Electromagnetic Materials and Devices

Edited by Man-Gui Han

Published in London, United Kingdom

IntechOpen

Supporting open minds since 2005

Electromagnetic Materials and Devices
http://dx.doi.org/10.5772/intechopen.79786
Edited by Man-Gui Han

Contributors
Vasa Radonic, Goran Kitic, Norbert Cselyuszka, Vesna Crnojevic-Bengin, Zion Menachem, Ming-Chun Tang, Yang Wang, Ting Shi, Man-Gui Han, Licinius Alcantara, Heng Luo, Hong Xiao, Jianying Chen, Meiwu Shi, Biao Zhao, Rui Zhang, Liying Zhang, Shuguang Bi, Ming Liu, Kuldeep Chand Verma, Ashish Sharma, Navdeep Goyal, Ravinder Kotnala, Vishal Kesari, Hitoshi Ohsato, Jobin Varghese, Heli Jantunen, Tsun-Hsu Chang, Zhuang Li, Rui Cao, Xiaohui Tao, Lihui Jiang, Dawei Rong, Víctor Hugo Méndez García, Irving Eduardo Cortes-Mestizo, Edgar Briones, Leticia Ithsmel Espinosa-Vega

First published in London, United Kingdom, 2020 by IntechOpen
IntechOpen is the global imprint of INTECHOPEN LIMITED, registered in England and Wales, registration number: 11086078, 7th floor, 10 Lower Thames Street, London, EC3R 6AF, United Kingdom
Printed in Croatia

British Library Cataloguing-in-Publication Data
A catalogue record for this book is available from the British Library

Additional hard and PDF copies can be obtained from orders@intechopen.com

Electromagnetic Materials and Devices
Edited by Man-Gui Han
p. cm.
Print ISBN 978-1-78985-226-4
Online ISBN 978-1-83880-101-4
eBook (PDF) ISBN 978-1-83880-102-1

For EU product safety concerns:
IN TECH d.o.o., Prolaz Marije Krucifikse Kozulić 3, 51000 Rijeka, Croatia, info@intechopen.com or visit our website at intechopen.com.

Meet the editor

Man-Gui Han obtained his PhD degree in Materials Science and Engineering from Iowa State University (USA) in 2004. Currently, he is a full professor at the University of Electronic Science and Technology of China. He has also been a journal editor of *IEEE Transactions on Magnetics* since from 2012. Up to 2019, he has published more than 70 peer-reviewed papers on magnetic materials. He established his academic reputation by finding a giant magnetostrictive effect in $Gd_5(Si_{0.5}Ge_{0.5})_4$, developing novel Fe-based nanocrystalline electromagnetic wave absorption materials, and pioneering the proposal of negative imaginary parts ($u'' < 0$) of permeability under the spin transfer torque effect.

Contents

Preface

With the advance of modern electronic technologies, many electronics work in high frequency, which requires many components to exhibit excellent electromagnetic properties. Discussions of electromagnetic properties mainly focus on high-frequency permeability and permittivity. Different applications have different requirements on these two electromagnetic parameters in the specific frequency range. Some require low magnetic loss and low dielectric loss. Some require large magnetic loss and large dielectric loss. While others require the real parts of permeability and/or permittivity to be negative. Electromagnetic materials nowadays can be naturally synthesized and artificially designed (such as metamaterials).

In this book, the contents discuss the following wide aspects of electromagnetic materials and applications: the measurements of permeability and permittivity; establishment ofthe relationships between crystal structures, microstructures, and permittivity (dielectric losses); approaches to tailor the high-frequency permeability of magnetic materials; the design of artificial materials to tailor the propagation of electromagnetic waves; and the application of electromagnetic materials (for instance, ferrites, ferromagnetic materials, multiferroic materials, magnetic-optical materials, etc.) in absorbing electromagnetic waves, weaving electromagnetic function textiles, and electromagnetic interference shielding. Some chapters also show the requirements of specific applications on electromagnetic parameters (such as antenna, circulator, and isolators). According to a survey, books with so many new advances in electromagnetic materials are rare.

Finally, I would like to express my thanks to the chapter authors for their willingness to share their expertise with readers. I am sure the research findings in this book will be interesting and helpful to graduate students, scientists, and engineers all around the world.

Man-Gui Han
Professor,
University of Electronic Science and Technology of China,
China

Section 1

High Frequency Dielectric Properties

Chapter 1

Dielectric Losses of Microwave Ceramics Based on Crystal Structure

Hitoshi Ohsato, Jobin Varghese and Heli Jantunen

Abstract

So far, many microwave dielectric materials have been investigated for a range of telecommunication applications. In dielectrics, the three main dielectric properties are quality factor (Q), dielectric constant and temperature coefficient of resonant frequency. Among these, the most essential dielectric property is Q. More specifically, Q is the inverse of the dielectric loss (tanδ); thus Q = 1/tanδ. There are two kinds of losses: those depending on crystal structure and losses due to external factors. The former is intrinsic losses such as ordering, symmetry, and phonon vibration. The latter is extrinsic losses due to factors such as grain size, defects, inclusions and distortion. In this chapter, the authors present the origin of dielectric losses based on the crystal structure. An ideal and well-proportional crystal structure constitutes a low loss material. Most dielectric materials are paraelectrics with inversion symmetry i and high symmetry. In general, it is believed that ordering gives rise to a high Q, on which many researchers are casting doubt. In the case of complex perovskites, the symmetry changes from cubic to trigonal. Ordering and symmetry should be compared with the structure. In this chapter, three essential conditions for the origin of high Q such as high symmetry, compositional ordering and compositional density are presented.

Keywords: microwave dielectrics, Q-factor, ordering, symmetry, indialite/cordierite, pseudo tungsten-bronze, complex perovskite

1. Introduction

Microwave and millimetre-wave dielectric materials [1–6] have been investigated for a wide range of telecommunication applications, such as mobile and smartphones, wireless local area network (LAN) modules and intelligent transport system (ITS). Millimetre-wave dielectric materials with high quality factor Q and low dielectric constant ε_r are required for the next 5G telecommunication applications used for noncondensed high data transfer on LAN/ personal area networks (PAN) and the higher frequency radar on autonomous cars.

In microwave dielectrics, there are three fundamental dielectric properties: quality factor (Q), dielectric constant (ε_r) and temperature coefficient of resonant frequency (TCf/τ_f) [1, 2, 6]. Microwave dielectrics have been used as the critical constituents of wireless communications [7–10], such as resonators, filters and temperature-stable capacitors with a near zero ppm/°C $TC\varepsilon_r$ (temperature coefficient of the dielectric constant). Among the dielectric properties, the most essential property is Q, the inversion of the dielectric loss ($\tan\delta$); thus $Q = 1/\tan\delta$. The dielectric losses of microwave dielectrics should be small. So, most of the microwave dielectrics are paraelectrics with inversion symmetry i, while most of the electronic materials are ferroelectrics with spontaneous polarity showing substantial dielectric losses [11–13]. The microwave dielectrics attract attention as a high potential material, which have an over-well-proportional rigid crystal structure with symmetry. That is, the structure should be without electric defects, nondistortion and without strain.

Under the influence of an electric field, four types of polarisation mechanisms can occur in dielectric ceramics, that is, interfacial, dipolar, ionic and electronic. In general, the microwave dielectric properties such as ε_r and Q are mostly influenced by ionic or electronic polarisation. The dielectric polarisation generates the dielectric losses in the presence of an electromagnetic wave. When the frequency is increased to millimetre-wave values, the dielectric losses may be increased or decreased depending on the polarisation mechanism. There are two kinds of losses: those depending on crystal structure and losses due to external factors. It was believed that the intrinsic losses are due to the ordering/disordering, symmetry and phonon vibration, while extrinsic losses are due to factors such as grain size, defects, inclusions, density and distortion from stress.

In this chapter, the origins of high Q are discussed based on the intrinsic factors related to the crystal structure, such as symmetry, compositional ordering and compositional density. Although it has previously been believed that ordering based on the order-disorder phase transition brings high Q [14], the authors propose that it is primarily a high symmetry that leads to high Q [15]. The following focused studies relate to specific examples; indialite with high symmetry showing higher Q than cordierite with an ordered structure [16–18]; pseudo tungsten-bronze solid solutions without phase transition showing high Q based on the compositional ordering [19–21]; complex perovskite compounds with order-disorder transitions depending on density and grain size [22, 23] and complex perovskites with composition deviated from the stoichiometric depending on the compositional density showing a high Q [24–29].

2. Focused studies

2.1 Indialite/cordierite glass ceramics

2.1.1 Indialite Q-factor improved by Ni-substitution

Cordierite ($Mg_2Al_4Si_5O_{18}$) has two polymorphs: cordierite and indialite, as shown in **Figure 1(a)** and **(c)**, respectively [30, 31]. Cordierite is of low symmetry form: orthorhombic crystal system *Cccm* (No. 66), which has $Si_4Al_2O_{18}$ six-membered tetrahedron rings with ordered SiO_4 and AlO_4 tetrahedra as shown in **Figure 1(b)**. On the other hand, indialite is of high symmetry form: hexagonal crystal system *P6/mcc* (No. 192), which has disordered $Si_4Al_2O_{18}$ equilateral hexagonal rings as shown in **Figure 1(c)**.

Cordierite shows a lower ε_r of 6.19 which depends on the silicates and a near-zero *TCf* of −24 ppm/°C [32] as compared to other silicates as shown in **Figure 2(a)**. Based on these properties, Terada et al. carried out initiative research on these microwave dielectrics [16]. They reported an excellent *Qf* by substituting Ni for Mg as shown in **Figure 2(b)**. The *Qf* was improved from 40×10^3 GHz to 100×10^3 GHz by Ni substitution of $x = 0.1$ in $(Mg_{1-x}Ni_x)_2Al_4Si_5O_{18}$. The Ni substitution did not change the ε_r value

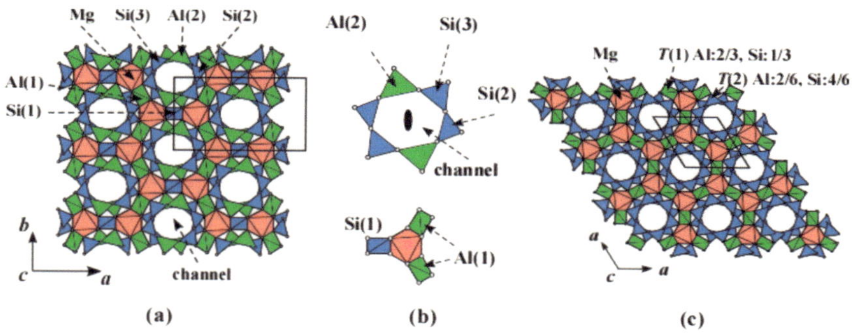

Figure 1.
Schematic representation of cordierite (a), six-membered tetrahedron ring with ordered SiO_4 and AlO_4 (b) and indialite (c).

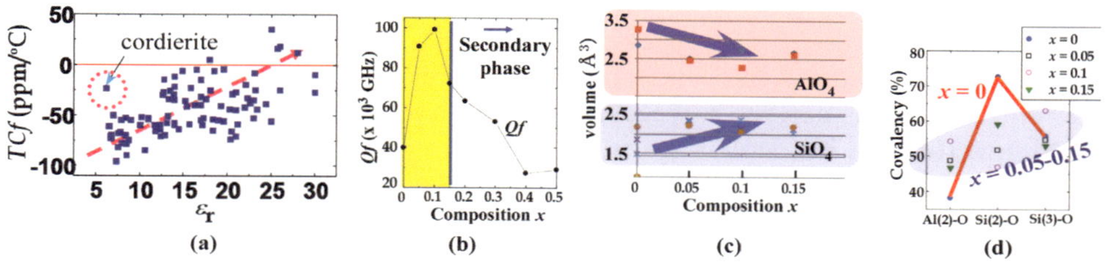

Figure 2.
Cordierite with near zero ppm/°C deviated from other compounds (a). Ni-substituted cordierite Qf (b), volume of AlO_4 and SiO_4 (c) and covalencies of Si-O and Al-O as a function of composition x (d).

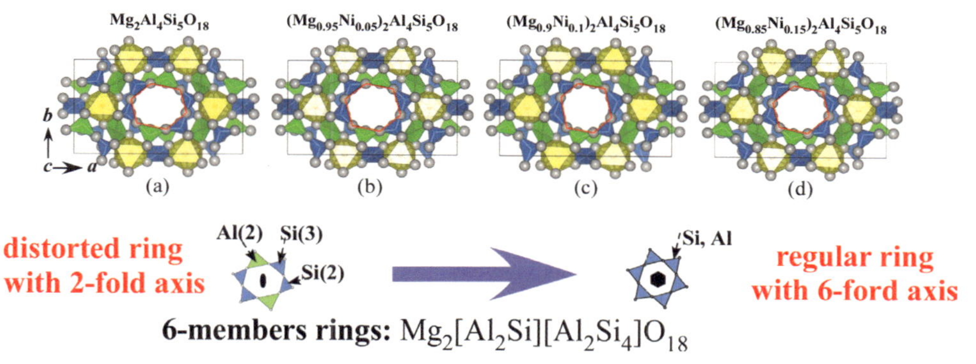

Figure 3.
Crystal structure of Ni-substituted cordierite: $(Mg_{1-x}Ni_x)_2Al_4Si_5O_{18}$ with composition x = 0 (a), 0.05 (b), 0.1 (c) and 0.15 (d).

considerably, but the *TCf* was degraded from −24 to −30 ppm/°C [16]. For $x > 0.1$, the properties were affected by the formation of the secondary phase of $NiAl_2O_4$.

Terada et al. also analysed the crystal structure by the Rietveld method [33] to clarify the origin of the improved *Qf* value. The X-ray powder diffraction (XRPD) pattern was obtained by a multi-detector system (MDS) [34] in the synchrotron radiation "Photon Factory" of the National Laboratory for High Energy Physics in Tsukuba, Japan. **Figure 3(a)–(d)** shows the crystal structures of Ni-substituted cordierite $(Mg_{1-x}Ni_x)_2Al_4Si_5O_{18}$ with $x = 0$, 0.05, 0.1 and 0.15. The crystal structure showed a tendency to deform to indialite with high symmetry on the hexagonal ring composed of corner-sharing of (Si, Al)O_4 tetrahedra in the *a-b* plane. Ni-substituted cordierite $(Mg_{1-x}Ni_x)_2Al_4Si_5O_{18}$ with composition $x = 0.1$

(**Figure 3(c)**) was obviously closer to equilateral hexagonal rings compared to $(Mg_{0.95}Ni_{0.05})_2Al_4Si_5O_{18}$ (**Figure 3(b)**) and $Mg_2Al_4Si_5O_{18}$ (**Figure 3(a)**).

The transformation from cordierite to indialite, represented by the ratio of disordering between the SiO_4 and AlO_4 tetrahedra, is based on the volumes and covalencies of the SiO_4 and AlO_4 tetrahedra [35]. The volume was calculated using atomic coordinates obtained by Rietveld crystal structural analysis as shown above. The covalency (f_c) of the cation-oxygen bond was estimated from the following equation [36].

$$f_c = as^M \tag{1}$$

The empirical constants a and M depending on the inner-shell electron number 10 are 0.54 v.u. and 1.64, respectively [37], where s is the bond length obtaining from the following equation:

$$s = (R/R1)^{-N} \tag{2}$$

where, R is defined as the bond length, and $R1$ and N are the measured parameter reliant on the cation site and each cation-anion pair, respectively.

Figure 2(c) and **(d)** depicts the calculated volume and covalency of SiO_4 and AlO_4 octahedra, respectively. These figures show the phase changing from cordierite to indialite as substitution of Ni in the Mg site. In the cordierite $Mg_2Al_4Si_5O_{18}$ (**Figure 1(a)**), Si/Al ions in the tetrahedra are ordered. Therefore, the volume and covalency of tetrahedra are different values, but the values are becoming similar to the substitution of Ni in the Mg site. This is due to the disordering of Si/Al ion phase transition in the cordierite (**Figure 1(a)**) to indialite (**Figure 1(c)**). In the indialite, the disordered $Si_4Al_2O_{18}$ equilateral hexagonal rings with 6-ford axis are the main framework as analysed by the Rietveld method as shown in **Figure 3(d)**. The improvement of *Qf* as shown in **Figure 2(b)** should be based on the disordering due to high symmetry instead of an ordering of SiO_4 and AlO_4 tetrahedra by order-disorder transition. It is one example of high symmetry bringing a higher *Q* than ordering by the order-disorder transition [18].

2.1.2 Indialite glass ceramics with high Q

As described in the previous section, the *Qf* value of indialite derived by substituting Ni for Mg was improved to three times that of cordierite. Based on the new knowledge, Ohsato et al. proposed the synthesis of indialite with superior microwave dielectric properties [17]. The indialite, being a high-temperature form, could not be synthesised by the solid-state reaction because the order-disorder phase transition is hindered by the incongruent melting to form mullite and liquid. On the other hand, indialite is an intermediate phase during the crystallisation process from glass with a cordierite composition to cordierite, as shown in **Figure 4**.
Therefore, fabrication of indialite glass ceramics has been attempted [17, 39]. Although the indialite is a metastable phase transforming to cordierite at higher temperatures, it is a relatively stable phase which occurs in nature formed by the crystallisation of natural glass. As this occurrence is in India, the mineral was named indialite. Another phase of μ-cordierite precipitating in the early stage of the crystallisation of cordierite glass is β-quartz solid solutions. The naming of μ-cordierite is not correct because of the different crystal structure, so the name that should be used is β-quartz solid solutions [38].

The cordierite composition was melted at 1550°C and was cast into a cylindrical rod with the diameter φ = 10 mm and l = 30 mm in a graphite mould. In order to avoid fracture due to internal strain, the cast glass rod was annealed at 760°C below

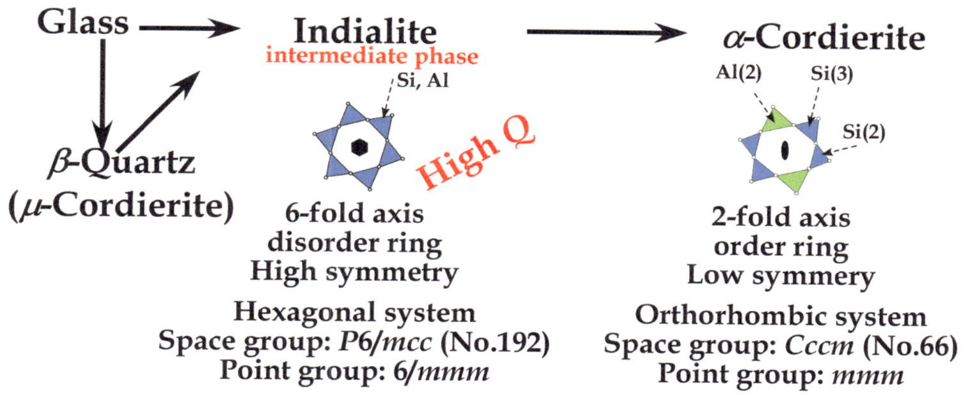

Figure 4.
Polymorphism of cordierite: indialite is the high temperature/high symmetry form, and cordierite is the low temperature/low symmetry form. In addition, indialite is an intermediate phase during crystallisation from glass to cordierite.

the glass transition point of 778°C [38]. The 10 mm diameter glass rod was cut to form a resonator with a height of 6 mm. The glass pellets were crystallised at temperatures in the range 1200–1470°C/10 and 20 h. The crystallised pellets had two problems: deformation by the formation of glass phase and cracking by anisotropic crystal growth from the surface (**Figure 5(a)**) [39]. **Figure 5(b)** and **(c)** shows photographs taken by a polarising microscope of a thin section of the crystallised samples. The needle-like crystals grown from the surface had an orientation with *c*-axis elongation. The microwave dielectric properties of the sample with cracking had a wide scattering range of the data [17, 39].

Figure 6(a) shows the volume of indialite/cordierite examined by the Rietveld method [40], which is estimated with two phases such as indialite and cordierite. Hereabout, the residual % is compared to that of cordierite. At 1200°C, the precipitated phase of indialite was about 96.7%. The volume of indialite reduced as the temperature and to 17.1% (82.9% for cordierite) at 1400°C. **Figure 6(b)** and **(c)** shows the microwave dielectric properties of indialite/cordierite glass ceramics and remarkably high *Qf* value of more than 200 × 10^3 GHz at 1300°C/20 h [17]. This is much better than the highest *Qf* value of 100 × 10^3 GHz obtained by substitution with Ni using the conventional solid-state reaction as previously described (**Figure 2(b)**) and is feasible for millimetre-wave dielectrics. The *Qf* values decreased as crystallisation temperature. In comparison with the amount of indialite as shown in **Figure 6(a)**

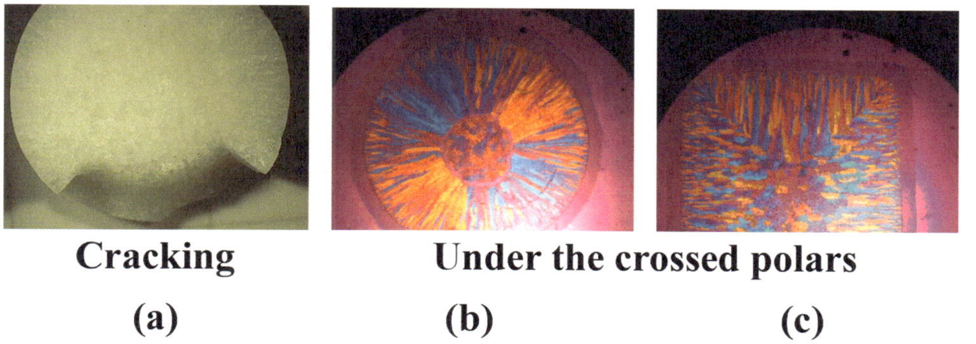

Figure 5.
Cracking of crystallised pellets (a) and anisotropic crystal growth of the pellets under the crossed polars with a sensitive test plate (b) and (c).

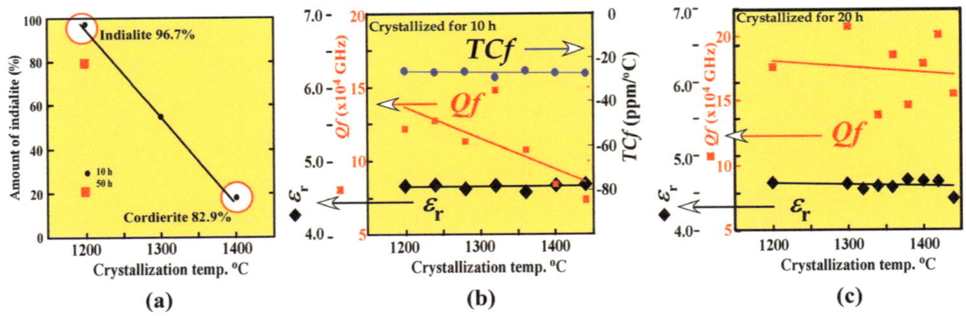

Figure 6.
Amount of indialite (a) and microwave dielectric properties of crystallised indialite at 1200–1440°C for 10 (b) and 20 (c) hours.

[39] and its *Qf* values as shown in **Figure 6(b)** and **(c)** [17], it is clear that the indialite glass ceramics present a higher *Qf* than that of cordierite. The ε_r was the lowest among the silicates, about 4.7 as shown in **Figure 6(b)** and **(c)**, and the *TCf* was −27 ppm/°C as shown in **Figure 6(b)**. Therefore, from these figures, indialite shows a higher *Qf* than cordierite. This *TCf* value of −27 ppm/°C is better than that of other silicates having a low *TCf* of approximately −60 ppm/°C [39].

2.1.3 Conclusions for indialite/cordierite glass ceramics

- Indialite/cordierite glass ceramics are one of the examples of high symmetry bringing a higher Q than ordering by order-disorder transition. Indialite glass ceramics with disordered high symmetry have higher *Qf* properties than cordierite with ordered low symmetry.

- Cordierite with substituted Ni for Mg synthesised by solid-state reaction exhibited an improved *Qf* from 40×10^3 to 100×10^3 GHz (**Figure 2(b)**). Rietveld crystal structure analysis showed that the cordierite was transformed to indialite [16].

- A novel idea from glass ceramics suggested the fabrication of indialite as an intermediate phase. Glass ceramics crystallised at 1200°C were almost completely indialite at 96.7% with a high *Qf* of 150×10^3 GHz, and those crystallised at 1400°C were cordierite at 82.9% with a lower *Qf* of 80×10^3 GHz. (**Figure 6**) [17, 39].

- Indialite/cordierite crystallised from cordierite glass at 1300°C/20 h showed good microwave dielectric properties of $\varepsilon_r = 4.7$, $Qf > 200 \times 10^3$ GHz and $TCf = -27$ ppm/°C (**Figure 6**) [17, 39].

2.2 Pseudo tungsten-bronze solid solutions: compositional ordering bringing high Q

2.2.1 Introduction

The pseudo tungsten-bronze solid solutions $Ba_{6-3x}R_{8+2x}Ti_{18}O_{54}$ (R = rare earth) located on the tie-line of $BaTiO_3$-$R_2Ti_3O_9$ are shown in **Figure 7(a)** and have been utilised in mobile phones because of their high dielectric constant of 80–90 [20, 21]. This solid solution was first reported by Varfolomeev et al. [41], based on Nd and Sm systems. The composition ranges $0.0 << x << 0.7$ for R = Nd and $0.3 << x << 0.7$ for

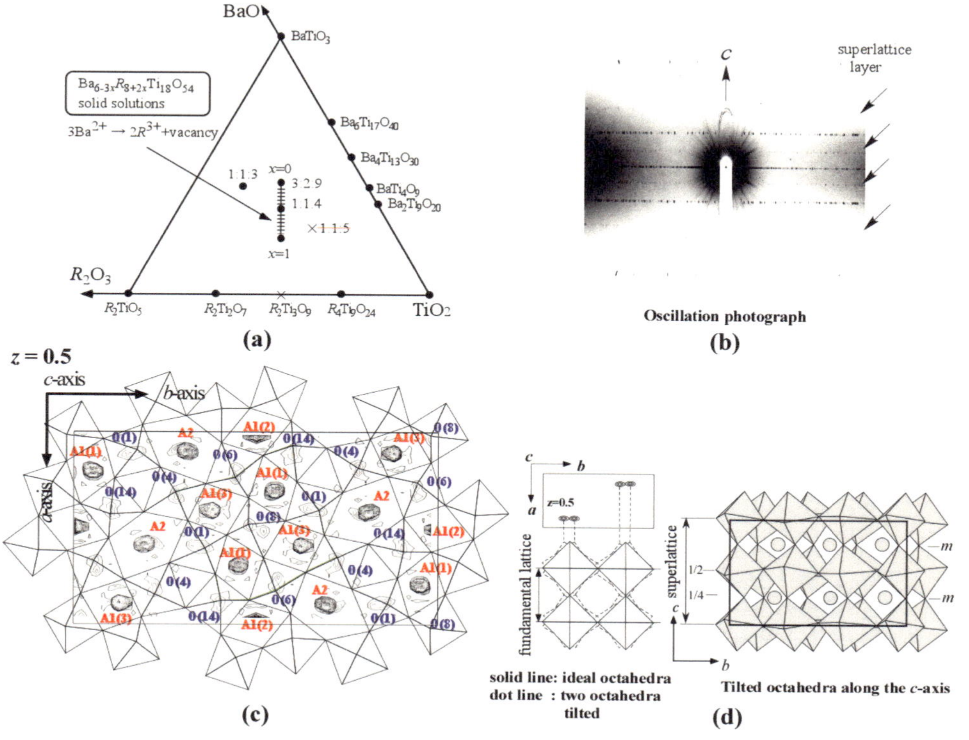

Figure 7.
A part of the BaO-R_2O_3-TiO_2 ternary phase diagram with pseudo tungsten-bronze type solid solutions (a). Oscillation photograph along c-axis of pseudo tungsten-bronze type solid solutions (b). Electron density map (Fourier map) of the fundamental structure superimposed on a superstructure framework (c) and TiO_6 tilting octahedra along the c-axis on the super-lattice (d) deduced from the splitting of oxygen in the fundamental structure (c), and the splitting of oxygen atoms based on the tilting of octahedra as shown in left side figure of the fundamental lattice (d). Right side schematic figure: super structure produced by tilting octahedral (d).

Sm [42] were reported by Ohsato et al. [19] and Negas et al. [43]. The composition range of the solid solutions becomes narrower with the decrease in the ionic radius of the *R*-ion, and Ga and Eu form only BaO·R_2O_3·4TiO_2 composition [44].

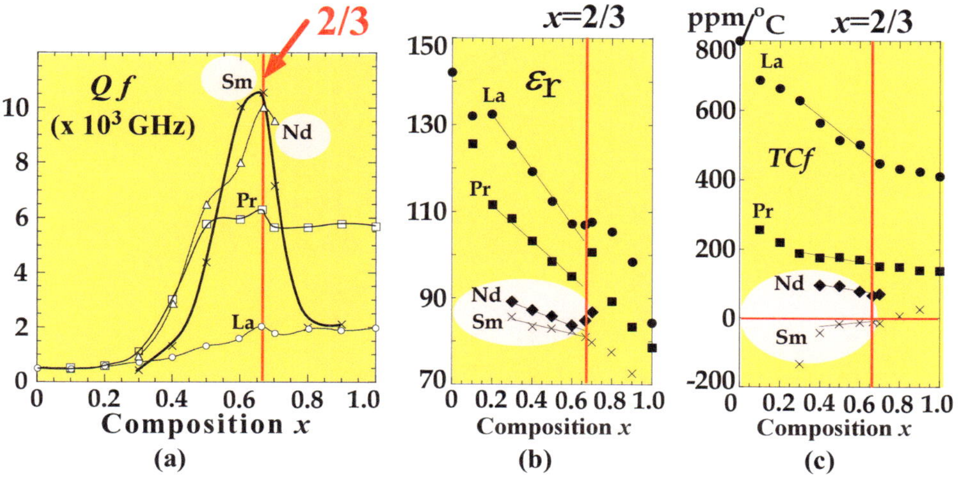

Figure 8.
Qf (a), ε_r (b) and TCf (c) of Sm, Nd, Pr and La system as a function of composition x.

Ohsato et al. and Negas et al. reported the microwave dielectric properties for the Sm, Nd, Pr and La systems as a function of composition x as shown in the **Figure 8(a)** [20, 43, 45] and Fukuda et al. reported the Pr system [46]. On the solid solutions, the composition with $x = 2/3$ was found by Ohsato et al. [42], at which the Qf value becomes the highest due to the ordering in the rhombic and pentagonal sites. The dielectric constants ε_r and TCf (**Figure 8(b)** and **(c)**) are decreased as a function of the composition x and are affected by volume and tilting angle of the TiO_6 octahedra and the polarizabilities of R and Ba ions [20]. The Clausius-Mosotti equation determined the temperature coefficient of the dielectric constant $TC\varepsilon_r$ as a function of the ratio of the mean radii (r_a/r_b) of A- and B-site ions by Valant et al. [47]. Hither, ra/rb is connected to the tilting of the TiO_6 octahedra. In this study, on the system without order-disorder phase transition that is without symmetry change, it is discussed that the ordering especially compositional ordering brings high Qf.

2.2.2 Crystal structure of pseudo tungsten-bronze solid solutions

2.2.2.1 Structure

The crystal structure of the pseudo tungsten-bronze $Ba_{6-3x}R_{8+2x}Ti_{18}O_{54}$ (R = rare earth) solid solutions [48–51] includes the perovskite blocks of 2 × 2 unit cells with rhombic ($A1$) sites and pentagonal ($A2$) sites, as shown in **Figure 9**, which are named after the tetragonal tungsten-bronze structure with 1 × 1 perovskite blocks and pentagonal sites [20, 48, 50]. On this compound, two large sites including Ba- and R-ions are placed such as $A1$ and $A2$. The Ba-ions engaged on the pentagonal $A2$-sites and R-ions $A1$-sites on the perovskite blocks. Two more sites, B and C are positioned on the tungsten-bronze crystal structure. The B-site is the same as the TiO_6 octahedral place in the perovskite, and the C-site is a triangular site which is usually empty. This crystal structure of this compound has a special relationship with the perovskite structure. If the two ions are the same size, the structure will change to perovskite with only an $A1$-site owing to the combination of the

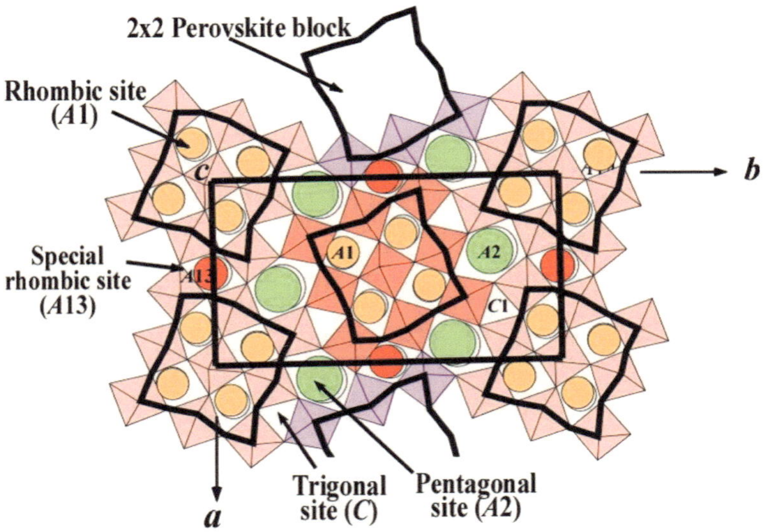

Figure 9.
Crystal structure of the pseudo tungsten-bronze solid solutions. Rhombic (A1) sites located in 2 × 2 unit cells of perovskite blocks, and pentagonal (A2) and trigonal sites (C).

pentagonal *A*2-site and the trigonal *C*-sites [20, 52]. The crystal data are as follows: orthorhombic crystal system of space group *Pbnm* (No. 62), point group *mmm* and lattice parameters a = 12.13, b = 22.27, c = 7.64 Å, Z = 2 and Dx = 5.91 g/cm^3.

2.2.2.2 Tilting

The structure has a super lattice along the c-axis with a double lattice of perovskite as shown in **Figure 7(b)** of an oscillation photograph with super diffraction lines [53, 54]. The crystal data of the fundamental lattice are as follows: orthorhombic crystal system of space group *Pbam* (No. 55), point group *mmm* and lattice parameters a = 12.13, b = 22.27, c = 3.82 Å, Z = 1 and Dx = 5.91 g/cm^3. The super lattice is depending on the tilting of TiO_6 octahedra as shown in **Figure 7(d)**. The tilting was endowed in the density map (**Figure 7(c)**) which is of the fundamental lattice superimposed on a superstructure framework. The top oxygen ions (O(1), O(4), O(6), O(8) and O(14)) of octahedra are separated into two along the c-axis. The left figure of **Figure 7(d)** shows the reason for splitting of the top oxygen [20]. However, this super lattice is not depending on the order-disorder phase transition as complex perovskite as explained at 2.3 section. The tilting of octahedra might be depending on the size of *A*-ion in the perovskite block.

2.2.2.3 Compositional ordering

The chemical formula of the solid solutions is $Ba_{6-3x}R_{8+2x}Ti_{18}O_{54}$, and the structural formula is $[Ba_4]_{A2}[Ba_{2-3x}R_{8+2x}]_{A1}Ti_{18}O_{54}$. Here, the amount of Ba in the *A*1-sites becomes zero ($2 - 3x = 0$), that is, $x = 2/3$. This composition is special as the following sentence: the structural formula is $[Ba_4]_{A2}[R_{8+4/3}]_{A1}Ti_{18}O_{54}$ which is occupied separately by Ba in *A*2 and by *R* in *A*1 as shown in **Figure 10(b)**. This special composition is called "compositional ordering" [20, 21, 42].

2.2.3 Microwave dielectric properties of pseudo tungsten-bronze solid solutions

Figure 8 shows the microwave dielectric properties of the solid solutions as a function of composition x of $Ba_{6-3x}R_{8+2x}Ti_{18}O_{54}$ [20, 21, 42]. The quality factor (Qf) changes nonlinearly and has the highest value at particular point $x = 2/3$ with compositional ordering specified above [55]. The highest Qf value might be depending on the internal strain. **Figure 11(a)** confers internal strain η obtained from the slope of equation $\beta\cos\theta = r/t + 2\eta\sin\theta$. The internal strain η of the special point $x = 2/3$ is the lowest with the compositional ordering as a function of composition x as shown

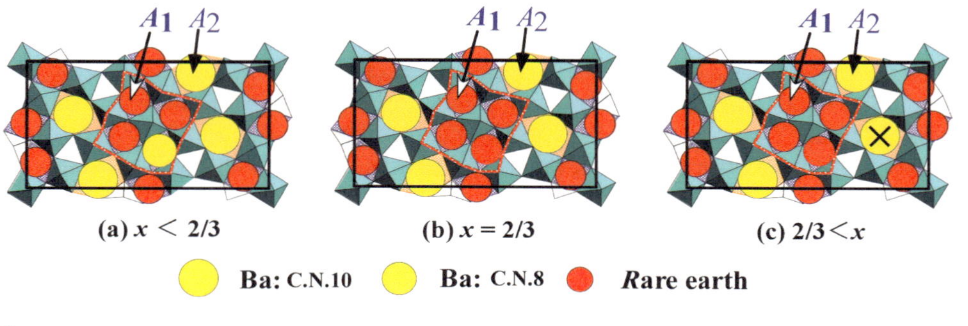

Figure 10.
Structure of disordering (a), compositional ordering (b) and defects in A2-sites (c), depending on the x values of $Ba_{6-3x}R_{8+2x}Ti_{18}O_{54}$.

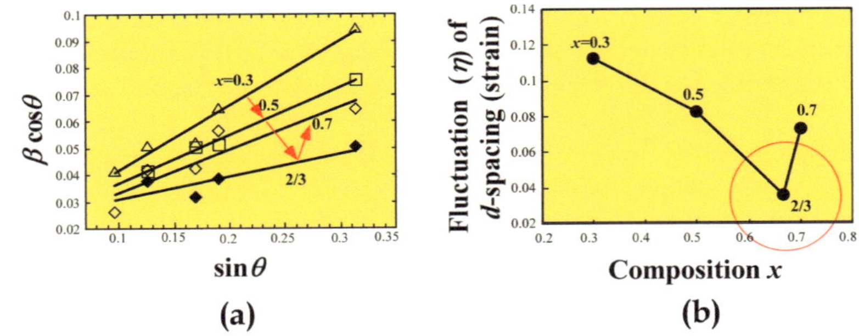

Figure 11.
Internal strain η values obtained from the slope of equation βcosθ = r/t + 2ηsinθ as a function of sinθ for x = 0.3, 0.5, 2/3 and 0.7 (a) and strain η (d-spacing) as a function of composition x (b).

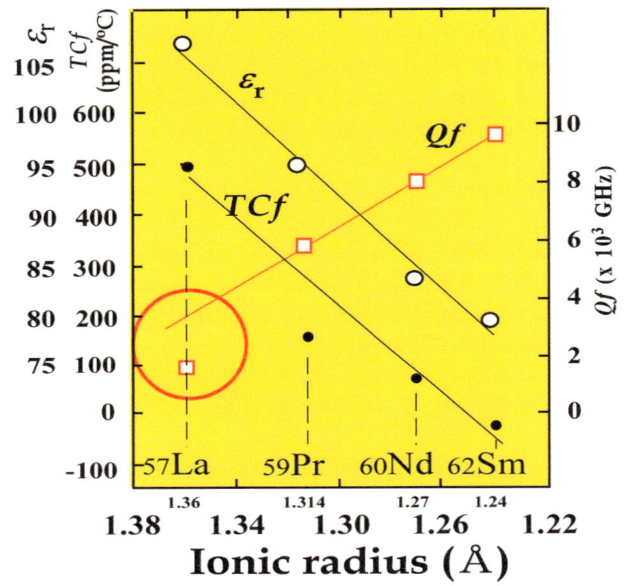

Figure 12.
Microwave dielectric properties as a function of ionic radius of R ion.

in **Figure 11(b)**. The internal strain comes from the fluctuation of *d*-spacing of the lattice broadening the full width at half maximum (FWHM) [20, 21, 56].

The *Qf* value at the special point $x = 2/3$ shows the highest of 10.5×10^3 GHz in the Sm system, 10.0×10^3 GHz in the Nd system and 2.0×10^3 GHz in the La system as depicted in **Figure 8(a)** [20, 21, 56]. The *Qf* values reducing in the order of Sm, Nd, Pr and La are depending on the ionic radius relating size difference between Ba and *R* [57], and that of La is deviating from the *Qf* line through the Sm, Nd and Pr as shown in **Figure 12**. If the sizes are similar, the crystal structure should become perovskite structure. In the case of Sm, the difference is maximum which introduces the stability of the crystal structure. The size of La ion is similar to Ba, so the structure might be unstable to be low *Qf*.

2.2.4 Symmetry and ordering for Q

On the microwave dielectrics, high Q has been brought by a high potential material, which has an over-well-proportional rigid crystal structure with symmetry

[11–13]. That is, the structure should be without electric defects, nondistortion and strain. Complex perovskites were described later, it is believed that ordering by long time sintering brings high *Q*, but we are pointing out symmetry is the predominant factor [14, 15]. In the case of indialite/cordierite, indialite with high symmetry shows higher *Q* than cordierite with ordering [17, 18, 39]. This case has an order-disorder phase transition. On the other hand, in the case of pseudo tungsten-bronze solid solutions which has no phase transition, one of ordering that is the compositional ordering brings high *Q* [20, 21]. In the case of no symmetry change, ordering is predominant.

2.2.5 Conclusions for pseudo tungsten-bronze

- The pseudo tungsten-bronze solid solutions have been used for mobile phones for miniaturisation based on their high *Qf* and high ε_r.

- The compound has a unique point of $x = 2/3$ on the $Ba_{6-3x}R_{8+2x}Ti_{18}O_{54}$ chemical formula which shows the highest *Qf* value.

- The special point of $x = 2/3$ on the structural formula of $[Ba_4]_{A2}[Ba_{2-3x}R_{8+2x}]_{A1}Ti_{18}O_{54}$ is the composition at which Ba-ions disappear on the *A*1-sites because $2 - 3x = 0$. That is the point of compositional ordering.

- The compositional ordering brings high *Q* by maintaining the stability of the crystal structure.

2.3 Complex perovskites

There are many kinds of complex perovskites such as 1:1, 1:2 and 1:3 type in *B*-site and 1:1 type in *A*-sites [21]. In this chapter, 1:2 type complex perovskite compounds $A^{2+}(B^{2+}{}_{1/3}B^{5+}{}_{2/3})O_3$ are presented such as $Ba(Zn_{1/3}Ta_{2/3})O_3$ (BZT), $Ba(Mg_{1/3}Ta_{2/3})O_3$ (BMT) and $Ba(Zn_{1/3}Nb_{2/3})O_3$ (BZN). These complex perovskite compounds have order-disorder phase transitions (**Figure 13(a)** and **(b)**) [58]. The ordered phase that appears at low temperatures is a trigonal (rhombohedral) structure of space group $P\bar{3}m1$ (No. 164), and the disordered phase appearing at high temperatures is a high symmetry cubic structure of $Pm\bar{3}m$ (No. 221), as shown

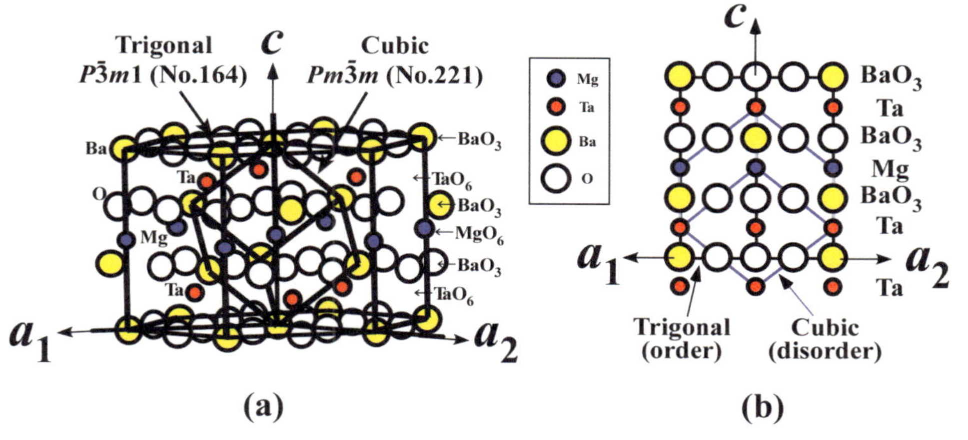

Figure 13.
Complex perovskite crystal structure composed by Mg/TaO₆ octahedra located between BaO₃ closed packing layer, showing relationship between cubic and trigonal crystal lattice. Perspective figure (a) and (110) plane (b).

in **Figure 13** [21]. In the ordered form of BMT, Mg^{2+} and Ta^{5+}-ions located among the adjacent packing layers of BaO_3 are ordering as -Mg-Ta-Ta-Mg-Ta-Ta-Mg-, as shown in **Figure 13**. On the other hand, in the disordered form, Mg^{2+} and Ta^{5+}-ions occupy *B*-sites statistically.

2.3.1 Introduction

Kawashima et al. [14] presented that ordering brings a high *Q* based on BMT with long duration sintering, which showed high *Qf* and ordering. This has previously been believed to be the case because long duration sintering samples generally show high *Qf* and ordering. However, some examples have arisen that contradict this relation, such as BMT-$Ba(Co_{1/3}Ta_{2/3})O_3$ [59], BMT-$BaZrO_3$ [60], BMT-$BaSnO_3$ [61] and BZT-(SrBa)$(Ga_{1/2}Ta_{1/2})O_3$ [62]. Koga et al. [23–26] presented the relationship between high *Qf* and the ordering ratio as determined by the Rietveld method, the high *Qf* samples with disordered structure synthesised by spark plasma sintering (SPS) [63] and the effects of annealing of disordered BZN with an order-disorder transition point of 1350°C [26]. HRTEM and Rietveld studies confirmed the ordering and disordering of BZN samples [64]. Partial ternary phase diagrams such as BaO-ZnO-Ta_2O_5, BaO-MgO-Ta_2O_5 and BaO-ZnO-Nb_2O_5 were studied on the composition with high *Qf* that deviated from the stoichiometric composition of BZT/BMT/BZN by Kugimiya et al. [22, 27], Koga et al. [24, 26] and Kolodiazhnyi [29]. Kugimiya pointed out that the solid solutions with high density and high *Qf* located on the tie-line BMT-$Ba_5Ta_4O_{15}$, which have completed the ideal chemical formula without oxygen defects. It is one of the conditions for high Q that the high compositional density brings high *Qf*.

2.3.2 Origin of high Q for microwave complex perovskite

In this section, it is explained that ordering has no relation with *Qf* based on the following three sets presented by Koga et al. [23, 25, 26, 63].

2.3.2.1 Ordering ratio and Qf

The ordering of BZT was observed on the samples with high *Qf* sintered at 1350°C [23] over 80 h. **Figure 14** presents the XRPD patterns (**a**) with super lattice lines (asterisked), and the high angle diffraction patterns (**b**) which depicts splitting of 420 cubic diffraction peak into two peaks, namely 226 and 422 in the trigonal system. These data are consistent with the report by Kawashima et al. [14].

Koga et al. investigated the amount of BZT ceramic as ordering ratio by the Rietveld method [23], which is shown in **Figure 15(a)**. The ordering ratio saturates at about 80%, but the *Qf* values increase up to 100 × 10^3 GHz. This shows that the effect of ordering on the *Qf* is not so significant. However, the *Qf* values are affected by density and grain size as shown in **Figure 15(b)** and **(c)**, respectively [15, 23].

2.3.2.2 BZN with a clear order-disorder transition

Many complex perovskites such as BMT and BZT have the order-disorder phase transition at high temperature, and the order-disorder transition is not so clear. On the other hand, BZN shows clearly the phase transition at lower temperature 1350°C [26]. **Figure 16(a)** shows *Qf* as a function of sintering temperature. Under the transition temperature such as 1200 and 1300°C, the sintered samples show order with under 50 × 10^3 GHz of *Qf*. Moreover, at 1400°C, higher than the transition

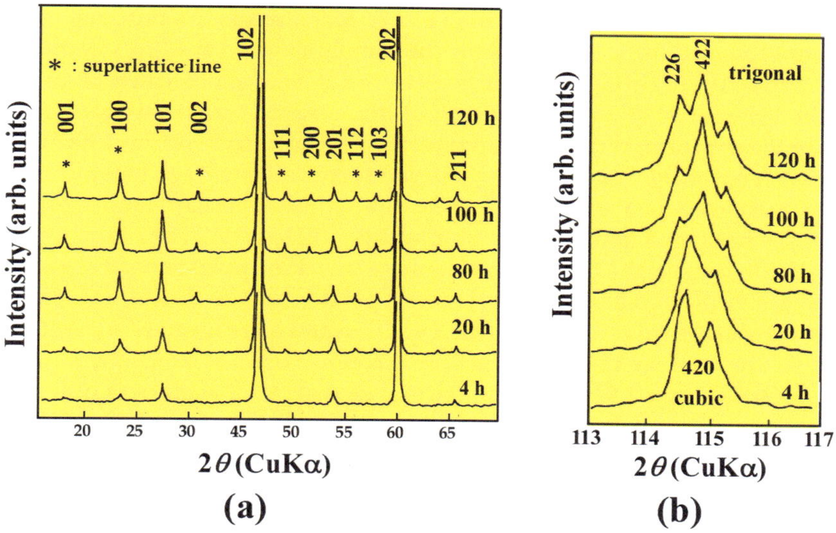

Figure 14.
XRPD patterns of BZT ceramics with different sintering times at 1350°C (a), here, asterisks are super lattice diffractions, and Magnified XRPD patterns around 2θ = 115° in which 420 diffraction peak split to 226 and 422(b).

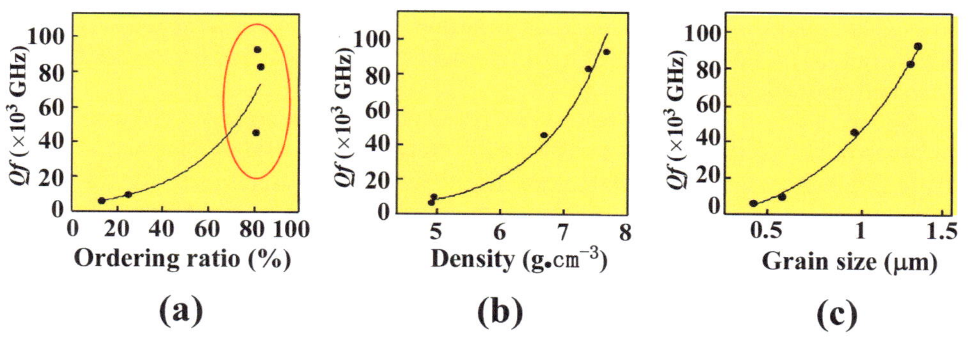

Figure 15.
The Qf as functions of ordering ratio (a), density (b) and grain size (c) of BZT ceramics.

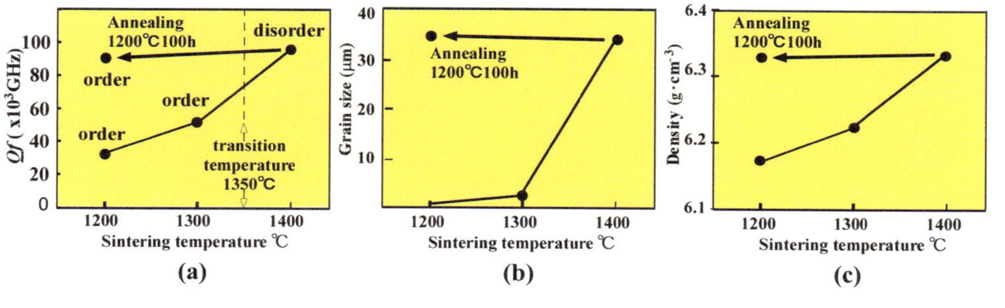

Figure 16.
Qf (a), grain size (b) and density (c) as a function of sintering temperature of BZN ceramics.

temperature, the *Qf* values increased to 90 × 10^3 GHz with disordering structure. This shows that the high symmetry form with disorder performs higher *Qf* than ordering form. Moreover, the sample annealed at 1200°C/100 h transformed to order form, but the *Qf* value did not improve and slightly decreased. Grain size and densities as shown in **Figure 16(b)** and **(c)** also increased as the sintering temperature from 1200 to 1400°C [15, 26]. As if the sample sintered at 1400°C annealed at 1200°C/100 h, the grain size and densities were not changed. Because of annealing, the slight decrease in *Qf* might be a result of the low symmetry that accompanies order. On the contrary, Wu et al. [65] presented annealing of BZN at 1300°C brings high *Qf* with ordering. The annealing temperature is high enough for sintering, so sintering was proceeded with ordering the same as Kawashima's results [14].

The BZN samples A and B are also studied by XRPD and HRTEM, which sintered at 1400°C/100 h above the order-disorder phase transition point and subsequently annealed at 1200°C/100 h below the transition point, respectively [26, 64]. The two samples were identified by conventional XRPD as shown in **Figure 17(a)**. As the super lattice lines are not clear, the high angle XRPD patterns around 2θ~115° were measured (**Figure 17(b)**). On the XRPD pattern, the sample A shows a single peak of the 420 diffraction, so it was confirmed as disorder phase. On the other hand, the sample B shows the peak splitting of 422 and 226 depending ordering. These results are comparable with Koga's data [23]. These two samples were analysed by the Rietveld method.

HRTEM figures as shown in **Figure 18** for most area of sample A (**Figure 18(a)**) and B (**Figure 18(c)**) are disordered and ordered area along the [111]c direction, respectively. A fast Fourier transform (FFT) image is inserted in **Figure 18(a)** of a disordered area without further reflections along the [111]c direction and in **Figure 18(c)** of a ordered area with additional two reflection points for super lattice. In the both sample A and B, mixed area of disordered and ordered area existed in **Figure 18(b)**, and in the sample B, ordered area showing twin-related anti-phase domain boundary also existed as shown in **Figure 18(d)**. The FFT image of twin area shows superimposed of ordered diffractions with four additional points.

Figure 19 depicts the high-resolution XRPD pattern of sample A and B using synchrotron radiation [64]. The super lattice diffraction 100 t peaks (reciprocal lattice plane 100 in the trigonal crystal system) are observed in both samples. The diffraction intensity of sample A is lower than that of sample B. These super lattice diffraction intensity peaks are comparable with the ordering ratios, that is the sample A and B have the value of 27.6 and 54.2%, respectively, obtained by the Rietveld method. Although the degree of ordering of sample B is large compared to that of sample A, it was assumed about 80% ordering for a whole sample, as in the case of BZT [23].

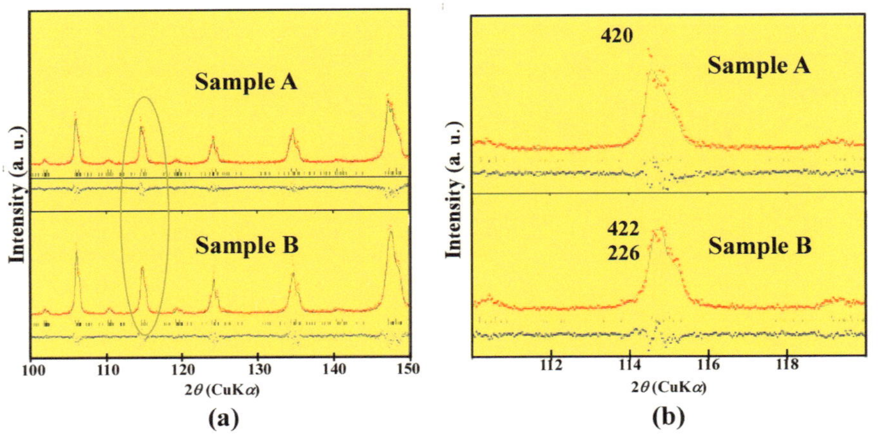

Figure 17.
XRPD patterns for BZN ceramics sintered at 1400°C (sample A) and annealed at 1200°C (sample B) (a) and magnified high angle XRPD patterns around 2θ~115° (b).

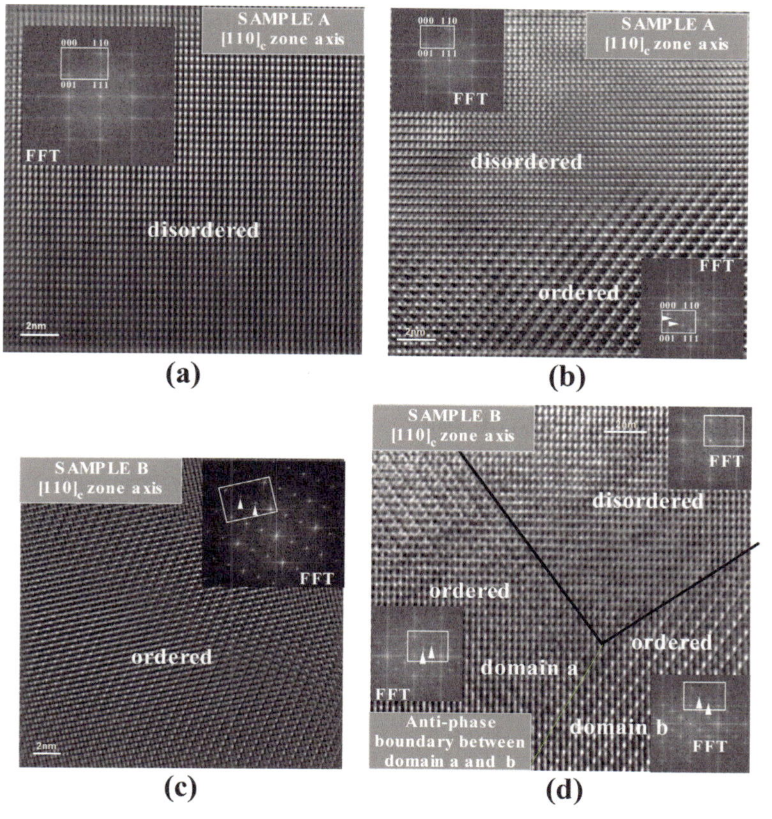

Figure 18.
HRTEM images of sample A and B with FFT image along the [111]c direction: disordered area in sample A (a), mixed area of disordered and ordered area in sample A (b), ordered area in sample B (c) and twin related anti-phase domain boundary in sample B (d).

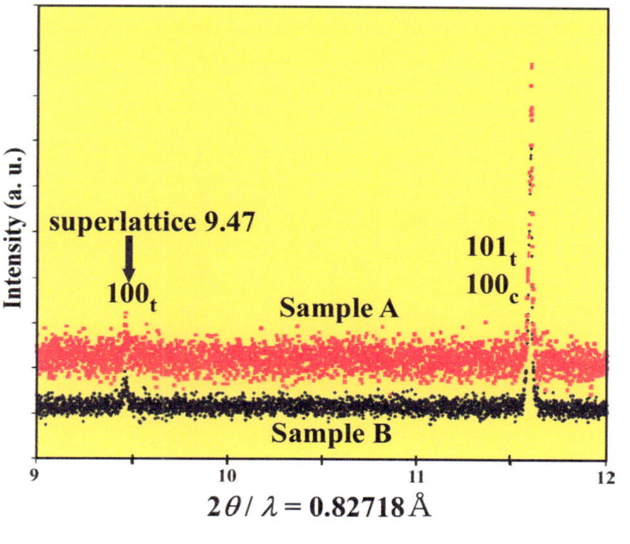

Figure 19.
High-resolution synchrotron XRPD patterns (λ = 0.82718 Å) for sample A and B with super lattice peak 100_t. Here, subscript t is trigonal, and c is cubic.

It is revealed that the degree of ordering increased from 27.6 to 54.2% due to the annealing. However, the *Qf* values, grain size and the density have no influence on the degree of ordering (**Figure 16**). While the disordered area of sample A (sintered above the transitional temperature) changes to the low-temperature phase with ordering by the annealing, the *Qf* values were expected to be increased. However, the *Qf* values changed only somewhat from 95.7 × 10^3 GHz to 95.0 × 10^3 GHz [64]. The effect of ordering is not acceptable to change the *Qf* value considerably.

2.3.2.3 BZT with disordering leaded high Qf by SPS

The ordered and disordered BZT ceramics can be achieved by varying the sintering duration in the conventional solid-state reaction (SSR). A high density and high *Q* ceramics of ordered BZT were obtained by SSR with a long sintering time of over 80 h, while the disordered BZT was not possible to fabricate by using SSR. Koga et al. [63] reported the high density disordered BZT ceramics for a short sintering time of 5 mins by using spark plasma sintering (SPS). **Figure 20(a)** presents the *Qf* as a function of the densities of BZT fabricated using SSR and SPS [15, 63]. The fabricated SPS samples were shown to be disordered cubic type of perovskite as depicted in the XRPD pattern (**Figure 20(b)**) with a peak of 420 reflection in compared with the ordered trigonal type with peaks separations of 422 and 226 when sintered using SSR (1400°C 100 h). The ceramics were sintered at the temperature between 1150 and 1300°C under 30 Mpa pressure [63].

This may result in the disordered BZT with a high density of 7.62 g/cm^3, which is approximately 20% higher than that of low-density samples of 5.0–6.0 g/cm^3 synthesized by conventional SSR. The full width at half maximum (FWHM) of the 420 peak became narrower with an increase in the temperature from 1100 to 1300°C (**Figure 20(b)**) indicates that the degree of crystallisation of the disordered cubic phase is improved without the need for conversion to the ordered trigonal phase. Regardless of the method of synthesis, *Qf* is strongly dependent on density, and *Qf* values were improved with increasing density. The dense disordered BZT ceramics synthesized by SPS showed a significantly high *Qf* (= 53.4 × 10^3 GHz) comparable to that of the ordered BZT with the same density (= ca. 7.5 g/cm^3) synthesized by SSR. The crystallisation with densification of BZT ceramics should play a more critical role in the improvement of the *Q* factor in the BZT system rather than the structural ordering.

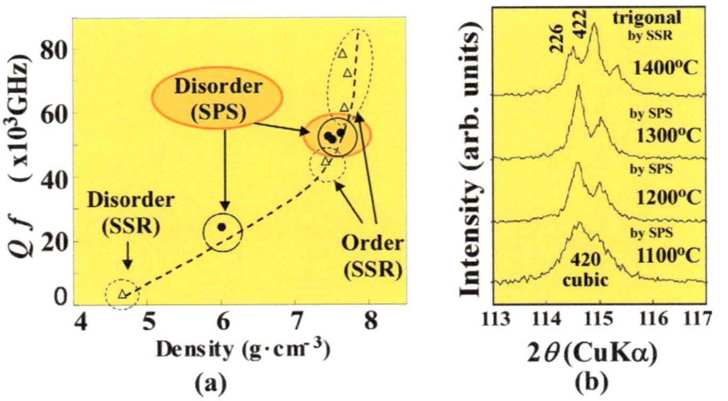

Figure 20.
Qf of BZT by solid-state reaction (SSR) and spark plasma sintering (SPS) as a function of density, Disordered BZT by SPS shows high Qf (a). Nonsplitting XRPD patterns around 420 diffraction of BZT sintering by SPS with different sintering temperatures compared with ordered sample by SSR with splitting pattern (b).

2.3.3 Deviated compositions with high Qf from stoichiometric complex perovskite composition

In a BaO-Mg/ZnO-Ta_2O_5 partial ternary ceramic (BMT/BZT system), complex perovskite such as BMT and BZT are forming solid solutions, and the *Qf* values varied intrinsically based on the crystal structure in the solid solutions depending on the density and defects. In this section, the crystal structure and properties on the varied compositions from the stoichiometric complex perovskite composition are reviewed for high *Qf* research.

2.3.3.1 The highest Qf composition with intrinsic compositional density by Kugimiya's research

Kugimiya [22, 27] presented the highest *Qf* composition with intrinsic compositional density on the Ta and Ba rich side near the BMT-$Ba_5Ta_4O_{15}$ tie-line in a BaO-MgO-$TaO_{5/2}$ partial system (BMT system), as shown in **Figure 21**. He presented three areas divided by the following two lines as shown in **Table 1** and **Figure 21**.

$$\alpha = 5\gamma/4 \quad (3)$$

$$\alpha = \gamma/2 \quad (4)$$

Here, α and γ are as written in the formula $\alpha BaO \cdot \gamma TaO_{5/2}$. On the $\alpha = 5\gamma/4$ line, $Ba_{1+\alpha}(Mg_{1/3}Ta_{2/3+4\alpha/5}V_{\alpha/5})O_{3+3\alpha}$ solid solutions are formed as the ideal compositions without vacancies in the *A*- and O-sites. In the *B*-site, the vacancy is neutralized and without charge.

In **Figure 21**, the composition with intrinsic compositional high density shows the highest *Q* of 50.0×10^3 on the tie-line between BMT and $Ba_5Ta_4O_{15}$ ($\alpha = 5\gamma/4$). The contour lines in **Figure 21** show *Q* values from 2.0×10^3 in the outer area to 25.0×10^3 in the centre. The contour is elongated parallel to the *Q* max line as drawn in **Figure 21,** and it changes steeply on the perpendicular to this line. So, the compositions without oxygen vacancy and with neutralised charge vacancies

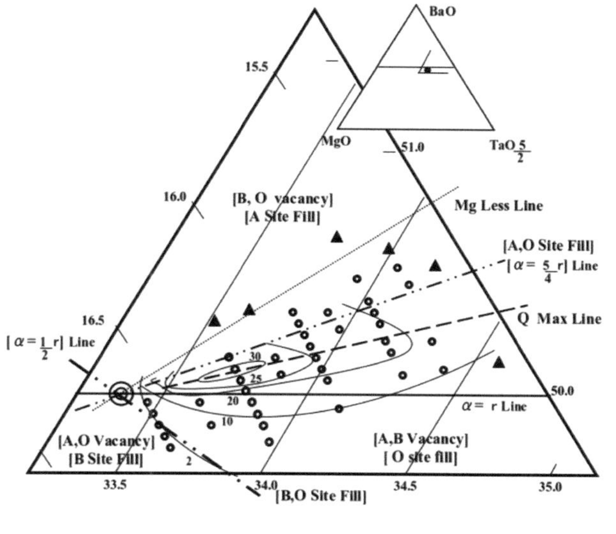

Figure 21.
BaO-MgO-$TaO_{5/2}$ partial system (BMT system).

α	Chemical formula	Vacancy
$\alpha > 5\gamma/4$	$Ba_{1+\alpha}(Mg_{1/3}Ta_{2/3+\gamma}V_{\alpha-\gamma})O_{3+\alpha+5\gamma/2}V_{2\alpha-5\gamma/2}$	*A*: fill, *B*, O: vacancy
$\alpha = 5\gamma/4$	$Ba_{1+\alpha}(Mg_{1/3}Ta_{2/3+4\alpha/5}V_{\alpha/5})O_{3+3\alpha}$	*A*, O: fill, *B*: vacancy
$5\gamma/4 > \alpha > \gamma/2$	$Ba_{1+\alpha}V_{5\gamma/6-2\alpha/3}(Mg_{1/3}Ta_{2/3+\gamma}V_{\alpha/3-\gamma/6})O_{3+\alpha+5\gamma/2}$	*A*, *B*: vacancy, O: fill
$\alpha = \gamma/2$	$Ba_{1+\alpha}V_{\alpha}(Mg_{1/3}Ta_{2/3+\gamma})O_{3+6\alpha}$	*A*: vacancy, *B*, O: fill
$\alpha < \gamma/2$	$Ba_{1+\alpha}V_{\gamma-\alpha}(Mg_{1/3}Ta_{2/3+\gamma})O_{3+\alpha+5\gamma/2}V_{\gamma/2-\alpha}$	*A*, O: vacancy, *B*: fill

Table 1.
*The chemical formula for three areas divided by two lines: $\alpha = 5\gamma/4$ and $\alpha = \gamma/2$, here, α and γ are in $Ba_{\alpha}Ta_{\gamma}O_{\alpha+5\gamma/2}$ and vacancies on the **A**-, **B**- and O-sites [22].*

are ideal for microwave dielectrics, and the density is high due to the partial substitution of Ta in the site of Mg, which is denoted as intrinsic compositional density [28]. Other regions have some defects degrading the *Qf* values, which were explained on the references [21, 22, 27, 28].

2.3.3.2 Phase conditions in the vicinity of BZT by Koga's research

Koga et al. [24, 25] showed the highest *Qf* composition shifted from the stoichiometric BZT composition. The ordering ratio of the deviated composition was not higher than that of the stoichiometric composition, which was calculated by the Rietveld method. These results were presented by the study of the phase relations in the vicinity of BZT in the BaO-ZnO-Ta_2O_5 ternary system, as shown in **Figure 22** [24, 25]. These samples were sintered at 1400°C/100 h as reported in Koga's paper. These diffraction patterns fit the Rietveld method well [23, 24]. Ordering ratios obtained are shown in **Figure 23(a)**. Three areas in the vicinity of BZT are presented as shown in **Figure 22**. 1st one (I) is ordering area with BZT single phase, the 2nd one (II) is ordering area with secondary phase and 3rd one (III) is disordering area with BZT single phase.

The first area (I) is characterised as a BZT single phase with an ordered structure and a high *Qf*. The varied compositions E and K have high *Qf* values about 50% higher than that of the stoichiometric BZT composition A. The ordering ratios at E and K are lower than that of stoichiometric BZT at A, but the density at E is the same as that of A [25]. The second (II) is composed by an ordered BZT

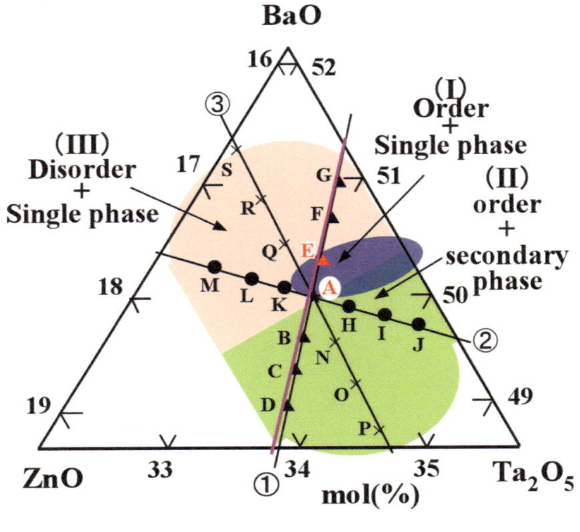

Figure 22.
Phase relations in the vicinity of BZT in the BaO-ZnO-Ta_2O_5 ternary system.

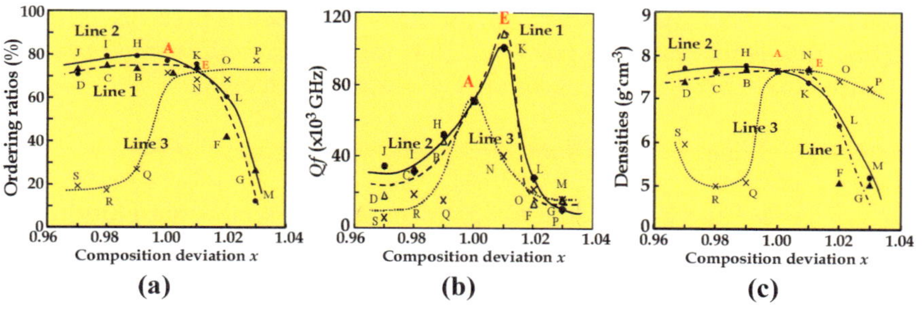

Figure 23.
Ordering ratio (a), Qf (b) and density (c) as a function of composition deviation from stoichiometric BZT.

accompanied by a secondary phase $BaTa_2O_6$ with a specific amount of Zn determined by X-ray microanalyser (XMA). The ordering ratio in this area is high at about 70–80% (**Figure 23(a)**). Although the structure is ordered, the *Qf* values decrease in the order of A-N-O-P from stoichiometric BZT (**Figure 23(b)**). The ordered BZT with the secondary phase is located on the Ta_2O_5 rich side as a eutectic phase diagram system. The third (III) with a disordered single phase shows low *Qf* and low density (**Figure 23(c)**). The low density comes from the numerous pores.

2.3.3.3 Phase conditions in the vicinity of BZT by Kolodiazhnyi's research

Kolodiazhnyi [29] also found the highest *Qf* of 330 × 10^3–340 × 10^3 GHz positions deviated from the stoichiometric BMT composition which is located in the BMT-$Ba_5Ta_4O_{15}$-$Ba_3Ta_2O_8$ compositional triangle (CT) as shown in **Figure 24**. The positions located in the single-phase BMT, which was indicated by green line. The position is close to the BMT-$Ba_5Ta_4O_{15}$ tie-line. A to H eight CTs are formed by BMT and five stable compounds, such as $Ba_5Ta_4O_{15}$, MgO, BaO, $Ba_9MgTa_{14}O_{45}$ and $Mg_4Ta_2O_9$, and three metastable compounds, $Ba_6Ta_2O_{11}$, $Ba_4Ta_2O_9$ and $Ba_3Ta_2O_8$. In A, B and C-CTs, although the samples demonstrated high density and a high degree

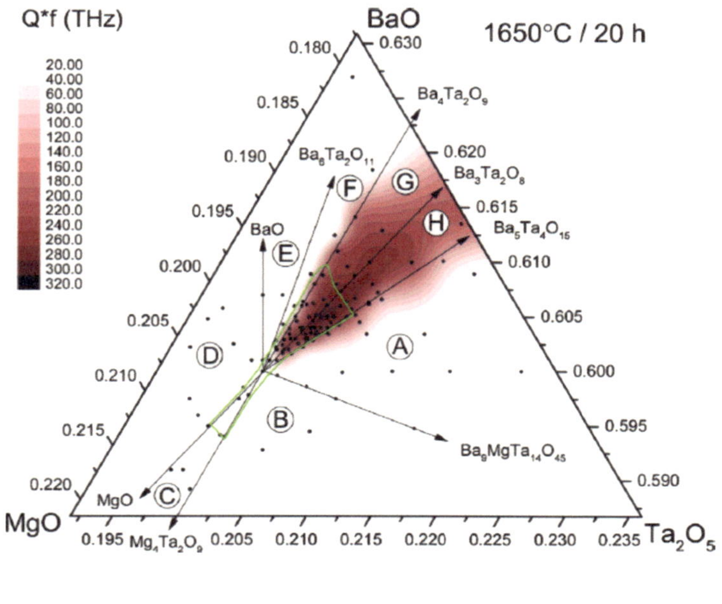

Figure 24.
Part of the BaO-MgO-Ta_2O_5 phase diagram in the vicinity of BMT divided into eight CTs. Small black dots indicate the target samples. Green line indicates an approximate boundary of the single-phase BMT.

of order, they showed low *Qf* values, attributed to the possible presence of the $Ba_9MgTa_{14}O_{45}$ second phase. Moreover, in D, E and F-CTs, as the samples were very low density, no electromagnetic resonance peaks were detected.

2.3.3.4 Koga's and Kolodiazhnyi's data comprehended in Kugimiya's data

Koga's data [24] and Kolodiazhnyi's [29] data are comparable with Kugimiya's BMT data [22]. The area (I) and the H-CT with the highest *Qf* as shown in **Figures 22** and **24**, respectively, are located on the opposite side of Kugimiya's data against the BMT-$Ba_5Ta_4O_{15}$ tie-line (**Figure 21**). These compositions will be comparable with that of the ideal crystal structure $Ba_{1+\alpha}(Mg_{1/3}Ta_{2/3+4\alpha/5}V_{\alpha/5})O_{3+3\alpha}$, as stated before in section (2.3.3.1) [22]. The formula is rewritten as $Ba(Mg_{1/3-\alpha/3}Ta_{2/3+2\alpha/15}V_{\alpha/5})O_3$ solid solutions on the tie-line BMT-$Ba_5Ta_4O_{15}$. The crystal structure in the composition region is ideal, without defects, and with an intrinsic high compositional density as described above. Surendran et al. [66] also reported a composition with high *Qf* deviated from stoichiometric BMT reviewed in detail in Intech Open Access Book [21].

2.3.4 Conclusions: important points concerning complex perovskites

- Ordering brings high *Qf* in the complex perovskite because of the long duration sintering. This situation has been bereaved for a long time. However, many examples contradicting this relation were presented.
- Koga et al. presented that *Qf* values of BZT did not depend on the ordering, preferably depending on the density and grain size.
- BZN with an order-disorder transition point at 1350°C (sample A) showed high *Qf* in the high-temperature disordered form. Moreover, annealing of the disordered sample B brings the ordered form, but the *Qf* does not improve. The both samples are analysed by the Rietveld method and HRTEM. The HRTEM presented the order form, disorder form and anti-phase domain by the FFT.
- Disordered samples with high density could not be synthesised by the solid-state reaction, but could be by SPS. The samples with disordered structure showed high *Q*. The ordering phenomenon is the only barometer of sintering in the solid-state reaction.
- Compositions deviated from stoichiometric complex perovskites such as BZT and BMT showed higher *Qf* and lower ordering than the stoichiometric composition. Based on these points, the ordering is not the reason for high *Qf*, and it is the only barometer of sintering.
- Intrinsic compositional density brings high *Qf*. On the BMT-$Ba_5Ta_4O_{15}$ tie-line, solid solutions are formed by the substitution Ta for Mg, which include high *Qf* compositions. The chemical composition with the highest *Qf* is $Ba_{1+\alpha}(Mg_{1/3}Ta_{2/3+4\alpha/5}V_{\alpha/5})O_{3+3\alpha}$, which is an ideal solid solutions without oxygen defects and neutralised vacancies (**Table 1**).
- Compositions deviated from stoichiometric BMT/BZT towards BaO and the Ta_2O_5 rich areas showing high *Qf*, as presented by Koga et al. [24], Kolodiazhny [29] and Surendran et al. [64], are comparable with intrinsic compositional density with high *Qf* as presented by Kugimiya [22].

3. Conclusions

The microwave dielectrics are the perfect, ideal and well-proportional crystal structures for low dielectric losses. Most of them belong to paraelectrics with inversion symmetry *i* and showing high symmetry and nondefects. In this chapter, the effects of ordering and symmetry were presented as follows: there are two types of ordering conditions. One is a case of nonphase transition such as pseudo tungsten-bronze solid solutions. These compounds show compositional ordering at a unique point of $x = 2/3$ on the $Ba_{6-3x}R_{8+2x}Ti_{18}O_{54}$ system, which shows the highest *Qf* without degradation of crystal symmetry. The other is a case of order-disorder phase transition such as indialite/cordierite. Indialite with a disordered structure and a high symmetry of 6/*mmm* point group has a higher *Qf* than cordierite with an ordered structure and low symmetry of *mmm* point group. It is clarified that the effect of high symmetry is predominant for high *Qf*. In the case of complex perovskite, a long sintering time of more than 80 h brings a high *Qf* accompanying ordering. It was clarified that the ordering is not necessary for high *Qf* and only a barometer of sintering in the solid-state reaction. Moreover, compositions deviated from stoichiometric complex perovskite showed higher *Qf* than the stoichiometric composition which has substituted Ta-ions for Mg/Zn-ions. The substitution brings a high density that is the compositional density. It was clarified that high compositional density brings high *Qf*.

Acknowledgements

The authors are grateful to Professors and graduate students of NIT, Meijo University and Hoseo University, and Doctors and researchers in the many companies, which collaborated with NIT. Visiting Professor Hitoshi Ohsato is grateful to the following projects: (1) support industries of Japan by Ministry of Economy, Trade and Industry (METI), Japan, (2) JSPS KAKENHI Grant Number 22560673, 25420721, JP16K06735 and (3) Nokia Foundation 2016 for Nokia Visiting Professors Project 201700003. Professor Heli Jantunen and Dr. Jobin Varghese are grateful to European Research Council Project No. 24001893 for financial assistance. The authors would also like to thank Honorary Research Professor Arthur E Hill of Salford University for valuable discussion and improving English during the preparation of this manuscript.

Author details

Hitoshi Ohsato[1,2]*, Jobin Varghese[1] and Heli Jantunen[1]

1 Microelectronics Research Unit, Faculty of Information Technology and Electrical Engineering, University of Oulu, Oulu, Finland

2 Department of Research, Nagoya Industrial Science Research Institute, Nagoya, Japan

*Address all correspondence to: ohsato.hitoshi@gmail.com

References

[1] Sebastian MT. Dielectric Materials for Wireless Communication. Amsterdam: Elsevier; 2008. ISBN 13:978-0-08-045330-9

[2] Sebastian MT, Ubic R, Jantunen H. Low-loss dielectric ceramic materials and their properties. International Materials Review. 2015;**60**:392-412

[3] Ohsato H. High frequency dielectric ceramics. In: Adachi G, editor. Materials Technology Hand Book for Rare-Earth Elements. Tokyo: NTS Inc; 2008. pp. 346-358 (Japanese)

[4] Ohsato H, Kagomiya I, Chae KW. Microwave dielectric ceramics with rare-earth elements (I). Journal of the Korean Physical Society. 2012;**61**:971-979

[5] Ohsato H, Kagomiya I, Kim JS. Microwave dielectric ceramics with rare-earth (II). Integrated Ferroelectrics. 2010;**115**:95-109

[6] Ohsato H. High frequency dielectrics. In: Shiosaki T, editor. Development and Applications of Ferroelectric Materials. Tokyo: CMC; 2001. pp. 135-147 (Japanese)

[7] Wakino K. Recent development of dielectric resonator materials and filters in Japan. Ferroelectrics. 1989;**91**:69-86

[8] Tamura H. Progress of ceramic dielectric materials for microwave and millimeterwave applications. In: MWE '99 Microwave Workshop Digest. 1999. pp. 175-180

[9] Kobayashi Y, Katoh M. Microwave measurement of dielectric properties of low-loss materials by the dielectric resonator rod method. IEEE Transaction on Microwave Theory and Technologies. 1985;**33**:586-592

[10] Ichinose N. High-frequency materials and their applications. New Ceramics & Electronic Ceramics. 1996;**9**(9):1-50

[11] Ohsato H. Microwave dielectrics. In: Fukunaga O, Haneda H, Makishima A, editors. Handbook of Multifunctional Ceramics. Tokyo: NTS; 2011. pp. 152-166 (Japanese)

[12] Ohsato H. Functional advances of microwave dielectrics for next generation. Ceramics International. 2012;**38**:S141-S146

[13] Ohsato H. Design of microwave dielectrics based on crystallography. In: Akedo J, Chen XM, Tseng T, editors. Advances in Multifunctional Materials and Systems II, Ceramic Transactions. Vol. 245. Hoboken, New Jersey: Wiley; 2014. pp. 87-100

[14] Kawashima S, Nishida M, Ueda I, Ouchi H. $Ba(Zn_{1/3}Ta_{2/3})O_3$ ceramics with low dielectric loss at microwave frequencies. Journal of the American Ceramic Society. 1983;**66**:421-423

[15] Ohsato H, Koga E, Kagomiya I, Kakimoto K. Origin of high *Q* for microwave complex perovskite. Key Engineering Materials. 2010;**421-422**:77-80

[16] Terada M, Kawamura K, Kagomiya I, Kakimoto K, Ohsato H. Effect of Ni substitution on the microwave dielectric properties of cordierite. Journal of the European Ceramic Society. 2007;**27**:3045-3148

[17] Ohsato H, Kim JS, Kim AY, Cheon CI, Chae KW. Millimeter-wave dielectric properties of cordierite/indialite glass ceramics. Japanese Journal of Applied Physics. 2011;**50**(9):09NF01-1-5

[18] Ohsato H. Millimeter-wave materials. In: Sebastian MT, Ubic R,

Jantunen H, editors. Microwave Materials and Applications. Wiley; 2017. pp. 203-265

[19] Ohsato H, Ohhashi T, Nishigaki S, Okuda T, Sumiya K, Suzuki S. Formation of solid solution of new tungsten bronze-type microwave dielectric compounds $Ba_{6-3x}R_{8+2x}Ti_{18}O_{54}$ (R = Nd and Sm, $0 \leq x \leq 1$). Japanese Journal of Applied Physics. 1993;**32**:4323-4326

[20] Ohsato H. Science of tungstenbronze-type like $Ba_{6-3x}R_{8+2x}Ti_{18}O_{54}$ (R = rare earth) microwave dielectric solid solutions. Journal of the European Ceramic Society. 2001;**21**:2703-2711

[21] Ohsato H. Microwave dielectrics with perovskite-type structure. In: Pan L, Zhu G, editors. Perovskite Materials—Synthesis, Characterization, Properties, and Applications. Rijeka, Croatia: INTECH; 2016. pp. 281-330. ISBN 978-953-51-4587-5. Available from: http://cdn.intechopen.com/pdfs-wm/49723.pdf [Accessed: 15-10-2018]

[22] Kugimiya K. Crystallographic study on the Q of $Ba(Mg_{1/3}Ta_{2/3})O_3$ dielectrics. In: Abstract for Kansai branch academic meeting. 5 September 2003; Senri-life science. B-20. In: Abstract for the 10th Meeting of Microwave/Millimeterwave Dielectrics and Related Materials on the Ceramic Soc. Japan: Nagoya Institute of Technology; 21st June 2004 (Japanese)

[23] Koga E, Moriwake H. Effects of superlattice ordering and ceramic microstructure on the microwave Q factor of complex Perovskite-type oxide $Ba(Zn_{1/3}Ta_{2/3})O_3$. Journal of the Ceramic Society of Japan. 2003;**111**:767-775 (Japanese)

[24] Koga E, Moriwake H, Kakimoto K, Ohsato H. Influence of composition deviation from stoichiometric $Ba(Zn_{1/3}Ta_{2/3})O_3$ on superlattice ordering and microwave quality factor Q. Journal of the Ceramic Society of Japan. 2005;**113**:172-178 (Japanese)

[25] Koga E, Yamagishi Y, Moriwake H, Kakimoto K, Ohsato H. Large Q factor variation within dense, highly ordered $Ba(Zn_{1/3}Ta_{2/3})O_3$ system. Journal of the European Ceramic Society. 2006;**26**:1961-1964

[26] Koga E, Yamagishi Y, Moriwake H, Kakimoto K, Ohsato H. Order-disorder transition and its effect on microwave quality factor Q in $Ba(Zn_{1/3}Nb_{2/3})O_3$ system. Journal of Electroceramics. 2006;**17**:375-379

[27] Fujimura T, Nishida M, Kugimiya K. Dielectric ceramics, United patent 5,246,898

[28] Ohsato H, Koga E, Kagomiya I, Kakimoto K. Dense composition with high-Q on the complex perovskite compounds. Ferroelectrics. 2009;**387**:28-35

[29] Kolodiazhnyi T. Origin of extrinsic dielectric loss in 1:2 ordered, single-phase $BaMg_{1/3}Ta_{2/3}O_3$. Journal of the European Ceramic Society. 2014;**34**:1741-1753

[30] Miyashiro A. Cordierite–indialite relations. American Journal of Science. 1957;**255**:43-62

[31] Gibbs GV. Polymorphism of cordierite. I. Crystal structure of low cordierite. American Mineralogist. 1966;**51**:1068-1087

[32] Wu JM, Huang HL. Effect of crystallization on microwave dielectric properties of stoichiometric cordierite glasses containing B_2O_3 and P_2O_5 glasses. Journal of Materials Research. 2000;**15**:222-227

[33] Izumi F, Ikeda T. A rietveld-analysis program RIETAN-98 and its applications to zeolites. Materials Science Forum. 2000;**321-324**:198-203

[34] Toraya H, Hibino H, Ohsumi K. A new powder diffractometer for synchrotron radiation with multiple-detector system. Journal of Synchrotron Radiation. 1996;**3**:75-83

[35] Ohsato H, Kagomiya I, Terada M, Kakimoto K. Origin of improvement of *Q* based on high symmetry accompanying Si–Al disordering in cordierite millimeter-wave ceramics. Journal of the European Ceramic Society. 2010;**30**:315-318

[36] Brown ID, Shannon RD. Empirical bond-strength–bond-length curves for oxides. Acta Cryst. 1973;**A29**:266-282

[37] Brown ID, Wu KK. Empirical parameters for calculating cation–oxygen bond valences. Acta Cryst. 1976;**B32**:1957-1959

[38] Ohsato H, Kim JS, Cheon CI, Kagomiya I. Crystallization of indialite/cordierite glass ceramics for millimeter-wave dielectrics. Ceramics International. 2015;**41**:S588-S595

[39] Ohsato H, Kim JS, Cheon CI, Kagomiya I. Millimeter-wave dielectrics of indialite/cordierite glass ceramics: Estimating Si/Al ordering by volume and covalency of Si/Al octahedron. Journal of the Ceramic Society of Japan. 2013;**121**:649-654

[40] Carvajal RJ. Fullprof Software. 2006. Available from: http://www-llb.cea.fr/fullweb/powder.htm [Accessed: 13-04-2018]

[41] Varfolomeev MB, Mironov AS, Kostomarov VS, Golubtsova LA, Zolotova TA. The synthesis and homogeneity ranges of the phases $Ba_{6-x}R_{8+2x/3}Ti_{18}O_{54}$. Russian Journal of Inorganic Chemistry. 1988;**33**:607-608

[42] Ohsato H, Mizuta M, Ikoma T, Onogi Z, Nishigaki S, Okuda T. Microwave dielectric properties of tungsten bronzetype $Ba_{6-3x}R_{8+2x}Ti_{18}O_{54}$ (*R* = La, Pr, Nd and Sm) solid solutions. Journal of the Ceramic Society of Japan International Edition. 1998;**106-185**:184-188

[43] Negas T, Davies PK. Influence of chemistry and processing on the electrical properties of $Ba_{6-3x}Ln_{8+2x}Ti_{18}O_{54}$ solid solutions. In: Material and Processes for Wireless Communications, Ceramic Transactions 53. Hoboken, New Jersey: Wiley; 1995. pp. 196-197

[44] Valant M, Suvorov D, Kolar D. X-ray investigations and dielectric property determination of the $Ba_{4.5}Gd_9Ti_{18}O_{54}$ compound. Japanese Journal of Applied Physics. 1996;**35**:144-150

[45] Ohsato H, Ohhashi T, Kato H, Nishigaki S, Okuda T. Microwave dielectric properties and structure of the $Ba_{6-3x}Sm_{8+2x}Ti_{18}O_{54}$ solid solutions. Japanese Journal of Applied Physics. 1995;**34**:187-191

[46] Fukuda K, Kitoh R, Awai I. Microwave characteristics of mixed phases of $BaPr_2Ti_4O_{12}$-$BaPr_2Ti_5O_{14}$ ceramics. Journal of Materials Research. 1995;**10**:312-319

[47] Valant M, Suvorov D, Rawn CJ. Intrinsic reasons for variations in dielectric properties of $Ba_{6-3x}R_{8+2x}Ti_{18}O_{54}$ (*R* = La-Gd) solid solutions. Japanese Journal of Applied Physics. 1999;**38**:2820-2826

[48] Matveeva RG, Varforomeev MB, Il'yuschenko LS. Refinement of the composition and crystal structure of $Ba_{3.75}Pr_{9.5}Ti_{18}O_{54}$. Zhurnal Neorganicheskoi Khimii. 1984;**29**:31-34 (Trans. Russ: Journal of Inorganic Chemistry. 1984;**29**:17-19)

[49] Roth RS, Beach F, Antoro A, Davis K, Soubeyroux JL. Structural of the nonstoichiometric solid solutions $Ba_2RE_4[Ba_x+RE_{2/3-2/3x}]Ti_9O_{27}$ (*RE* = Nd, Sm). In: 14 Int Congress Crystallog: Mat

Sci. C-138: Collected Abstract 07. 9-9; Perth, Australia. 1987

[50] Ohsato H, Ohhashi T, Okuda T. Structure of $Ba_{6-3x}Sm_{8+2x}Ti_{18}O_{54}$ (0<x<1). In: Asian Crystallographic Association Conference (AsCA '92): Ext. Abstract 14U-50; November 1992; Singapore. 1992

[51] Kolar D, Gabrscek S, Suvorov D. Structural and dielectric properties of perovskite-like rare earth titanates. In: Duran P, Fernandes JF, editors. Third Euro-Ceramics V.2. Vol. 2. Spein: Faenza Editrice Lberica; 1993. pp. 229-234

[52] Lundberg M, Sundberg M, Magneli A. The "pentagonal column" as a building unit in crystal and defect structure of some groups of transition metal compounds. Journal of Solid State Chemistry. 1982;**44**:32-40

[53] Ohsato H, Nishigaki S, Okuda T. Superlattice and dielectric properties of dielectric compounds. Japanese Journal of Applied Physics. 1992;**31**(9B):3136-3138

[54] Ohsato H. Crystallography of dielectrics. In: Crystallography in Japan (II)–The glorious development. Tokyo, Japan: The Crystallographic Society of Japan; 2014. pp. 184-185 (Japanese)

[55] Ohsato H, Imaeda M, Komura A, Okuda T. Non-linear microwave quality factor change based on the site occupancy of cations on the tungstenbronze-type $Ba_{6-3x}R_{8+2x}Ti_{18}O_{54}$ (R = rare earth) solid solutions. In: Nair KM, Bhalla AS, editors. Dielectric Ceramic Materials, Ceramic Transactions 100. Hoboken, New Jersey: Wiley; 1998. pp. 41-50

[56] Ohsato H, Imaeda M, Takagi Y, Komura A, Okuda T. Microwave quality factor improved by ordering of Ba and rare-earth on the tungstenbronze-type $Ba_{6-3x}R_{8+2x}Ti_{18}O_{54}$ (R = La, Nd and Sm) solid solutions. In: Proceeding of the XIth IEEE International Symposium on Applications of Ferroelectrics; IEEE catalog number 98CH36245. 1998. pp. 509-512

[57] Ohsato H, Mizuta M, Okuda T. Crystal structure and microwave dielectric properties of Tungstenbronze-type $Ba_{6-3x}R_{8+2x}Ti_{18}O_{54}$ (R = La, Nd and Sm) solid solutions. In: Morawiec H, Stroz D, editors. Applied Crystallography. World Scientific Publishing; 1998. pp. 440-447

[58] Galasso F, Pyle J. Ordering in compounds of the $A(B'_{0.33}Ta_{0.67})O_3$ type. Inorganic Chemistry. 1963;**2**:482-484

[59] Yokotani Y, Tsuruta T, Okuyama K, Kugimiya K. Low-dielectric loss ceramics for microwave uses. National Technical Report. 1994;**40**:11-16 (Japanese)

[60] Tamura H, Konoike T, Sakabe Y, Wakino K. Microwave dielectric properties of Ba-Nd-Ti-O system doped with metal oxides. Journal of the American Ceramic Society. 1984;**67**:C-59

[61] Matsumoto H, Tamura H, Wakino K. $Ba(Mg,Ta)O_3$-$BaSnO_3$ high-Q dielectric resonator. Japanese Journal of Applied Physics. 1991;**30**:2347-2349

[62] Kageyama K. Crystal structure and microwave dielectric properties of $Ba(Zn_{1/3}Ta_{2/3})O_3$-$(Sr,Ba)(Ga_{1/3}Ta_{1/2})O_3$ ceramics. Journal of the American Ceramic Society. 1992;**75**:1767-1771

[63] Koga E, Moriwake H, Kakimoto K, Ohsato H. Synthesis of disordered $Ba(Zn_{1/3}Ta_{2/3})O_3$ by spark plasma sintering and its microwave Q factor. Japanese Journal of Applied Physics. 2006;**45**(9B):7484-7488

[64] Ohsato H, Azough F, Koga E, Kagomiya I, Kakimoto K, Freer R. High symmetry brings high Q instead of ordering in $Ba(Zn_{1/3}Nb_{2/3})O_3$: A HRTEM

study. In: Akedo J, Ohsato H, Shimada T, editors. Advances in Multifunctional Materials and Systems, Ceramic Transactions. Vol. 216. Hoboken, New Jersey: Wiley; 2010. pp. 129-136

[65] Wu H, Davies PK. Influence of non-stoichiometry on the structure and properties of $Ba(Zn_{1/3}Nb_{2/3})O_3$ microwave dielectrics: I. Substitution of $Ba_3W_2O_9$. Journal of the American Ceramic Society. 2006;**89**(7):2239-2249. DOI: 10.1111/j.1551-2916.2006.01007.x

[66] Surendran KP, Sebastian MT, Mohanan P, Moreira RL, Dias A. Effect of nonstoichiometry on the structure and microwave dielectric properties of $Ba(Mg_{0.33}Ta_{0.67})O_3$. Chemistry of Materials. 2005;**17**:142-151

Chapter 2

Dielectric Responses in Multilayer C_f/Si_3N_4 as High-Temperature Microwave-Absorbing Materials

Heng Luo, Lianwen Deng and Peng Xiao

Abstract

High-temperature microwave-absorbing materials are in great demand in military and aerospace vehicles. The high-temperature dielectric behavior of multilayer C_f/Si_3N_4 composites fabricated by gelcasting has been intensively investigated at temperature coverage up to 800°C in the X-band (8.2–12.4 GHz). Experimental results show that the permittivity of Si_3N_4 matrix exhibits excellent thermo-stability with temperature coefficient lower than 10^{-3}°C^{-1}. Taking temperature-dependent polarized bound charge and damping coefficient into consideration, a revised dielectric relaxation model with Lorentz correction for Si_3N_4 ceramics has been established and validated by experimental results. Besides, a general model with respect to permittivity as a function of temperature and frequency has been established with the help of nonlinear numerical analysis to reveal mechanisms of temperature-dependent dielectric responses in C_f/Si_3N_4 composites. Temperature-dependent permittivity has been demonstrated to be well distributed on circular arcs with centers actually kept around the real (ε') axis in the Cole-Cole plane. Furthermore, space charge polarization and relaxation are discussed. These findings point to important guidelines to reveal the mechanism of dielectric behavior for carbon fiber functionalized composites including but not limited to C_f/Si_3N_4 composites at high temperatures, and pave the way for the development of high-temperature radar absorbing materials.

Keywords: high temperature, microwave-absorbing material, dielectric, relaxation

1. Introduction

Wireless electronic devices and communication instruments have found wide application in our daily life. Their efficient operation depends strongly on transmission behavior of alternating electromagnetic wave with frequency ranging from kilohertz (KHz) to gigahertz (GHz), and vice versa, are very sensitive to interference from external electromagnetic wave. Driven by the demand for both adequate interference rejection and controlled radiation, more and more efforts have been devoted to high-performance electromagnetic compatibility/interference (EMC/EMI) materials. As we all know, the propagation behavior of electromagnetic wave when encountering a material could be divided into three types in principle: reflection, absorption, and transmission. As typical EMI materials, metals or materials with high electrical conductivity could prevent external electromagnetic wave from penetration due to the large amount of free electrons. Last decades have witnessed intensive efforts toward exploring lightweight and cost-effective electromagnetic interference (EMI) materials with adequate shielding effectiveness [1–5], involving carbonaceous fillers-enabled polymers, novel lightweight metal composites, etc.

However, the primary function of EMI shielding is to reflect radiation using charge carriers that interact directly with incident electromagnetic field, and the back-radiation would in turn affect the surrounding environment and devices. What is more, the reflected radiation may also be caught by radar observation systems and lead to exposure of moving trace, which is extremely undesirable from the defense-oriented point of view. As a result, electromagnetic wave-absorbing materials with reduced reflection on the surface as well as enhanced internal attenuation are more favorable candidates for EMI shielding, especially in the GHz range. Polymers modified by carbon nanomaterials (e.g., carbon nanotubes [6–8], carbon nanofibers [9, 10], graphene [11, 12], etc.), metal powders [13, 14], and ferrite [15] have been demonstrated to be excellent microwave-absorbing/ shielding materials especially in the X-band (8.2–12.4 GHz) [16, 17], and have achieved successful application [18–20]. However, due to their inferior temperature stability and mechanical properties, their application is limited toward application under high temperature. For example, the temperature on the windward side of high-speed aircraft (>3 Ma) could reach up to 1000°C due to the aerodynamic heating effect. As a result, ceramics and their derivative architecture (r-GOs/SiO_2, CNT/SiO_2, ZnO/$ZrSiO_4$, SiC_f/SiC, etc.) [21–30] with the integration of desirable dielectric responses, high strength, oxidation resistance, thermo-stability, and low density have attracted growing attention for high-temperature-absorbing materials. Besides, Si_3N_4 ceramics are one of the most intensively studied ceramics in high-temperature applications due to their superior antioxidation (>1200°C) and mechanical and chemical stabilization properties [31–38]. More importantly, owing to the excellent electrical insulation property and low dielectric constant, Si_3N_4 ceramics are expected to be a promising candidate matrix as high-temperature microwave-absorbing materials [25, 29, 30, 39–42]. However, previous work mainly focused on experimental evolution of complex dielectric responses with temperature and qualitative analyses according to the Debye theory. Still, modeling for high-temperature dielectric behaviors is relatively limited and remains a great challenge due to the complexity of the components and microstructures for high-temperature microwave-absorbing materials, as well as high-temperature measurement system. It also should be noted that microwave dissipation capacity of a composite is strongly dependent on the structural design. Many investigations have shown that incorporation of reasonable structural design, involving multilayer structure [43–45] and periodic structure in metamaterials [46, 47], is an effective way to regulate dielectric response and guarantee desirable attenuation performance. Moreover, taking full advantage of tunable electromagnetic parameters in each layer, optimal microwave impedance matching as well as absorbing capability could be achieved. This fact means that it is essential to explore the mechanism of dielectric behavior of laminate-structure materials from new viewpoints.

In this chapter, we mainly focus on the microwave dielectric responses in laminate-structure or multilayer-structure C_f/Si_3N_4 composites from both experimental and theoretical points of view. Furthermore, a general model with respect to permittivity as a function of temperature and frequency would be established to reveal mechanisms of temperature-dependent dielectric responses for C_f/Si_3N_4 composites. These findings point to important guidelines to reveal the mechanism of dielectric behavior for carbon fiber functionalized composites including but not limited to C_f/Si_3N_4 composites at high temperatures, and pave the way for the development of high-temperature radar absorbing materials.

2. Experiments

2.1 Preparation of multilayer C_f/Si_3N_4 composites

Commercially available carbon fibers (T700, 12 K, TohoTenax Inc., Japan) were used as starting materials in this work. In order to avoid damage at high-temperature sintering, pyrolytic carbon(PyC)/SiC dual-coating on carbon fibers was prepared by chemical vapor deposition based on Methyltrichlorosilane (MTS)-H2-Ar system at 1150°C. The powder mixture of 85 wt% α-Si_3N_4 (purity > 93%, d_{50} = 0.5 μm, Beijing Unisplendor Founder High Technology Ceramics Co. Ltd., China), 5 wt% Al_2O_3, and 10 wt% Y_2O_3 was mixed with solvent-based acrylamide-N,N′-methylenebisacrylamide (AM-MBAM) system, and consolidated via gelcasting and pressureless-sintering route (as illustrated in **Figure 1**). Details of the multilayer C_f/Si_3N_4 samples' preparation are given in our previous work [35]. Note that each layer of carbon fiber plays dominant role to attenuate microwave energy, which could be adjudged the surface density of short carbon fiber.

2.2 High-temperature electromagnetic measurements

In order to evaluate the high-temperature permittivity, specimens with the size of 22.86 × 10.16 × 1.5 mm^3 were polished and determined in X-band through the wave-guide method with a vector network analyzer (Agilent N5230A, USA). As shown in **Figure 2**, the as-prepared Si_3N_4 ceramic sample was heated by an inner heater with a ramp rate of 10°C/min up to 800°C in air. For accuracy of measurement, the device was carefully calibrated with the through-reflect-line (TRL) approach, and a period of 10 min was applied to guarantee system stability at each evaluated temperature.

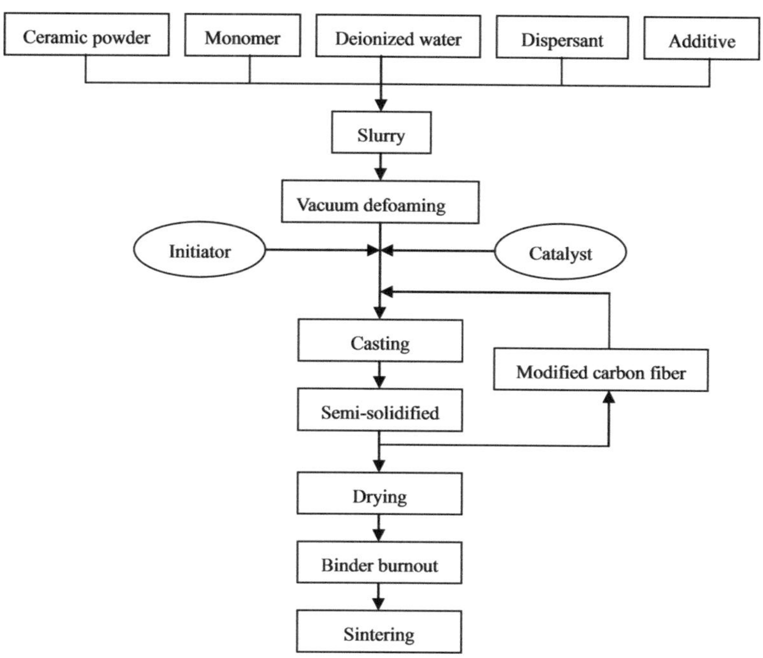

Figure 1.
The gelcasting process for preparation of multilayer C_f/Si_3N_4 composites (reprinted with permission from Ref. [35]).

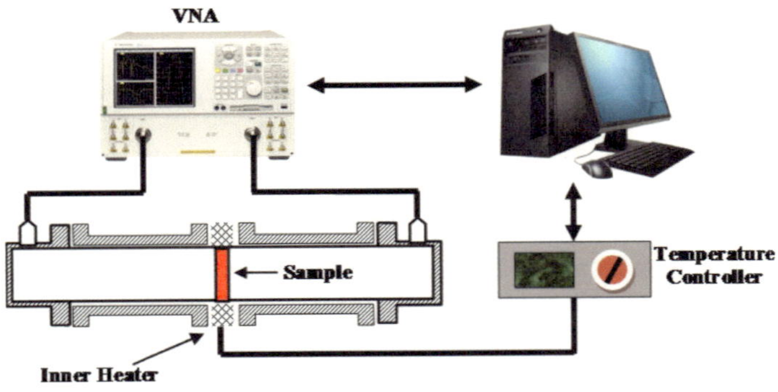

Figure 2.
Schematic diagram of complex permittivity test apparatus (reprinted with permission from Ref. [48]).

3. Microwave dielectric properties

3.1 Structure of multilayer C_f/Si_3N_4 composites

The optical image of cross-section of multilayer C_f/Si_3N_4 composites is shown in **Figure 3(a)**. As expected, three layers filled with short carbon fibers are uniformly embedded in the Si_3N_4 matrix. The microstructure of Si_3N_4 ceramic was formed by rod-like particles, which are evenly distributed and intercross with each other to form the main pores. Energy dispersive spectroscopy (EDS) analysis at spots A and B in **Figure 3(c)** demonstrates that the PyC/SiC interphase could effectively promote the chemical compatibility between carbon fibers and Si_3N_4 ceramic at high-temperature circumstance, which could be further proved by the XRD investigations (see **Figure 4**). As seen in **Figure 4**, in addition to the main β-Si_3N_4 peaks

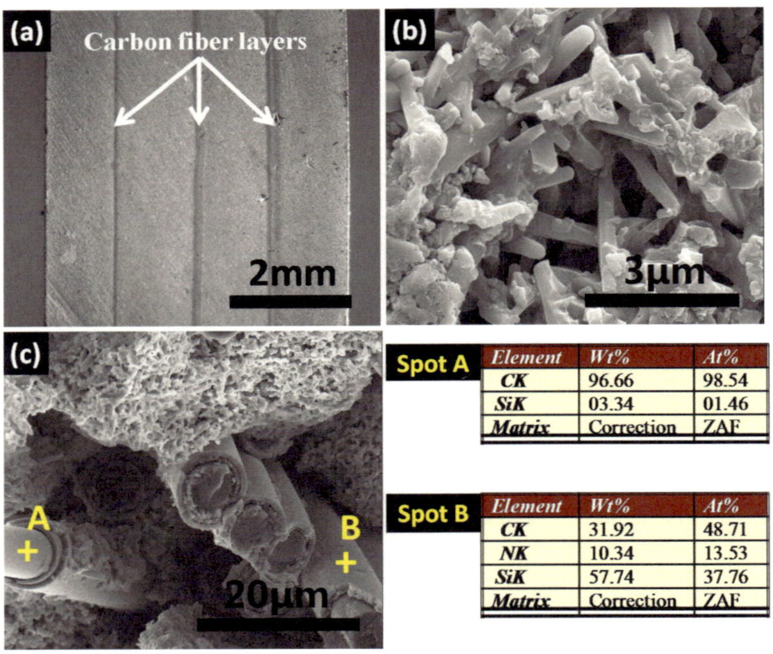

Figure 3.
(a) Cross-section of multilayer C_f/Si_3N_4 composites, facture surface located at (b) Si_3N_4 matrix and (c) carbon fibers (reprinted with permission from Ref. [39]).

and Y-Al oxide peaks, additional C and β-SiC peaks, corresponding to the carbon fiber and modification coating, were detected for C_f/Si_3N_4 composites.

3.2 Room-temperature dielectric properties of multilayer C_f/Si_3N_4 composites

According to the classical transmission line theory, microwave complex permittivity ($\varepsilon = \varepsilon' - j\varepsilon''$) is an important parameter to determine the absorbing performance. **Figure 5(a)** shows the real and imaginary permittivity of multilayer C_f/Si_3N_4 composites at X-band, as well as as-prepared Si_3N_4 ceramics. Clearly, the dielectric constant of Si_3N_4 ceramics presents frequency-independent behavior. The mean real and imaginary parts of permittivity and dielectric loss ($\tan\delta = \varepsilon''/\varepsilon'$) of pure Si_3N_4 ceramic were 7.7, 0.04, and 5.3×10^{-3}, respectively. The relatively low dielectric constant is considered to be helpful for microwave impedance matching with free space, which tends to reduce reflection of electromagnetic wave from the surface of material and enhance energy propagating in the material.

However, both the real permittivity and imaginary permittivity of C_f/Si_3N_4 sandwich composites decrease markedly as frequency increases at X-band, varying from 12.3 and 5.1 to 7.9 and 1.2, respectively. This phenomenon is usually called frequency dispersion characteristic, which is acknowledged to be beneficial to broaden the microwave absorption bandwidth. The reflection loss (R) of Si_3N_4 and C_f/Si_3N_4 sandwich composites was calculated according to the formula as follows:

$$R(dB) = 20 \log\left|\frac{Z_{in} - 1}{Z_{in} + 1}\right| \tag{1}$$

and

$$Z_{in} = \sqrt{\frac{\mu_r}{\varepsilon_r}} \tanh\left[j\left(\frac{2\pi}{c}\right)\sqrt{\mu_r \varepsilon_r} fd\right] \tag{2}$$

where Z_{in} refers to input impedance, j is the imaginary unit (i.e., equals to $\sqrt{-1}$), c is the velocity of electromagnetic waves in free space, f is the microwave frequency,

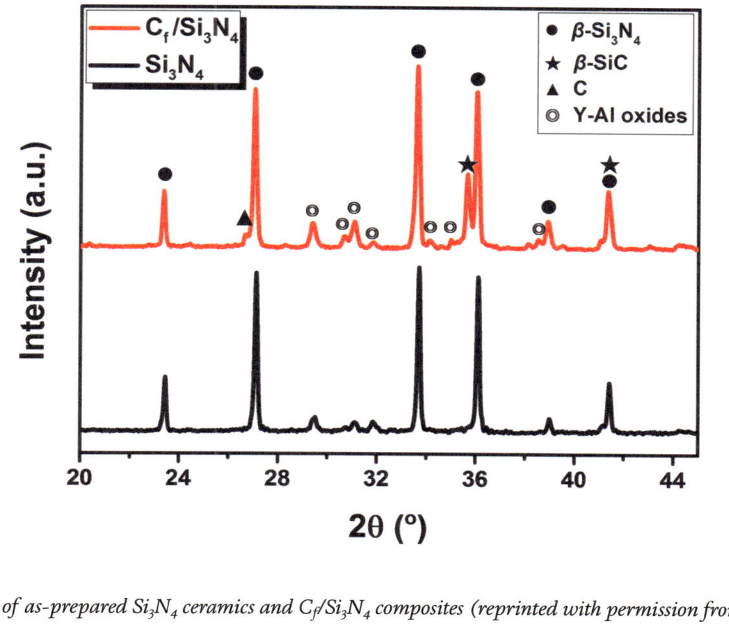

Figure 4.
XRD patterns of as-prepared Si_3N_4 ceramics and C_f/Si_3N_4 composites (reprinted with permission from Ref. [39]).

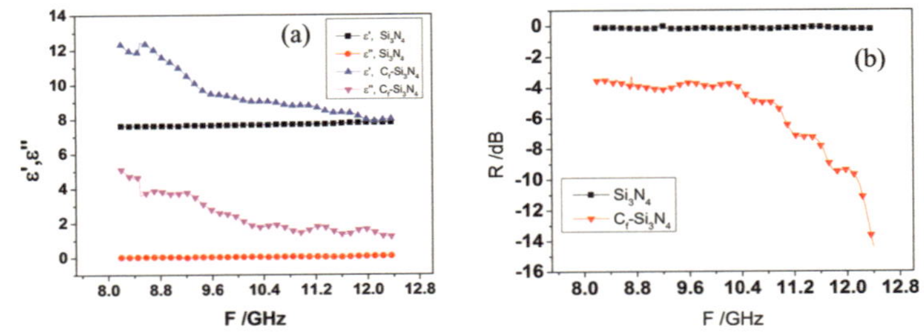

Figure 5.
(a) The permittivity and (b) reflection loss curves of Si_3N_4 and C_f/Si_3N_4 sandwich composites over X-band (reprinted with permission from Ref. [35]).

and d is the thickness of the samples. As depicted in **Figure 5(b)**, the microwave absorption ability of the C_f/Si_3N_4 sandwich composites was significantly enhanced compared with pure Si_3N_4 ceramic. The reflection loss of the C_f/Si_3N_4 sandwich composites gradually decreases from −3.5 dB to −14.4 dB with the increase of frequency, while that of the pure Si_3N_4 ceramic remains at −0.1 dB.

The enhanced microwave-absorbing performance could be mainly attributed to polarization relaxation. As we know, there exists migration of free electrons inside the electro-conductive carbon fibers, as well as charge accumulation at interfaces between short carbon fibers and insulated matrix when subjected to external electric field. As a result, the chopped carbon fibers are more inclined to be equivalent to micro-dipoles. With increase of frequency, the orientation of these dipoles could not keep up with change of electric field gradually, resulting in the real part of permittivity (ε') of C_f/Si_3N_4 sandwich composites decrease gradually. Furthermore, the scattering effect from defects and the crystal lattice on the back-and-forth movement of electrons under alternating electromagnetic waves predominately contributes to the dissipation of EM energy, which results in thermal energy.

For a deep-seated investigation of frequency-dependent dielectric responses of multilayer C_f/Si_3N_4 composites (**Figure 6(a)**), here we proposed an equivalent RC circuit model, where each layer of carbon fiber plays a role of one electrode in a plane-parallel capacitor, while each layer of Si_3N_4 ceramic plays the role of the dielectric (**Figure 6(b)**). Considering the existence of leakage current, leakage resistances were applied in equivalent circuit (**Figure 6(c)**).

According to the circuit theory knowledge, the relationship between permittivity and frequency ω follows:

$$\begin{aligned} \varepsilon' &= \frac{Q(C_1+C_2)^2}{R_1^2 C_1 C_2 (C_1+C_2)} \cdot \frac{1}{\omega^2} + \frac{Q R_1^2 C_1^2 C_2^2}{R_1^2 C_1 C_2 (C_1+C_2)} \\ \omega \cdot \varepsilon'' &= P \frac{(C_1+C_2)^2 + R_1^2 C_1^2 C_2^2 \cdot \omega^2}{(R_1+R_2)(C_1+C_2)^2 + R_1^2 R_2 C_1^2 C_2^2 \cdot \omega^2} \end{aligned} \quad (3)$$

where P, Q, and C are constants and are determined by the surface density of carbon fiber layers and thickness of Si_3N_4 layers. We have analyzed our experimental permittivity based on Eq. (3) using Trust-Region algorithm, which is illustrated in **Figure 7**. The points in **Figure 7** indicate the experimental data, while the results predicted by equivalent circuit model are given as solid line. Clearly, for multilayer C_f/Si_3N_4 composites, both ε' and ($\omega \cdot \varepsilon''$) are inversely proportional to the frequency square ω^2, and the predicted results agree quite well with the measured data. Additionally, the experimental data show oscillation phenomena at high frequency, which may result from charge and discharge processes between C_1 and C_2 (**Figure 6**) with the increase of frequency. Note that, even though the imaginary part of

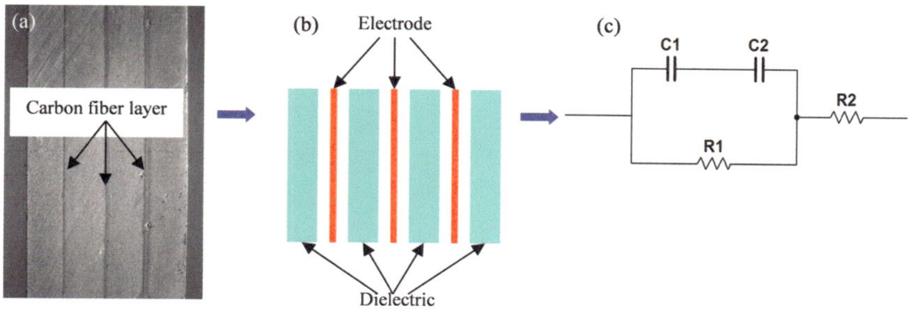

Figure 6.
(a) Cross-section morphology, (b) structural schematic diagram, and (c) equivalent circuit diagram of C_f/Si_3N_4 sandwich composites (reprinted with permission from Ref. [35]).

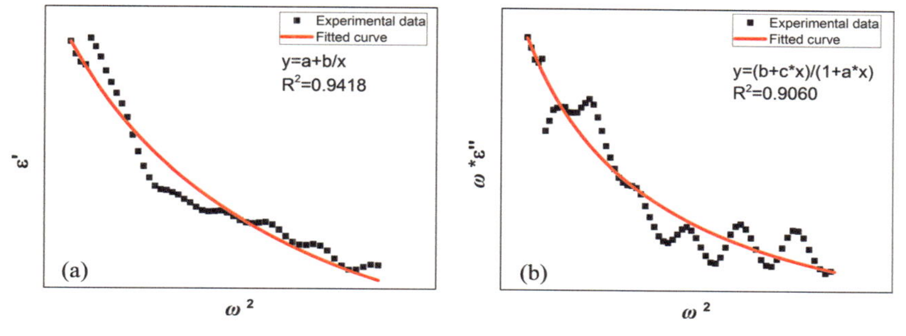

Figure 7.
Experimental data and curves of (a) ε′ and (b) ω · ε″ versus ω 2 (reprinted with permission from Ref. [35]).

permittivity declines faster than the real part with frequency increase, the reflection loss presents enhanced trend with frequency increase. This phenomenon is mainly attributed to the fact that microwave-absorbing efficiency is the combination of reflection from the material surface and attenuation inside the material. The lower the permittivity, the better its impedance matching between air and absorber. As a result, in order to achieve optimal reflection loss, one must lower reflection as much as possible and keep a modest loss tangent simultaneously.

3.3 High-temperature dielectric behaviors of Si_3N_4 ceramics

Due to the fact that there will inevitably be some variation of electromagnetic performance or even the mechanical property of materials served in high-temperature condition, the dielectric property would be supposed to dynamically change with temperature. How and to what extent does the permittivity dynamically change with temperature (increase or decrease)? All these are quite critical in parameter modification strategy for improving the accuracy of radar detection and guidance. Consequently, it is of utmost importance to explore the evolution of dielectric properties of Si_3N_4 ceramics used in high-temperature circumstances. Three-dimensional (3D) plots of the effect of temperature on permittivity of Si_3N_4 ceramics over X-band are shown in **Figure 8**. Clearly, the real permittivity (ε') shows no obvious change even though temperature rises up to 800°C. Likewise, thanks to the excellent electrical insulation of Si_3N_4 ceramics, the loss tangent (as shown in **Figure 8(b)**) is almost independent of frequency and temperature and remains lower than 0.06.

Herein, temperature coefficient κ is used to explicate the impact of temperature on dielectric response of as-prepared Si_3N_4 ceramics:

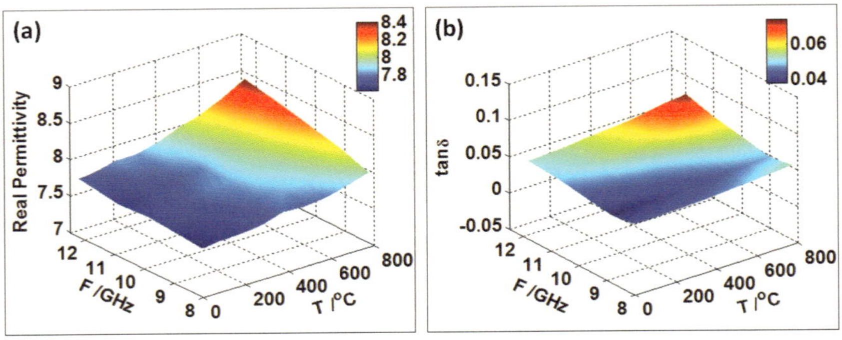

Figure 8.
Three-dimensional plots of complex permittivity of Si_3N_4 ceramics versus frequency and temperature (reprinted with permission from Ref. [39]).

$$\kappa = \frac{1}{\varphi} \cdot \frac{\Delta\varphi}{\Delta T}, \tag{4}$$

where T is the temperature and φ refers to either the dielectric constant or loss tangent. As summarized in **Table 1**, the temperature coefficients of both dielectric constant and loss tangent remain around 10^{-4}°C^{-1}. From this perspective, the as-prepared Si_3N_4 ceramics exhibit excellent thermo-stability of dielectric response within the range of evaluated temperatures. This weak temperature dependence further corroborates as-prepared Si_3N_4 ceramics to be a competitive candidate as the matrix of high-temperature microwave-absorbing materials.

It should be noted that the real permittivity increases slightly with frequency increase, which is contrary to the ordinary frequency dispersion effect described by the Debye model [49–55]. In order to further expound this peculiar frequency dispersion characteristic, it is essential to explore the details of electronic polarizing processing of Si_3N_4 ceramics. Considering the covalent bonding, the electronic polarization in Si_3N_4 ceramics mainly results from the bound charge's displacement deviated from the equilibrium position. The motion equation of bound charge driven by an external electric field $E_0e^{j\omega t}$ can be expressed as:

$$m\frac{\partial^2 x}{\partial t^2} = qE_0e^{j\omega t} - fx - 2\eta\frac{\partial x}{\partial t} \tag{5}$$

Temperature (°C)	$\kappa_{\varepsilon'}$ (×10^{-4}°C^{-1})		$\kappa_{\tan\delta}$ (×10^{-4}°C^{-1})	
	8.2 GHz	12.4 GHz	8.2 GHz	12.4 GHz
25	0.46	0.45	2.08	2.84
100	0.22	0.93	1.79	3.11
200	0.45	0.031	2.83	4.83
300	1.04	1.02	3.02	3.91
400	0.16	1.31	2.87	4.89
500	0.40	1.11	4.98	6.92
600	0.97	1.70	4.68	5.97
700	1.16	1.94	3.49	4.34

Table 1.
Temperature coefficient of permittivity and loss tangent at selected frequency (reprinted with permission from Ref. [39]).

where m, q, and x are the mass, charge, and displacement of single bound charge, respectively; f is the coefficient of restoring force; and η is the damping coefficient. Taking Lorentz correction [56–58] into consideration, the real permittivity of Si_3N_4 ceramic containing N polarized bound charges could be obtained as:

$$\varepsilon = \varepsilon_s + \frac{Nq^2}{\varepsilon_0 m} \cdot \frac{\omega_0{}^2 - \omega^2}{\left(\omega_0{}^2 - \omega^2\right)^2 + 4\eta^2\omega^2} \tag{6}$$

where ω_0 is the resonant frequency of Si_3N_4 ceramics, and ε_0 and ε_s are the vacuum permittivity and static dielectric constant of Si_3N_4 ceramics, respectively. Theoretical results have shown that the order of magnitude of resonant frequency ω_0 is around 10 eV (~10^{15} Hz) [59, 60] which is considerably larger than the tested frequency (~10^{10}Hz). Combining with Eq. (4), the real permittivity increases slightly with frequency increase, which is closely coincident with the experimental results. Furthermore, taking the effect of temperature into consideration, N and η should follow:

$$N \propto \exp\left(-E_a/RT\right) \tag{7}$$

$$\eta \propto \exp\left(-E_b/RT\right) \tag{8}$$

where E_a and E_b are the activation energy of electrons and lattice, respectively. The dependence of real permittivity of Si_3N_4 ceramics on temperature at three representative frequencies and the best fitting diagram according to Eqs. (6)–(8) are shown by solid lines in **Figure 9**.

It can be clearly seen that the dielectric constant gradually increases as temperature increased, starting from room temperature to 800°C, and results show that the real permittivity is well distributed on the predicted curves with coefficient of determination (R^2) ranging from 0.91 to 0.93. The characteristic parameters fitted from the temperature dependence of permittivity at three representative frequencies by the Trust-Region algorithm are also listed in **Figure 8**. The activation energy of electrons E_a is distributed between 15.46 and 17.49 KJ/mol, while the activation energy of

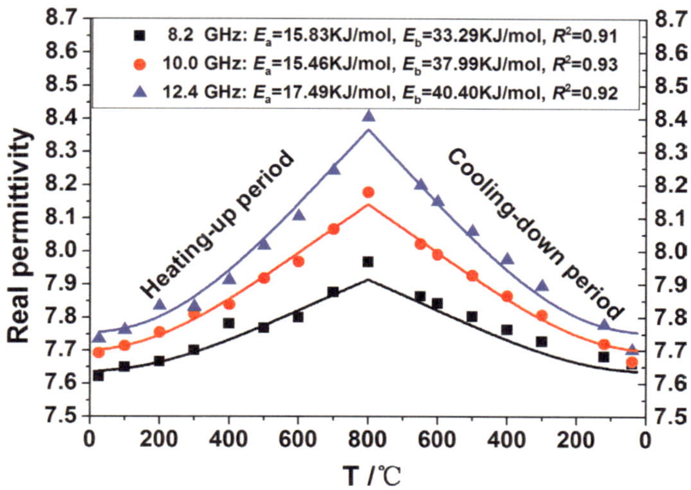

Figure 9.
Dependences of real permittivity of Si_3N_4 ceramics on temperature (reprinted with permission from Ref. [48]).

lattice E_b is distributed between 33.29 and 40.40 KJ/mol. The activation energy of electronic E_a is less than that of lattice E_b, which is mainly attributed to the binding force between the electrons and nucleus being lower than the covalent bonding force of lattice. Another important feature to be noticed is that the real permittivity of Si_3N_4 ceramics shows symmetrical features between the heating-up and cooling-down periods. The excellent thermo-stability of dielectric properties of Si_3N_4 ceramics has established the foundation for high-temperature radar absorbing materials.

3.4 High-temperature dielectric behaviors of multilayer C_f/Si_3N_4 composites

The evolution of ε' and ε'' for multilayer C_f/Si_3N_4 composites with temperature and frequency is illustrated in **Figure 10(a)** and **(b)**, respectively. It can be clearly seen that both ε' and ε'' of C_f/Si_3N_4 composites are enhanced with temperature. Similar phenomenon was also observed in other microwave-absorbing/shielding materials [24–26, 61–63]. Besides, as shown in **Figure 10(c)**, the loss tangent of C_f/Si_3N_4 composites remains higher than 0.6 over X-band and increases with increase in temperature. After comparison with low-loss β-SiC [64] as well as Si_3N_4, it is fairly clear that short carbon fibers are dominantly responsible for the frequency and temperature-dependent permittivity of C_f/Si_3N_4 composites. This is attributed to the unique microstructure (named skin-core structure), which was confirmed from transmission electron microscope analysis and selected-area electron-diffraction patterns previously [65, 66]. As illustrated in **Figure 11**, the core region consists of graphitic basal planes with random orientation, whereas graphitic basal planes are parallel to the fiber axis in the skin region [66, 67]. Therefore, electron migration behaviors both within and between graphitic basal planes inside carbon fibers would lead to electric charge accumulation at interfaces between carbon fibers and Si_3N_4 matrix (usually referred to as space charge polarization) when exposed to electromagnetic field. Furthermore, this space charge polarization is supposed to be

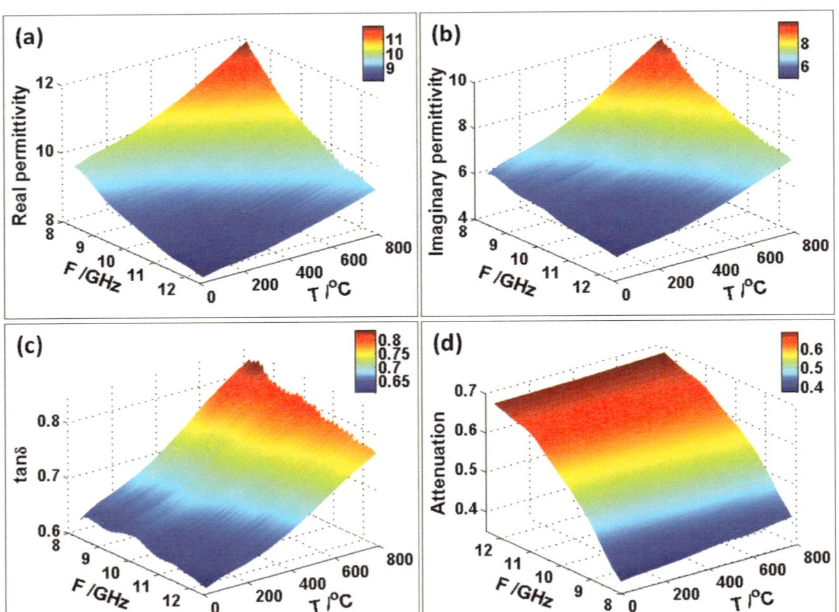

Figure 10.
Three-dimensional plots of dielectric properties of C_f/Si_3N_4 composites versus frequency and temperature: (a) real part, (b) imaginary part, (c) loss tangent, and (d) attenuation coefficient (reprinted with permission from Ref. [39]).

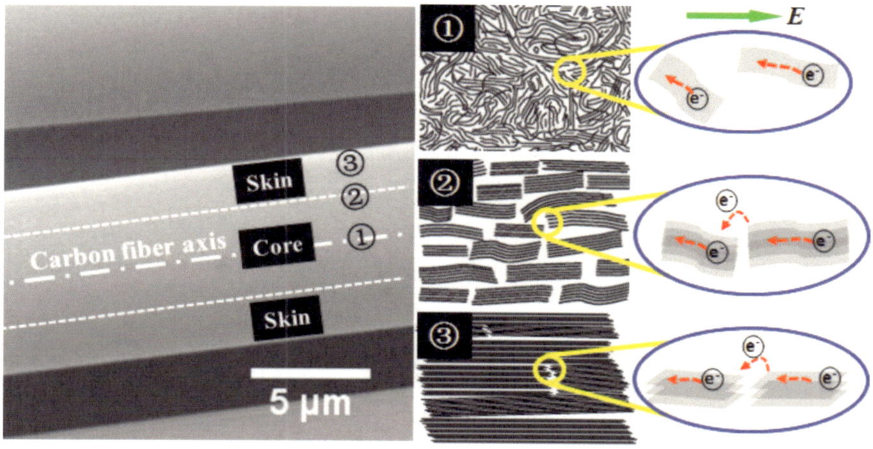

Figure 11.
Sketches of microstructure arrangement and electronic motion in carbon fiber (reprinted with permission from Ref. [39]).

enhanced with temperature rise since more and more electrons would be excited, which is in accordance with results in **Figure 10(a)** and **(b)**.

It should also be noted here that the lattice vibration also enhances with temperature rise, corresponding to the enhancement of scattering effect on migration of electrons. Fortunately, the energy of lattice vibration, which is described by phonons, is so small [68] that the scattering effect on electron migration could be neglected reasonably.

Attenuation coefficient is a quantity that characterizes how easily a material can be penetrated by incident microwave. A large attenuation coefficient means that the beam is quickly "attenuated" (weakened) as it passes through the material. The intrinsic attenuation coefficient of multilayer C_f/Si_3N_4 composites along with temperature and frequency derived from S parameters (i.e. $A = 1\text{-}S_{11}^2\text{-}S_{21}^2$) is illustrated in **Figure 10(d)**. It can be clearly observed that attenuation coefficient of C_f/Si_3N_4 composites enhances continuously with increase in frequency, and reaches a maximum value around 0.7 at 12.4 GHz, which is almost twice as much as that reported for C_f/SiO_2 composites [25]. This superior absorption performance suggests the multilayer C_f/Si_3N_4 composites to be competitive high-temperature microwave-absorbing materials. Another noteworthy phenomenon in **Figure 10(d)** is that the absorption coefficient remains steady in the investigated temperature range regardless of the loss tangent. As discussed previously, the overall absorbing performance results from reflection on the surface and attenuation inside the materials. However, the increasing permittivity means more severe impedance mismatch between air and the absorbing material, corresponding to reduction of penetrated electromagnetic energy. It could be explained by the reduced part of energy, which compensates the enhanced loss tangent, and consequently leads to a relatively steady absorption coefficient of C_f/Si_3N_4 composites with temperature rise.

3.5 Modeling of temperature-dependent dielectric responses for C_f/Si_3N_4 composites

As explicated in Section 3.2, the dielectric responses for multilayer C_f/Si_3N_4 composites have been evaluated experimentally over X-band. Additionally, the corresponding theoretical relationship of complex permittivity versus frequency at room temperature has been successfully established, which could be expressed as:

$$\varepsilon'(\omega) = \frac{c_1}{\omega^2} + c_2 \tag{9}$$

$$\varepsilon''(\omega) = \frac{\omega^2 + c_3}{c_4 \cdot \omega^3 + c_5 \cdot \omega} \tag{10}$$

where ω refers to the angular frequency (i.e., $\omega = 2\pi f$), and c_1, c_2, c_3, c_4, and c_5 are all pre-experimental parameters that are mainly associated with the surface density of short carbon fiber layers and thickness of Si_3N_4 layers. Herein, we first demonstrate this "room-temperature model" is still available at each evaluated temperature coverage up to 800°C or even higher with the help of nonlinear fitting technology. The best fitting curves of experimental data based on Eqs. (9) and (10) are depicted in **Figure 12**.

As expected, the measured results at each evaluated temperature agree quite well with the theoretical curve with coefficient of determination (R^2) above 0.98. These observed results suggest that both ε' and $\omega\varepsilon''$ of C_f/Si_3N_4 composites are still inversely proportional to the frequency square ω^2 within the temperature range of 25–800°C. However, a universal model coupled with frequency as well as temperature is urgently needed. Actually, a great deal of effort has been made to model frequency dispersive behaviors of permittivity for dielectrics [45]. It is well established that the development of all dielectric relaxation models that came after classical Debye's could be explicated as:

$$\varepsilon = \varepsilon_\infty + \frac{\varepsilon_s - \varepsilon_\infty}{1 + j\omega\tau} \tag{11}$$

where ε_s and ε_∞ are "static" and "infinite frequency" dielectric constants, respectively; j is the imaginary unit (i.e. equals to $\sqrt{-1}$); ω is the angular frequency; and τ is called the relaxation time. However, it is well known that the Debye model is originally developed for spherical polarizable molecules with a single relaxation time and without interaction between them [69]. Nevertheless, the short carbon fibers act as dipoles when exposed to altering electromagnetic fields oscillating in the microwave band, and electromagnetic field around each dipole is supposed to be coupled with that of neighboring dipoles. Furthermore, the relaxation time of chopped carbon fibers is supposed to be distributed over an interval rather than taking a constant value due to the variation of the fibers' length and the

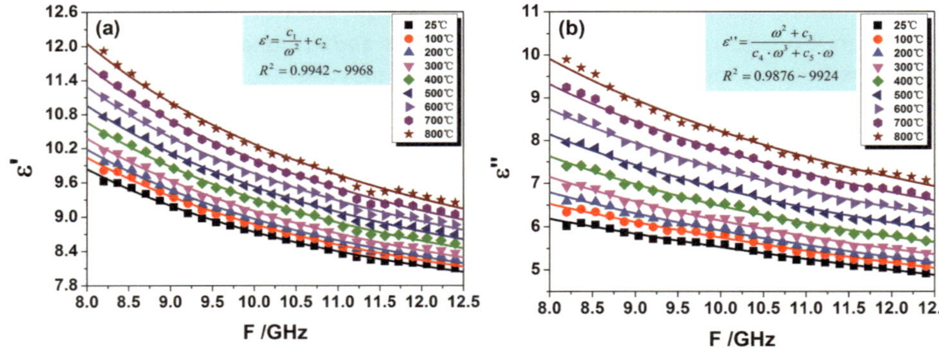

Figure 12.
(a) The real and (b) imaginary parts of permittivity for C_f/Si_3N_4 composites in X-band at the evaluated temperature (reprinted with permission from Ref. [39]).

heterogeneity of fibers' internal structure. This distribution property brings the Debye model (Eq. (11)) into more empirical and given by [25, 26, 70]:

$$\varepsilon = \varepsilon_\infty + \frac{\varepsilon_s - \varepsilon_\infty}{1 + (j\omega\tau)^{1-\alpha}} \tag{12}$$

where α is a parameter which determines the width of the distribution of relaxation time. All parameters in Eq. (12) should be a function of temperature. Substitution of Euler's formula $e^{jx} = \cos(x) + j\sin(x)$ in Eq. (12) yields

$$\varepsilon' = \varepsilon_\infty + \frac{(\varepsilon_s - \varepsilon_\infty)\cdot\left[1 + (\omega\tau)^{1-\alpha}\cdot\sin\frac{\alpha\pi}{2}\right]}{1 + 2(\omega\tau)^{1-\alpha}\cdot\sin\frac{\alpha\pi}{2} + (\omega\tau)^{2(1-\alpha)}} \tag{13}$$

$$\varepsilon'' = \frac{(\varepsilon_s - \varepsilon_\infty)\cdot(\omega\tau)^{1-\alpha}\cdot\cos\frac{\alpha\pi}{2}}{1 + 2(\omega\tau)^{1-\alpha}\cdot\sin\frac{\alpha\pi}{2} + (\omega\tau)^{2(1-\alpha)}} \tag{14}$$

We finally obtain the relationship between ε' and ε'', which could be expressed as

$$\left(\varepsilon' - \frac{\varepsilon_s + \varepsilon_\infty}{2}\right)^2 + \left(\varepsilon'' + \frac{\varepsilon_s + \varepsilon_\infty}{2}\tan\frac{\alpha\pi}{2}\right)^2 = \frac{(\varepsilon_s - \varepsilon_\infty)^2}{4\cos^2\frac{\alpha\pi}{2}} \tag{15}$$

Eq. (15) suggests that the locus of the permittivity in the $(\varepsilon', \varepsilon'')$ complex plane should still be a circular arc with different radius. To put this into perspective, the experimental data are re-plotted in **Figure 13**, in which the best fitted circles based on Eq. (15) are marked as solid curves. As seen, high agreement between the proposed model with experimental data is observed. Besides, the center tends to shift toward greater values of the ε' axis with increase in temperature, suggesting enhanced dielectric strength with temperature according to the Eq. (15). This may be attributed to the Enhanced electron concentration which participated in the polarization process occurs with temperature.

In addition, it is important to highlight some additional details. Firstly, the y-coordinates of fitted circular centers remain so small ($\sim 10^{-3}$) that Eq. (12) would be reduced to the classical Debye expression. From this point of view, the electronic polarization of short carbon fibers still follows the classic Debye relaxation process. Besides, the effects of temperature and electromagnetic interaction between neighboring short carbon fibers on relaxation time could be neglected reasonably.

Relaxation time also is one of key factors to analyze the dielectric behaviors for composites. After leaving out the effect of α, Eqs. (10) and (11) could be rewritten as follows:

$$\varepsilon' = \frac{1}{\tau}\cdot\frac{\varepsilon''}{\omega} + \varepsilon_\infty \tag{16}$$

It can be clearly seen from Eq. (16) that the real and imaginary parts of permittivity in (ε', ε''/ω) coordinate should be linear, which is also confirmed in **Figure 13**. What is more, the slope of each line is exactly the inverse of relaxation time τ at a certain temperature evaluated. In this case, we can come to a conclusion that the relaxation time for multilayer C_f/Si_3N_4 composites is weakly dependent on the temperature since the differences between fitted lines are quite modest. For clarity, a detailed plot of τ as a function of temperature is also illustrated as an inset in **Figure 14**. An increase from 216.1 to 250.2 ps has been derived when samples are

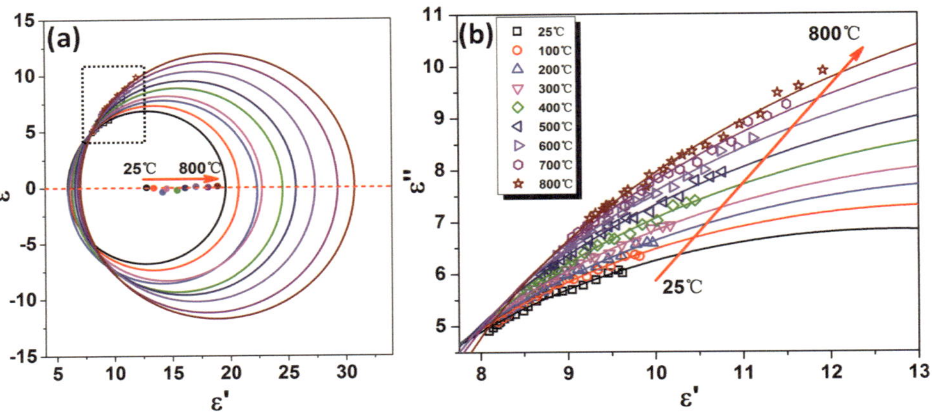

Figure 13.
(a) Argand diagram of C_f/Si_3N_4 composites at different temperature, (b) detailed view for the region marked by black box in (a) (reprinted with permission from Ref. [39]).

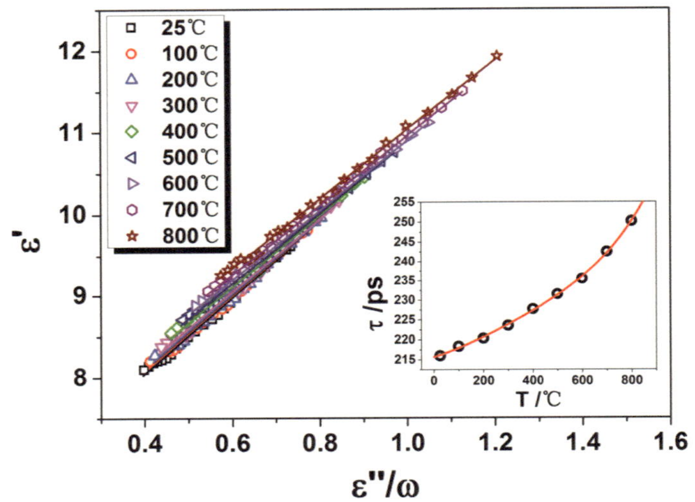

Figure 14.
Plot of ε' as a function of ε''/ω with an inset of temperature-dependent relaxation time (reprinted with permission from Ref. [39]).

heated from room temperature to 800°C. This gradually increasing trend may be ascribed to the enhancement of scattering effect between electrons for increase both the number and energy of electrons. Furthermore, the relaxation time τ (216.1–250.2 ps) is almost twice as much as a single cycle for time-harmonic electromagnetic wave in X-band ($t_0 = 1/f =$ 80.65–121.95 ps). As a result, electronic migration could not keep up with the pace of external alternating electronic field, which leads to continuous decrease of permittivity with frequency increase.

4. Conclusion

In this chapter, microwave dielectric properties of multilayer C_f/Si_3N_4 composites fabricated via gelcasting and pressureless sintering were intensively studied in X-band. Firstly, a strong frequency dependence of the real and imaginary parts of permittivity at room temperature was observed at the X-band. Particularly, an equivalent RC circuit model concerning the frequency-dependent permittivity of multilayer C_f/Si_3N_4 composites has been established. The predicted results reveal that both ε' and $(\omega \cdot \varepsilon'')$ are inversely proportional to the frequency square, and agree quite well with the measured data. Secondly, high-temperature dielectric behaviors of Si_3N_4 ceramic show that both permittivity and loss tangent exhibit excellent thermo-stability with temperature coefficient lower than 10^{-3}°C^{-1} within the temperature range of 25–800°C. A revised dielectric relaxation model with Lorentz correction for as-prepared Si_3N_4 ceramics was established and validated by experimental data. The activation energy of electrons (15.46–17.49 KJ/mol) was demonstrated to be slightly smaller than that of lattice (33.29–40.40 KJ/mol). Finally, the microwave attenuation coefficient of multilayer C_f/Si_3N_4 composites was inclined to be independent of temperature, and a maximum value of 0.7 could be achieved. The obvious positive temperature coefficient characteristic for permittivity is mainly attributed to the enhancement of electric polarization and relaxation of electron migration for graphitic basal planes of short carbon fibers. In addition to the fact that the room-temperature model concerning frequency-dependent permittivity is still available when extended into the full range of temperature coverage up to 800°C, an empirical equation with respect to temperature-dependent permittivity of multilayer C_f/Si_3N_4 composites has been established. It is concluded that the measured complex permittivity of multilayer C_f/Si_3N_4 composites is well distributed on circular arcs with centers actually kept around the real (ε') axis in $(\varepsilon', \varepsilon'')$ complex plane. Furthermore, the relaxation time as a function of temperature also has been derived. Results suggest that the relaxation time for multilayer C_f/Si_3N_4 composites increases from 216.1 to 250.2 ps when heated from room temperature to 800°C, and is almost twice as much as a single cycle for electromagnetic wave in X-band which leads to continuous decrease in permittivity with frequency increase. These findings point to important guidelines for analyzing high-temperature dielectric behaviors and revealing fundamental mechanisms for carbon fiber functionalized composites including but not limited to C_f/Si_3N_4 composites.

Acknowledgements

The authors would like to acknowledge the generous funding from the National Key Research and Development Program of China (Grant No. 2017YFA0204600), the State Key Development Program for Basic Research of China (Grant No. 2011CB605804), and the National Natural Science Foundation of China (Grant No. 51802352).

Conflict of interest

The authors declared that they have no conflicts of interest to this work.

Author details

Heng Luo[1,2]*, Lianwen Deng[1] and Peng Xiao[2]

1 School of Physics and Electronics, Central South University, Changsha, China

2 State Key Laboratory of Powder Metallurgy, Central South University, Changsha, China

*Address all correspondence to: luohengcsu@csu.edu.cn

References

[1] Mohan RR, Varma SJ, Sankaran J. Impressive electromagnetic shielding effects exhibited by highly ordered, micrometer thick polyaniline films. Applied Physics Letters. 2016;**108**(15):154101. DOI: 10.1063/1.4945791

[2] Zhao Y-T et al. Transparent electromagnetic shielding enclosure with CVD graphene. Applied Physics Letters. 2016;**109**(10):103507. DOI: 10.1063/1.4962474

[3] Zeng Z et al. Thin and flexible multi-walled carbon nanotube/ waterborne polyurethane composites with high-performance electromagnetic interference shielding. Carbon. 2016;**96**:768-777. DOI: 10.1016/j. carbon.2015.10.004

[4] Wen B et al. Reduced graphene oxides: Light-weight and high-efficiency electromagnetic interference shielding at elevated temperatures (Adv. Mater. 21/2014). Advanced Materials. 2014;**26**(21):3357-3357. DOI: 10.1002/ adma.201470138

[5] Shahzad F et al. Electromagnetic interference shielding with 2D transition metal carbides (MXenes). Science. 2016;**353**(6304):1137-1140. DOI: 10.1126/science.aag2421

[6] Zhang Y et al. Additive manufacturing of carbon nanotube-photopolymer composite radar absorbing materials. Polymer Composites. 2016. DOI: 10.1002/ pc.24117

[7] Hayashida K, Matsuoka Y. Electromagnetic interference shielding properties of polymer-grafted carbon nanotube composites with high electrical resistance. Carbon. 2015;**85**:363-371. DOI: 10.1016/j. carbon.2015.01.006

[8] Li N et al. Electromagnetic interference (EMI) shielding of single-walled carbon nanotube epoxy composites. Nano Letters. 2006;**6**(6):1141-1145. DOI: 10.1021/ nl0602589

[9] Micheli D et al. X-Band microwave characterization of carbon-based nanocomposite material, absorption capability comparison and RAS design simulation. Composites Science and Technology. 2010;**70**(2):400-409. DOI: 10.1016/j.compscitech.2009.11.015

[10] Al-Saleh MH, Sundararaj U. A review of vapor grown carbon nanofiber/polymer conductive composites. Carbon. 2009;**47**(1):2-22. DOI: 10.1016/j.carbon.2008.09.039

[11] Han M et al. Hierarchical graphene/SiC nanowire networks in polymer-derived ceramics with enhanced electromagnetic wave absorbing capability. Journal of the European Ceramic Society. 2016;**36**(11):2695-2703. DOI: 10.1016/j. jeurceramsoc.2016.04.003

[12] Huang X et al. A three-dimensional graphene/Fe3O4/carbon microtube of sandwich-type architecture with improved wave absorbing performance. Scripta Materialia. 2016;**120**:107-111. DOI: 10.1016/j.scriptamat.2016.04.025

[13] Olmedo L, Hourquebie P, Jousse F. Microwave absorbing materials based on conducting polymers. Advanced Materials. 1993;**5**(5):373-377. DOI: 10.1002/adma.19930050509

[14] Huang Y et al. The microwave absorption properties of carbon-encapsulated nickel nanoparticles/ silicone resin flexible absorbing material. Journal of Alloys and Compounds. 2016;**682**:138-143. DOI: 10.1016/j.jallcom.2016.04.289

[15] Chen Y et al. Enhanced electromagnetic interference shielding efficiency of polystyrene/graphene composites with magnetic Fe_3O_4 nanoparticles. Carbon. 2015;**82**:67-76. DOI: 10.1016/j.carbon.2014.10.031

[16] Al-Saleh MH, Saadeh WH, Sundararaj U. EMI shielding effectiveness of carbon based nanostructured polymeric materials: A comparative study. Carbon. 2013;**60**:146-156. DOI: 10.1016/j.carbon.2013.04.008

[17] Qin F, Brosseau C. A review and analysis of microwave absorption in polymer composites filled with carbonaceous particles. Journal of Applied Physics. 2012;**111**(6):061301. DOI: 10.1063/1.3688435

[18] Chung DDL. Carbon materials for structural self-sensing, electromagnetic shielding and thermal interfacing. Carbon. 2012;**50**(9):3342-3353. DOI: 10.1016/j.carbon.2012.01.031

[19] Liu L, Das A, Megaridis CM. Terahertz shielding of carbon nanomaterials and their composites – A review and applications. Carbon. 2014;**69**:1-16. DOI: 10.1016/j.carbon.2013.12.021

[20] Qin F, Peng H-X. Ferromagnetic microwires enabled multifunctional composite materials. Progress in Materials Science. 2013;**58**(2):183-259. DOI: 10.1016/j.pmatsci.2012.06.001

[21] Yin X et al. Electromagnetic properties of Si–C–N based ceramics and composites. International Materials Reviews. 2014;**59**(6):326-355. DOI: 10.1179/1743280414Y.0000000037

[22] Duan W et al. A review of absorption properties in silicon-based polymer derived ceramics. Journal of the European Ceramic Society. 2016;**36**(15):3681-3689. DOI: 10.1016/j.jeurceramsoc.2016.02.002

[23] Yin X et al. Dielectric, electromagnetic absorption and interference shielding properties of porous yttria-stabilized zirconia/silicon carbide composites. Ceramics International. 2012;**38**(3):2421-2427. DOI: 10.1016/j.ceramint.2011.11.008

[24] Wen B et al. Reduced graphene oxides: Light-weight and high-efficiency electromagnetic interference shielding at elevated temperatures. Advanced Materials. 2014;**26**(21):3484-3489. DOI: 10.1002/adma.201400108

[25] Cao M-S et al. The effects of temperature and frequency on the dielectric properties, electromagnetic interference shielding and microwave-absorption of short carbon fiber/silica composites. Carbon. 2010;**48**(3):788-796. DOI: 10.1016/j.carbon.2009.10.028

[26] Wen B et al. Temperature dependent microwave attenuation behavior for carbon-nanotube/silica composites. Carbon. 2013;**65**:124-139. DOI: 10.1016/j.carbon.2013.07.110

[27] Kong L et al. High-temperature electromagnetic wave absorption properties of $ZnO/ZrSiO_4$ composite ceramics. Journal of the American Ceramic Society. 2013;**96**(7):2211-2217. DOI: 10.1111/jace.12321

[28] Iveković A et al. Current status and prospects of SiCf/SiC for fusion structural applications. Journal of the European Ceramic Society. 2013;**33**(10):1577-1589. DOI: 10.1016/j.jeurceramsoc.2013.02.013

[29] Liu H, Tian H, Cheng H. Dielectric properties of SiC fiber-reinforced SiC matrix composites in the temperature range from 25 to 700°C at frequencies between 8.2 and 18GHz. Journal of Nuclear Materials. 2013;**432**(1-3):57-60. DOI: 10.1016/j.jnucmat.2012.08.026

[30] Tian H, Liu H-T, Cheng H-F. A high-temperature radar absorbing

structure: Design, fabrication, and characterization. Composites Science and Technology. 2014;**90**:202-208. DOI: 10.1016/j.compscitech.2013.11.013

[31] Azarhoushang B, Soltani B, Zahedi A. Laser-assisted grinding of silicon nitride by picosecond laser. The International Journal of Advanced Manufacturing Technology. 2017;**93**(5):2517-2529. DOI: 10.1007/s00170-017-0440-9

[32] Xu Y et al. Shear-thinning behavior of the $CaO-SiO_2-CaF_2-Si_3N_4$ system mold flux and its practical application. International Journal of Minerals, Metallurgy, and Materials. 2017;**24**(10):1096-1103. DOI: 10.1007/s12613-017-1500-8

[33] Zuo K-H, Zeng Y-P, Jiang D-L. The mechanical and dielectric properties of Si_3N_4-based sandwich ceramics. Materials & Design. 2012;**35**:770-773. DOI: 10.1016/j.matdes.2011.09.019

[34] Yang J, Yu J, Huang Y. Recent developments in gelcasting of ceramics. Journal of the European Ceramic Society. 2011;**31**(14):2569-2591. DOI: 10.1016/j.jeurceramsoc.2010.12.035

[35] Luo H et al. Dielectric properties of Cf–Si3N4 sandwich composites prepared by gelcasting. Ceramics International. 2014;**40**(6):8253-8259. DOI: 10.1016/j.ceramint.2014.01.023

[36] Bocanegra-Bernal MH, Matovic B. Mechanical properties of silicon nitride-based ceramics and its use in structural applications at high temperatures. Materials Science and Engineering A. 2010;**527**(6):1314-1338. DOI: 10.1016/j.msea.2009.09.064

[37] Dambatta YS et al. Ultrasonic assisted grinding of advanced materials for biomedical and aerospace applications—a review. The International Journal of Advanced Manufacturing Technology. 2017;**92**(9):3825-3858. DOI: 10.1007/s00170-017-0316-z

[38] Tatli Z, Thompson DP. Low temperature densification of silicon nitride materials. Journal of the European Ceramic Society. 2007;**27**(2-3):791-795. DOI: 10.1016/j.jeurceramsoc.2006.04.010

[39] Luo H et al. Modeling for high-temperature dielectric behavior of multilayer Cf/Si3N4 composites in X-band. Journal of the European Ceramic Society. 2017;**37**(5):1961-1968. DOI: 10.1016/j.jeurceramsoc.2016.12.028

[40] Peng C-H, Chen PS, Chang C-C. High-temperature microwave bilayer absorber based on lithium aluminum silicate/lithium aluminum silicate-SiC composite. Ceramics International. 2014;**40**(1 Part A):47-55. DOI: 10.1016/j.ceramint.2013.05.101

[41] Yuan J et al. High dielectric loss and microwave absorption behavior of multiferroic $BiFeO_3$ ceramic. Ceramics International. 2013;**39**(6):7241-7246. DOI: 10.1016/j.ceramint.2013.01.068

[42] Qing YC et al. Microwave absorbing ceramic coatings with multi-walled carbon nanotubes and ceramic powder by polymer pyrolysis route. Composites Science and Technology. 2013;**89**:10-14. DOI: 10.1016/j.compscitech.2013.09.007

[43] Liu H, Cheng H, Tian H. Design, preparation and microwave absorbing properties of resin matrix composites reinforced by SiC fibers with different electrical properties. Materials Science and Engineering B. 2014;**179**:17-24. DOI: 10.1016/j.mseb.2013.09.019

[44] Chen M et al. Gradient multilayer structural design of $CNTs/SiO_2$ composites for improving microwave absorbing properties. Materials & Design. 2011;**32**(5):3013-3016. DOI: 10.1016/j.matdes.2010.12.043

[45] Luo H, Xiao P, Hong W. Dielectric behavior of laminate-structure Cf/Si_3N_4 composites in X-band. Applied Physics Letters. 2014;**105**(17):172903. DOI: 10.1063/1.4900932

[46] Cui TJ, Smith D, Liu R. Metamaterials: Theory, Design, and Applications. 233 Spring Street, New York, USA Springer Publishing Company, Incorporated; 2009. p. 368

[47] Liu Y, Zhang X. Metamaterials: A new frontier of science and technology. Chemical Society Reviews. 2011;**40**(5):2494-2507. DOI: 10.1039/C0CS00184H

[48] Shao S et al. Effect of temperature on dielectric response in X-band of silicon nitride ceramics prepared by gelcasting. AIP Advances. 2018;**8**(7):075127. DOI: 10.1063/1.5033965

[49] Amirat Y, Shelukhin V. Homogenization of time harmonic Maxwell equations and the frequency dispersion effect. Journal de Mathématiques Pures et Appliquées. 2011;**95**(4):420-443. DOI: 10.1016/j.matpur.2010.10.007

[50] Haijun Z et al. Complex permittivity, permeability, and microwave absorption of Zn- and Ti-substituted barium ferrite by citrate sol–gel process. Materials Science and Engineering B. 2002;**96**(3):289-295. DOI: 10.1016/S0921-5107(02)00381-1

[51] Joshi A, Kumar S, Verma NK. Study of dispersion, absorption and permittivity of an synthetic insulation paper—with change in frequency and thermal aging. NDT & E International. 2006;**39**(1):19-21. DOI: 10.1016/j.ndteint.2005.05.004

[52] Kamenetsky FM. Frequency dispersion of rock properties in equations of electromagnetics. Journal of Applied Geophysics. 2011;**74**(4):185-193. DOI: 10.1016/j.jappgeo.2011.04.004

[53] Razzitte AC, Fano WG, Jacobo SE. Electrical permittivity of Ni and NiZn ferrite–polymer composites. Physica B: Condensed Matter. 2004;**354**(1-4):228-231. DOI: 10.1016/j.physb.2004.09.054

[54] Rica RA, Jiménez ML, Delgado AV. Electric permittivity of concentrated suspensions of elongated goethite particles. Journal of Colloid and Interface Science. 2010;**343**(2):564-573. DOI: 10.1016/j.jcis.2009.11.063

[55] Tan YQ et al. Giant dielectric-permittivity property and relevant mechanism of $Bi_{2/3}Cu_3Ti_4O_{12}$ ceramics. Materials Chemistry and Physics. 2010;**124**(2-3):1100-1104. DOI: 10.1016/j.matchemphys.2010.08.041

[56] Moysés Araújo C et al. Electrical resistivity, MNM transition and band-gap narrowing of cubic GaN: Si. Microelectronics Journal. 2002;**33**(4):365-369. DOI: 10.1016/S0026-2692(01)00133-1

[57] Fernandez JRL et al. Electrical resistivity and band-gap shift of Si-doped GaN and metal-nonmetal transition in cubic GaN, InN and AlN systems. Journal of Crystal Growth. 2001;**231**(3):420-427. DOI: 10.1016/S0022-0248(01)01473-7

[58] Schuurmans FJP, Vries Pd, Lagendijk A. Local-field effects on spontaneous emission of impurity atoms in homogeneous dielectrics. Physics Letters A. 2000;**264**(6):472-477. DOI: 10.1016/S0375-9601(99)00855-5

[59] Hao Wang YC, Kaneta Y, Iwata S. First-principles investigation of the structural, electronic and optical properties of olivine-Si3N4 and olivine-Ge3N4. Journal of Physics: Condensed Matter. 2006;**18**(47):10663-10676. DOI: 10.1088/0953-8984/18/47/012

[60] Xu M et al. Theoretical prediction of electronic structures and optical properties of Y-doped γ-Si3N4. Physica B: Condensed Matter. 2008;**403** (13-16):2515-2520. DOI: 10.1016/j.physb.2008.01.042

[61] Yang H-J et al. Silicon carbide powders: Temperature-dependent dielectric properties and enhanced microwave absorption at gigahertz range. Solid State Communications. 2013;**163**:1-6. DOI: 10.1016/j.ssc.2013.03.004

[62] Wang H et al. High temperature electromagnetic and microwave absorbing properties of polyimide/multi-walled carbon nanotubes nancomposites. Chemical Physics Letters. 2015;**633**:223-228. DOI: 10.1016/j.cplett.2015.05.048

[63] Dou Y-K et al. The enhanced polarization relaxation and excellent high-temperature dielectric properties of N-doped SiC. Applied Physics Letters. 2014;**104**(5):052102. DOI: 10.1063/1.4864062

[64] Liu Y et al. Transmission electron microscopy study of the microstructure of unidirectional C/C composites fabricated by catalytic chemical vapor infiltration. Carbon. 2013;**51**:381-389. DOI: 10.1016/j.carbon.2012.08.070

[65] Koziel S, Bekasiewicz A. Multi-objective optimization of expensive electromagnetic simulation models. Applied Soft Computing. 2016;**47**: 332-342. DOI: 10.1016/j.asoc.2016.05.033

[66] Zhou G et al. Microstructure difference between core and skin of T700 carbon fibers in heat-treated carbon/carbon composites. Carbon. 2011;**49**(9):2883-2892. DOI: 10.1016/j.carbon.2011.02.025

[67] Qin X et al. A comparison of the effect of graphitization on microstructures and properties of polyacrylonitrile and mesophase pitch-based carbon fibers. Carbon. 2012;**50**(12):4459-4469. DOI: 10.1016/j.carbon.2012.05.024

[68] Cai Y et al. First-principles study of vibrational and dielectric properties of β-Si3N4. Physical Review B. 2006;**74**(17):174301. DOI: 10.1103/PhysRevB.74.174301

[69] Kirkwood JG, Fuoss RM. Anomalous dispersion and dielectric loss in polar polymers. The Journal of Chemical Physics. 1941;**9**(4):329-340. DOI: 10.1063/1.1750905

[70] Cole KS, Cole RH. Dispersion and absorption in dielectrics I. Alternating current characteristics. The Journal of Chemical Physics. 1941;**9**(4):341-351. DOI: 10.1063/1.1750906

Chapter 3

The Influence of the Dielectric Materials on the Fields in the Millimeter and Infrared Wave Regimes

Zion Menachem

Abstract

This chapter presents the influence of the dielectric materials on the output field for the millimeter and infrared regimes. This chapter presents seven examples of the discontinuous problems in the cross section of the straight waveguide. Two different geometrical of the dielectric profiles in the cross section of the straight rectangular and circular waveguides will be proposed to understand the behavior of the output fields. The two different methods for rectangular and circular waveguides and the techniques to calculate any geometry in the cross section are very important to understand the influence of the dielectric materials on the output fields. The two different methods are based on Laplace and Fourier transforms and the inverse Laplace and Fourier transforms. Laplace transform on the differential wave equations is needed to obtain the wave equations and the output fields that are expressed directly as functions of the transmitted fields at the entrance of the waveguide. Thus, the Laplace transform is necessary to obtain the comfortable and simple *input-output* connections of the fields. The applications are useful for straight waveguides in the millimeter and infrared wave regimes.

Keywords: wave propagation, dielectric profiles, rectangular and circular waveguides, dielectric materials

1. Introduction

Various methods for analysis of waveguides have been studied in the literature. The review for the modal analysis of general methods has been published [1]. The important methods, such as the finite difference method and integral equation method, and methods based on series expansion have been described. An analytical model for the corrugated rectangular waveguide has been extended to compute the dispersion and interaction impedance [2]. The application of analytical method based on the field equations has been presented to design corrugated rectangular waveguide slow-wave structure TH_z amplifiers.

A fundamental technique has been proposed to compute the propagation constant of waves in a lossy rectangular waveguide [3]. An important consequence of this work is the demonstration that the loss computed for degenerate modes propagating simultaneously is not simply additive. The electromagnetic fields in rectangular conducting waveguides filled with uniaxial anisotropic media have been characterized [4].

A full-vectorial boundary integral equation method for computing guided modes of optical waveguides has been proposed [5]. The integral equations are used to compute the Neumann-to-Dirichlet operators for sub-domains of constant refractive index on the transverse plane of the waveguide. Wave propagation in an inhomogeneous transversely magnetized rectangular waveguide has been studied with the aid of a modified Sturm-Liouville differential equation [6].

An advantageous finite element method for the rectangular waveguide problem has been developed [7] by which complex propagation characteristics may be obtained for arbitrarily shaped waveguide. The characteristic impedance of the fundamental mode in a rectangular waveguide was computed using finite element method. The finite element method has been used to derive approximate values of the possible propagation constant for each frequency. A new structure has been proposed for microwave filters [8]. This structure utilizes a waveguide filled by several dielectric layers. The relative electric permittivity and the length of the layers were optimally obtained using least mean square method.

An interesting method has been introduced for frequency domain analysis of arbitrary longitudinally inhomogeneous waveguides [9]. In this method, the integral equations of the longitudinally inhomogeneous waveguides are converted from their differential equations and solved using the method of moments. A general method has been introduced to frequency domain analysis of longitudinally inhomogeneous waveguides [10]. In this method, the electric permittivity and also the transverse electric and magnetic fields were expanded in Taylor's series. The field solutions were obtained after finding unknown coefficients of the series. A general method has been introduced to analyze aperiodic or periodic longitudinally inhomogeneous waveguides [11]. The periodic longitudinally inhomogeneous waveguides were analyzed using the Fourier series expansion of the electric permittivity function to find their propagation constant and characteristic impedances.

Various methods for the analysis of cylindrical hollow metallic or metallic with inner dielectric coating waveguide have been studied in the literature. A review of the hollow waveguide technology [12, 13] and a review of IR transmitting, hollow waveguides, fibers, and integrated optics [14] were published. Hollow waveguides with both metallic and dielectric internal layers have been proposed to reduce the transmission losses. A hollow waveguide can be made, in principle, from any flexible or rigid tube (plastic, glass, metal, etc.) if its inner hollow surface (the core) is covered by a metallic layer and a dielectric overlayer. This layer structure enables us to transmit both the TE and TM polarizations with low attenuation [15].

A transfer matrix function for the analysis of electromagnetic wave propagation along the straight dielectric waveguide with arbitrary profiles has been proposed [16]. This method is based on the Laplace and Fourier transforms. This method is based on Fourier coefficients of the transverse dielectric profile and those of the input-wave profile. Laplace transform is necessary to obtain the comfortable and simple input–output connections of the fields. The transverse field profiles are computed by the inverse Laplace and Fourier transforms.

The influence of the spot size and cross section on the output fields and power density along the straight hollow waveguide has been proposed [17]. The derivation is based on Maxwell's equations. The longitudinal components of the fields are developed into the Fourier-Bessel series. The transverse components of the fields are expressed as functions of the longitudinal components in the Laplace plane and are obtained by using the inverse Laplace transform by the residue method.

These are two kinds of different methods that enable us to solve practical problems with different boundary conditions. The calculations in all methods are based on using Laplace and Fourier transforms, and the output fields are computed by the inverse Laplace and Fourier transforms. Laplace transform on the differential wave equations is needed to obtain the wave equations (and thus also the output fields) that are expressed directly as functions of the transmitted fields at the entrance of the waveguide at $z = 0^+$. Thus, the Laplace transform is necessary to obtain the comfortable and simple *input-output* connections of the fields.

All models that are mentioned refer to solve interesting wave propagation problems with a particular geometry. If we want to solve more complex discontinuous problems of coatings in the cross section of the dielectric waveguides, then it is important to develop for each problem an improved technique for calculating the profiles with the dielectric material in the cross section of the straight waveguide.

This chapter presents two techniques for two different geometries of the straight waveguide. The two proposed techniques are very important to solve discontinuous problems with dielectric material in the cross section of the straight rectangular and circular waveguides. The proposed technique relates to the method for the propagation along the straight rectangular metallic waveguide [16]. The examples will be demonstrated for the rectangular and circular dielectric profiles in the straight rectangular waveguide.

In this chapter, we present seven dielectric structures as shown in **Figure 1(a)–(g)**. **Figure 1(a)–(e)** shows five examples of the discontinuous dielectric materials in the cross section of the rectangular straight waveguide. **Figure 1(f)–(g)** shows two examples of the discontinuous dielectric materials in the cross section of the circular straight waveguide.

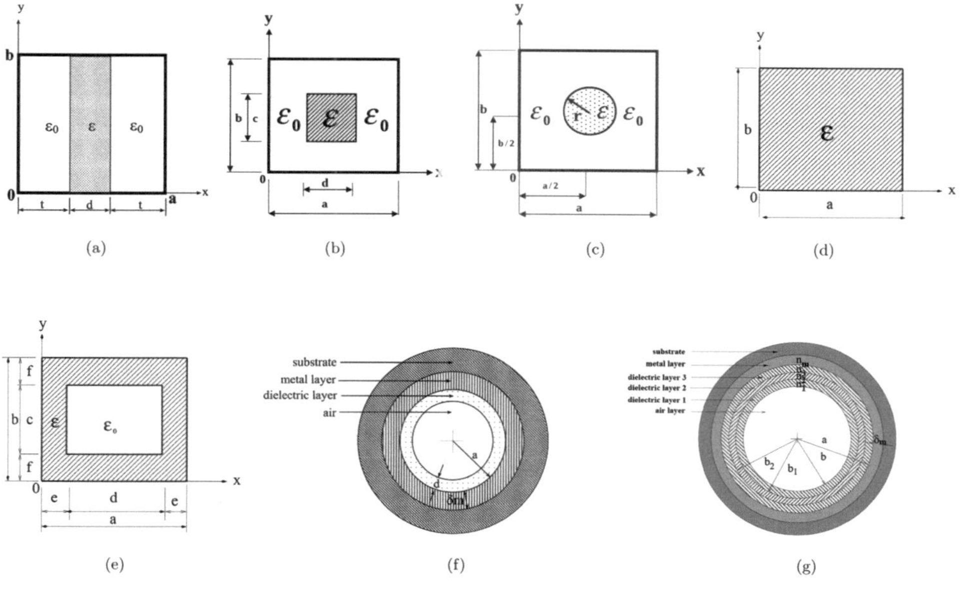

Figure 1.
Seven examples of dielectric materials in the cross section of the straight waveguides. (a) Slab profile in the cross section of the straight waveguides. (b) Rectangular profile in the rectangular waveguide. (c) Circular profile in the rectangular waveguide. (d) Full dielectric material in the rectangular waveguide. (e) Hollow rectangular waveguide with one dielectric material between the hollow rectangle and the metal. (f) Hollow waveguide with one dielectric coating. (g) Hollow waveguide with three dielectric coatings.

2. Two proposed techniques for discontinuous problems in the cross section of the rectangular and circular waveguides

The wave equations for the components of the electric and magnetic field are given by

$$\nabla^2 E + \omega^2 \mu \varepsilon E + \nabla \left(E \cdot \frac{\nabla \varepsilon}{\varepsilon} \right) = 0 \tag{1}$$

and

$$\nabla^2 H + \omega^2 \mu \varepsilon H + \frac{\nabla \varepsilon}{\varepsilon} \times (\nabla \times H) = 0 \tag{2}$$

where ε_0 represents the vacuum dielectric constant, χ_0 is the susceptibility, and g is its dielectric profile function in the waveguide.

Let us introduce the dielectric profile function for the examples as shown in **Figure 1(a)–(g)** for the inhomogeneous dielectric materials.

3. The derivation for rectangular straight waveguide

The wave Eqs. (1) and (2) are given in the case of the rectangular straight waveguide, where

$$\varepsilon(x, y) = \varepsilon_0 (1 + \chi_0 g(x, y)),\ g_x = [1/\varepsilon(x, y)][\partial \varepsilon(x, y)/\partial x], \text{ and}$$
$$g_y = [1/\varepsilon(x, y)][\partial \varepsilon(x, y)/\partial y].$$

3.1 The first technique to calculate the discontinuous structure of the cross section

Figure 2(a) shows an example of the cross section of the straight waveguide (**Figure 1(a)**) for $g(x)$ function. In order to solve inhomogeneous dielectric profiles, we use the ω_ε function, with the parameters ε_1 and ε_2 (**Figures 1(a)** and **(b)**).

The ω_ε function [18] is used in order to solve discontinuous problems in the cross section of the straight waveguide. The ω_ε function is defined according to **Figure 2(b)** as

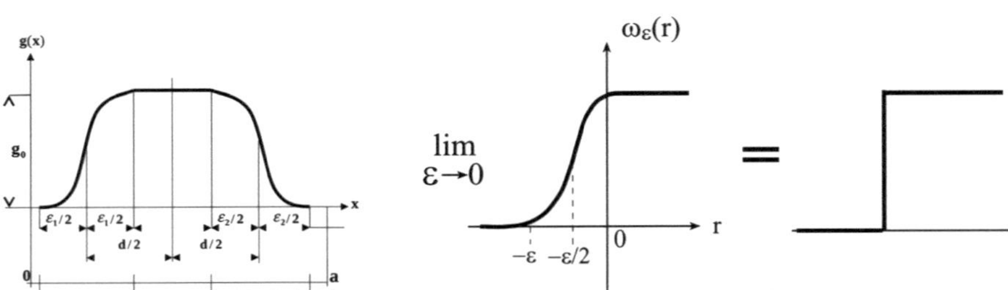

Figure 2.
(a) An example of the discontinuous problem of the slab dielectric profile in the straight rectangular waveguide. (b) The ω_ε function in the limit $\varepsilon \to 0$.

$$\omega_\varepsilon(r) = \begin{cases} C_\varepsilon e^{-\frac{\varepsilon^2}{\varepsilon^2-|r|^2}} & |r| \le \varepsilon \\ 0 & |r| > \varepsilon \end{cases}, \tag{3}$$

where C_ε is a constant and $\int \omega_\varepsilon(r)dr = 1$.

In order to solve inhomogeneous dielectric profiles (e.g., in **Figure 1(a)–(b)**) in the cross section of the straight waveguide, the parameter ε is used according to the ω_ε function (**Figure 2(b)**), where $\varepsilon \to 0$. The dielectric profile in this case of a rectangular dielectric material in the rectangular cross section (**Figure 1(b)**) is given by

$$g(x) = \begin{cases} 0 & 0 \le x < (a-d-\varepsilon)/2 \\ g_0 \exp\left[1 - \dfrac{\varepsilon^2}{\varepsilon^2 - [x-(a-d+\varepsilon)/2]^2}\right] & (a-d-\varepsilon)/2 \le x < (a-d+\varepsilon)/2 \\ g_0 & (a-d+\varepsilon)/2 < x < (a+d-\varepsilon)/2 \\ g_0 \exp\left[1 - \dfrac{\varepsilon^2}{\varepsilon^2 - [x-(a+d-\varepsilon)/2]^2}\right] & (a+d-\varepsilon)/2 \le x < (a+d+\varepsilon)/2 \\ 0 & (a+d+\varepsilon)/2 < x \le a \end{cases}, \tag{4}$$

and

$$g(y) = \begin{cases} 0 & 0 \le y < (b-c-\varepsilon)/2 \\ g_0 \exp\left[1 - \dfrac{\varepsilon^2}{\varepsilon^2 - [y-(b-c+\varepsilon)/2]^2}\right] & (b-c-\varepsilon)/2 \le y < (b-c+\varepsilon)/2 \\ g_0 & (b-c+\varepsilon)/2 < y < (b+c-\varepsilon)/2 \\ g_0 \exp\left[1 - \dfrac{\varepsilon^2}{\varepsilon^2 - [y-(b+c-\varepsilon)/2]^2}\right] & (b+c-\varepsilon)/2 \le y < (b+c+\varepsilon)/2 \\ 0 & (b+c+\varepsilon)/2 < y \le b \end{cases}. \tag{5}$$

The elements of the matrices are given according to **Figure 1(b)**, in the case of b $\neq$ c by

$$\begin{aligned} g(n,m) = \frac{g_0}{ab} &\left\{ \int_{(a-d-\varepsilon)/2}^{(a-d+\varepsilon)/2} \exp\left[1 - \frac{\varepsilon^2}{\varepsilon^2 - [x-(a-d+\varepsilon)/2]^2}\right] \cos\left(\frac{n\pi x}{a}\right) dx \right. \\ &\left. + \int_{(a-d+\varepsilon)/2}^{(a+d-\varepsilon)/2} \cos\left(\frac{n\pi x}{a}\right) dx + \int_{(a+d-\varepsilon)/2}^{(a+d+\varepsilon)/2} \exp\left[1 - \frac{\varepsilon^2}{\varepsilon^2 - [x-(a+d-\varepsilon)/2]^2}\right] \cos\left(\frac{n\pi x}{a}\right) dx \right\} \\ &\left\{ \int_{(b-c-\varepsilon)/2}^{(b-c+\varepsilon)/2} \exp\left[1 - \frac{\varepsilon^2}{\varepsilon^2 - [y-(b-c+\varepsilon)/2]^2}\right] \cos\left(\frac{m\pi y}{b}\right) dy \right. \\ &\left. + \int_{(b-c+\varepsilon)/2}^{(b+c-\varepsilon)/2} \cos\left(\frac{m\pi y}{b}\right) dy + \int_{(b+c-\varepsilon)/2}^{(b+c+\varepsilon)/2} \exp\left[1 - \frac{\varepsilon^2}{\varepsilon^2 - [y-(b+c-\varepsilon)/2]^2}\right] \cos\left(\frac{m\pi y}{b}\right) dy \right\}, \end{aligned} \tag{6}$$

where

$$\int_{(a-d+\varepsilon)/2}^{(a+d-\varepsilon)/2} \cos\left(\frac{n\pi x}{a}\right)dx = \begin{cases} (2a/n\pi)\sin((n\pi/2a)(d-\varepsilon))\cos((n\pi)/2) & n \neq 0 \\ d-\varepsilon & n=0 \end{cases}$$

and

$$\int_{(b-c+\varepsilon)/2}^{(b+c-\varepsilon)/2} \cos\left(\frac{m\pi y}{b}\right)dy = \begin{cases} (2b/m\pi)\sin((m\pi/2b)(c-\varepsilon))\cos((m\pi)/2) & n \neq 0 \\ c-\varepsilon & n=0 \end{cases}.$$

The elements of the matrices are given according to **Figure 1(a)**, in the case of b = c by

$$g(n,m) = \frac{g_0}{ab}\left\{\int_{(a-d-\varepsilon)/2}^{(a-d+\varepsilon)/2} \exp\left[1-\frac{\varepsilon^2}{\varepsilon^2-[x-(a-d+\varepsilon)/2]^2}\right]\cos\left(\frac{n\pi x}{a}\right)dx\right.$$
$$\left.+\int_{(a-d+\varepsilon)/2}^{(a+d-\varepsilon)/2}\cos\left(\frac{n\pi x}{a}\right)dx+\int_{(a+d-\varepsilon)/2}^{(a+d+\varepsilon)/2}\exp\left[1-\frac{\varepsilon^2}{\varepsilon^2-[x-(a+d-\varepsilon)/2]^2}\right]\cos\left(\frac{n\pi x}{a}\right)dx\right\}$$
$$\left\{\int_0^b \cos\left(\frac{m\pi y}{b}\right)dy\right\},$$

where

$$\int_0^b \cos\left(\frac{m\pi y}{b}\right)dy = (b/m\pi)\sin(m\pi) = \begin{cases} b & m=0 \\ 0 & m\neq 0 \end{cases}.$$

3.2 The second technique to calculate the discontinuous structure of the cross section

The second technique to calculate the discontinuous structure of the cross section as shown in **Figure 1(a)** and **(b)**.

The dielectric profile $g(x,y)$ is given according to $\varepsilon(x,y) = \varepsilon_0(1+g(x,y))$. According to **Figure 3** and for $g(x,y) = g_0$, we obtain

$$g(n,m) = \frac{g_0}{4ab}\int_{-a}^{a}dx\int_{-b}^{b}\exp\left[-j(k_x x+k_y y)\right]dy$$
$$= \frac{g_0}{4ab}\left\{\int_{x_{11}}^{x_{12}}dx\int_{y_{11}}^{y_{12}}\exp\left[-j(k_x x+k_y y)\right]dy+\int_{-x_{12}}^{-x_{11}}dx\int_{y_{11}}^{y_{12}}\exp\left[-j(k_x x+k_y y)\right]dy\right.$$
$$\left.+\int_{-x_{12}}^{-x_{11}}dx\int_{-y_{12}}^{-y_{11}}\exp\left[-j(k_x x+k_y y)\right]dy+\int_{x_{11}}^{x_{12}}dx\int_{-y_{12}}^{-y_{11}}\exp\left[-j(k_x x+k_y y)\right]dy\right\}. \quad (7)$$

If y_{11} and y_{12} are functions of x, then we obtain

$$g(n,m) = \frac{g_0}{abk_y}\int_{x_{11}}^{x_{12}}\left[\sin(k_y y_{12})-\sin(k_y y_{11})\right]\cos(k_x x)dx$$
$$= \frac{2g_0}{am\pi}\int_{x_{11}}^{x_{12}}\sin\left[\frac{k_y}{2}(y_{12}-y_{11})\right]\cos\left[\frac{k_y}{2}(y_{12}+y_{11})\right]\cos(k_x x)dx. \quad (8)$$

The dielectric profile for **Figure 1(b)** is given by

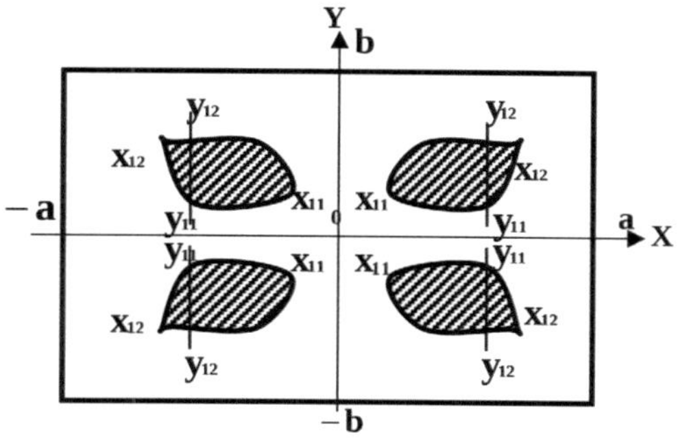

Figure 3.
The arbitrary profile in the cross section.

$$g(n,m) = \begin{cases} \frac{g_0}{4ab}(4cd) & n=0, m=0 \\ \frac{g_0}{4ab}\left(\frac{8d}{k_{0y}m}\sin\left(\frac{k_{0y}mc}{2}\right)\cos\left(\frac{k_{0y}mb}{2}\right)\right) & n=0, m\neq 0 \\ \frac{g_0}{4ab}\left(\frac{8c}{k_{0x}n}\sin\left(\frac{k_{0x}nd}{2}\right)\cos\left(\frac{k_{0x}na}{2}\right)\right) & n\neq 0, m=0 \\ \frac{g_0}{4ab}\left(\frac{16}{k_{0x}k_{0y}nm}\sin\left(\frac{k_{0x}nd}{2}\right)\cos\left(\frac{k_{0x}na}{2}\right)\sin\left(\frac{k_{0y}mc}{2}\right)\cos\left(\frac{k_{0y}mb}{2}\right)\right) & n\neq 0, m\neq 0 \end{cases} \quad (9)$$

3.3 The dielectric profile for the circular profile in the cross section

The dielectric profile for the circular profile in the cross section of the straight rectangular waveguide is given by (**Figure 1(c)**)

$$g(x,y) = \begin{cases} g_0 & 0 \le r < r_1 - \varepsilon_1/2 \\ g_0 \exp\left[1 - q_\varepsilon(r)\right] & r_1 - \varepsilon/2 \le r < r_1 + \varepsilon/2 \end{cases}, \quad (10)$$

where

$$q_\varepsilon(r) = \frac{\varepsilon^2}{\varepsilon^2 - \left[r - (r_1 - \varepsilon/2)\right]^2},$$

else $g(x,y) = 0$. The radius of the circle is given by $r = \sqrt{(x - a/2)^2 + (y - b/2)^2}$.

3.4 The dielectric profile for the waveguide filled with dielectric material in the entire cross section

The dielectric profile (**Figure 1(d)**) is given by

$$g(n,m) = \begin{cases} g_0 & n=0, m=0 \\ \frac{g_0}{4ab}\left(\frac{8a}{k_{0y}m}\sin\left(\frac{k_{0y}mb}{2}\right)\cos\left(\frac{k_{0y}mb}{2}\right)\right) & n=0, m\neq 0 \\ \frac{g_0}{4ab}\left(\frac{8b}{k_{0x}n}\sin\left(\frac{k_{0x}na}{2}\right)\cos\left(\frac{k_{0x}na}{2}\right)\right) & n\neq 0, m=0 \\ \frac{g_0}{4ab}\left(\frac{16}{k_{0x}k_{0y}nm}\sin\left(\frac{k_{0x}na}{2}\right)\cos\left(\frac{k_{0x}na}{2}\right)\sin\left(\frac{k_{0y}mb}{2}\right)\cos\left(\frac{k_{0y}mb}{2}\right)\right) & n\neq 0, m\neq 0 \end{cases} \quad (11)$$

3.5 The hollow rectangular waveguide with the dielectric material between the hollow rectangle and the metallic

The dielectric profile of the hollow rectangular waveguide with the dielectric material between the hollow rectangle and the metal (**Figure 1(e)**) is calculated by subtracting the dielectric profile of **Figure 1(b)** from the dielectric profile of **Figure 1(d)**.

The matrix **G** is given by the form

$$G = \begin{bmatrix} g_{00} & g_{-10} & g_{-20} & \cdots & g_{-nm} & \cdots & g_{-NM} \\ g_{10} & g_{00} & g_{-10} & \cdots & g_{-(n-1)m} & \cdots & g_{-(N-1)M} \\ g_{20} & g_{10} & \ddots & \ddots & \ddots & & \\ \vdots & g_{20} & \ddots & \ddots & \ddots & & \\ g_{nm} & \ddots & \ddots & \ddots & g_{00} & \vdots & \\ \vdots & & & & & & \\ g_{NM} & \cdots & \cdots & \cdots & \cdots & \cdots & g_{00} \end{bmatrix}. \quad (12)$$

Similarly, the G_x and G_y matrices are obtained by the derivatives of the dielectric profile. These matrices relate to the method that is based on the Laplace and Fourier transforms and the inverse Laplace and Fourier transforms [16]. Laplace transform is necessary to obtain the comfortable and simple input-output connections of the fields. The output transverse fields are computed by the inverse Laplace and Fourier transforms.

This method becomes an improved method by using the proposed technique and the particular application also in the cases of discontinuous problems of the hollow rectangular waveguide with dielectric material between the hollow rectangle and the metal (**Figure 1(e)**), in the cross section of the straight rectangular waveguide. In addition, we can find the thickness of the dielectric layer that is recommended to obtain the desired behavior of the output fields.

Several examples will demonstrate in the next section in order to understand the influence of the hollow rectangular waveguide with dielectric material in the cross section (**Figure 1**) on the output field. All the graphical results will be demonstrated as a response to a half-sine (TE_{10}) input-wave profile and the hollow rectangular waveguide with dielectric material in the cross section of the straight rectangular waveguide.

4. The derivation for circular straight waveguide

The wave Eqs. (1) and (2) are given in the case of the circular straight waveguide, where

$$\varepsilon(r) = \varepsilon_0[1 + \chi_0 g(r)] \text{ and } g_r(r) = [1/\varepsilon(r)][\partial\varepsilon(r)/\partial r].$$

The proposed technique to calculate the refractive index for discontinuous problems (**Figure 1(f)** and **(g)**) is given in this section for the one dielectric coating (**Figure 1(f)**) and for three dielectric coatings (**Figure 1(g)**).

4.1 The refractive index for the circular hollow waveguide with one dielectric coating in the cross section

The cross section of the hollow waveguide (**Figure 1(f)**) is made of a tube of various types of one dielectric layer and a metallic layer. The refractive indices of the air, dielectric, and metallic layers are $n_{(0)} = 1$, $n_{(AgI)} = 2$, and $n_{(Ag)} = 10 - j60$, respectively. The value of the refractive index of the material at a wavelength of λ = 10.6 μm is taken from the table performed by Miyagi et al. [19]. The refractive indices of the air, dielectric layer (AgI), and metallic layer (Ag) are shown in **Figure 1(f)**.

The refractive index (n(r)) is dependent on the transition's regions in the cross section between the two different materials (air-AgI, AgI-Ag).

The refractive index is calculated as follows:

$$n(r) = \begin{cases} n_0 & 0 \le r < b - \varepsilon_1/2 \\ n_0 + (n_d - n_0)\exp\left[1 - \dfrac{{\varepsilon_1}^2}{{\varepsilon_1}^2 - [r - (b + \varepsilon_1/2)]^2}\right] & b - \varepsilon_1/2 \le r < b + \varepsilon_1/2 \\ n_d & b + \varepsilon_1/2 \le r < a - \varepsilon_2/2, \\ n_d + (n_m - n_d)\exp\left[1 - \dfrac{{\varepsilon_2}^2}{{\varepsilon_2}^2 - [r - (a + \varepsilon_2/2)]^2}\right] & a - \varepsilon_2/2 \le r < a + \varepsilon_2/2 \\ n_m & \textit{else} \end{cases}$$

where the internal and external diameters are denoted as 2b, 2a, and 2(a+δ_m), respectively, where δ_m is the metallic layer. The thickness of the dielectric coating (d) is defined as [a − b], and the thickness of the metallic layer (δ_m) is defined as [(a+δ_m) − a]. The parameter ε is very small [ε = [a − b]/50]. The refractive indices of the air, dielectric, and metallic layers are denoted as n_0, n_d, and n_m, respectively.

4.2 The refractive index for the circular hollow waveguide with three dielectric coatings in the cross section

The cross section of the hollow waveguide (**Figure 1(g)**) is made of a tube of various types of three dielectric layers and a metallic layer. The internal and external diameters are denoted as 2b, 2 b_1, 2 b_2, 2a, and 2(a + δ_m), respectively, where δ_m is the thickness of the metallic layer. In addition, we denote the thickness of the dielectric layers as d_1, d_2, and d_3, respectively, where $d_1 = b_1 - $ b, $d_2 = b_2 - b_1$, and d_3 = a − b_2. The refractive index in the particular case with the three dielectric layers and the metallic layer in the cross section of the straight hollow waveguide (**Figure 1(g)**) is calculated as follows:

$$n(r) = \begin{cases} n_0 & 0 \le r < b - \varepsilon/2 \\ n_0 + (n_1 - n_0)\exp\left[1 - \dfrac{\varepsilon^2}{\varepsilon^2 - [r - (b + \varepsilon/2)]^2}\right] & b - \varepsilon/2 \le r < b + \varepsilon/2 \\ n_1 & b + \varepsilon/2 \le r < b_1 - \varepsilon/2 \\ n_1 + (n_2 - n_1)\exp\left[1 - \dfrac{\varepsilon^2}{\varepsilon^2 - [r - (b_1 + \varepsilon/2)]^2}\right] & b_1 - \varepsilon_2/2 \le r < b_1 + \varepsilon_2/2 \\ n_2 & b_1 + \varepsilon/2 \le r < b_2 - \varepsilon/2 \\ n_2 + (n_3 - n_2)\exp\left[1 - \dfrac{\varepsilon^2}{\varepsilon^2 - [r - (b_2 + \varepsilon/2)]^2}\right] & b_2 - \varepsilon/2 \le r < b_2 + \varepsilon/2 \\ n_3 & b_2 + \varepsilon/2 \le r < a - \varepsilon/2 \\ n_3 + (n_m - n_3)\exp\left[1 - \dfrac{\varepsilon^2}{\varepsilon^2 - [r - (a + \varepsilon/2)]^2}\right] & a - \varepsilon/2 \le r < a + \varepsilon/2 \\ n_m & \textit{else} \end{cases},$$

where the parameter ε is very small $[\varepsilon = [a - b]/50]$. The refractive indices of the air, dielectric, and metallic layers are denoted as n_0, n_1, n_2, n_3, and n_m, respectively. In this study we suppose that $n_3 > n_2 > n_1$.

The proposed technique to calculate the refractive indices of the dielectric profile of one dielectric coating (**Figure 1(f)**) or three dielectric coatings (**Figure 1(g)**), and the metallic layer in the cross section relate to the method that is based on Maxwell's equations, the Fourier-Bessel series, Laplace transform, and the inverse Laplace transform by the residue method [17]. This method becomes an improved method by using the proposed technique also in the cases of discontinuous problems of the hollow circular waveguide with one dielectric coating (**Figure 1(f)**), three dielectric coatings (**Figure 1(g)**), or more dielectric coatings.

5. Numerical results

Several examples for the rectangular and circular waveguides with the discontinuous dielectric profile in the cross section of the straight waveguide are demonstrated in this section according to **Figure 1(a)–(g)**.

Figure 4(a)–(c) demonstrates the output field as a response to a half-sine (TE_{10}) input-wave profile in the case of the slab profile (**Figure 1(a)**), where a = b = 20 mm, c = 20 mm, and d = 2 mm, for ε_r = 3, 4, and 5, respectively. **Figure 4 (c)** shows the output field for ε_r = 3, 4, and 5, respectively, where y = b/2 = 10 mm.

By increasing only the value of the dielectric profile from ε_r = 3 to ε_r = 5, the width of the output field decreased, and also the output amplitude decreased.

Figure 5(a)–(e) demonstrates the output field as a response to a half-sine (TE_{10}) input-wave profile in the case of the rectangular dielectric profile in the rectangular waveguide (**Figure 1(b)**), where a = b = 20 mm and c = d = 2 mm, for ε_r = 3, 5, 7, and 10, respectively. **Figure 5(e)** shows the output field for ε_r = 3, 5, 7, and 10, respectively, where y = b/2 = 10 mm.

By increasing only the dielectric profile from ε_r = 3 to ε_r = 5, the width of the output field increased, and also the output amplitude increased.

The output fields are strongly affected by the input-wave profile (TE_{10} mode), the location, and the dielectric profile, as shown in **Figure 4(a)–(c)** and **Figure 5(a)–(e)**.

Figure 6(a)–(e) shows the output field as a response to a half-sine (TE_{10}) input-wave profile in the case of the circular dielectric profile (**Figure 1(c)**), for ε_r = 3, 5, 7 and 10, respectively, where a = b = 20 mm, and the radius of the circular dielectric profile is equal to 1 mm. **Figure 6(e)** shows the output field for ε_r =3, 5, 7, and 10, respectively, where y = b/2 = 10 mm. The other parameters are z = 0.15 m, k_0 = 167 $1/m$, λ = 3.75 cm, and β = 58 $1/m$.

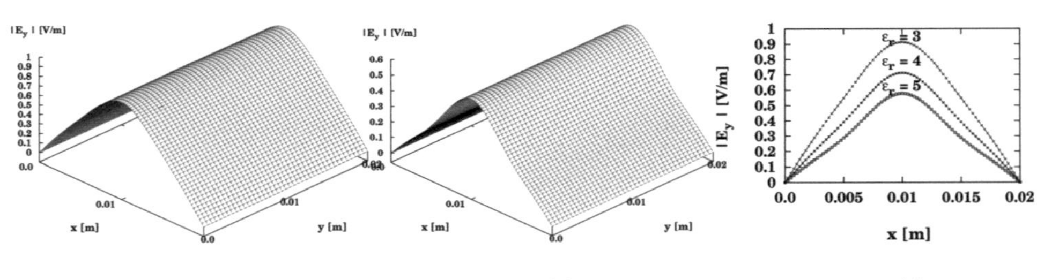

Figure 4.
*The output field as a response to a half-sine (TE_{10}) input-wave profile in the case of the slab dielectric profile (**Figure 1(a)**), where a = b = 20 mm, c = 20 mm, and d = 2 mm where (a) ε_r = 3 and (b) ε_r = 5. (c). The output field for ε_r =3, 4, and 5, respectively, where y = b/2 = 10 mm.*

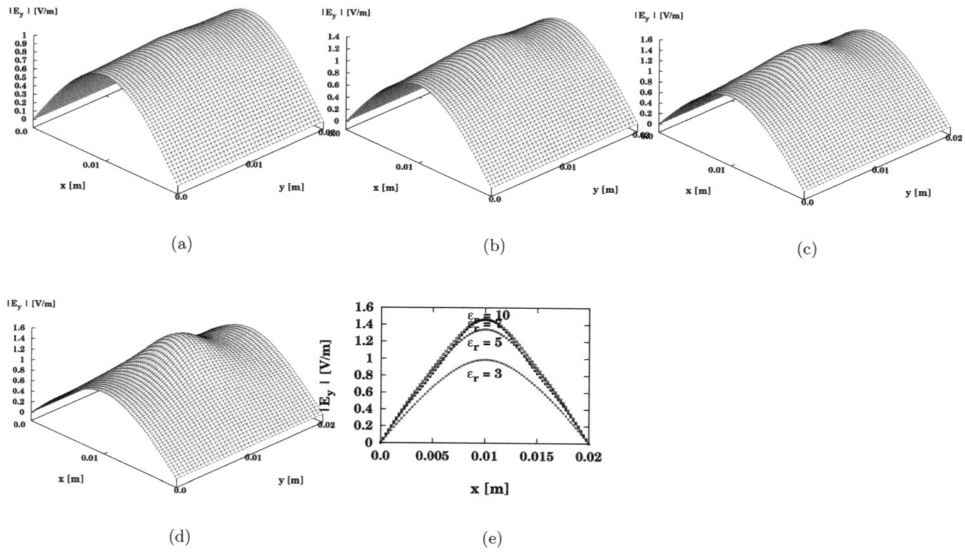

Figure 5.
*The output field as a response to a half-sine (TE_{10}) input-wave profile in the case of the rectangular dielectric material (**Figure 1(b)**), where a = b = 20 mm and c = d = 2 mm: (a) $\varepsilon_r = 3$, (b) $\varepsilon_r = 5$, (c) $\varepsilon_r = 7$, and (d) $\varepsilon_r = 10$. The other parameters are a = b = 20 mm, z = 0.15 m, $k_0 = 167$ 1/m, $\lambda = 3.75$ cm, and $\beta = 58$ 1/m. (e) The output field for $\varepsilon_r = 3$, 5, 7, and 10, respectively, where y = b/2 = 10 mm.*

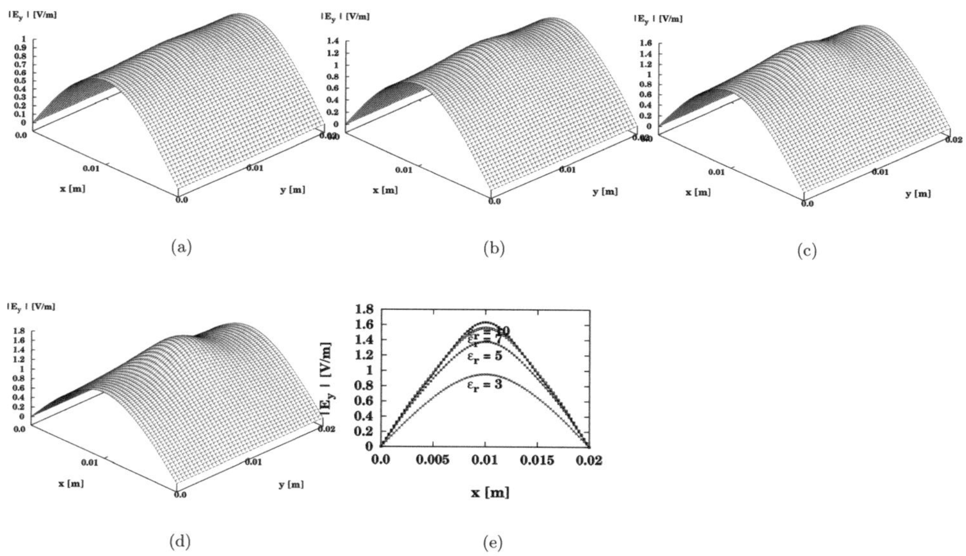

Figure 6.
*The output field as a response to a half-sine (TE_{10}) input-wave profile in the case of the circular dielectric profile (**Figure 1(c)**), where a = b = 20 mm, and the radius of the circular dielectric profile is equal to 1 mm: (a) $\varepsilon_r = 3$, (b) $\varepsilon_r = 5$, (c) $\varepsilon_r = 7$, and (d) $\varepsilon_r = 10$. The other parameters are z = 0.15 m, $k_0 = 167$ 1/m, $\lambda = 3.75$ cm, and $\beta = 58$ 1/m. (e) The output field for $\varepsilon_r = 3$, 5, 7, and 10, respectively, where y = b/2 = 10 mm.*

The proposed technique in Section 3.3 is also effective to solve discontinuous problems of periodic circular profiles in the cross section of the straight rectangular waveguides, and some examples were demonstrated in Ref. [20].

The behavior of the output fields (**Figures 5(a)–(e)** and **6(a)–(e)**) is similar when the dimensions of the rectangular dielectric profile (**Figure 1(b)**) and the circular profile (**Figure 1(c)**) are very close. The output field (**Figure 5(a)–(e)**) is shown for c = d = 2 mm as regards to the dimensions a = b = 20 mm. The output field (**Figure 6(a)–(e)**) is shown where the radius of circular profile is equal to 1 mm (viz., the diameter 2 mm), as regards to the dimensions a = b = 20 mm.

Figure 7(a)–(e) shows the output field as a response to a half-sine (TE_{10}) input-wave profile in the case of the circular dielectric profile (**Figure 1(c)**), where a = b = 20 mm, and the radius of the circular dielectric profile is equal to 2 mm for ε_r = 3, 5, 7, and 10, respectively. The other parameters are z = 0.15 m, k_0 = 167 $1/m$, λ = 3.75 cm, and β = 58 $1/m$. **Figure 7(e)** shows the output field for ε_r =3, 5, 7, and 10, respectively, where y = b/2 = 10 mm.

By changing only the value of the radius of the circular dielectric profile (**Figure 1(c)**) from 1 mm to 2 mm, as regards to the dimensions of the cross section of the waveguide (a = b = 20 mm), the output field of the Gaussian shape increased, and the half-sine (TE_{10}) input-wave profile decreased.

The dielectric profile of the hollow rectangular waveguide with the dielectric material between the hollow rectangle and the metal (**Figure 1(e)**) is calculated by subtracting the dielectric profile of the waveguide with the dielectric material in the core (**Figure 1(b)**) from the dielectric profile according to the waveguide entirely with the dielectric profile (**Figure 1(d)**).

Figure 8(a)–(c) shows the output field as a response to a half-sine (TE_{10}) input-wave profile in the case of the hollow rectangular waveguide with one dielectric material between the hollow rectangle and the metal (**Figure 1(e)**), where a = b = 20 mm, c = d = 14 mm, and d = 14 mm, namely, e = 3 mm and f = 3 mm. **Figure 8(a)–(b)** shows the output field for ε_r = 2.5 and ε_r = 4, respectively. **Figure 8(c)** shows the output field for ε_r = 2.5, 3, 3.5, and 4, respectively, where y = b/2 = 10 mm. The other parameters are z = 0.15 m, k_0 = 167 $1/m$, λ = 3.75 cm, and β = 58 $1/m$.

Figure 9(a)–(c) shows the output power density in the case of the hollow circular waveguide with one dielectric coating (**Figure 1(f)**), where a = 0.5 mm. **Figure 9(a)–(b)** shows the output power density for w_0 = 0.15 mm and w_0 = 0.25 mm, respectively. The output power density of the central peak is shown

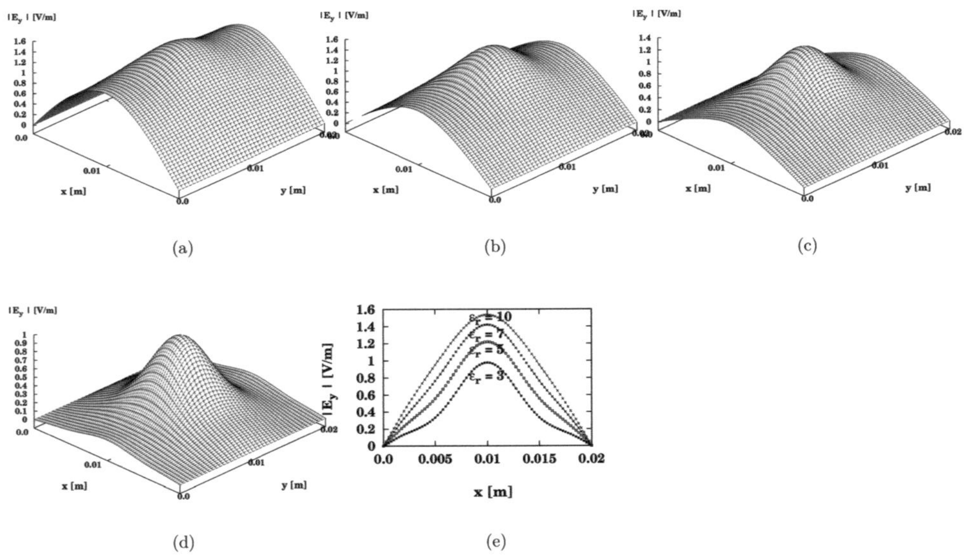

Figure 7.
*The output field as a response to a half-sine (TE_{10}) input-wave profile in the case of the circular dielectric profile (**Figure 1(c)**), where a = b = 20 mm, and the radius of the circular dielectric profile is equal to 2 mm: (a) ε_r = 3, (b) ε_r = 5, (c) ε_r = 7, and (d) ε_r = 10. The other parameters are z = 0.15 m, k_0 = 167 $1/m$, λ = 3.75 cm, and β = 58 $1/m$. (e) The output field for ε_r = 3, 5, 7, and 10, respectively, where y = b/2 = 10 mm.*

DOI: http://dx.doi.org/10.5772/intechopen.80943

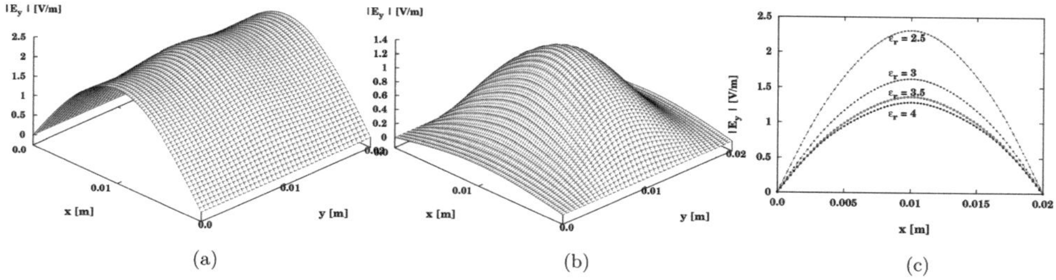

Figure 8.
*The output field as a response to a half-sine (TE_{10}) input-wave profile in the case of the hollow rectangular waveguide with one dielectric material between the hollow rectangle and the metal (**Figure 1(e)**), where a = b = 20 mm, c = 14 mm, and d = 14 mm, namely, e = 3 mm and f = 3 mm: (a) ε_r = 2.5, and (b) ε_r = 4. (c). The output field for ε_r = 2.5, 3, 3.5, and 4, respectively, where y = b/2 = 10 mm. The other parameters are z = 0.15 m, k_0 = 167 1/m, λ = 3.75 cm, and β = 58 1/m.*

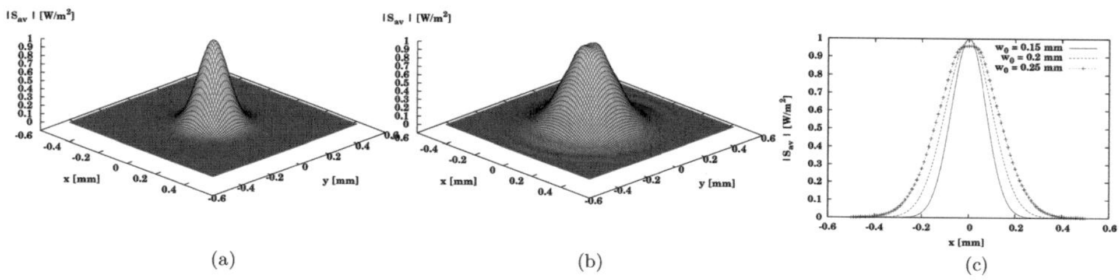

Figure 9.
*The output power density in the case of the hollow circular waveguide with one dielectric coating (**Figure 1(e)**), where a = 0.5 mm: (a) w_0 = 0.15 mm, and (b) w_0 = 0.25 mm. (c). The output power density of the central peak for w_0 = 0.15 mm, w_0 = 0.2 mm, and w_0 = 0.25 mm, respectively, where y = b/2. The other parameters are z = 1 m, n_d =2.2, and $n_{(Ag)}$ = 13.5 - j 75.3.*

for w_0 = 0.15 mm, w_0 = 0.2 mm, and w_0 = 0.25 mm, respectively, where y = b/2. The other parameters are z = 1 m, n_d = 2.2, and $n_{(Ag)}$ = 13.5 - j 75.3.

Figure 1(f) and **(g)** shows two examples of discontinuous problems for circular waveguides. The practical results are demonstrated for **Figure 1(f)**.

Figure 10(a)–(c) shows also the output power density in the case of the hollow circular waveguide with one dielectric coating (**Figure 1(f)**), where a = 0.5 mm, but for other values of the spot size. **Figure 10(a)–(b)** shows the output power density for w_0 = 0.26 mm and w_0 = 0.3 mm, respectively. The output power density of the central peak is shown for w_0 = 0.26 mm, w_0 = 0.28 mm, and w_0 = 0.3 mm, respectively, where y = b/2. The other parameters are z = 1 m, n_d = 2.2, and $n_{(Ag)}$ = 13.5 − j 75.3.

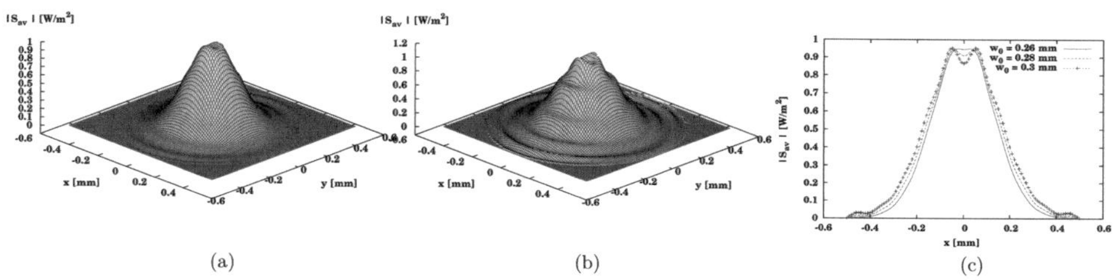

Figure 10.
*The output power density in the case of the hollow circular waveguide with one dielectric coating (**Figure 1(e)**), where a = 0.5 mm: (a) w_0 = 0.26 mm, and (b) w_0 = 0.3 mm. (c). The output power density of the central peak for w_0 = 0.26 mm, w_0 = 0.28 mm, and w_0 = 0.3 mm, respectively, where y = b/2. The other parameters are z = 1 m, n_d = 2.2, and $n_{(Ag)}$ = 13.5 − j 75.3.*

By changing only the values of the spot size from w_0 = 0.15 mm, w_0 = 0.2 mm, and w_0 = 0.25 mm to w_0 = 0.26 mm, w_0 = 0.28 mm, and w_0 = 0.3 mm, respectively, the results of the output power density for a = 0.5 mm are changed as shown in **Figure 10(a)–10(c)**.

The output modal profile is greatly affected by the parameters of the spot size and the dimensions of the cross section of the waveguide. **Figure 10(a)–(c)** demonstrates that in addition to the main propagation mode, several other secondary modes and symmetric output shape appear in the results of the output power density for the values of w_0 = 0.26 mm, w_0 = 0.28 mm, and w_0 = 0.3 mm, respectively.

The proposed technique in Section 4.2 is also effective to solve discontinuous problems of the straight hollow circular waveguide with three dielectric layers (**Figure 1(g)**), and some examples were demonstrated in Ref. [21].

6. Conclusions

Several examples for the rectangular and circular waveguides with the discontinuous dielectric profile in the cross section of the straight waveguide were demonstrated in this research, according to **Figure 1(a)–(g)**.

Figure 4(a)–(c) demonstrates the output field as a response to a half-sine (TE_{10}) input-wave profile in the case of the slab profile (**Figure 1(a)**), where a = b = 20 *mm*, c = 20 *mm*, and d = 2 *mm* for ε_r = 3 and 5, respectively. By increasing only the value of the dielectric profile from ε_r = 3 to ε_r = 5, the width of the output field decreased, and also the output amplitude decreased.

Figure 5(a)–(e) demonstrates the output field as a response to a half-sine (TE_{10}) input-wave profile in the case of the rectangular dielectric profile in the rectangular waveguide (**Figure 1(b)**), where a = b = 20 mm and c = d = 2 *mm*, for ε_r = 3, 5, 7, and 10, respectively. By increasing only the dielectric profile from ε_r = 3 to ε_r = 5, the width of the output field increased, and also the output amplitude increased. The output fields are strongly affected by the input-wave profile (TE_{10} mode), the location, and the dielectric profile, as shown in **Figure 4(a)–(c)** and **Figure 5(a)–(e)**.

The behavior of the output fields (**Figures 5(a)–(e)** and **6(a)–(e)**) is similar when the dimensions of the rectangular dielectric profile (**Figure 1(b)**) and the circular profile (**Figure 1(c)**) are very close. The output field (**Figure 5(a)–(e)**) is shown for c = d = 2 *mm* as regards to the dimensions a = b = 20 mm. The output field (**Figure 6(a)–(e)**) is shown where the radius of circular profile is equal to 1 *mm* (viz., the diameter 2 *mm*), as regards to the dimensions a = b = 20 mm.

Figures 6(a)–(e) and **7(a)–(e)** show the output field as a response to a half-sine (TE_{10}) input-wave profile in the case of the circular dielectric profile (**Figure 1(c)**), for ε_r = 3, 5, 7, and 10, respectively, where a = b = 20 mm, and the radius of the circular dielectric profile is equal to 1 *mm*. By changing only the value of the radius of the circular dielectric profile (**Figure 1(c)**) from 1 *mm* to 2 *mm*, as regards to the dimensions of the cross section of the waveguide (a = b = 20 *mm*), the output field of the Gaussian shape increased, and the half-sine (TE_{10}) input-wave profile decreased.

Figure 8(a)–(c) shows the output field as a response to a half-sine (TE_{10}) input-wave profile in the case of the hollow rectangular waveguide with one dielectric material between the hollow rectangle and the metal (**Figure 1(e)**), where a = b = 20 *mm*, c = d = 14 *mm*, and d = 14 *mm*, namely, e = 3 *mm* and f = 3 *mm*.

Figures 9(a)–(c) and **10(a)–(c)** show the output power density in the case of the hollow circular waveguide with one dielectric coating (**Figure 1(f)**), where a = 0.5 mm. By changing only the values of the spot size from w_0 = 0.15 mm, w_0 = 0.2 mm, and w_0 = 0.25 to w_0 = 0.26 mm, w_0 = 0.28 mm, and w_0 = 0.3 mm,

DOI: http://dx.doi.org/10.5772/intechopen.80943

respectively, the results of the output power density for a = 0.5 mm are changed as shown in **Figure 10(a)–(c)**.

The output modal profile is greatly affected by the parameters of the spot size and the dimensions of the cross section of the waveguide. **Figure 10(a)–(c)** demonstrates that in addition to the main propagation mode, several other secondary modes and symmetric output shape appear in the results of the output power density for the values of w_0 = 0.26 mm, w_0 = 0.28 mm, and w_0 = 0.3 mm, respectively.

The two important parameters that we studied were the spot size and the dimensions of the cross section of the straight hollow waveguide. The output results are affected by the parameters of the spot size and the dimensions of the cross section of the waveguide.

Author details

Zion Menachem
Department of Electrical Engineering, Sami Shamoon College of Engineering, Beer Sheva, Israel

*Address all correspondence to: zionm@post.tau.ac.il

References

[1] Chiang KS. Review of numerical and approximate methods for the modal analysis of general optical dielectric waveguides. Optical and Quantum Electronics. 1993;**26**: S113-S134

[2] Mineo M, Carlo AD, Paoloni C. Analytical design method for corrugated rectangular waveguide SWS THZ vacuum tubes. Journal of Electromagnetic Waves and Applications. 2010;**24**:2479-2494

[3] Yeap KH, Tham CY, Yassin G, Yeong KC. Attenuation in rectangular waveguides with finite conductivity walls. Radioengineering. 2011;**20**: 472-478

[4] Liu S, Li LW, Leong MS, Yeo T Sr. Rectangular conducting waveguide filled with uniaxial anisotropic media: A modal analysis and dyadic Green's function. Progress in Electromagnetics Research. 2000;**25**:111-129

[5] Lu W, Lu YY. Waveguide mode solver based on Neumann-to-Dirichlet operators and boundary integral equations. Journal of Computational Physics. 2012;**231**:1360-1371

[6] Chen TT. Wave propagation in an inhomogeneous transversely magnetized rectangular waveguide. Applied Scientific Research. 1960;**8**: 141-148

[7] Vaish A, Parthasarathy H. Analysis of a rectangular waveguide using finite element method. Progress in Electromagnetics Research C. 2008;**2**: 117-125

[8] Khalaj-Amirhosseini M. Microwave filters using waveguides filled by multi-layer dielectric. Progress in Electromagnetics Research, PIER. 2006; **66**:105-110

[9] Khalaj-Amirhosseini M. Analysis of longitudinally inhomogeneous waveguides using the method of moments. Progress in Electromagnetics Research, PIER. 2007;**74**:57-67

[10] Khalaj-Amirhosseini M. Analysis of longitudinally inhomogeneous waveguides using Taylor's series expansion. Journal of Electromagnetic Waves and Applications. 2006;**16**: 1093-1100

[11] Khalaj-Amirhosseini M. Analysis of longitudinally inhomogeneous waveguides using the Fourier series expansion. Journal of Electromagnetic Waves and Applications. 2006;**20**: 1299-1310

[12] Harrington JA, Matsuura Y. Review of hollow waveguide technology. SPIE. 1995;**2396**:4-14

[13] Harrington JA, Harris DM, Katzir A, editors. Biomedical Optoelectronic Instrumentation, 4–14; 1995

[14] Harrington JA. A review of IR transmitting, hollow waveguides. Fiber and Integrated Optics. 2000;**19**:211-228

[15] Marhic ME. Mode-coupling analysis of bending losses in IR metallic waveguides. Applied Optics. 1981;**20**: 3436-3441

[16] Menachem Z, Jerby E. Transfer matrix function (TMF) for propagation in dielectric waveguides with arbitrary transverse profiles. IEEE Transactions on Microwave Theory and Techniques. 1998;**46**:975-982

[17] Menachem Z, Tapuchi S. Influence of the spot-size and cross-section on the output fields and power density along the straight hollow waveguide. Progress in Electromagnetics Research. 2013;**48**: 151-173

[18] Vladimirov V. Equations of Mathematical Physics. New York: Marcel Dekker, Inc; 1971

[19] Miyagi M, Harada K, Kawakami S. Wave propagation and attenuation in the general class of circular hollow waveguides with uniform curvature. IEEE Transactions on Microwave Theory and Techniques. 1984;**MTT-32**: 513-521

[20] Menachem Z, Tapuchi S. Circular and periodic circular profiles in a rectangular cross section along the straight waveguide. Applied Physics Research. 2015;**7**:121-136

[21] Menachem Z, Tapuchi S. Straight hollow waveguide with three dielectric layers in the cross section. Journal of Electromagnetic Waves and Applications. 2016;**30**:2125-2137

Chapter 4

Phase-Shift Transmission Line Method for Permittivity Measurement and Its Potential in Sensor Applications

Vasa Radonic, Norbert Cselyuszka, Vesna Crnojevic-Bengin and Goran Kitic

Abstract

This chapter offers a detailed insight into a dielectric characterization of the materials based on the phase-shift measurements of the transmission signal. The chapter will provide in-depth theoretical background of the phase-shift transmission line measurement in the microstrip architecture and determination of dielectric permittivity of design under test for several measurement configurations. Potential of the phase-shift method will be demonstrated through applications in the characterization of an unknown dielectric constant in multilayered structure, realization of the soil moisture sensor, and sensor for determination of the dielectric constant of a fluid in microfluidic channel. Moreover, specific techniques for increasing the phase shift based on the electromagnetic bandgap structure, the aperture in the ground plane and the left-handed effect will be presented. In the end, the realization of simple in-field detection device for determination of permittivity based on the phase-shift measurement will be demonstrated.

Keywords: phase-shift method, transmission line, microstrip, permittivity, sensors

1. Introduction

The measurement of dielectric properties of the materials found application in different fields, such as material science, absorber development, biomedical research, tissue engineering, wood industry, food quality control, etc. [1–5]. A number of methods have been developed over a time for characterization of the dielectric properties of the materials such as time domain method [6, 7], capacitive method [3, 8], transmission line (TL) methods [3, 9, 10], resonant method [3, 9], etc. The selection of the appropriate method depends on the measured frequency range, expected values of permittivity, measurement accuracy, form of material (solid, powder, and liquid), sample shape and size, temperature, etc. [8, 11]. Moreover, depending on some important aspects, the measurement methods can be divided into contact or noncontact methods (depends on whether the sample is touched or not), destructive or nondestructive (depends on whether sample can be destroyed or not), narrowband and wideband (depends on frequency range), etc. Since each method has its own advantages and limitations, the selection of the appropriate one depends on a particular application, required accuracy, sample, and other factors. There are a number of commercially available holders, kits, and probes that operate on different principles and allow measurement of the dielectric constant of the material in different forms on different temperatures and frequency ranges [8, 11]. However, the most of them are designed to be connected with expensive instruments such as network analyzers, LCR meters, or impedance analyzers.

One of the commonly used methods suitable for the material characterization in a wide frequency range, from around 10 MHz to 75 GHz, is the TL method. The TL method includes both measurements of the reflection and/or transmission characteristic [3–10]. This method can be used for characterization of permittivity as well as permeability of hard solid materials with medium losses. High-loss materials can be also characterized using this method, if the sample is kept relatively thin. The TL holders are usually made of a coaxial, a waveguide, or a microstrip line section. However, the specific design of the holder or specific multilayered configurations can be used for characterization of powders, liquid, or gases. Usually, the method requires initial sample preparation to fit into the section of the TL, typically the waveguide or the coaxial line. For accurate permittivity measurement, the sample has to be exposed to the maximal electric field, and therefore the position of the sample is very important. A typical measurement configuration of this method consists of the TL section with a sample placed inside, a vector network analyzer (VNA) used to measure the two ports complex scattering parameters (S-parameters), and a software that converts the measured S-parameters to the complex permittivity or permeability. In addition, the TL method requires initial calibration with various terminations before the measurement.

The measurement of the phase shift of the transmitted signal represents relatively fast and simple method for determination of the dielectric properties of the material. It is characterized by fast time response, and in comparison with other methods, it is less sensitive to the noise [10, 12]. Furthermore, this method allows characterization at a single frequency which simplifies the development of a supporting electronic, allows easy integration with sensor element, and allows realization of low-cost in-field sensing devices. Therefore, it found application in the realization of different types of sensors such as soil moisture sensors [13], microfluidic sensor for detention fluid mixture concentration [14], etc.

In this chapter, the phase-shift method will be explained on the example of a microstrip line configuration, and the permittivity of the materials will be determined by measuring the phase shift of the transmitted signal. Theoretical background of the phase-shift method and mathematical equations for determination of real and imaginary part of complex permittivity based on the phase of the transmitted signal will be presented in Section 2. The unknown permittivity of material will be determined for several measurement configurations in Section 3. Potential of the phase-shift method will be demonstrated through several applications in the characterization of an unknown dielectric constant in multilayered structure, a soil moisture sensor, and sensor for determination of fluid properties in microfluidic channel. Advance techniques for increasing the sensitivity of the phase-shift measurement will be presented in Section 4, while the simple in-field detection device for determination of the permittivity based on the phase measurement will be presented in Section 5. The conclusions are given in Section 6.

2. Phase-shift method

2.1 Determination of real part of the dielectric constant

Phase-shift method is based on the measurement of the phase shift of a sinusoidal signal that propagates along a transmission line.

Phase shift $\Delta\varphi$ is defined by velocity and frequency of the propagating signal as well as physical properties of the transmission line:

$$\Delta\varphi = \frac{\omega L_{TL}}{v_p}, \tag{1}$$

where ω is the angular frequency, v_p is the phase velocity, and L_{TL} is the length of transmission line.

In order to determine a phase velocity of electromagnetic wave, we will start with the expression for imaginary part of the complex propagation constant for lossy medium [15]:

$$\beta = \frac{\omega\sqrt{\mu\varepsilon}}{\sqrt{2}}\sqrt{1+\sqrt{1+\frac{\sigma^2}{\omega^2\varepsilon^2}}}, \tag{2}$$

where μ, ε, and σ are the real parts of the permeability, permittivity, and electrical conductivity of the medium through which the signal is propagating, respectively. If the imaginary part of the complex propagation constant is known, the phase velocity can be determined as

$$v_p = \frac{\omega}{\beta} = \frac{\sqrt{2}}{\sqrt{\mu\varepsilon}}\frac{1}{\sqrt{1+\sqrt{1+\frac{\sigma^2}{\omega^2\varepsilon^2}}}}. \tag{3}$$

Based on Eq. (3), it can be seen that phase velocity is dominantly influenced by permittivity, permeability and signal frequency, and then electrical conductivity.

The main advantage of the phase-shift method lies on the fact that on the frequencies high enough, the influence of conductivity can be neglected:

$$\frac{\sigma^2}{\omega^2\,\varepsilon^2} \ll 1; \tag{4}$$

therefore, expression for velocity on high frequencies can be reduced to

$$v_p = \frac{1}{\sqrt{\mu\varepsilon}}, \tag{5}$$

where phase velocity is determined by permeability and permittivity only. This is especially important for soil moisture sensor which will be discussed later since most of the materials in the soil are diamagnetic or paramagnetic. If we assume that electromagnetic wave propagates through a nonmagnetic medium, that is, the magnetic permeability is equal to $\mu_0 = 4\pi \bullet 10^{-7}$ H/m, the phase velocity and the phase shift are dependent on dielectric permittivity only:

$$\Delta\varphi = \omega L_{TL}\sqrt{\mu_0\varepsilon} = \omega L_{TL}\sqrt{\mu_0\varepsilon_0\varepsilon_r} = \frac{\omega L_{TL}}{c_0}\sqrt{\varepsilon_r}, \tag{6}$$

where c_0 is the speed of the light in the vacuum and ε_r is a relative dielectric constant of the medium.

Unlike the coaxial line or the waveguide, the microstrip structure is particularly interesting for a simple and low-cost fabrication of compact sensors based on dielectric permittivity change which can be easily integrated with supporting electronics. In this way, standard printed circuit board technology used for microstrip manufacturing offers fabrication of a complete solution of the sensor in a single substrate. Therefore, the concept of the phase-shift measurement will be explained on the example of microstrip line. In the microstrip, the influence of the change in permittivity is reflected in effective permittivity which is determined by permittivity of a medium above the microstrip line and permittivity of the dielectric substrate, **Figure 1**.

Effective permittivity of the microstrip line shown in **Figure 1** can be expressed as

$$\varepsilon_{eff} = \frac{\varepsilon_s + \varepsilon_m}{2} + \frac{\varepsilon_s - \varepsilon_m}{2} \frac{1}{\sqrt{1 + 12\frac{h}{w}}}, \tag{7}$$

where ε_r and ε_m are the permittivity of dielectric substrate and the medium above the line, respectively, h is the height of the substrate, and w is the width of the microstrip line [16].

It can be seen that a variation in the permittivity of the medium causes a variation in effective permittivity that results in the change of the phase velocity. The change in the phase velocity changes the phase shift of the signal. It is evident that phase shift is determined by the value of the permittivity of the medium above the microstrip line. Therefore, the real part of dielectric constant of unknown medium can be detected with simple measurement of the phase shift of the transmitted signal.

The range of the phase shift $\Delta\Phi$, is determined by the upper and lower values of the measured permittivity:

$$\Delta\Phi = \Delta\varphi_{max} - \Delta\varphi_{min} = \omega L_{TL}\sqrt{\mu_0\varepsilon_0}\left(\sqrt{\varepsilon_{eff\,max}} - \sqrt{\varepsilon_{eff\,min}}\right). \tag{8}$$

This parameter needs to be adjusted to the supporting electronics that measure the phase shift. From Eq. (8) it can be seen that with appropriate choice of the operating frequency and the proper optimization of the geometrical parameters (mostly the length of the transmission line), the range of the phase shift can be optimized to the maximal measurable value.

As stated above, the phase shift depends on the properties of the transmission line. Since the microstrip line can be characterized by inductance per unit length, L',

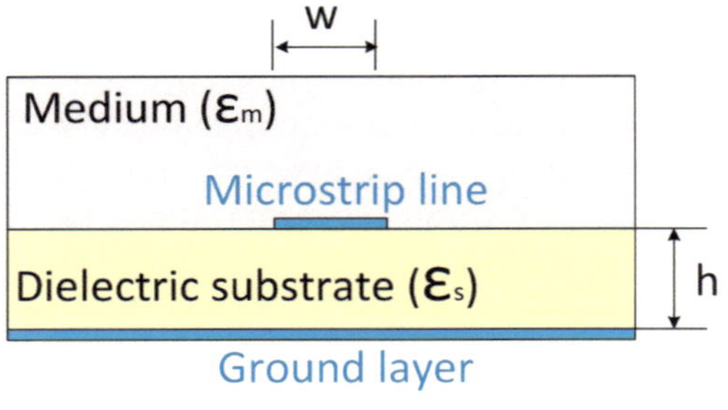

Figure 1.
Configuration of the microstrip line.

and capacitance per unit length, C', on high frequencies, the phase velocity of the signal that propagates along microstrip can be defined as [17]

$$v_p = \frac{1}{L'C'}. \tag{9}$$

Combining Eqs. (1) and (9), phase shift of the signal propagating along microstrip line with length L_{TL} can be expressed as

$$\Delta\varphi = \omega L_{TL}\sqrt{L'C'} = \omega\sqrt{LC}, \tag{10}$$

while the phase-shift range is

$$\Delta\Phi = \Delta\varphi_{max} - \Delta\varphi_{min} = \omega\left(\sqrt{L_{max}C_{max}} - \sqrt{L_{min}C_{min}}\right). \tag{11}$$

The capacitance can be written in the form of the vacuum capacitance C_0, which represents the capacitance of the microstrip line when both substrate and medium relative permittivities are equal to 1, and effective permittivity

$$C = C_0\varepsilon_{eff}. \tag{12}$$

In addition, if we assume that the electromagnetic wave propagates through a nonmagnetic medium, the inductivity of the transmission line does not depend on the sample under test ($L_{max} = L_{min} = L$). Therefore, the expression for the phase-shift range can be reduced to

$$\Delta\Phi = \omega\sqrt{LC_0}\left(\sqrt{\varepsilon_{eff\,max}} - \sqrt{\varepsilon_{eff\,min}}\right). \tag{13}$$

Based on Eq. (13), it can be concluded that the phase-shift range can be increased by performing the measurements on higher frequencies or by increasing the total microstrip line inductance or capacitance. Therefore, the phase-shift range can be optimized to the capabilities of the supporting electronics for measurement of the phase shift.

To validate the proposed method for the characterization of the material properties in wider frequency range, we compare the results of the dielectric constant calculated using proposed method with a predetermined value of dielectric constant, **Figure 2**. For the comparison, the microstrip transmission line was used where the unknown medium was placed above the microstrip line and dielectric substrate, as shown in **Figure 1**. The calculated dielectric constant was extracted from the phase shift of the transited signal using described procedure. The proposed result reveals that the phase-shift detection method provides high accuracy with the relative errors lower than 0.5% for the real parts of the dielectric constant in the measured frequency range.

2.2 Determination of the imaginary part of the dielectric constant

From the previous analysis, we demonstrate how the real part of the complex permittivity can be determined from the phase shift of the transmitted signal. However, the imaginary part of the complex permittivity cannot be directly calculated from the phase shift. It can be estimated using Kramers-Kronig (K-K) relation [18–22].

The real and imaginary part of the complex dielectric constant is correlated with K-K relation [18–20]. This relation is a direct consequence of the principle of

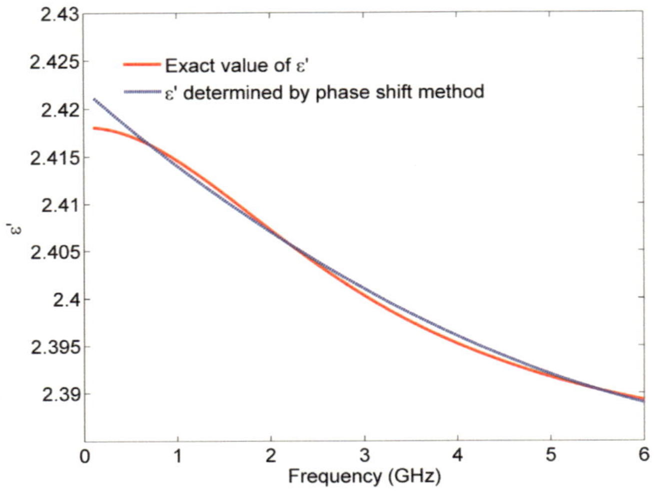

Figure 2.
Real part of the dielectric constant calculated using the phase-shift method compared with the predetermined value.

causality [21]. K-K relation describes a fundamental correlation between the real and imaginary part of the complex dielectric constant and allows us to retrieve imaginary part of the dielectric constant from the measured real part or vice versa.

The imaginary part of the complex dielectric constant can be calculated as

$$\varepsilon''(\omega) = -\frac{2\omega}{\pi}\wp\int_0^{\infty}\frac{\varepsilon'(\Omega)-1}{\Omega^2-\omega^2}d\Omega, \tag{14}$$

where ω is the angular frequency, ε' is the frequency-dependent real part of the dielectric constant, and $\wp$ is the Cauchy principal value [22].

The imaginary part of the complex dielectric constants retrieved using K-K relation from the real part of the dielectric constant (**Figure 2**) measured by phase-shift method is shown in **Figure 3**. The calculated imaginary part shows a good

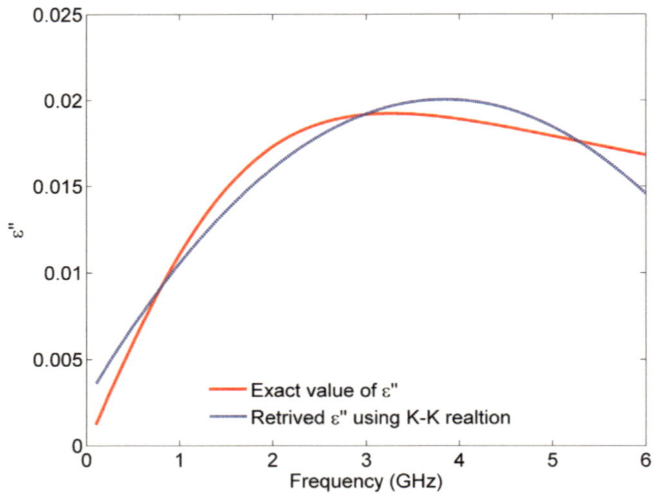

Figure 3.
Imaginary part of the complex dielectric constant retrieved by K-K relation from real part of the dielectric constant measured by phase-shift method.

agreement with the predetermined value of imaginary parts. However, the relative error is about 9% since the accuracy of Eq. (14) depends on the measured frequency range. It can be noted that the range from 0 to ∞ in the integral in Eq. (14) is not achievable in practice. Therefore, the limited range of frequencies affects the accuracy of the results and causes the error that can be observed in **Figure 3**.

3. Multilayered substrate configurations

In a previous section, we presented how the real and imaginary parts of the dielectric constant of the conventional microstrip configuration can be determined using the phase-shift method based on the phase of the transmitted signal. The same method can be applied for nonhomogeneous substrate such as multilayered or heterogeneous substrate. In this section we will analyze different microstrip multilayered substrate configurations interesting for the realization of different sensor topologies. The phase-shift method will be used for the calculation of the effective dielectric constant and determination of the real part of dielectric constant of individual layers. Similarly, the imaginary part of the complex permittivity can be determined from real part using procedure described in Section 2.2.

If we assume that the microstrip transmission line is realized on a multilayered substrate consisting of N layers with different dielectric constants, the effective dielectric constant, ε_s, of the multilayer substrate with N layers can be calculated using [14, 23]

$$\varepsilon_s = \frac{\sum_{i=1}^{N} |d_i|}{\sum_{i=1}^{N} \left|\frac{d_i}{\varepsilon_i}\right|}, \tag{15}$$

where N is the number of the layers, ε_i is the dielectric constant of the i-th layer, and d_i is a coefficient which can be calculated using the following equation:

$$d_i = \frac{K(k_i)}{K'(k_i)} - \frac{K(k_{i-1})}{K'(k_{i-1})} \cdots - \frac{K(k_1)}{K'(k_1)}, \tag{16}$$

where K and $K' = K(k_i)$ are the complete elliptical integrals of the first kind [24] and k_i is

$$k_i = \frac{1}{\cosh\left(\frac{\pi w}{4\sum_{i=1}^{N} h_i}\right)}, \tag{17}$$

where w is the width of the microstrip line and h_i is the thickness of the i-th layer. If all geometrical parameters are known as well as the dielectric constants of all layers except of one arbitrary layer, this unknown dielectric constant can be determined based on effective dielectric constant of the multilayered substrate. Previously, the effective dielectric constant has to be determined from the phase shift of the transmitted signal.

If the parameters of several layers are unknown, the unknown values of the dielectric constants can be found by solving system of equations. This procedure requires several measurements with different sets of geometrical parameters, typically the length of the microstrip line. In that case, the number of the essential measurements required for the determination of all parameter depends on the number of the unknown materials.

Beside conventional microstrip line, next three examples present typical microstrip configurations commonly used in the sensor design. Therefore, the determination of the dielectric constant will be explained on bilayered, tri-layered, and the embedded substrate configurations. In the following section, the practical applications of the analyzed configurations will be demonstrated.

The simplest case is a microstrip line realized on a bilayer substrate, shown in **Figure 4**, when the property of one layer is known (ε_{r1}) and the other one is unknown (ε_{r2}). This presents a typical configuration of the microstrip sensor with thin sensitive film deposited on the dielectric substrate.

The effective dielectric constant of the substrate combination bellow microstrip line (ε_s) can be obtained using procedure explained in Section 2, while the unknown dielectric constant in the bilayered configuration can be expressed as

$$\varepsilon_{r2} = d_2 \left[\pm \left(\frac{|d_1| + |d_2|}{\varepsilon_s} - \frac{|d_1|}{|\varepsilon_{r1}|} \right) \right]^{-1}, \tag{18}$$

where coefficients d_i can be obtained using Eq. (16).

The example of the calculated unknown dielectric constant in the bilayered configuration is shown in **Figure 5**. The preset values for the dielectric constants were set to constant value of ε_{r1} = 3.6 and ε_{r2} = 2.4, while the geometrical parameters of the microstrip line were set to h_1 = 0.1 mm, h_2 = 0.2 mm, and w = 1.5 mm, and the length of the microstrip line was set to L = 10 mm. It can be seen that the

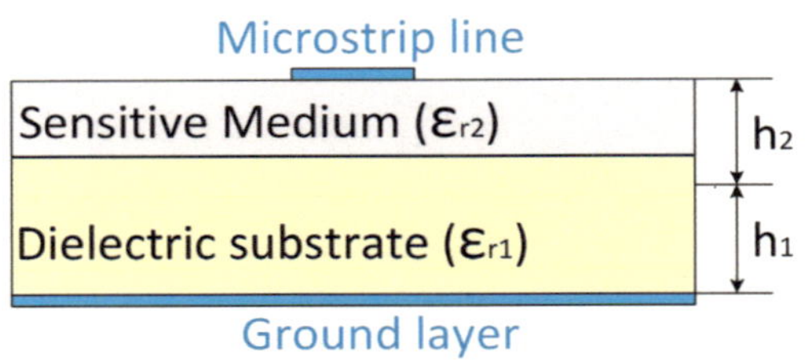

Figure 4.
Microstrip line on bilayered substrate.

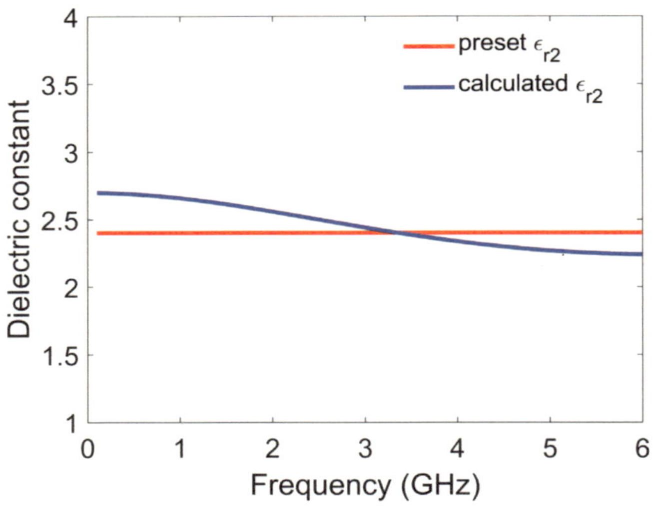

Figure 5.
Calculated unknown dielectric constant in the bilayered substrate combination.

calculated dielectric constant is in a good agreement with the preset value in the calculated frequency range.

Another example presents tri-layered substrate where middle layer is the one with an unknown dielectric constant, **Figure 6**. The bottom and top layers have the same known dielectric contestants, while the middle one is with an unknown material and with known geometrical parameters. This geometrical configuration is typical for the realization of the sensor for the characterization of a fluid or gasses in which the reservoir is placed between two dielectrics bellow microstrip line [25, 26].

Using above-described procedure, the effective dielectric constant of the tri-layer substrate (ε_s) can be obtained from the phase shift, while the unknown dielectric constant, ε_{r2}, can be calculated using the following equation:

$$\varepsilon_{r2} = d_2 \left[\pm \left(\frac{\sum_{i=1}^{3} |d_i|}{\varepsilon_s} - \frac{|d_1| + |d_3|}{|\varepsilon_{r1}|} \right) \right]^{-1}, \tag{19}$$

where coefficients d_i can be determined using Eq. (16).

The example of the calculated unknown dielectric constant in tri-layered configuration is shown in **Figure 7**. The dielectric constants used for this configuration

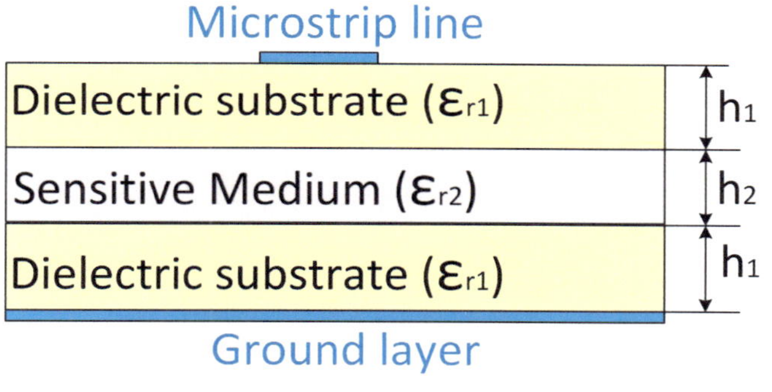

Figure 6.
Microstrip line on tri-layered substrate.

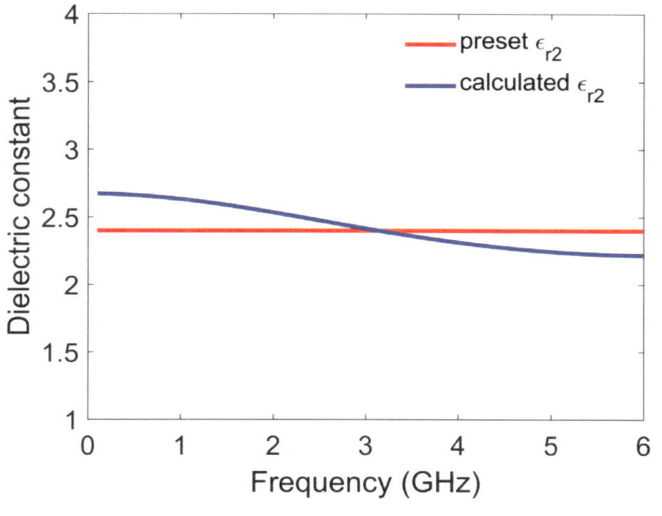

Figure 7.
Calculated unknown dielectric constant in the tri-layered substrate configuration.

were set to constant values of ε_{r1} = 3.6 and ε_{r2} = 2.4, while the geometrical parameters of the microstrip line were set to h_1 = 0.1 mm, h_2= 0.2 mm, w = 1.5 mm, and L = 10 mm. Good agreement is obtained with the maximal relative error of 10% in the observed frequency range.

The third configuration in which the substrate with an unknown dielectric constant is embedded into the substrate with known properties and dimensions is shown in **Figure 8**. This configuration is particularly interesting for the microfluidic applications, for the characterization of the fluid inside the channel [21].

This configuration of the substrate can be observed as a tri-layered substrate in which the middle layer is composed of two materials with different dielectric constants, ε_{r1} and ε_{r2}. The first and third layers are with known dielectric constant, ε_{r1}. If we assume that all geometrical parameters are known, the calculation of the effective dielectric constants of the middle layer ($\varepsilon_{r1_{eff}}$) can be calculated using Eq. (19). This formula does not determine the unknown dielectric constant, just the effective dielectric constant of the middle layer. Therefore, the unknown dielectric constant ε_{r2} can be calculated using Bruggeman formalism [27]:

$$\varepsilon_{r2e} = V\varepsilon_{r2} + (1 - V)\varepsilon_{r1}, \tag{20}$$

where V is the volumetric fraction of the microfluidic channel in the surrounding substrate. The dielectric constant of the unknown embedded material can be

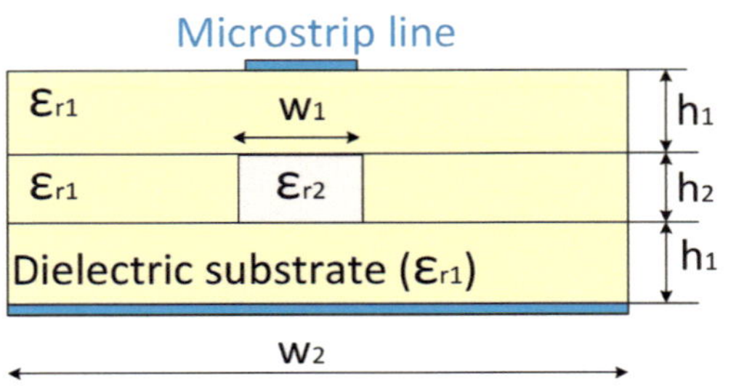

Figure 8.
Microstrip line on the substrate with embedded rectangular channel.

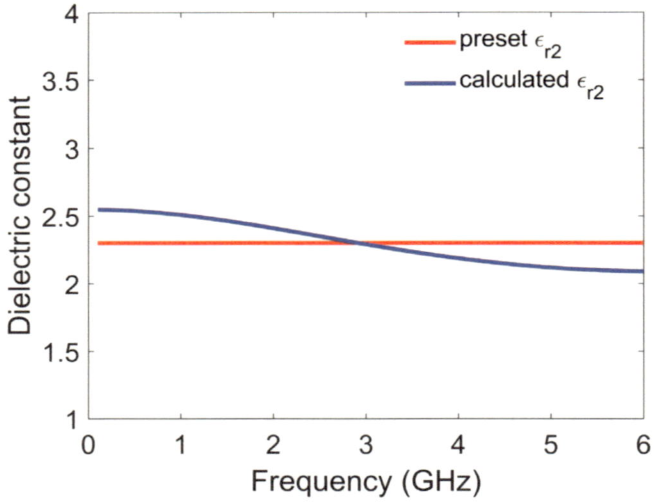

Figure 9.
Calculated unknown dielectric constant in substrate configuration with embedded channel.

determined using the procedure for the extraction of the effective dielectric constant of the tri-layered substrate from the phase shift combined with Eqs. (19) and (20).

The example of the calculated unknown dielectric constant in the embedded substrate configuration is shown in **Figure 9**. The preset values in this case were set to constant values of ε_{r1} = 3.6 and ε_{r2} = 2.35, while the geometrical parameters of the microstrip line were set to h_1 = 0.1 mm, h_2 = 0.2 mm, w = 1.5 mm, and L = 10 mm. The good agreement between the calculated dielectric constant and the preset value is obtained with the relative error lower than 8.5%.

From the previous examples, it can be seen that the phase-shift method has a potential for the characterization of the unknown dielectric materials in different configurations with a good accuracy, and therefore it presents good choice for the rapid characterization of material.

4. Techniques for increasing the phase shift

This section summarized various innovative techniques that can be used for improvement of the sensitivity of phase-shift measurement in the microstrip architecture and their advantages over conventional design.

The first technique is based on increasing the effective length of the transmission line, that is, increasing the inductance and capacitance of the microstrip line. However, this technique results in increased length and reduction in the compactness of the structure. To preserve compactness and satisfy the requirements for good performances, different shapes of the microstrip line have been recently used [28–31]. Since the space-filling property of a fractal offers high potentials for miniaturization of microwave circuits [28–31], the application of the fractal curves theoretically allows the design of infinite-length lines on finite substrate area. In that manner compact transmission line with improved phase response can be obtained using fractal curves. Based on that principle, we have designed a compact soil moisture sensor [28] that consists of two parallel fractal line segments, where each segment comprises two Hilbert fractal curves of the fourth order connected in serial, **Figure 10a**. The additional analyses confirm that the increase of the iteration of the Hilbert fractal curve

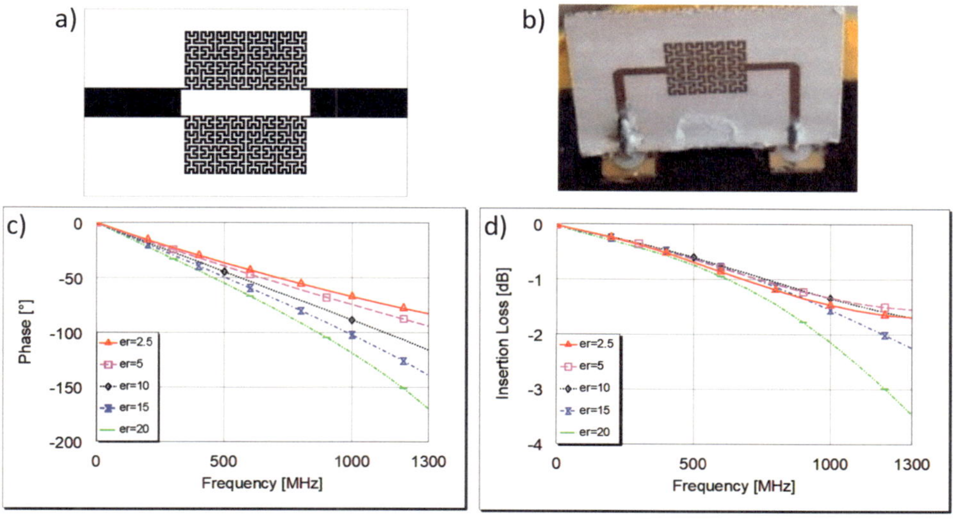

Figure 10.
Soil moisture sensor based on Hilbert fractal curve: (a) layout of the sensor, (b) fabricated sensor, (c) phase-shift response, and (d) insertion loss [28].

increases the range of phase shift [32], but accordingly the insertion loss too. Therefore, in the proposed configuration, two transmission lines are connected in parallel to reduce insertion losses. When Hilbert fractal curves are connected in parallel, the range of the phase shift is approximately the same, but insertion losses are reduced. The Hilbert curve itself is realized with the line width and the spacing between the lines of 100 μm. The results of the proposed sensor placed in the medium with different values of the dielectric constants are shown in **Figure 10**. The range of the phase shift for this configuration at the frequency of 1.2 GHz is 66.64°, while the insertion loss in the worst case is 2.98 dB. The consequence of the line modification is 0.4 dB greater insertion loss comparing to the conventional microstrip line. Although the insertion loss in this case is larger, it is still within acceptable range of 3 dB. However, for the same line length, the range of the phase shift is increased for more than three times. Stated results show that the usage of Hilbert fractal curve leads to more compact sensor characterized by higher sensitivity.

Another technique to increase the phase shift is based on an aperture in the ground plane, where the part of the ground plane, positioned under Hilbert curves of the sensor, shown in **Figure 11**, was removed. In this manner a certain passage for the lines of the electric field is made, so they can pass through it into the soil under the sensor and end up at the bottom side of the ground plane, **Figure 11**. In this manner, soil moisture has larger effect on the sensor characteristics. The results for the sensor with the aperture in the ground plane are presented in **Figure 11**. It can be seen that the range of the phase shift for the sensor with the aperture in the ground plane is 70.76° at the frequency of 1.2 GHz, and the insertion loss is 2.98 dB. Modification in ground plane improved the range of the phase shift for additional 6%, while insertion loss did not change. Quartz sand was used to validate the sensor performances, and the phase-shift measurements were performed for different moisture levels, at operating frequencies in the range of 500–2500 MHz [33, 34]. To investigate the influence of the conductivity, that is, of the soil type, on the calibration curves, the measurement procedure was repeated for the sand moisturized with water with two different salinities (conductivities): 29 and 70‰. If the operating frequency is sufficiently high, Eq. (4) will be satisfied. Therefore, the electrical conductivity can be neglected, and the expression for the phase velocity can be

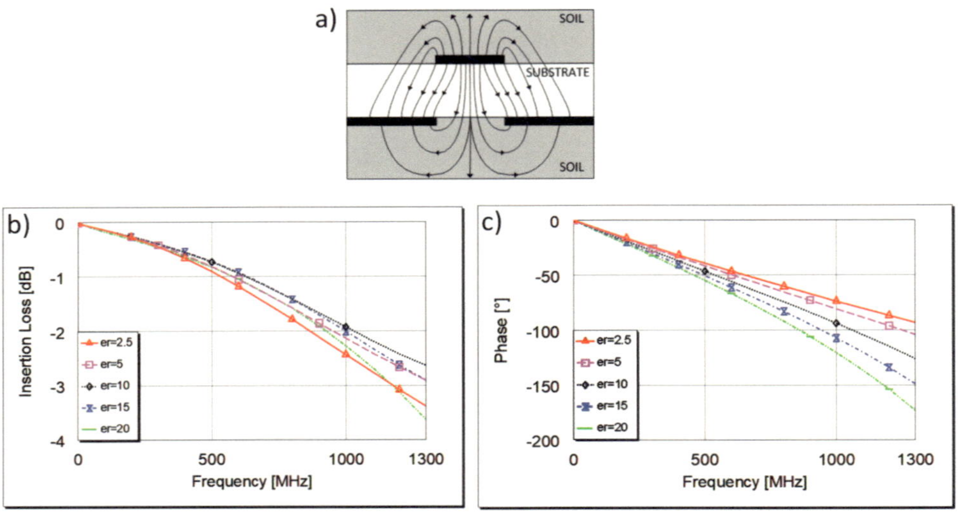

Figure 11.
Soil moisture sensor with the aperture in the ground plane: (a) cross section, (b) phase-shift response, and (c) insertion loss [28].

simplified to Eq. (5). In (**Figure 12**), fitted calibration curves (volumetric water content in the function of the phase shift) for the measurement performed at 500 and 2500 MHz are presented. A significant difference between the calibration curves is obtained at 500 MHz from differently treated samples. However, if the operating frequency is 2500 MHz, the phase shifts obtained from all three samples are almost identical. These results indicate that the Hilbert sensor with the aperture in the ground plane can be successfully used to measure soil moisture independently of the soil type (**Figure 12**).

Third technique to improve sensitivity is based on a defected electromagnetic bandgap (EBG) structure, periodical structure realized as a pattern in the microstrip ground plane. A concept to improve microstrip sensor sensitivity based on the EBG structure was firstly proposed in [35], where it is demonstrated that the sensor sensitivity can be increased by reducing the wave group velocity of the propagating signal. Illustration of this technique is shown in **Figure 13** in the realization of the 3D-printed microfluidic sensor for the determination of the characteristics of different fluids in the microfluidic channel [14]. The bottom layer of the sensor, shown in **Figure 13c**, represents the ground plane realized using defected EBG structure, periodical structure that consists of etched holes. The introduction of the uniform EBG structure in the ground plane forms a frequency region where propagation is forbidden, that is, bandgap in the transmission characteristic [36], while the defect in the EBG results in a resonance in the bandgap, which frequency is determined by the size of the defect. Introduction of the defected EBG structure improves phase shift of the microstrip line in comparison to the conventional microstrip line. Moreover, in comparison with a conventional microstrip line, the intensity of the electric field is stronger in the vicinity of the defect in the EBG [14]. Therefore, the changes of the dielectric constant of the liquid that flows in the channel will have the highest impact to the phase response. In the proposed configuration, the EBG structure is designed to provide bandgap between 5 and 9 GHz, while the defect in the EBG causes the resonance at 6 GHz. By introducing defected EBG structure, the phase change significantly increases, especially at the frequencies that are close to the bandgap edges and at the resonance in the bandgap, due to decrease in the wave phase velocity. From the transmission characteristic, **Figure 14**, it can be seen that the resonant frequency of the defect in the bandgap slightly shifts due to the change of the dielectric constant of the fluid in microchannel. The effect of the EBG structure is predominant at the frequency of 6 GHz where the wave phase velocity is minimal. The results show that the change of the fluid permittivity from 1 (air) to

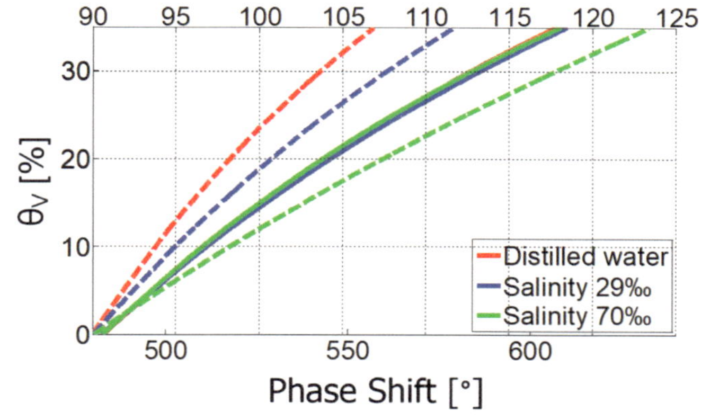

Figure 12.
Volumetric water content for 500 MHz (dashed lines, upper abscissa) and 2500 MHz (full lines, lower abscissa) [33, 34].

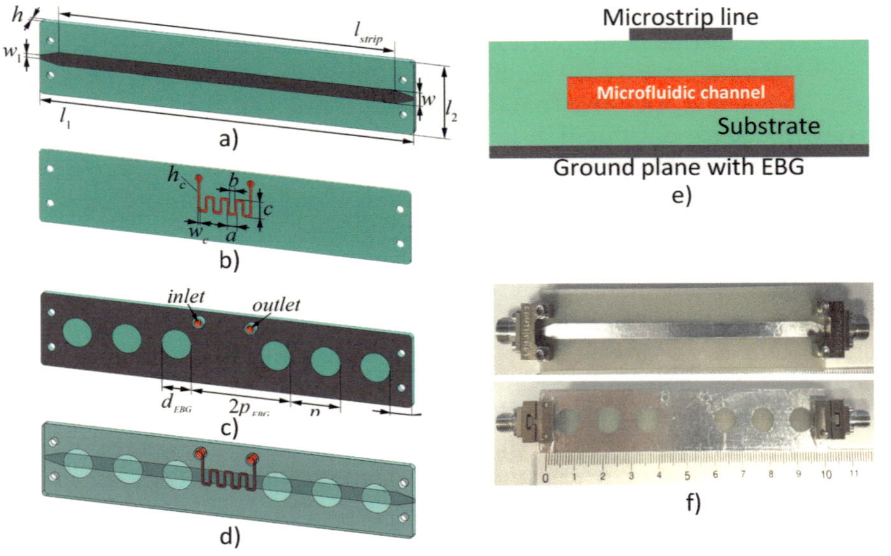

Figure 13.
Microfluidic EBG sensor: (a) top layer, (b) substrate with embedded channel, (c) bottom layer with defected EBG, (d) 3D view, (e) cross section, and (f) top and bottom side of the fabricated sensor [14].

81 (water) causes the phase-shift difference of 84°, **Figure 14**. Compared to the phase shift of the conventional microstrip line without EBG which is only 10.2° at 6 GHz, the proposed design shows eight times higher phase shift. Furthermore, the proposed sensor shows relatively high and almost linear dependence for the fluid materials with permittivity lower than 30. Therefore, the proposed sensor is characterized by relatively high sensitivity and linearity, which makes it a suitable candidate for monitoring small concentrations of a specific fluid in different mixtures. The potential application has been demonstrated in the realization of the microfluidic sensor for detection of toluene concentration in toluene-methanol mixture [14], **Figure 15**.

The fourth technique is based on metamaterials and a left-handed (LH) transmission line approach [37, 38]. Metamaterials are artificial structures that can be designed to exhibit specific electromagnetic properties not commonly found in nature. Metamaterials with simultaneously negative permittivity and

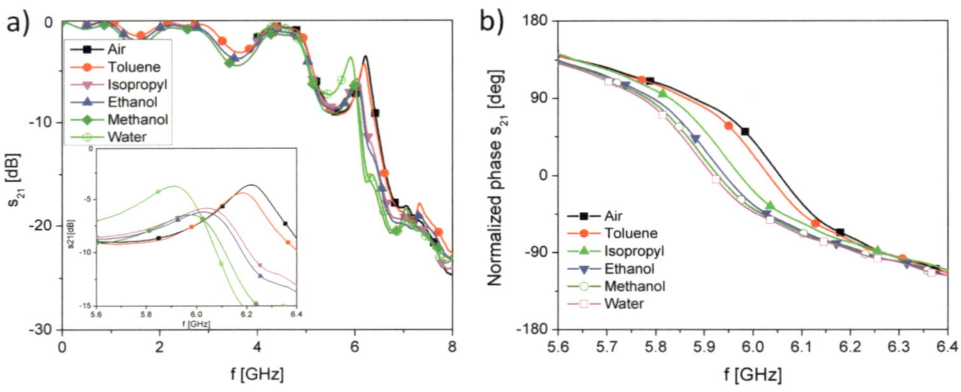

Figure 14.
Measured results of the microfluidic EBG sensor with different fluids inside the microfluidic channel: (a) transmission characteristic and (b) phase characteristic [14].

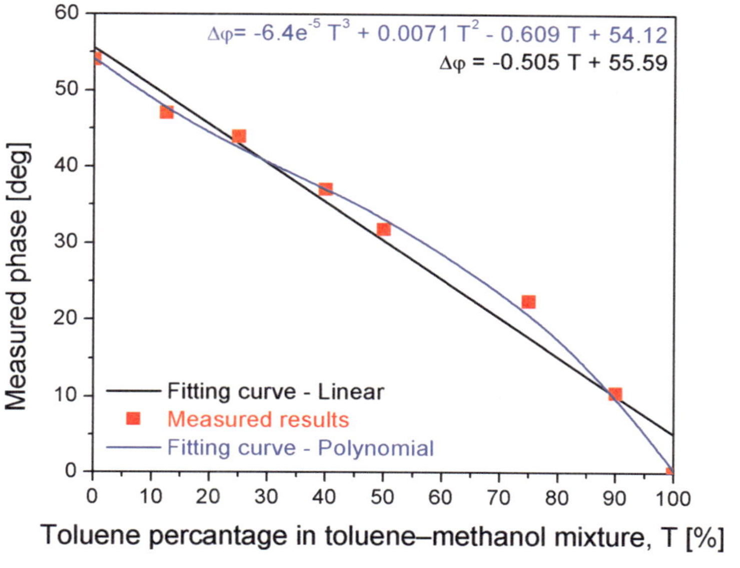

Figure 15.
Measurement of the toluene concentration in toluene-methanol mixture: sensor measured response (dots) and the corresponding fitting curves (lines) [14].

permeability, more commonly referred to as LH materials, have a propagation constant equal to zero at non-zero frequency [37]. Therefore, they can support electromagnetic waves with a group and phase velocities that are antiparallel, known as backward waves [39]. Consequently, energy will travel away from the source, while wave fronts travel backward toward the source. However, the ideal LH structure does not exist in the nature and can be formed as a combination of the LH section and conventional transmission line (RH). This structure known as a composite left-/right-handed (CLRH) transmission line can be formed using a capacitance in series with shunted inductance. In the case of CLRH due to the backward wave propagation, the phase "advance" occurs in the LH frequency range, while phase delay occurs in the RH frequency range [38]. This concept was used in the design of microfluidic sensor for the measurement of the characteristic of the fluid that flows in the microfluidic reservoir placed under CLRH transmission line. The conventional microstrip line in phase comparator was replaced with CRLH transmission line, **Figure 16**. Since the proposed line consists of one CRLH unit cell, it provides passband response at frequency of 1.9 GHz that is characterized with narrowband LH behavior. Therefore, the phase advance at Output 2 is obtained comparing to the signal that propagates true the conventional RH line (Output 1), **Figure 17b**. By changing the properties of the fluid that flows in the microfluidic channel, the central resonance of the LH band slightly shifts to the lower frequency, **Figure 17a**, while the slope of the phase characteristics intensely changes, **Figure 17b**. It can be mentioned that the level between two signals is lower than 8 dB in the worst case at 1.9 GHz; therefore the standard phase detector can be used for the phase-shift measurement. In the proposed configuration, the conventional RH line was used as a referent one, and therefore it was bent in the meander shape to provide maximal measurement phase-shift range at 1.9 GHz. The proposed sensor is characterized with maximal achievable sensitivity, good linearity, and design flexibility, since the maximum extent of the phase can be obtained for the arbitrary range of the dielectric constants by simple modification of the CLRH transmission line.

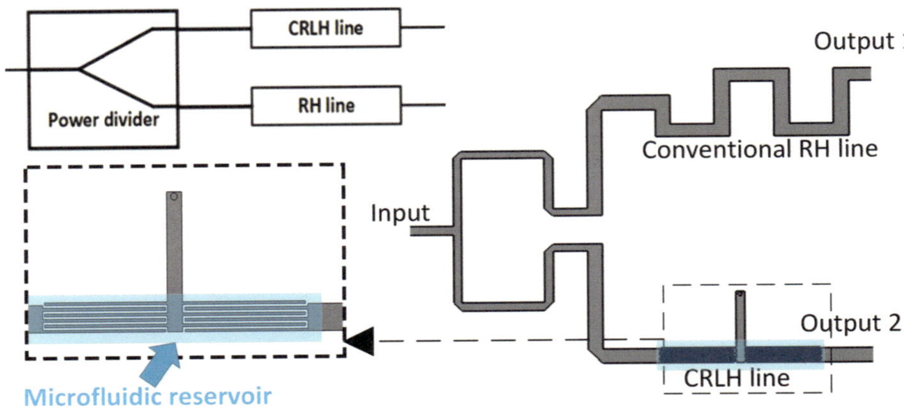

Figure 16.
Blok diagram and layout of the CRLH sensor.

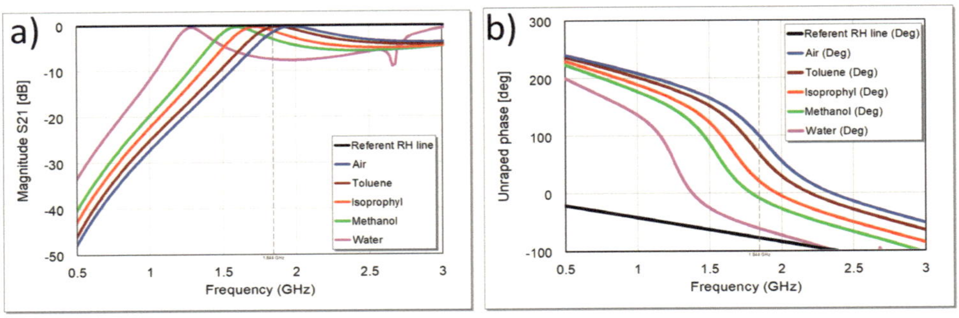

Figure 17.
Response of the CLRH sensor: (a) transmission characteristic and (b) phase characteristic.

5. Simple in-field detection circuit

The phase-shift method, as it was stated above, is the simple method that can be used for the determination of the complex permittivity. Therefore, the expensive instrument for the determination of permittivity can be replaced with a simple detection circuit. The block diagram, and detailed electrical scheme for measurement of the phase shift at single operating frequency, is presented in **Figure 18** together with the fabricated prototype realized in LTCC (low-temperature cofired ceramics) technology.

Detection circuit consists of a microwave oscillator, a quadrature hybrid, a sensor element, and a phase detector functional blocks. The operating principle can be described in the following way: sinusoidal signal generated by source, that is, microwave oscillator, is divided by quadrature hybrid on measurement and referent signal. The measurement signal propagates along sensor element where phase shift occurs according to the dielectric properties of the surrounding medium. The signal from the sensor element is fed into the input of the phase detector which compares the phase of the measurement and the referent signal. The output of the phase detector is the voltage proportional to the phase shift of the input signals which can be related to the measured dielectric constant. Referent signal propagates through the phase-shifter module which has a purpose to provide the calibration of the sensor and to overcome deviations of the nominal properties of the materials and

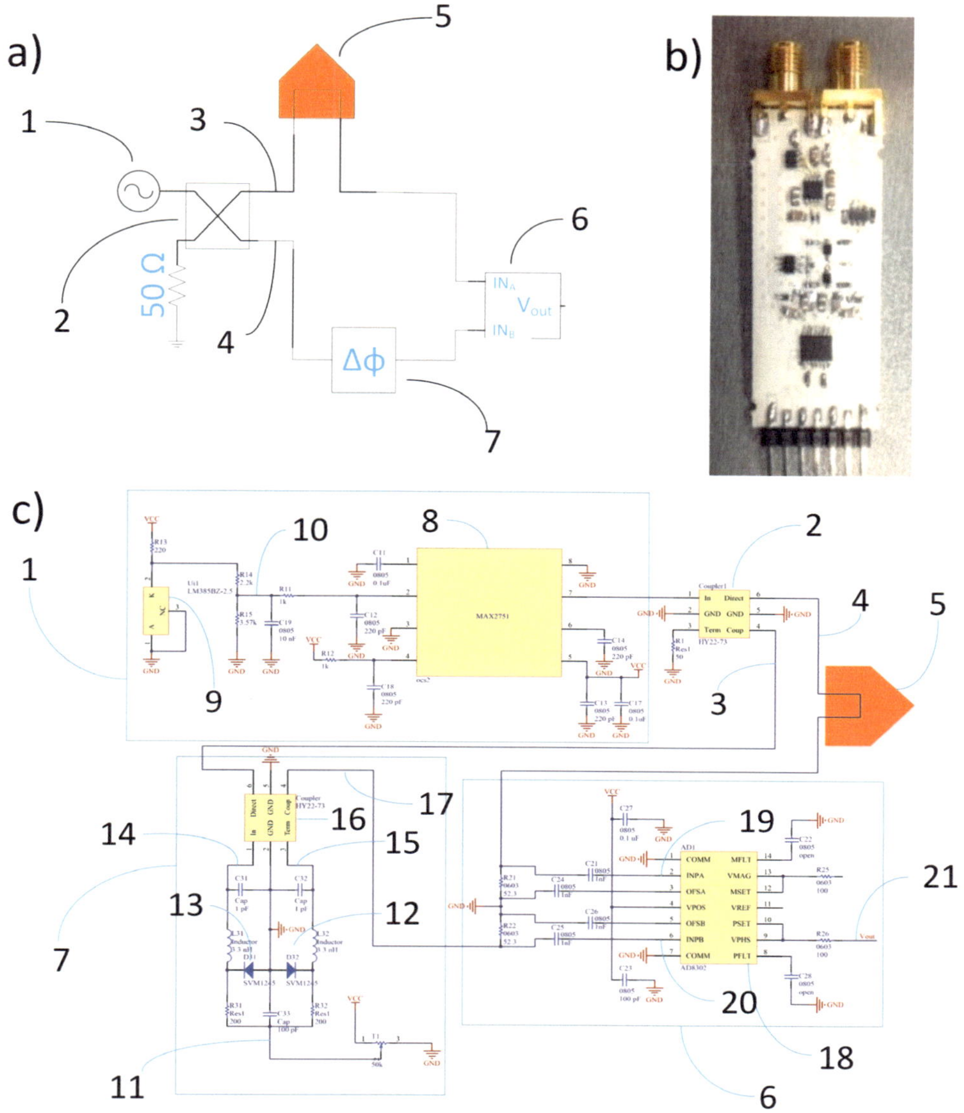

1-Microwave oscillator	8-Voltage controlled oscillator	16-Quadrature coupler
2-Quadrature hybrid	9-Referent voltage	17-Output from phase shifter
3-Measurement signal	10-Voltage divider	18-integraged circuit AD8302
4-Referent signal	11-Controlled voltage	19-signal at input A of AD8302
5-Sensor	12/13- Varactor Diodes	20- signal at input B of AD8302
6- Phase detector	14-Superimposed signal 1	21-output signal from AD8302
7-Phase shifter module	15-Superimposed signal 2	

Figure 18.
Phase-shift measurement device: (a) block diagram, (b) fabricated circuit, and (c) detail electronic circuit of the phase-shift measurement device [13].

components used for sensor fabrication. In this way the repeatability of measurement of the different sensors will be secured.

The phase-shift measurement electrical circuit is detailly presented in **Figure 18c**, indicating abovementioned functional blocks of the circuit, **Figure 18a**. The oscillator is realized by the voltage-controlled oscillator MAX2751 (8), which is adjusted by the referent voltage (9) and the voltage divider (10) to operating frequency of 2.2 GHz. The signal from the oscillator is divided into measurement and referent signals using commercially available quadrature coupler circuit HY22-73 (2). The phase-shifter module allows control of the phase shift and enables fine-tuning of the difference in the phase between the measurement and referent signals. The phase shift can be achieved by varying the control voltage (11) that affects the capacitance of the varactor diodes (12 and 13). The change in capacitance affects the signals (14) and (15) which superimpose with the input referent signal (3) through the quadrature coupler (16) and in this way changes the phase of the resultant signal at the output of the module (17). The measurement of the phase shift between the measurement and referent signal is done using a phase detector module implemented with the integrated circuit AD8302 Analogue Devices (18). The phase detector circuit is set to the phase difference measurement mode according to the manufacturer's recommendation. Integrated circuit AD8302 on its output (21) gives a voltage signal that is proportional to the phase difference of the signal on its inputs (19) and (20).

The accuracy of the designed phase-shift measurement device was experimentally verified by comparing the results of the measurement of the phase shift of the signal, induced by phase-shift circuit (functional block (7), **Figure 18c**), in the range from 0 to 90° with the results of the measurement obtained using vector network analyzer (VNA). The results of the comparison are shown in **Figure 19**. It can be seen that the results agree well and the relative error in respect of the full-scale output is 5.56% which confirms performance of the designed phase-shift circuit.

Additional module for different sensor topologies can be designed to provide a conversion of the measured voltage to the corresponding value of the dielectric constant.

6. Conclusions

In this chapter dielectric characterization technique based on the phase-shift method has been presented. It has been shown that both the real and imaginary part of a complex permittivity can be calculated solely by using the phase shift of the transmitted signal. The exact formulas for the calculation of the unknown dielectric properties in the microstrip and multilayered substrate configurations such as bilayered, tri-layered, and embedded configurations were presented.

To enhance sensitivity of sensors that operate on phase-shift principle, various techniques for increasing the phase-shift range have been presented. The first technique uses space-filling property of Hilbert fractal curve to increase the phase-shift range and yet preserve the compactness of the structure. The technique that increases sensitivity of the sensor by introducing an aperture in the ground layer has been presented on the example of the soil moisture sensor. The third technique based on EBG effect has been illustrated on the example of 3D-printed microfluidic sensor for detection of toluene concentration in toluene-methanol mixture. Metamaterial and CRLH transmission line approach is used as the technique in the realization of microfluidic sensor with a maximum extent of the phase-shift range.

In the end a simple in-field detection circuit for determination of permittivity based on the phase-shift measurement on single operating frequency has been described. With the help of described concepts, a complete set of tools has been introduced which enables the design and optimization of the phase-shift-based sensors.

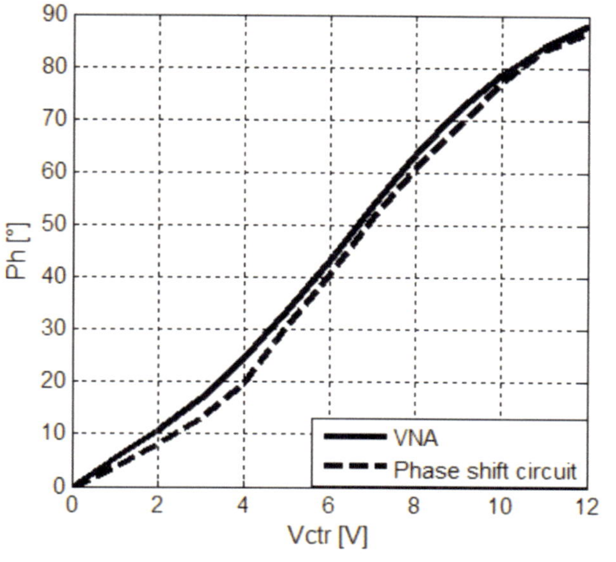

Figure 19.
Comparisons between VNA and designed phase-shift measurement circuit.

Acknowledgements

This result is part of a project that has received funding from the European Union's Horizon 2020 research and innovation programme under the grant agreement No. 664387.

Author details

Vasa Radonic*, Norbert Cselyuszka, Vesna Crnojevic-Bengin and Goran Kitic
University of Novi Sad, BioSense Institute–Research and Development Institute for Information Technologies in Biosystems, Novi Sad, Serbia

*Address all correspondence to: vasarad@uns.ac.rs

References

[1] Foster KR, Schwan HP. Dielectric properties of tissues and biological materials: A critical review. Critical Reviews in Biomedical Engineering. 1989, 1989;**17**(1):25-104

[2] Nelson SO. Fundamentals of dielectric properties measurements and agricultural applications. Journal of Microwave Power and Electromagnetic Energy. 2010;**44**(2): 98-113. DOI: 10.1080/08327823. 2010.11689778

[3] Jarvis JB, Janezic MD, Riddle BF, Johnk RT, Kabos P, Holloway CL, et al. Measuring the Permittivity and Permeability of Lossy Materials: Solids, Liquids, Metals, Building Materials, and Negative-Index Materials. NIST Technical Note 1536; 2005

[4] Krraoui H, Mejri F, Aguili T. Dielectric constant measurement of materials by a microwave technique: Application to the characterization of vegetation leaves. Journal of Electromagnetic Waves and Applications. 2016;**30**(12):1643-1660. DOI: 10.1080/09205071.2016.1208592

[5] Brodie G, Jacob MV, Farrell P. Techniques for measuring dielectric properties. In: Microwave and Radio-Frequency Technologies in Agriculture. 1st ed. Warsaw, Poland: Sciendo; 2015. pp. 52-77. DOI: 10.1515/ 9783110455403-007

[6] Nozaki R, Bose TK. Broadband complex permittivity measurements by time-domain spectroscopy. IEEE Transactions on Instrumentation and Measurement. 1990;**39**(6):945-951. DOI: 10.1109/19.65803

[7] Cataldo A, Tarricone L, Attivissimo F, Trotta A. A TDR method for real-time monitoring of liquids. IEEE Transactions on Instrumentation and Measurement. 2007;**56**(5):1616-1625. DOI: 10.1109/TIM.2007.903596

[8] Keysight Technologies. Basics of Measuring the Dielectric Properties of Materials, Application note, Literature number 5989-2589EN, 2017

[9] Stuchly MA, Stuchly SS. Coaxial line reflection methods for measuring dielectric properties of biological substances at radio and microwave frequencies—A review. IEEE Transaction of Instrumentation and Measurement. 1980;**M-29**:176-183. DOI: 10.1109/TIM.1980.4314902

[10] Chen L, Varadan VV, Ong CK, Neo CP. Microwave Electronics: Measurement and Materials Characterization. 1st ed. Chichester, England: John Wiley & Sons; 2005. DOI: 10.1002/0470020466

[11] Rode & Swartz. Measurement of Dielectric Material Properties. Application Note. RAC-0607-0019, 2012

[12] You KY. Effects of sample thickness for dielectric measurements using transmission phase-shift method. International Journal of Advances in Microwave Technology. 2016;**1**(3): 64-67

[13] Kitic G. Microwave soil moisture sensor based on phase shift method independent of electrical conductivity of the soil. The Intellectual Property Office Republic of Serbia; 2018. П-2018/ 0253—March 2018

[14] Radonic V, Birgermajer S, Kitic G. Microfluidic EBG sensor based on phase-shift method realized using 3D Printing technology. Sensors. 2017; **17**(4):892. DOI: 10.3390/s17040892

[15] Balanis CA. Advanced Engineering Electromagnetics. New York: John Wiley & Sons, Publisher Inc; 1989

[16] Liu R, Zhang Z, Zhong R, Chen X, Li J. Nanotechnology synthesis study: Technical Report 0-5239-1; April 2007. pp. 76-79

[17] Pozar DM. Microwave Engineering. 4th ed. New Jersey: John Wiley and Sons; 2011

[18] Kuzuoglu M, Mittra R. Frequency dependence of the constitutive parameters of causal perfectly matched anisotropic absorbers. IEEE Microwave and Guided Wave Letters. 1996;**6**(12): 447-449. DOI: 10.1109/75.544545

[19] Steeman PAM, Turnhout JV. A numerical Kramers-Kronig transform for the calculation of dielectric relaxation losses free from Ohmic conduction losses. Colloid and Polymer Science. 1997;**275**(2):106-115. DOI: 10.1007/s003960050059

[20] Lucarini V, Saarinen JJ, Peiponen K-E, Vartiainen EM. Kramers–Kronig Relations in Optical Materials Research. Berlin/Heidelberg: Springer-Verlag; 2005. DOI: 10.1007/b138913

[21] Landau LD, Lifshitz EM. Electrodynamics of Continuous Media. Moscow, Russia: Pergamon Press; 1960

[22] Arfken GB, Weber HJ. Mathematical Methods for Physicists. 6th ed. Burlington, MA, USA: Elsevier Academic Press; 2005, 2005. DOI: 10.1016/C2009-0-30629-7

[23] Jha KR, Singh G. Terahertz Planar Antennas for Next Generation Communication. Cham, Switzerland: Springer International Publishing; 2014. DOI: 10.1007/978-3-319-02341-0

[24] Whittaker ET, Watson GN. A Course in Modern Analysis. 4th ed. Cambridge, England: Cambridge University Press; 1990. DOI: 10.1017/CBO9780511608759

[25] Lee K, Hassan A, Lee CH, Bae J. Microstrip patch sensor for salinity determination. Sensors. 2017;**17**(12): 2941. DOI: 10.3390/s17122941

[26] Jankovic N, Radonic V. A microwave microfluidic sensor based on a dual-mode resonator for dual-sensing applications. Sensors. 2017;**17**(12):2713. DOI: 10.3390/s17122713

[27] Lakhtakia A, Michel B, Weiglhofer WS. Bruggeman formalism for two models of uniaxial composite media; dielectric properties. Composites Science and Technology. 1997;**57**: 185-196. DOI: S0266353896001224

[28] Radonic V, Kitic G, Crnojevic-Bengin V. Novel Hilbert soil moisture sensor bassed on the phase shift method, In: Proceedings of Mediterranean Microwave Symposium (MMS 2010); 25–27 August 2010; Guzelyurt, Cyprus; 2010

[29] Crnojevic-Bengin V, Radonic V, Jokanovic B. Fractal geometries of split-ring resonators. IEEE Transactions of Microwave Theory and Techniques. 2008;**56**(10):2312-2321. DOI: 10.1109/TMTT.2008.2003522

[30] Jarry P, Beneat J. Design and Realizations of Miniaturized Fractal Microwave and RF Filters. 1st ed. Hoboken, NJ, USA: Wiley; 2009. ISBN: 978-0-470-48781-5

[31] Stojanović G, Radovanović M, Radonic V. A New Fractal-Based Design of Stacked Integrated Transformers, Active and Passive Electronic Components; 2008. Article ID 134805. DOI: 10.1155/2008/134805

[32] Kitic G. Microwave soil moisture sensors based on distributed elements [thesis]. University of Novi Sad, Faculty of Technical Sciences; 2015

[33] Will B, Crnojević-Bengin V, Kitić G. Microwave soil moisture sensors. In: Proceedings of the 43rd European Microwave Conference (EuMA); 6–10 October 2013; Nuremberg, Germany. IEEE; 2013. DOI: 10.23919/EuMC.2013.6686793

[34] Kitić G, Crnojević Bengin V. The influence of conductivity on the microstrip soil moisture sensor. In: 3rd Global Workshop on Proximal Soil Sensing; 26–29 May 2013; Potsdam, Germany

[35] García-Baños B, Cuesta-Soto F, Griol A, Catalá-Civera JM, Pitarch J. Enhancement of sensitivity of microwave planar sensors with EBG structures. IEEE Sensors Journal. 2006; **6**:1518-1522. DOI: 10.1109/JSEN.2006.884506

[36] Griol A, Mira D, Martinez A, Marti J. Multiple frequency photonic bandgap microstrip structures based on defects insertion. Microwave and Optical Technology Letters. 2003;**36**:479-481. DOI: 10.1002/mop.10795

[37] Christophe Caloz C, Itoh T. Electromagnetic Metamaterials: Transmission Line Theory and Microwave Applications. 1st ed. Wiley-IEEE Press; 2005. DOI: 10.1002/0471754323

[38] Lai A, Itoh T, Caloz C. Composite right/left-handed transmission line metamaterials. IEEE Microwave Magazine. 2004;**5**(3):34-50. DOI: 10.1109/MMW.2004.1337766

[39] Ramo S, Whinnery JR, Duzer TV. Fields and Waves in Communication Electronics. 2nd ed. New York, NY, USA: Wiley; 1984

Section 2

High Frequency Magnetic Properties

Chapter 5

Ferromagnetism in Multiferroic $BaTiO_3$, Spinel MFe_2O_4 (M = Mn, Co, Ni, Zn) Ferrite and DMS ZnO Nanostructures

Kuldeep Chand Verma, Ashish Sharma, Navdeep Goyal and Ravinder Kumar Kotnala

Abstract

Multiferroic magnetoelectric material has significance for new design nanoscale spintronic devices. In single-phase multiferroic $BaTiO_3$, the magnetism occurs with doping of transition metals, TM ions, which has partially filled d-orbitals. Interestingly, the magnetic ordering is strongly related with oxygen vacancies, and thus, it is thought to be a source of ferromagnetism of TM:$BaTiO_3$. The nanostructural MFe_2O_4 (M = Mn, Co, Ni, Cu, Zn, etc.) ferrite has an inverse spinel structure, for which M^{2+} ions in octahedral site and Fe^{3+} ions are equally distributed between tetrahedral and octahedral sites. These antiparallel sublattices (cations M^{2+} and Fe^{3+} occupy either tetrahedral or octahedral sites) are coupled with O_{2-} ion due to superexchange interaction to form ferrimagnetic structure. Moreover, the future spintronic technologies using diluted magnetic semiconductors, DMS materials might have realized ferromagnetic origin. A simultaneous doping from TM and rare earth ions in ZnO nanoparticles could increase the antiferromagnetic ordering to achieve high-Tc ferromagnetism. The role of the oxygen vacancies as the dominant defects in doped ZnO that must involve bound magnetic polarons as the origin of ferromagnetism.

Keywords: magnetically ordered material, oxygen vacancies, spin glass

1. Introduction

Multiferroics are combining multiple order parameters, offer an exciting way of coupling phenomena such as electronic and magnetic order. However, because simultaneous electric and magnetic order is difficult to achieve, multiferroics-especially those that function at or approaching room temperature-are extremely rare. For a crystal, when electrons are surrounded atomic nuclei to orient themselves in a same fashion then the crystal induces macroscopic ferromagnetism and electrical polarization. Since induction of ferromagnetism is essential in the technologies that involved sensors, computer hard drives, power generation, etc. Recently, the diluted magnetic semiconductors (DMSs) such as ZnO, SnO_2, TiO_2, etc. have generated potential in spintronics because DMS has the collective ordering that mediated via semiconductor charge carriers, as well as electron scattering at localized magnetic impurities and electron–electron interactions. To use DMSs for practical spintronic devices, a relatively high concentration of magnetic elements needed in the semiconductor host, and a large ferromagnetism is required with a Curie temperature (T_c) above room temperature. The transition metal (TM) ferrites with a spinel structure (MFe_2O_4; M = Co^{2+}, Ni^{2+}, Cu^{2+}, Zn^{2+}, etc.) are used in a wide variety of technological applications such as magnetic memory devices and biomedicine. However, these spinel ferrites are the candidate materials for multiferroic heterostructure due to their excellent magnetic response. For such multiferroic heterostructures, the perovskite ($BaTiO_3$, $PbTiO_3$, $BiFeO_3$, etc.) has an opportunity of higher piezoelectric coefficient that may pool with magnetostrictive materials ($CoFe_2O_4$, $NiFe_2O_4$, $ZnFe_2O_4$, etc.) via lattice strain effect.

1.1 Multiferroic $BaTiO_3$

The magnetoelectric (ME) effect—the induction of magnetization by an electric field and the induction of electric polarization by a magnetic field [1]. Multiferroic nanostructures have given recent advances in new type of memory devices including multistate data storage and spintronics [2]. $BaTiO_3$ (BTO) is a rare single-phase multiferroic. In multiferroics, the magnetic order is due to exchange interactions between magnetic dipoles, which themselves originate from unfilled shells of electron orbitals. Similarly, the electric order is due to the ordering of local electric dipoles, elastic order is due to the ordering of atomic displacements due to strain. The three crystallographic phases of BTO are ferroelectric: rhombohedral <190 K, orthorhombic for 190 K< T < 278 K and tetragonal for 278 K < T < 395 K. At higher temperatures, BTO is a paraelectric. For tetragonal BTO (**Figure 1(b)**) having lattice constants: $a(\text{Å}) = b(\text{Å}) = 3.99$; $c(\text{Å}) = 4.03$; space group = P4*mm*, the displacement of Ti^{4+} ion along *c*-axis might be induced electrical polarization (ferroelectricity). It involved hybridization of charge among Ti cation (3d states) with O anions (2p states). In cubic phase, the Ba^{2+} is located at the centre of the cube with coordination number 12.

Since to the formation of ME random access memories (MERAMs), the main thing that required ME coupling via interfacial exchange coupling among a multiferroic and a ferromagnet, which can change the magnetization of the ferromagnetic coating with respect to a voltage (**Figure 1(a)**) [3]. For such MERAMs, an electric field is enabled by ME coupling could control the exchange coupling between multiferroic and ferromagnetic at the interface. This exchange coupling at the interface reins the magnetization of the ferromagnetic layer, and therefore the magnetization might be change with multiferroic electrical polarization. Therefore, for perovskite (ABO_3) BTO, Ba^{2+} (A-site cation) induce the required distortion for ferroelectricity, while magnetism can be achieved by the doping such as TM = Cr, Mn, Fe, Co, Ni, Cu along B-site cation [4, 5].

1.2 Spinel ferrites

The spinel structure typically represented as AB_2O_4, where 'A' indicates four-fold coordinated tetrahedral sub-lattice sites and 'B' indicates six-fold coordinated octahedral sub-lattice (**Figure 1(c)**). There are 8 A-sites in which the metal cations are tetrahedrally coordinated with oxygen, and 16 B-sites, which possess octahedral coordination. Normal ferrites have divalent cations residing solely as the central ion on the tetrahedral sub-lattice with only trivalent Fe cations occupying octahedral sub-lattice sites. Harrisa and Sepelak [6] suggested superexchange interaction, the J_{BB} is strong and negative indicating antiferromagnetic coupling between Fe^{3+}–O^{2-}–Fe^{3+} with octahedral sub-lattice cation spins largely canceling out. When

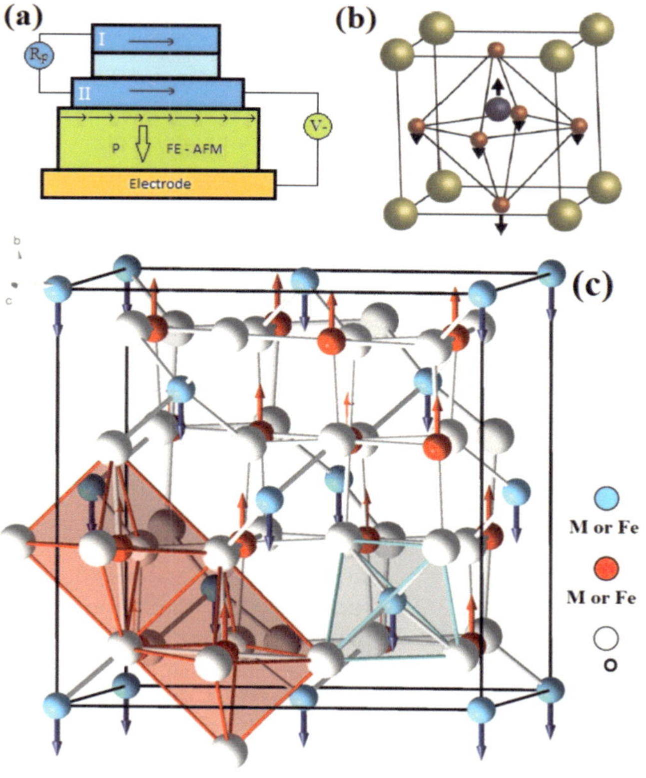

Figure 1.
(a) Schematic MERAM device in which the binary information is stored by the magnetization direction of the ferromagnetic layer (I & II), read by the resistance, R_p of the magnetic trilayer, and written by applying a voltage across the multiferroic ferroelectric-antiferromagnetic layer (FE-AFM). (b) Tetragonal crystalline structure of BTO, with a lattice ferroelectric distortion. (c) Spinel unit cell. Arrows represent magnetic moments and their antiparallel alignment. Adopted from Harrisa and Sepelak [6].

the divalent cation resides on the [B] site and A and B ions share the balance of [B] and the totality of (A), the spinel is inverse: $(A_{1-\delta}B_\delta)[A_\delta B_{2-\delta}]O_4$, where δ is the inversion degree, *i.e.*, $NiFe_2O_4$. Ferrite of the type $NiFe_2O_4$ (NFO), $CoFe_2O_4$ (CFO) and $MnFe_2O_4$ (MFO) with the spinel structure are magnetic ceramics which have potential in electronic and magnetic components [7]. The NFO has an inverse spinel structure for which Fe^{3+} ions occupied tetrahedral A-sites; whereas Fe^{3+} and Ni^{2+} ions are sit on octahedral B-sites. This NFO is ferrimagnetic material has magnetization originated by antiparallel spins on A- and B-sites. However, for normal spinel structure of CFO, Co is a divalent atom, occupying tetrahedral A-sites, while Fe is a trivalent atom, sitting on the octahedral B-sites. For MFO, the inverted spinel structure is partial for which the Mn^{2+} and Fe^{3+} ions with half-filled 3d shell (ground state is singlet ^6S with spin = 5/2 with zero orbital momentum) and the crystal field is not sufficient to split it [8]. The magnetic moment of MFO agrees well with Neel's coupling scheme and has lower resistivity than CFO and NFO ferrites [9].

1.3 Diluted magnetic semiconductors

Recently, the realization of spin in DMS, semiconductor with substituted magnetic impurities (Fe, Co, Ni, Cu) has attracted great interest in design of spintronics devices like spin field-effect transistors, non-volatile memory devices, and programmable logic gates [10–13]. Among various DMS, ZnO is a promising spin source, since it epitomizes DMS with T_c well above room temperature. The DMS ZnO is a wide band gap that gained recent research in spintronics, due to an ability to change its optical and magnetic behavior with doping of TM = Fe, Co, Mn, Ni, Cu, Cr, or V ions and/or by intrinsic defects, such as oxygen vacancy (V_O) and zinc vacancy (V_{Zn}). The DMS ZnO has wide applications included:

1.3.1 Magnetic recording

For read heads in magnetic disk recorders (computer components), read head senses the magnetic bits that are stored on the media, which is stored as magnetized regions of the media, called magnetic domains, along tracks (**Figure 2(a)**) [13]. Magnetization is stored as a "0" in one direction and as a "1" in the other. Although, there is no magnetic field emanating from the interior of a magnetized domain itself, uncompensated magnetic poles in the vicinity of the domain walls generate magnetic fields (sensed by the GMR element) that extend out of the media.

1.3.2 Nonvolatile memories

The "Nonvolatile" refers to information storage that does not "evaporate" when power is removed from a system, *i.e.*, magnetic disks and tapes. Prinz [13] has recently demonstrated that GMR elements can be fabricate in arrays with standard lithographic processes to obtain memory that has speed and density approaching that of semiconductor memory, but is nonvolatile. A schematic representation of RAM that is constructed of GMR elements is shown in **Figure 2(b)**. The spin-dependent scattering of the carriers (electrons & holes) is minimized for parallel magnetic moment of the ferromagnetic layer, to induce lowest value of resistance. However, the highest resistance is the result of maximized spin-dependent scattering carriers via anti-aligned ferromagnetic layers. An external magnetic field could give the direction of magnetic moments, applied to the materials. The spin-valve structures of GMR set into series using lithographic (wires) termed as a sense line. This sense line has information storage due to resistance (resistance of elements). The sense line runs the current which is detected at the end by amplifiers because resistance changes in the elements.

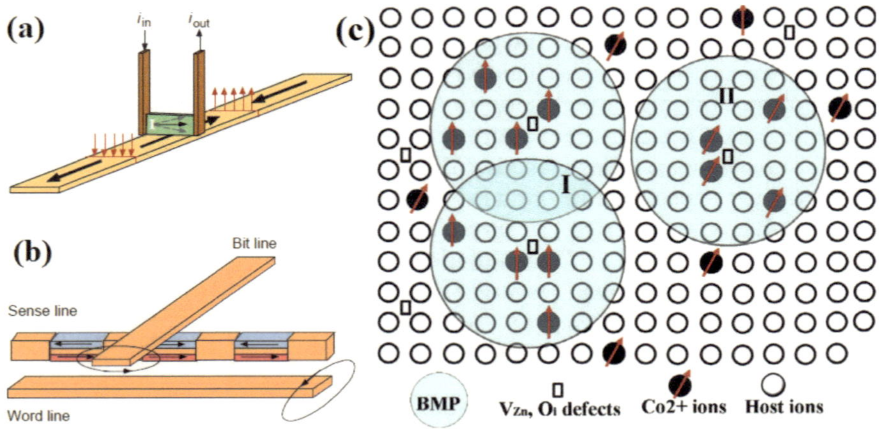

Figure 2.
A schematic representation of (a) GMR read head (I) that passes over recording media containing magnetized regions, (b) RAM that is constructed of GMR elements. Adopted from Prinz [13], (c) bound magnetic polaron (BMP), V_{Zn}, O_i trapped carriers couple with the 3d shell spins of TM ions within its hydrogenic orbit.

1.3.3 Ferromagnetism in DMS ZnO due to bound magnetic polarons

Generally, the room temperature ferromagnetism (RTFM) in DMS ZnO is given because of bound magnetic polaron (BMP) model [14]. The localized spins of the dopant ion interact with the charge carriers that are bound to a small number of defects such as oxygen vacancies, resulting into a magnetic polarization of the surrounding local moments. Since the magnetism in TM ions doped ZnO nanoparticles relates with exchange interactions between unpaired electron spins, that arising from the lattice imperfections such as oxygen vacancies, V_O at the surface of the nanoparticles. Pal et al. [15] described BMP formation in Co doped ZnO and shown in **Figure 2(c)**. The electrons trapped in the defect vacancies undergo orbital coupling with the *d* shells of the adjacent divalent dopant ion and form BMP. In BMP model, the bound electrons (holes) hold in defect states that coupled through TM ions to overlap ferromagnetic regions, which responsible into high T_C (due to formation of long-range ferromagnetic ordering) [16]. When the dopant ions are donors or acceptors, the exchange interactions are sp-d that would lead BMPs formation.

2. Experimental methods

The multiferroic, DMS and ferrites materials might be synthesized by methods such as a sol-gel [17], chemical combustion [18], hydrothermal [19], metallo-organic decomposition (MOD) [20], conventional solid-state reaction [21], sol-gel precipitation [22], thermal evaporation [23], etc.

3. Result and discussion

3.1 Multiferroic systems of $BaTiO_3$

3.1.1 X-ray diffraction for $BaTiO_3$ and $BaTM_{0.01}Ti_{0.99}O_3$ nanoparticles

The sol-gel method is used to prepare pure $BaTiO_3$ and $BaTM_{0.01}Ti_{0.99}O$ [TM = Cr (BTO:Cr), Mn (BTO:Mn), Fe (BTO:Fe), Co (BTO:Co), Ni (BTO:Ni), Cu (BTO:Cu)] nanoparticles [4]. The TM ions in perovskite BTO structure highly influenced lattice constants to induce lattice strain and unit cell expansion, which responsible into defects vacancies formation. **Figure 3** shows the Rietveld refinement (Full-Prof program) of X-ray diffraction (XRD) patterns for pure and $BaTM_{0.01}Ti_{0.99}O$ nanoparticles measured at room temperature. A polycrystalline with tetragonal BTO phase (space group:P4*mm*) is detected. The fitting parameters, $R_p(\%)$ = 6.7, 7.1, 6.7, 4.9, 9.1, 8.2 and 9.8, $R_{wp}(\%)$ = 3.2, 9.9, 9.2, 10.0, 12.5, 1.2 and 14.6, χ^2 = 1.1, 1.4, 0.81, 1.4, 3.3, 3.3 and 0.9 and distortion ratio, (c/a) = 1.00959, 1.00932, 1.00688, 1.00909, 1.00776, 1.00625 and 1.00508, respectively, refined for pure BTO and BTO with Cr, Mn, Fe, Co, Ni, Cu doping. Also, **Figure 3(a)** shows the tetragonal splitting of (200) diffraction peak at 2θ = 44.3–45.7°. The diffraction peak of pure BTO has shifted towards a lower diffraction angle with TM doping which supports the lattice strain in BTO. Such splitting of (200) peak might be confirmed the tetragonal phase formation. It is due to electrostatic repulsions between 3d electrons of Ti^{4+} ions and 2p electrons of O^{2-} ions, the structure becomes distorted. It is also reported in Ref. [4] that the average particles size, from TEM, D_{TEM} (nm) = 20 ± 3, 13 ± 1, 33 ± 5, 35 ± 3, 17 ± 1 and 47 ± 7, the value of saturation magnetization, M_s (emu g^{-1}) = 0.056, 0.042, 0.066, 0.035, 0.013 and 0.021, and the ME coupling constant, α_{ME} (mV cm^{-1} Oe^{-1}) = 25.91, 11.27, 31.15, 16.58, 11.61 and 16.48, respectively, measured with Cr, Mn, Fe, Co, Ni, Cu doping into BTO.

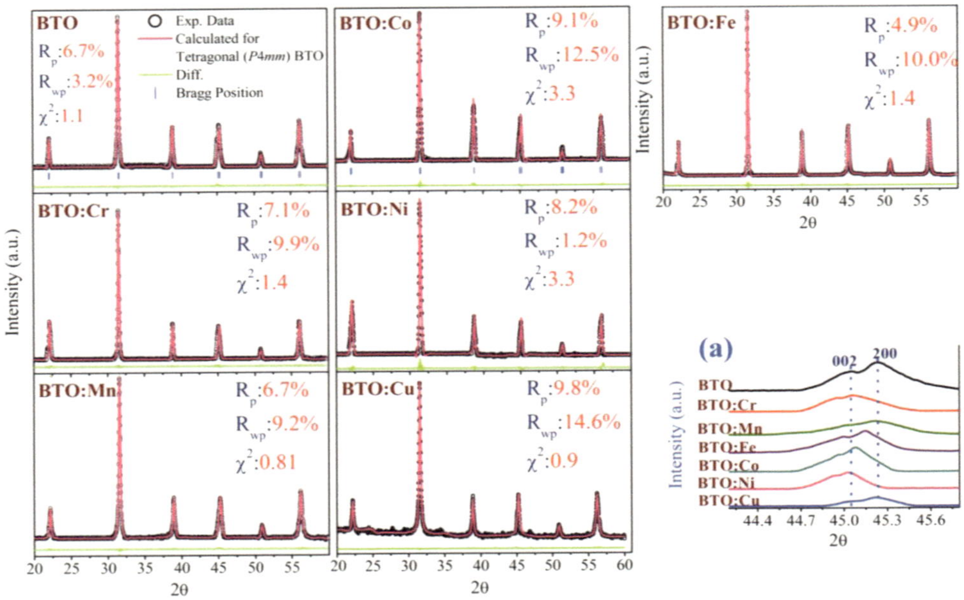

Figure 3.
XRD pattern for pure $BaTiO_3$ and $BaTM_{0.01}Ti_{0.99}O_3$ nanoparticles. (a) Showing the splitting of the (200) peak. Adopted from Verma and Kotnala [4].

3.1.2 Nanostructure of $BaTiO_3$ (BTO) and $BaFe_{0.01}Ti_{0.99}O_3$ (BFTO) multiferroic

The BTO and BFTO were prepared by a hydrothermal method of processing temperature 180°C/48 h [5]. The XRD pattern shows the coexistence of cubic/tetragonal/hexagonal phases of BTO and cubic/tetragonal of BFTO. **Figure 4(a** and **b)** reveals the TEM images of BTO and BFTO nanostructure. It shown that BTO (**Figure 4(a)**) is the product that consist of nanorods structure having hexagonal like face of average diameter 50 nm and length 75 nm. However, **Figure 4(b)** shows the nanowires formation of BFTO with average diameter ~45 nm and the length >1.5 μm. It is also reported in Ref. [5] that the room temperature *M-H* hysteresis shows diamagnetism in BTO and ferromagnetism in BFTO with M_s ~ 82.23 memu g^{-1}, M_r ~ 31.91 memu g^{-1} with H_c ~ 122.68 Oe and ME coupling coefficient, α_{ME} = 16 mV Oe^{-1} cm^{-1}.

3.1.3 Ferromagnetism/ferroelectricity in Ce,La:$BaFe_{0.01}Ti_{0.99}O_3$ nanostructures

The nano-aggregation type $Ba(Fe_{0.67}Ce_{0.33})_{0.01}Ti_{0.99}O_3$ (BFTO:Ce) and $Ba(Fe_{0.67}La_{0.33})_{0.01}Ti_{0.99}O_3$ (BFTO:La) product is synthesized by a hydrothermal process [2]. Rietveld refinement of XRD pattern indicates polycrystalline phase with tetragonal BFTO. It is reported that the Ce and La ions in BFTO improved lattice distortion, c/a ratio. These dopant Ce and La in BFTO forms nano-aggregation type product with average value of aggregation diameter, D = 40 and 22 nm, respectively, for BFTO:Ce and BFTO:La. The formation of tetragonal BTO phase and lattice defects due to vacancies is attributed by Raman active modes. These defects and vacancies are also confirmed with photoluminescence measurement that might be altered due to higher surface-to-volume ratio in nano-aggregation. **Figure 4(d** and **e)** shows the ferromagnetic behavior of Ce, La doped BFTO, respectively, by magnetization versus field (M-H_{dc} hysteresis) measured at room temperature [2]. The values of M_s (emu g^{-1}) = 0.15 and 0.08, and M_r (emu g^{-1}) = 0.039 and 0.015 with H_c (Oe) = 242 and 201, respectively, for BFTO:Ce and BFTO:La. It is well predicted that the formation of Fe^{4+}–O^{2-}–Fe^{4+} interaction is ferromagnetic, which dominate in BFTO over the antiferromagnetic Fe^{3+}–O^{2-}–Fe^{4+}

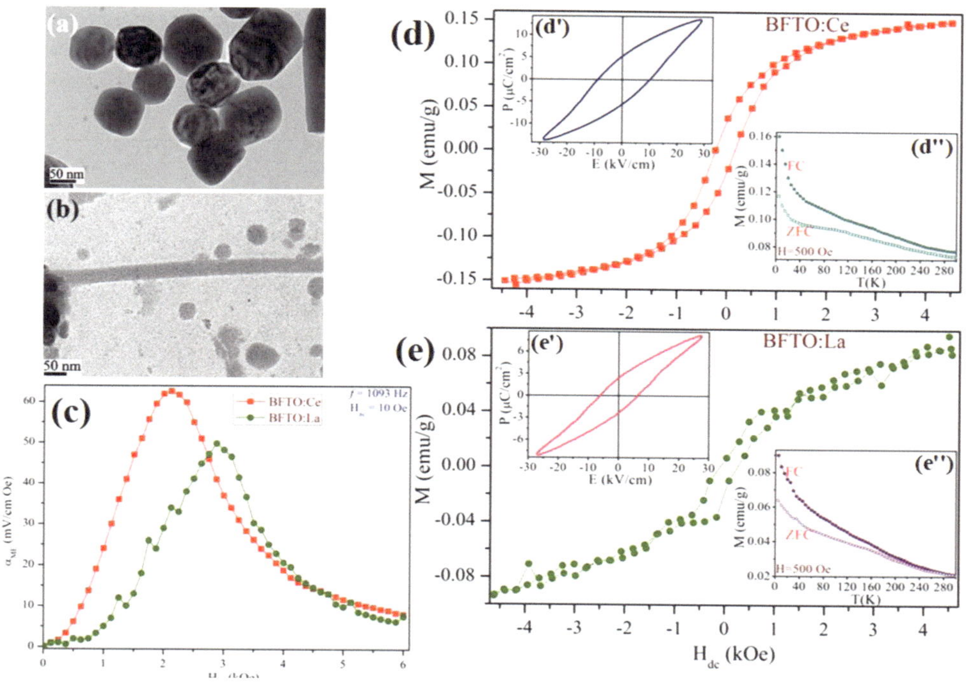

Figure 4.
Transmission electron microscopy (TEM) of (a) pure $BaTiO_3$ *(b)* $Ba(Fe_{0.01}Ti_{0.99})O_3$*. (c) ME voltage coefficient (*α_{ME}*), (d and e) M-*H_{dc} *hysteresis at room temperature, (d' and e') P-E hysteresis, and (d" and e") M(T) following ZFC/FC at H = 500 Oe for Ce, La doped BFTO. Adopted from Refs. [2, 5].*

and Fe^{3+}–O^{2-}–Fe^{3+} interactions, producing weak ferromagnetism. Generally, the ferromagnetism in Fe-doped $BaTiO_3$ is explained into two ways: the partially filled inner shells (d- or f-levels) and formation of nanostructures. An F-centre exchange (FCE) mechanism describes the required ferromagnetism [24, 25]. Such a mechanism expects $Fe^{3+}-V_O^{2-}-Fe^{3+}$ transition that generally exists in the structure where an electron trapped in oxygen vacancies, V_O to form F-centre. For this, the electron occupies an orbital, p_z that overlaps d_z^2 of d shells in iron neighbors. The Fe^{3+} ions have $3d^5$ electronic configurations for which spin down trapped electron and spin up in two iron Neighbors. Therefore, the F-centre has the exchange interaction among two iron ions that would leads to ferromagnetism. The insets of **Figure 4(d′** and **e′)** show the polarization-electric field (P-E) hysteresis at room temperature. With Ce doping into BFTO, the value of spontaneous polarization, P_s ($\mu C\ cm^{-2}$) = 13.83 and remanent polarization, P_r ($\mu C\ cm^{-2}$) = 5.45 with electric coercivity, E_c ($kV\ cm^{-1}$) = 9.53. However, with La doping, P_s ($\mu C\ cm^{-2}$) = 8.28, and P_r ($\mu C\ cm^{-2}$) = 2.46, with E_c ($kV\ cm^{-1}$) = 6.14. These values of polarization have an improvement over reported work [26–28]. This is due to nano-aggregation formation and lattice distortion enhancement in BTO lattice. The smaller polarization in BFTO:La is the nano-size effect that involved compensation of polarization-induced surface charges [29].

3.1.4 Magnetization at low temperature measurement

The origin of observed ferromagnetism at room temperature in Ce- and La-doped BFTO is described by measuring their magnetization from zero-field cooling (ZFC) and field cooling (FC) at 500 Oe (**Figure 4(d″** and **e″)**). A clear separation between FC and ZFC retains up to low temperature without blocking temperature is observed. This is an indication of weak antiferromagnetic interactions.

An upward curvature observed in M-T curve suggests a Curie-Weiss like behavior. It is attributed with short-range ferromagnetism, or a spin-cluster within a matrix of spin disorder [30]. Li et al. [31] suggested that the oxygen vacancy might be mediate antiferromagnetic-ferromagnetic interactions in multiferroics.

3.1.5 ME coupling in Ce, La-substituted BFTO

The longitudinal ME coupling coefficient, α_{ME} is measured for Ce and La-doped BFTO and shown in **Figure 4(c)**. The samples are biased with *ac* magnetizing field, H_{ac} = 10 Oe at 1093 Hz, and a *dc* magnetic field, H_{dc} is applied collinear to it. The value of α_{ME} was determined as a function of *dc* magnetic field using: $\alpha_{ME} = \frac{\partial E}{\partial H} = \frac{1}{t}\frac{\partial V}{\partial H} = \frac{V_{out}}{t \times H_{ac}}$. In **Figure 4(c)**, the value of α_{ME} increases rapidly to a maximal value (due to an enhancement of elastic interactions) and then slowly decreasing in the higher region of H_{dc}. The maximum value of α_{ME} (mV cm^{-1} Oe^{-1}) is 62.65 and 49.79, respectively, for BFTO:Ce and BFTO:La.

3.2 Spinel ferrites MFe_2O_4 (M = Mn, Co, Ni, Zn, Cu, etc.)

3.2.1 Structural studies of $Co_{0.65}Zn_{0.35}Fe_2O_4$

Figure 5 (a) shows the Rietveld refinement (space group: $Fd\bar{3}m$) of $Co_{0.65}Zn_{0.35}Fe_2O_4$ (CZFO) ferrite, which indicate spinel phase [32]. The value of χ^2 is 1.648. The lattice parameter of CZFO is found to be 8.4183 Å. A three-dimensional sketch of CZFO unit cell projected along the *c*-axis (**Figure 5(b)**). The inset of **Figure 5(a)** shows the SEM image of a CZFO pellet sample to displays densely packed grains with few scattered pores and voids. It is also reported in Ref. [32] that the high dielectric permittivity value is obtained. The ferrimagnetic-paramagnetic phase transition is ~640 K.

3.2.2 Lattice constant, grains size and magnetism in pure NFO, CFO and MFO thin films

Table 1 shows the experimental results of lattice constant (*a*), grain's size [*x* (SEM) & *x'* (AFM)], M_s, M_r and H_c measured at room temperature (300 K) and 10 K, and magnetic phase transition temperature (T_{pm}) of $NiFe_2O_4$ (NFO),

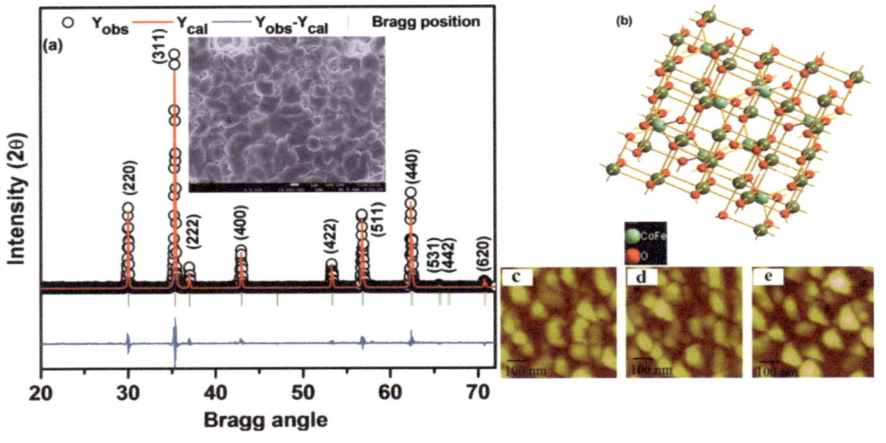

Figure 5.
(a) XRD patterns and SEM image (inset) of $Co_{0.65}Zn_{0.35}Fe_2O_4$ (CZFO). (b) CZFO unit cell with a spinel structure. AFM of (c) NFO, (d) CFO (e) MFO thin films. Adopted from Refs. [9, 32].

DOI: http://dx.doi.org/10.5772/intechopen.82437

Sample	*a* (Å)	*x* (nm)	*x'* (nm)	M_s (emu cc^{-1})		M_r (emu cc^{-1})		H_c (Oe)		T_{pm} (K)
				300 K	**10 K**	**300 K**	**10 K**	**300 K**	**10 K**	
NFO	8.161	44	46	50.60	76.42	14.33	63.31	265.33	73.90	687
CFO	8.312	60	61	33.50	54.57	15.50	8.90	1292.00	69.33	693
MFO	8.425	74	75	5.40	13.35	1.10	7.20	113.30	39.28	581

Adopted from Verma et al. [9].

Table 1.
*Values of lattice constant (**a**), grain's size (**x**) (SEM), grain's size (**x'**) (AFM), M_s, M_r and H_c at 300 K and 10 K, and magnetic phase transition (T_{pm}) of pure NFO, CFO and MFO thin films.*

$CoFe_2O_4$ (CFO) and $MnFe_2O_4$ (MFO) thin films [9]. These ferrites were prepared by a MOD method using spin coating. The thickness of all the film is ~700 nm. It is also reported that the miller indices of cubic spinel ferrites structure are (2 2 0), (3 1 1), (2 2 2), (4 0 0), (3 3 1), (4 2 2) and (5 1 1), respectively, detected with diffraction angle 2θ = 30.48, 34.99, 37.48, 42.58, 48.02, 51.23 and 55.85° for NFO, for CFO, 2θ = 30.25, 34.87, 37.36, 42.58, 47.20, 51.23 and 55.84° and 2θ = 29.77, 34.64, 37.36, 41.75, 47.08, 51.23 and 57.04° for MFO. The higher coercivity value of CFO than NFO and MFO is the effect of higher magneto-crystalline anisotropy of Co cations than Ni and Mn. The observed values of M_s, M_r and H_c of the NFO, CFO and MFO films are quite smaller than bulk form [9]. For this, the decrease in M_s in ferrite nanoparticles is the canted spins in the surface layers [33]. Also, it observed very smaller values of M_r, M_s and H_c of MFO because lowering number of magnetic domains and the rate of alignment of the spins with the applied field [34, 35].

3.2.3 AFM images of NFO, CFO and MFO thin films

The value of average grain's size from AFM (**Figure 5(c–e)**) is 46, 61 and 75 nm, and the surface roughness 2.5, 4 and 2 nm, respectively, for NFO, CFO and MFO ferrite thin films, were synthesized by MOD method [9].

3.2.4 Magnetic ordering of ferrites in $MFe_2O_4/BaTiO_3$ nanocomposite

The multiferroic $MFe_2O_4/BaTiO_3$ (MFO/BTO) thin films were fabricated by a MOD method [20]. The addition of ferrite MFe_2O_4 in BTO results into lattice strain due to tetragonal distortion, unit cell expansion/contraction and lattice mismatch. It enhances ME coupling. The lattice distortion, *c*/*a*, tetragonal phase of BTO, spinel ferrite MFO and the lattice strain in composite phases is studied by XRD pattern [20]. These reported results also shown the AFM images to evaluate the value of average grain size for MnFO/BTO, CFO/BTO, NFO/BTO and ZFO/BTO, which is 25, 102, 24 and 133 nm, respectively. XPS analysis indicate that Fe exist into mixed +2/+3 valence states and O in both O^{2-} and deficient states. **Figure 6(a)** shows the room temperature ferromagnetic behavior of MFO/BTO thin films by measuring ferromagnetic hysteresis (M-H_{dc}). The measured value of saturation magnetization, M_s (kJ T^{-1} m^{-3}) = 1.29, 20.25, 27.64 and 6.77, remanent magnetization M_r (kJ T^{-1} m^{-3}) = 0.03, 3.75, 6.76 and 1.49 and with magnetic coercivity, H_c ($\times 10^5$ Am^{-1}) = 0.013, 0.079, 0.167 and 0.135, respectively, for MnFO/BTO, CFO/BTO, NFO/BTO and ZFO/BTO nanocomposite. An enlarged view of M-H_{dc} as the inset of **Figure 6(a)** is also shown for more clarification. This magnetic measurement of MFO/BTO is compared with magnetization of a single phase MnFO (5.40 kJ T^{-1} m^{-3}), CFO (33.50 kJ T^{-1} m^{-3}), NFO (50.60 kJ T^{-1} m^{-3}) and ZFO (230 kJ T^{-1} m^{-3}), which indicate abrupt reduction in nanocomposite sample. This is because non-magnetic BTO phase exist in nanocomposite, which reduced the magnetization of ferrite. Since the non-magnetic elements weaken the A-B superexchange interaction which result into an increase the distance between the magnetic moments in A and B sites in spinel structure. Since the magnetization decreases/increases in ferrites is a relaxation process that might be concerned with the redistribution of oxygen vacancies reported by Wang et al. [36]. The presence of a magnetic dead or antiferromagnetic coating on the nanostructural surface have reducible value of saturation magnetization of ferrite nanoparticles [37]. Vamvakidis et al. [38] suggested that the reduction in inversion degree (~0.22) of $MnFe_2O_4$ due to the partial oxidation of Mn^{2+} to Mn^{3+} ions results into weaker superexchange interactions between the tetrahedral and octahedral sites within the spinel structure. Bullita et al. [39] reported the cation distribution of $ZnFe_2O_4$ at the nanoscale level, which is contributed by partial inverted spinel structure results into

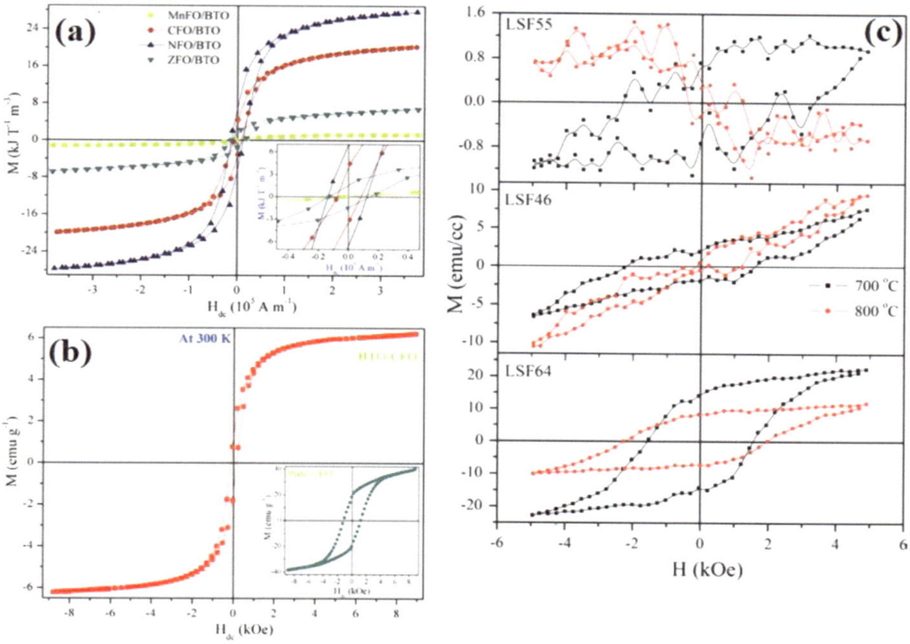

Figure 6.
(a) M-H$_{dc}$ hysteresis of MFe$_2$O$_4$/BaTiO$_3$ nanocomposite thin films at 300 K. (b) Room temperature M-H$_{dc}$ measurement for pure CFO and BTO-CFO nanoparticles. (c) the M-H measurement of LSF thin films. Adopted from Refs. [19, 20, 35].

an increase magnetization. Peddis et al. [40] indicates inversion degree of CoFe$_2$O$_4$ nanoparticles, which has a better correlation between spin canting and cationic distribution to get competitively higher M$_s$.

3.2.5 Magnetization in CFO and 0.25BTO–0.75CFO nanoparticles

The multiferroic 0.25BTO–0.75CFO nanocomposite is prepared by hydrothermal process at 180°C/48 h [19]. The spinel CFO and tetragonal BTO structure is studied by XRD pattern. The 0.25BTO–0.75CFO nanocomposite has nanoparticles formation with average particles size from FESEM is 80 ± 10 nm and from TEM is 76 ± 13 nm. **Figure 6(b)** shows the M-H$_{dc}$ hysteresis for pure CFO and 0.25BTO–0.75CFO nanoparticles, measured at room temperature. The value of M$_s$ (emu g^{-1}) = 39.027 and 6.199, M$_r$ (emu g^{-1}) = 19.724 and 1.545 with H$_c$ (Oe) = 1226 and 149, respectively, for CFO and 0.25BTO–0.75CFO nanoparticles. These measured values of M$_s$ and H$_c$ for pure CFO are smaller than reported for bulk CoFe$_2$O$_4$ (M$_s$ (emu g^{-1}) =73 and H$_c$(kOe) = 5, at 300 K). However, the nanocomposite of 0.25BTO–0.75CFO has abrupt decrement in M$_s$ value, which is explained on the basis of BTO coating over CFO. For nanostructural CFO, the inversion degree must be swung with surface spin-canting and finite size-effect, to execute magnetic disorder. The change in bond angle and bond length may also influence the magnetization of CFO in composite. Since the spin coupling due to 3d unpaired electrons in Co^{2+} and Fe^{3+} along A and B-sites of spinel structure would lead to the ferrimagnetism of CoFe$_2$O$_4$. Due to such spin arrangement by unpaired electrons, there exist Fe$_B$–O–Fe$_B$, Fe$_B$–O–Co$_B$, Fe$_B$–O–Fe$_A$, Co$_B$–O–Fe$_A$ and Co$_A$–O–Fe$_B$ superexchange interactions to induce magnetic ordering. In this, the antiferromagnetic interactions are strongest via A–O–B superexchange (inter-sublattice) and the weak ferromagnetism is attributed with A–O–A and B–O–B superexchange (intra-sublattice). It is reported for an ideal spinel CFO, the bond angle is 90° of B–O–B, 120° of A–O–B and 80° of A–O–A magnetic exchange interactions. In the present case, the Rietveld analysis [19], calculated the values of bond-angle/length between A and B sites of CFO in 0.25BTO–0.75CFO nanocomposite, *i.e.*, bond angle = 91.07° for Fe_B–O–Fe_B, 133.57° for Fe_A–O–Co_B and 76.13° of Fe_A–O–Fe_A. The values of bond length (l), $l_{A\text{-}A}$ = 0.3626(1) nm, $l_{B\text{-}B}$ = 0.2961(2) nm and $l_{A\text{-}B}$ = 0.34713 (2) nm. It is reported that the superexchange interactions might be related with bond-angle/length among bonds in Fe, Co and O atoms [41]. For the bond angle of 90° (Fe_B—O–Fe_B bonds), the ferromagnetic super-exchange interaction are resulted. When this bond angle is increased, the antiferromagnetic super-exchange interactions are there and generally, bond 120° of super-exchange Fe_A–O–Fe_B is the dominant interaction of magnetite. Above 120° of bond angle, the antiferromagnetic strength is increased which is maximum at 180° [42]. Hence the calculated values of bond angles for Co and Fe ions along A and B-sites are deviated with theoretical one that might be indicated the distorted spinel lattice to influence resulting magnetic behavior.

3.2.6 *Ferromagnetism in x$Li_{0.5}Fe_{2.5}O_4$-(1 − x)$SrFe_2O_4$ thin films*

The role of hard/soft ferrites composite of $x$$Li_{0.5}Fe_{2.5}O_4$-(1 − x)$SrFe_2O_4$ (LSF) (x = 0.4, 0.5, and 0.6) on the magnetic exchange-spring systems have been studied [35]. LSF thin films were prepared by MOD method. XRD result shows the polycrystalline behavior of ferrite with cubic phase of $Li_{0.5}Fe_{2.5}O_4$ and orthorhombic $SrFe_2O_4$. The magnetic behavior of the three series of LSF films is investigated by measuring M-H hysteresis (**Figure 6(c)**). The Li doping into LSF enhanced the value of M_s and M_r and decreased coercivity value. At equal content of Li:Sr., the magnetic behavior is unpredicted. At annealing temperature of 800°C, the LSF55 thin film have anti S-type (diamagnetism), LSF64 has reducible magnetization, while LSF46 has a similar magnetism that seen at 700°C of annealing temperature. The values of saturation magnetization, M_s (emu cc^{-1}) = 1.21, 8.01, and 22.78, remanent magnetization, M_r (emu cc^{-1}) = 0.64, 2.13, and 14.20, with magnetic coercivity, H_c(kOe) = 2.32, 2.18, and 1.54, respectively, measured for LSF55, LSF46, and LSF64 sample, annealed at 700°C. Also from **Figure 6(c)**, the magnetic hysteresis is involved two-step processes, which might have typically exchange-spring

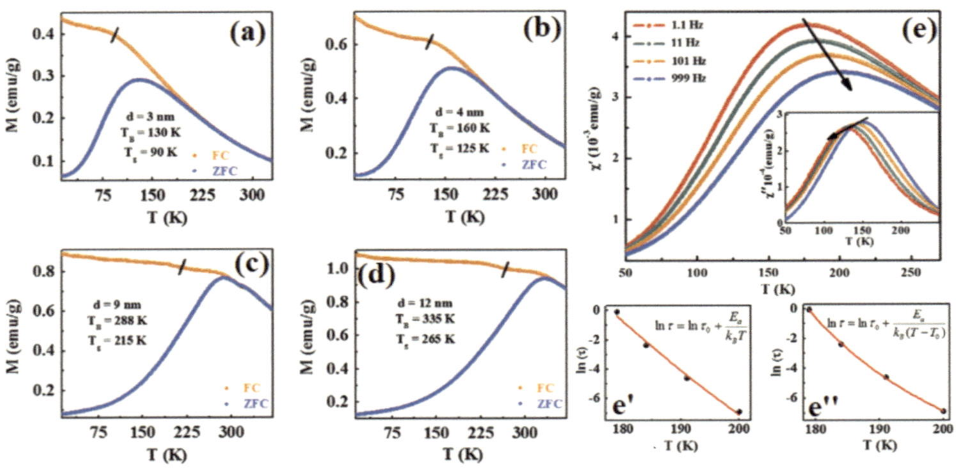

Figure 7.
(a–d) ZFC-FC magnetization of $CoFe_2O_4$ nanoparticles. (e) Temperature dependent $\chi'(T)/\chi''(T)$ of ac susceptibility for 4 nm $CoFe_2O_4$. (e') Arrhenius law (e") Vogel-Fulcher law of $\chi'(T)$. Adopted from Mohapatra et al. [45].

regime. The positive nucleation field is increased with Li doping. This positive nucleation field occurrence is predicted with a micro-magnetic model, where the perpendicular bilayer of shape anisotropy contribution is there [43].

3.2.7 Super-spin glass formation in $CoFe_2O_4$ nanoparticles

The magnetic $CoFe_2O_4$ nanoparticles of mean size 2–16 nm have been synthesized through a solventless thermolysis technique [44]. **Figure 7(a–d)** shows the ZFC/FC magnetization of 3, 4, 9 and 12 nm $CoFe_2O_4$ nanoparticles in a field of 5 Oe [45]. The ZFC magnetization of these $CoFe_2O_4$ nanoparticles has observed maxima, which represents the blocking temperature, T_B. However, the FC magnetization is continuously increased below T_B, but at a very low temperature, the FC magnetization slightly become constant. Expectedly, the FC magnetization below T_B is abruptly increased with a decrease in temperature, which is because the superparamagnetic nanoparticles have no inter-particle interactions. This is typically a signature of super-spin glass transition [44]. The temperature dependent real $\chi'(T)$ and the imaginary $\chi''(T)$ components of the *ac* magnetic susceptibility for 4 nm $CoFe_2O_4$ nanoparticles are shown in **Figure 7(e)**. The $\chi'(T)$ curve for 1.1 Hz displays a sharp peak at the blocking temperature, T_B = 179 K and the position of peak is frequency dependent. By increasing frequency from 1.1 to 999 Hz, there is shifting of peak position from 179 to 200 K that calculated the value of ΔT_m = 21 K. However, $\chi''(T)$ in the inset of **Figure 7(e)** has also peak shifting behavior by changing peak position from 130 to 151 K with frequency 1.1–999 Hz. This is analyzed with relation: $\varphi = \Delta T_m/T_m\Delta\log(f)$, where ΔT_m is the difference between T_m measured in $\Delta\log(f)$ frequency interval. For non-interacting nanoparticles, this parameter 'φ' is usually more than 0.13, for nanoparticle based superspin-glasses, the range is $0.005 < \varphi < 0.05$ and for intermediate interactions, $0.05 < \varphi < 0.13$) [46]. From **Figure 7**, the φ values calculated to be ~0.015, lies for super-spin-glass systems. The super-spin glass behavior is further studied with Neel-Arrhenius law and Vogel-Fulcher model as shown in **Figure 7(e′ and e″)**.

3.3 ZnO: diluted magnetic semiconductor

Recently, DMS ZnO has a technological potential due to its large direct band gap (3.37 eV) and exciton binding energy about 60 meV, which is comparably high [47]. The room temperature ferromagnetism, RTFM in 3d ions doped ZnO is reported by Dietl et al. [48]. But the pure ZnO is also give RTFM when the crystalline product have small sized nanoparticles [49, 50]. The oxygen vacancies are located on the nanoparticles surface and responsible in major participation of RTFM [51]. Garcia et al. [49] found that the ZnO nanoparticles had absorbed certain organic molecules to modify the electronic structure to give RTFM without any magnetic impurity. Xu et al. [50] suggested singly charged oxygen vacancies that depend upon nano-size and heating condition and located mainly near on ZnO nanoparticles surface to induce ferromagnetism. During last decade, the magnetism of ZnO with TM doping is extensively studied. Sato et al. [52] have used local density approximation (LDA) to discuss ferromagnetic ordering which is more stable in Fe-doped ZnO. Karmakar et al. [53] have found antiferromagnetism, which prefers to stabilize Fe-doped ZnO without any native defects. Spaldin [54] had found ferromagnetic ordering which is not possible when Zn sites are substituted with Co or Mn, unless additional hole carriers are incorporated. However, rare earth (RE) atoms have partially filled *f*-orbitals which carry high magnetic moments and form magnetic coupling as for TM ions with partially filled *d*-orbitals [55]. Deng et al. [56] investigated the effect of La doping on the electronic structure and optical properties of ZnO using the first-principle calculation. It is also recently reported that the RE Ce ions are responsible in support antiferromagnetic interactions in TM = Co, Fe doped ZnO [57].

3.3.1 XRD for La, Gd substituted $Zn_{0.95}Co_{0.05}O$ nanostructure

The $Zn_{0.95}Co_{0.05}O$ (ZCO5), $Zn_{0.92}Co_{0.05}La_{0.03}O$ (ZCLO53) and $Zn_{0.92}Co_{0.05}Gd_{0.03}O$ (ZCGO53) nanostructures were synthesized by a sol-gel process [11]. The structural information of La and Gd doped ZCO5 nanostructure using the Rietveld method for XRD pattern of wurtzite structure (space group $P6_3mc$) refined and shown in **Figure 8(a)**. The miller indices of wurtzite structure (100), (002), (101), (102) and (110), respectively, observed with diffraction angle 2θ = 31.88, 34.49, 36.34, 47.59 and 49.65° of ZnO. The XRD peak intensity of $Zn_{0.95}Co_{0.05}O$ is reduced with La and Gd doping into it. This is because the ionic radii of La^{3+} and Gd^{3+} ions is much larger than TM Zn^{2+} and Co^{2+} ions, which results into lattice deformation of ZnO. Due to this, the lattice parameters such as lattice distortion, Zn–O bond length/angle and per unit cell volume are affected to induce lattice defects. The value of lattice constant, a = 3.252(1) Å, 3.253(2) Å and 3.255(1) Å, c = 5.204(1) Å, 5.218(3) Å and 5.212(3) Å, V = 47.660(1) Å^3, 47.818(3) Å^3 and 47.823(2) Å^3, and χ^2 = 3.37, 0.893 and 2.9, respectively, extracted for ZCO5, ZCLO53 and ZCGO53. It is also reported that the average value of particles size, D of nano-aggregation is 142 and 86 nm, respectively, for ZCLO53 and ZCGO53 [12]. The wurtzite ZnO structure and lattice defects (vacancies/interstitials) are also found with Raman study. Photoluminescence spectra have near band edge emission (shown energy band gap, E_g = 3.26 eV for ZCLO53 and for ZCGO53, E_g = 3.27 eV) and the defects/

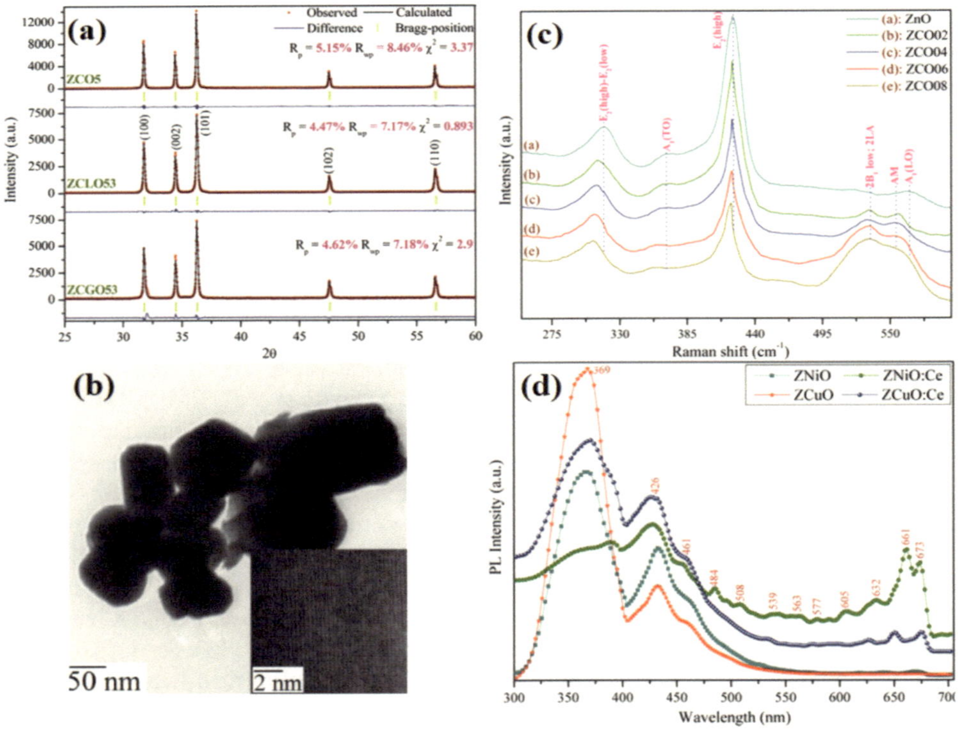

Figure 8.
(a) XRD pattern of Co, La and Gd doped ZnO nanostructure. (b) TEM image of $Zn_{0.94}Fe_{0.03}Ce_{0.03}O$ nanoparticles. (c) Raman spectra of pure and low Co concentrated ZnO nanoparticles. (d) Photoluminescence (PL) of Ni, Cu, Ce substituted ZnO nanoparticles. Adopted from Refs. [10, 11, 57, 58].

V_O evolution with variation in visible PL spectra. The $Zn_{0.95}Co_{0.05}O$ nanoparticles are paramagnetic at room temperature and involved superparamagnetic transition at low temperature. The antiferromagnetic interactions are enhanced with La and Gd doping into $Zn_{0.95}Co_{0.05}O$, is confirmed with ZFC/FC magnetization.

3.3.2 TEM of $Zn_{0.94}Fe_{0.03}Ce_{0.03}O$ (ZFCeO) nanoparticles

The DMS ZFCeO nanoparticles were synthesized by a sol-gel process [57]. **Figure 8(b)** shows the TEM image and the average particles size, D, of nanoparticles is 97 ± 4 nm. The inset of TEM is the HRTEM to show the crystalline formation and lattice spacing after doping into ZnO. It can be noted that the distorted lattice has an enhanced interplanar spacing *d* [corresponding to (101) planes] ~ 0.247 nm. For pure ZnO, the value of *d* ~ 0.237 nm. In HRTEM image, some little spots that covered the lattice fringes of spacing are also observed. This is an indication towards ferromagnetic clusters or structural inclusions formation.

3.3.3 Raman study evaluated lattice structure inducing defects

The $Zn_{1-x}Co_xO$ [x = 0.002 (ZCO02), 0.004 (ZCO04), 0.006 (ZCO06) and 0.008 (ZCO08)] nanoparticles were synthesized by a sol-gel process [58]. The XRD pattern results into wurtzite ZnO structure. The ZnO with Co doping has nanorods type morphology with diameter, D (nm) = 18 ± 2, 23 ± 3, 41 ± 5 and 53 ± 3, and length L (nm) = 39 ± 3, 57 ± 5, 95 ± 3 and 127 ± 5, respectively, for ZCO02, ZCO04, ZCO06 and ZCO08. **Figure 8(c)** shows Raman vibrational modes, which are located at around 314, 368, 422 and 533 cm^{-1} attributed to E_2(high)-E_2(low), A_1(TO), E_2(high) and (2B_1 low; 2LA) phonon modes, respectively [59]. The sharpest and strongest peak at about 422 cm^{-1} can be attributed to nonpolar high frequency mode, E_2(high), involved motion of oxygen, which is the characteristic of wurtzite lattice. With increasing Co concentration, a pronounced weakening in peak height, E_2(high) mode, than pure ZnO, has been observed. The intensity of E_2 mode of pure ZnO is shifted towards lower frequencies with increasing Co doping. This happens because decreasing binding energy of Zn-O bonds and a tensile strain in nanograins. An additional strong peak known as additional mode (AM) is observed at 554 cm^{-1} whose intensity is increased with Co concentration. This AM mode is the quasi-longitudinal optical mode formed with abundant shallow donor defects (Zn interstitial, oxygen vacancies, etc.). It is also reported in Ref. [59] that the RTFM is enhanced in low Co concentrated ZnO nanoparticles due to lattice defects. The low temperature ZFC/FC magnetic measurement indicates long-range antiferromagnetic-ferromagnetic ordering to form BMPs.

3.3.4 Photoluminescence spectra extract ZnO lattice defects

The $Zn_{0.95}Ni_{0.05}O$ (ZNiO), $Zn_{0.91}Ni_{0.05}Ce_{0.04}O$ (ZNiO:Ce), $Zn_{0.95}Cu_{0.05}O$ (ZCuO) and $Zn_{0.91}Cu_{0.05}Ce_{0.04}O$ (ZCuO:Ce) nanoparticles were synthesized by a sol-gel process [10]. **Figure 8(d)** show the PL emission for Ni, Cu, Ce substituted ZnO nanoparticles at room temperature. The peak at 369 (3.36 eV) is correlated with surface exciton recombination, which is the near band edge emission of ZnO [59]. The visible emission is formed due to radiative recombination of a photogenerated hole for which an electron occupied oxygen vacancies. The violet emission at 426 nm is the effect of radiative defects related oxygen and Zn vacancies. The peak at 461 and 484 nm is the blue emission, which have two defect level formed due to transition from Zn_i to valance band or bottom of the conduction band to O interstitial. The peak at 632, 661 and 673 nm is the red emission formed with intrinsic defect of O [60]. The Ce-doping into ZNiO and ZCuO is the production of oxygen vacancies defects in PL spectra [61].

3.3.5 Ferromagnetism at 300 K, 10 K of $Zn_{0.95}Ni_{0.05}O$ and $Zn_{0.91}Ni_{0.05}Ce_{0.04}O$ nanoparticles

Figure 9 shows M-H hysteresis at 300 and 10 K, respectively, for ZNiO, ZNiO:Ce nanoparticles [10]. At 300 K, the values of M_s (emu g^{-1}) = 0.073 and 0.085, and M_r (emu g^{-1}) = 0.0066 and 0.0187 with H_c (Oe) = 150 and 558, respectively, for ZNiO and ZNiO:Ce. At 10 K, the value of M_s (emu g^{-1}) = 0.096 and 0.198 and M_r (emu g^{-1}) = 0.0129 and 0.0642 with H_c (Oe) = 76 and 165, respectively, for ZNiO and ZNiO:Ce. The RTFM in TM substituted ZnO has a great influence of the intrinsic defects such as oxygen vacancies. This is since the BMP model ascribed that the bound electrons (holes) in the defect states can couple with TM ions and cause the ferromagnetic regions to overlap, giving long range ferromagnetic ordering. A theoretical prediction based on first principle calculation in Ni-doped ZnO shows that the ferromagnetic ordering is energetically favorable through double or superexchange mechanisms [10]. The defects such as oxygen vacancies in DMS are responsible in generating carriers for ferromagnetic ordering. The observed values of M_s and M_r at 10 K, are enhanced than at room temperature. However, at low temperature (~10 K), the value of H_c is varies abruptly. This is due to the existence of some ferromagnetic cluster assemblies in the samples, which enlarged H_c, may an indication towards superparamagnetic/spin-glass formation [62].

3.3.6 FC/ZFC magnetization of $Zn_{0.95}Ni_{0.05}O$ and $Zn_{0.91}Ni_{0.05}Ce_{0.04}O$ nanoparticles

The origin of observed RTFM in Ni, Ce substituted ZnO is evaluated by the temperature dependent magnetization [M(T)] with FC at 500 Oe and ZFC measurements (**Figure 9**) [47]. The FC/ZFC curves are separated with decrease in temperature, which usually appears with a coexistent system of antiferromagnetic and ferromagnetic phases. The steep increase of magnetization in FC curve with decreasing temperatures below 50 K is the characteristic of DMS [63]. Decreasing

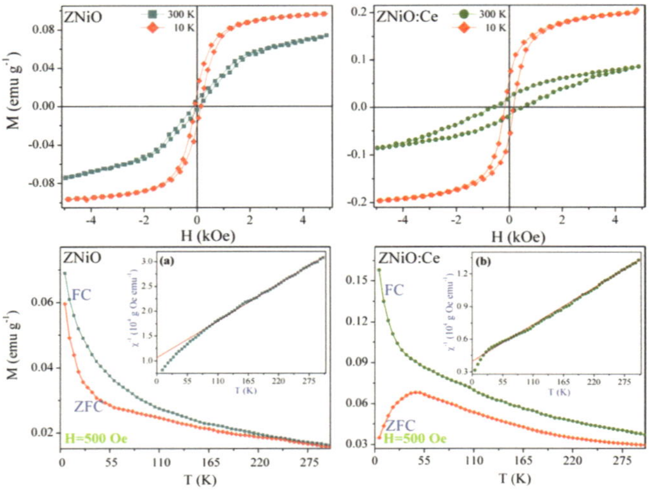

Figure 9.
M-H hysteresis at 300 and 10 K, and M(T) measurement following FC/ZFC at H = 500 Oe for ZNiO, ZNiO:Ce nanoparticles. Inset shows their χ^{-1}-T of Curie-Weiss law. Adopted from Verma and Kotnala [10].

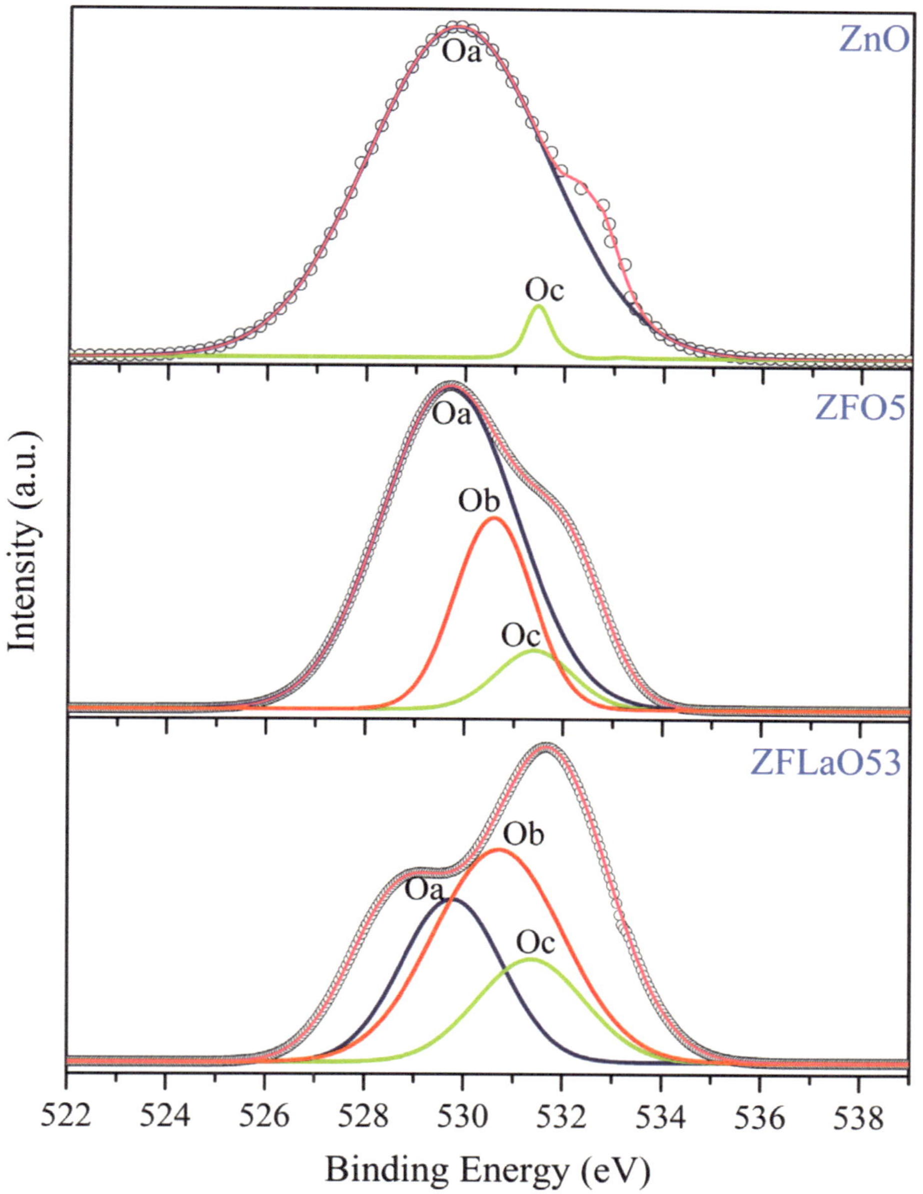

Figure 10.
O 1s XPS spectra of pure ZnO, $Zn_{0.95}Fe_{0.05}O$ and $Zn_{0.92}Fe_{0.05}La_{0.03}O$ nanoparticles. Adopted from Verma and Kotnala [47].

temperatures result into an increase in polarons interaction distance, leading to overlap between neighboring polarons and so allowing them to interact through the magnetic impurities, forming correlated polarons clusters. Though the direct interaction among localized carriers induced antiferromagnetism, and the ferromagnetic interaction due to BMP is possible with large magnetic impurities. When the unoccupied 3d states overlap the impurity band might be attributed high T_c value of ZnO. This is explained on the basis of donor impurity band for which oxygen vacancies involved F-centers into host ZnO [64]. The dip in ZFC curve indicates the existence of the blocking temperature, T_B. The upward curvature observed in the FC M(T) measurements (inset of **Figure 9**) of Ni and Ce substituted ZnO nanoparticles, involve a Curie–Weiss behavior related with susceptibility (χ) as: $\chi = \frac{C}{T-\theta}$; where C is the material specific Curie constant, T is the absolute temperature, and θ is the Weiss constant (in K) [10]. The estimated value of θ is found to be negative, which indicate antiferromagnetic interactions. But $\chi^{-1}(T)$ displays a notable deviation from the Curie-Weiss law that might be attributed by short-range ferromagnetism, antiferromagnetism, or a spin-glass system [65].

3.3.7 O 1s XPS spectra

In order to examined the presence of oxygen vacancies, V_O, the O 1s XPS spectra of pure ZnO, $Zn_{0.95}Fe_{0.05}O$ (ZFO5) and $Zn_{0.92}Fe_{0.05}La_{0.03}O$ (ZFLaO53) nanoparticles are given in **Figure 10** [47]. It deconvoluted into three peaks (O_a, O_b, O_c). The peak located on the low binding energy side (O_a) around 529.73 eV is attributed to O^{2-} ions on the wurtzite structure. The O_b peak at about 530.65 eV is associated with O^{2-} ions in oxygen-deficient regions within ZnO matrix, which indicate the formation of defects. The O_c peak at about 531.37 eV is usually attributed to chemisorbed oxygen on the nanostructural surface of the ZnO, such as adsorbed $-CO_3$, H_2O or adsorbed O_2 [66, 67].

4. Conclusion

The ferromagnetism in multiferroic, DMS and ferrites is a topic of intensive research due to their superficial potential in spintronic and magnetic memory devices. The lattice defects related to these multiferroic, DMS and ferrites have an important role in mediating observed ferromagnetism. The lattice defects mainly are oxygen vacancies and their formation might be related with crystalline structure, heating condition during syntheses and types of nanostructural product.

Acknowledgements

This work is financially supported by Dr. D.S. Kothari Postdoctoral Fellowship (No.F.4-2/2006 (BSR)/PH/16-17/0066) under University Grants Commission, India.

Author details

Kuldeep Chand Verma[1*], Ashish Sharma[1], Navdeep Goyal[1] and Ravinder Kumar Kotnala[2]

1 Department of Physics, Panjab University, Chandigarh, India

2 CSIR-National Physical Laboratory, New Delhi, India

*Address all correspondence to: dkuldeep.physics@gmail.com

References

[1] Buurma AJC, Blake GR, Palstra TTM. Multiferroic Materials: Physics and Properties, from Encyclopedia of Materials: Science and Technology (2nd edn.), 2006. Reference Module in Materials Science and Materials Engineering. Amsterdam: Elsevier; 2016. pp. 1-7

[2] Verma KC, Kotnala RK. Lattice defects induce multiferroic responses in Ce, La-substituted $BaFe_{0.01}Ti_{0.99}O_3$ nanostructures. Journal of the American Ceramic Society. 2016;**99**:1601-1608

[3] Bibes M, Barthelemy A. Towards a magnetoelectric memory. Nature Materials. 2008;**7**:425-426

[4] Verma KC, Kotnala RK. Multiferroic approach for Cr, Mn, Fe, Co, Ni, Cu substituted $BaTiO_3$ nanoparticles. Materials Research Express. 2016;**3**:055006

[5] Verma KC, Vinay G, Kaur J, Kotnala RK. Raman spectra, photoluminescence, magnetism and magnetoelectric coupling in pure and Fe doped $BaTiO_3$ nanostructures. Journal of Alloys and Compounds. 2013;**578**:5-11

[6] Harrisa VG, Sepelak V. Mechano-chemically processed zinc ferrite nanoparticles: Evolution of structure and impact of induced cation inversion. Journal of Magnetism and Magnetic Materials. 2018;**465**:603-610

[7] Sugimoto M. The past, present, and future of ferrites. Journal of the American Ceramic Society. 1999;**82**:269-279

[8] Price GD, Price SL, Burdett JK. The factors influencing cation site-preferences in spinels a new mendelyevian approach. Physics and Chemistry of Minerals. 1982;**8**:69-76

[9] Verma KC, Singh VP, Ram M, Shah J, Kotnala RK. Structural, microstructural and magnetic properties of $NiFe_2O_4$, $CoFe_2O_4$ and $MnFe_2O_4$ nanoferrite thin films. Journal of Magnetism and Magnetic Materials. 2011;**323**:3271-3275

[10] Verma KC, Kotnala RK. Understanding lattice defects to influence ferromagnetic order of ZnO nanoparticles by Ni, Cu, Ce ions. Journal of Solid State Chemistry. 2017;**246**:150-159

[11] Verma KC, Kotnala RK. Defects due to lattice distortion and nano-size intermediate ferromagnetism in La, Gd substituted $Zn_{0.95}Co_{0.05}O$. Current Applied Physics. 2016;**16**:175-182

[12] Kaur J, Kotnala RK, Gupta V, Verma KC. Anionic polymerization in Co and Fe doped ZnO: Nanorods, magnetism and photoactivity. Current Applied Physics. 2014;**14**:749-756

[13] Prinz GA. Magnetoelectronics. Science. 1998;**282**:1660-1663

[14] Verma KC, Kotnala RK. Realizing ferromagnetic ordering in SnO_2 and ZnO nanostructures with Fe, Co, Ce ions. Physical Chemistry Chemical Physics. 2016;**18**:17565-17574

[15] Pal B, Dhara S, Giri PK, Sarkar D. Room temperature ferromagnetism with high magnetic moment and optical properties of Co doped ZnO nanorods synthesized by a solvothermal route. Journal of Alloys and Compounds. 2014;**615**:378-385

[16] McCabe GH, Fries T, Liu MT, Shapira Y, Ram-Mohan LR, Kershaw R, et al. Bound magnetic polarons in p-type $Cu_2Mn_{0.9}Zn_{0.1}SnS_4$. Physical Review B. 1997;**56**(**11**):6673-6680

[17] Verma KC, Kotnala RK, Negi NS. Improved dielectric and ferromagnetic properties in Fe-doped $PbTiO_3$ nanoparticles at room temperature.

Applied Physics Letters. 2008;**92**:152902-152903

[18] Verma KC, Kotnala RK. Tailoring the multiferroic behavior in $BiFeO_3$ nanostructures by Pb doping. RSC Advances. 2016;**6**:57727-57738

[19] Verma KC, Singh M, Kotnala RK, Goyal N. Magnetic field control of polarization/capacitance/voltage/resistance through lattice strain in $BaTiO_3$-$CoFe_2O_4$ multiferroic nanocomposite. Journal of Magnetism and Magnetic Materials. 2019;**469**:483-493

[20] Verma KC, Singh D, Kumar S, Kotnala RK. Multiferroic effects in MFe_2O_4/$BaTiO_3$ (M = Mn, Co, Ni, Zn) nanocomposites. Journal of Alloys and Compounds. 2017;**709**:344-355

[21] Shen Y, Sun J, Li L, Yao Y, Zhou C, Su R, et al. The enhanced magnetodielectric interaction of $(1-x)BaTiO_3$–$xCoFe_2O_4$ multiferroic composites. Journal of Materials Chemistry C. 2014;**2**:2545-2551

[22] Jiang K, Zhu JJ, Wu JD, Sun J, Hu ZG, Chu JH. Influences of oxygen pressure on optical properties and interband electronic transitions in multiferroic bismuth ferrite nanocrystalline films grown by pulsed laser deposition. ACS Applied Materials & Interfaces. 2011;**3**:4844-4852

[23] Li H, Huang Y, Zhang Q, Qiao Y, Gu Y, Liu J, et al. Facile synthesis of highly uniform Mn/Co-codoped ZnO nanowires: Optical, electrical, and magnetic properties. Nanoscale. 2011;**3**:654-660

[24] Kaur J, Kotnala RK, Verma KC. Multiferroic properties of $Ba(Fe_xTi_{1-x})O_3$ nanorods. Materials Letters. 2011;**65**:3160-3163

[25] Verma KC, Kaur J, Negi NS, Kotnala RK. Multiferroic and magnetoelectric properties of nanostructured $BaFe_{0.01}Ti_{0.99}O_3$ thin films obtained under polyethylene glycol conditions. Solid State Communications. 2014;**178**:11-15

[26] Liu S, Akbashev AR, Yang X, Liu X, Li W, Zhao L, et al. Hollandites as a new class of multiferroics. Scientific Reports. 2014;**4**:6203

[27] Zhao H, Kimura H, Cheng Z, Osada M, Wang J, Wang X, et al. Large magnetoelectric coupling in magnetically short-range ordered $Bi_5Ti_3FeO_{15}$ film. SC Reports. 2014;**4**:5255

[28] Bretos I, Jimenez R, Lazaro CG, Montero I, Calzada ML. Defect-mediated ferroelectric domain depinning of polycrystalline $BiFeO_3$ multiferroic thin films. Applied Physics Letters. 2014;**104**:092905

[29] Spanier JE, Kolpak AM, Urban JJ, Grinberg I, Ouyang L, Yun WS, et al. Ferroelectric phase transition in individual single-crystalline $BaTiO_3$ nanowires. Nano Letters. 2006;**6**:735-739

[30] Beltran JJ, Barrero CA, Punnoose A. Evidence of ferromagnetic signal enhancement in Fe and Co codoped ZnO nanoparticles by increasing superficial Co^{3+} content. Journal of Physical Chemistry C. 2014;**118**:13203-13217

[31] Li W et al. Oxygen-vacancy-induced antiferromagnetism to ferromagnetism transformation in $Eu_{0.5}Ba_{0.5}TiO_3$-delta multiferroic thin films. Scientific Reports. 2013;**3**:618-616

[32] Pradhan DK, Kumari S, Puli VS, Das PT, Pradhan DK, Kumar A, et al. Correlation of dielectric, electrical and magnetic properties near the magnetic phase transition temperature of cobalt zinc ferrite. Physical Chemistry Chemical Physics. 2017;**19**:210-218

[33] Clavel G, Marichy C, Willinger MG, Ravaine S, Zitoun D, Pinna N. $CoFe_2O_4$-TiO_2 and $CoFe_2O_4$-ZnO thin film nanostructures elaborated from colloidal chemistry and atomic layer deposition. Langmuir. 2010;**26**(23):18400

[34] Ahmed MA, Okasha N, Mansour SF, El-dek SI. Bi-modal improvement of the physico-chemical characteristics of PEG and MFe_2O_4 subnanoferrite. Journal of Alloys and Compounds. 2010;**496**:345-350

[35] Verma KC, Kotnala RK. Spring like ferromagnetic behavior of $x Li_{0.5}Fe_{2.5}O_4$-$(1 - x)SrFe_2O_4$ nanoferrite thin films. Journal of Nanoparticle Research. 2011;**13**:4437-4444

[36] Wang P, Jin C, Zheng D, Li D, Gong J, Li P, et al. Strain and ferroelectric-field effects co-mediated magnetism in (011)-$CoFe_2O_4$/Pb $(Mg_{1/3}Nb_{2/3})_{0.7}Ti_{0.3}O_3$ multiferroic heterostructures. ACS Applied Materials & Interfaces. 2016;**8**:24198-24204

[37] Bateer B, Tian C, Qu Y, Du S, Yang Y, Ren Z, et al. Synthesis, size and magnetic properties of controllable $MnFe_2O_4$ nanoparticles with versatile surface functionalities. Dalton Transactions. 2014;**43**:9885-9891

[38] Vamvakidis K, Katsikini M, Sakellari D, Paloura EC, Kalogirou O, Samara CD. Reducing the inversion degree of $MnFe_2O_4$ nanoparticles through synthesis to enhance magnetization: Evaluation of their ^1H NMR relaxation and heating efficiency. Dalton Transactions. 2014;**43**:12754-12765

[39] Bullita S, Casu A, Casula MF, Concas G, Congiu F, Corrias A, et al. $ZnFe_2O_4$ nanoparticles dispersed in a highly porous silica aerogel matrix: A magnetic study. Physical Chemistry Chemical Physics. 2014;**16**: 4843-4852

[40] Peddis D, Yaacoub N, Ferretti M, Martinelli A, Piccaluga G, Musinu A, et al. Cationic distribution and spin canting in $CoFe_2O_4$ nanoparticles. Journal of Physics. Condensed Matter. 2011;**23**:426004-426008

[41] Nedelkoski Z, Kepaptsoglou D, Lari L, Wen T, Booth RA, Oberdick SD, et al. Origin of reduced magnetization and domain formation in small magnetite nanoparticles. Scientific Reports. 2017;**7**:45997

[42] Robinson DW. Magnetism and the chemical bond. In: John B, editor. Goodenough Interscience. New York: Wiley; 1963

[43] Casoli F, Nasi L, Albertini F, Fabbrici S, Bocchi C, Germini F, et al. Morphology evolution and magnetic properties improvement in FePt epitaxial films by in situ annealing after growth. Journal of Applied Physics. 2008;**103**:043912

[44] Mohapatra J, Mitra A, Bahadur D, Aslam M. Surface controlled synthesis of MFe_2O_4 (M = Mn, Fe, Co, Ni and Zn) nanoparticles and their magnetic characteristics. CrystEngComm. 2013;**15**:524-532

[45] Mohapatra J, Mitra A, Bahadur D, Aslam M. Superspin glass behavior of self-interacting $CoFe_2O_4$ nanoparticles. Journal of Alloys and Compounds. 2015;**628**:416-423

[46] Jonsson PE. Superparamagnetism and spin glass dynamics of interacting magnetic nanoparticle systems. Advances in Chemical Physics. 2004;**128**:191-248

[47] Verma KC, Kotnala RK. Oxygen vacancy induced by La and Fe into ZnO nanoparticles to modify ferromagnetic ordering. Journal of Solid State Chemistry. 2016;**237**:211-218

[48] Dietl T, Ohno H, Matsukura F, Cibert J, Ferrand D. Zener model description of ferromagnetism in zinc-blende magnetic semiconductors. Science. 2000;**287**:1019-1022

[49] Garcia MA, Merino JM, Pinel EF, Quesada A, dela Venta J, Ruız Gonzalez ML, et al. Magnetic properties of ZnO nanoparticles. Nano Letters. 2007;7:1489-1492

[50] Xu X, Xu C, Dai J, Hu J, Li F, Zhang S. Size dependence of defect-induced room temperature ferromagnetism in undoped ZnO nanoparticles. Journal of Physical Chemistry C. 2012;**116**:8813-8818

[51] Gao D, Zhang Z, Fu J, Xu Y, Qi J, Xue D. Room temperature ferromagnetism of pure ZnO nanoparticles. Journal of Applied Physics. 2009;**105**:113928-113924

[52] Sato K, Katayama-Yoshida H. Magnetic interactions in transition-metal-doped ZnO: An ab initio study. Ferromagnetism in a transition metal atom doped ZnO. Physica E. 2001;**10**:251-255

[53] Karmakar D, Mandal SK, Kadam RM, Paulose PL, Rajarajan AK, Nath TK, et al. Ferromagnetism in Fe-doped ZnO nanocrystals: Experiment and theory. Physical Review B. 2007;**75**:144404-144414

[54] Spaldin NA. Search for ferromagnetism in transition-metal-doped piezoelectric ZnO. Physical Review B. 2004;**69**:125201-125207

[55] El Hachimi AG, Zaari H, Benyoussef A, El Yadari M, El Kenz A. First-principles prediction of the magnetism of 4f rare-earth-metal-doped wurtzite zinc oxide. Journal of Rare Earths. 2014;**32**:715-721

[56] Deng SH, Duan MY, Xu M, He L. Effect of La doping on the electronic structure and optical properties of ZnO. Physica B. 2011;**406**:2314-2318

[57] Verma KC, Kotnala RK. Defects-assisted ferromagnetism due to bound magnetic polarons in Ce into Fe, Co:ZnO nanoparticles and first-principle calculations. Physical Chemistry Chemical Physics. 2016;**18**:5647-5657

[58] Verma KC, Bhatia R, Kumar S, Kotnala RK. Vacancies driven magnetic ordering in ZnO nanoparticles due to low concentrated Co ions. Materials Research Express. 2016;**3**:076103

[59] Zeng H, Duan G, Li Y, Yang S, Xu X, Cai W. Blue luminescence of ZnO nanoparticles based on non-equilibrium processes: Defect origins and emission controls. Advanced Functional Materials. 2010;**20**:561-572

[60] Kumar S, Basu S, Rana B, Barman A, Chatterjee S, Jha SN, et al. Structural, optical and magnetic properties of sol-gel derived ZnO:Co diluted magnetic semiconductor nanocrystals: An EXAFS study. Journal of Materials Chemistry C. 2014;**2**:481-495

[61] Dar MI, Arora N, Singh NP, Sampath S, Shivashankar SA. Role of spectator ions in influencing the properties of dopant-free ZnO nanocrystals. New Journal of Chemistry. 2014;**38**:4783-4790

[62] Kovaleva NN, Kugel KI, Bazhenov AV, Fursova TN, Loser W, Xu Y, et al. Formation of metallic magnetic clusters in a Kondo-lattice metal: Evidence from an optical study. Scientific Reports. 2012;**2**:890-897

[63] Singhal A, Achary SN, Manjanna J, Chatterjee S, Ayyub P, Tyagi AK. Chemical synthesis and structural and magnetic properties of dispersible cobalt- and nickel-doped ZnO

Nanocrystals. Journal of Physical Chemistry C. 2010;**114**:3422-3430

[64] Lu ZL, Hsu HS, Tzeng YH, Zhang FM, Du YW, Huang JCA. The origins of ferromagnetism in Co-doped ZnO single crystalline films: From bound magnetic polaron to free carrier-mediated exchange interaction. Applied Physics Letters. 2009;**95**:102501-102503

[65] Yang CY, Lu YH, Lin WH, Lee MH, Hsu YJ, Tseng YC. Structural imperfections and attendant localized/itinerant ferromagnetism in ZnO nanoparticles. Journal of Physics D: Applied Physics. 2014;**47**:345003

[66] Janotti A, Van de Walle CG. Oxygen vacancies in ZnO. Applied Physics Letters. 2005;**87**:122102-122103

[67] Verma KC, Kotnala RK, Goyal N. Multi-Functionality of Spintronic Materials, from Nanoelectronics Devices, Circuits and Systems. Amsterdam, Netherlands: Elsevier; 2019. pp. 153-215

Chapter 6

High-Frequency Permeability of Fe-Based Nanostructured Materials

Mangui Han

Abstract

Frequency dependence of permeability of magnetic materials is crucial for their various electromagnetic applications. In this chapter, the approaches to tailor the high-frequency permeability of Fe-based nanostructured materials are discussed: shape controlling, particle size distribution, coating treatments, phase transitions and external excitations. Special attention is paid on the electromagnetic wave absorbing using these Fe-based magnetic materials (Fe-Si-Al alloys, Fe-Cu-Nb-Si-B nanocrystalline alloys and Fe nanowires). Micromagnetic simulations also have been presented to discuss the intrinsic high-frequency permeability. Negative imaginary part of permeability ($\mu'' < 0$) is proved possible under the spin transfer torque effect.

Keywords: complex permeability, electromagnetic wave absorbing, negative imaginary part of permeability, micromagnetic simulation, natural resonance

1. Introduction

The high-frequency permeability values are crucial for many applications of electromagnetic (EM) materials. At low frequency region, a typical application is magnetic cores for transformers or inductors; these devices demand the larger real parts of permeability ($\mu' > 0$) and lower imaginary parts of permeability ($\mu'' > 0$ and $\mu'' \to 0$) for magnetic materials. With the working frequency increases, many electronic devices work above gigahertz (GHz). The unwanted electromagnetic wave pollution is problematic, which can be overcome by using the electromagnetic wave attenuation composites with the absorbing frequency band falling into the GHz zone. Therefore, the electromagnetic wave absorbing materials require their working frequencies to be above 1 GHz and larger μ'' values for efficient absorption via the typical magnetic loss mechanism (natural resonance). Due to the small magnetocrystalline anisotropy constants of spinel ferrites, the magnetic loss peaks of spinel ferrites are below 1 GHz. Although the natural resonance frequencies of hexagonal ferrites can be a few GHz, the shortcoming of both ferrites (spinel and hexagonal) is the poor temperature stability; in other words, the permeability decreases faster with increasing environmental temperature, making these ferrites not suitable candidates for EM attenuation applications. The physical law governing the relationship of initial permeability, working frequency and magnetization is the well-known "Snoek's law" [1]. This law tells us that if we want to increase the working frequency of a specific ferrite (*Ms* is constant), we have to sacrifice their permeability and vice versa. People often tailor the $\mu \sim f$ spectrum of a ferrite by cation substitution or microstructures by changing the sintering conditions. We think this technique is time-consuming and inefficient for mass production. In this chapter, we propose Fe-based nanostructured materials (Fe-Cu-Nb-Si-B alloys, also called "FINEMET") as ideal EM attenuation materials, which offer us greater freedom to tailor their high-frequency permeability spectra and possess good temperature stability. Firstly, we will discuss how to increase the permeability by controlling the shape of particles, and then we move to discuss the impacts of size distribution, phase transitions and coating on the permeability. Next, we will discuss the cause of frequently observed broad permeability spectra using the micromagnetic simulations. Finally, we will prove how we can obtain negative imaginary parts of permeability via spin transfer torque effect. Our discussions mainly focus on the electromagnetic wave absorbing (attenuation) applications, which require large imaginary parts of permeability and working frequency above 1 GHz.

2. Shape controlling

Soft magnetic materials can give large initial permeability, but their working frequency is far below 1 GHz due to their small magnetocrystalline anisotropic constants. As per Snoek's law, if the working frequency is shifted to GHz range, the permeability has to be compromised. To increase the working frequency, we have to look at other anisotropy fields, such as shape anisotropy and stress (or external field)-induced anisotropy. Shape anisotropy is most commonly employed. Materials in forms of thin films, microwires, nanowires and flakes can possess a strong shape anisotropy. Here, we present how to enhance the permeability of a soft ferromagnetic material (Fe-Si-Al alloys) above 1 GHz via shape controlling. The Fe-Si-Al alloy ingot (composition: $Fe_{84.94}Si_{9.68}Al_{5.38}$) was prepared using the hydrogen reduction method in a furnace with the starting materials (Si, Al and Fe with high purity). Subsequently, the Fe-Si-Al ingot was first pulverized into particles with irregular shapes; later these particles were further milled into flakes under different milling times (10, 20 and 30 hours, respectively). Our traditional ball milling process description can be found in our published paper [2]. The scanning electron microscopy images of Fe-Si-Al particles are shown in **Figure 1**. Clearly, the preliminarily pulverized particles have irregular shapes. The traditional ball milling process can transform them into flakes after 10, 20 or 30 hours of milling. The typical milling result is illustrated in **Figure 1b** showing flaky particles milled for 30 hours. The high-frequency complex permeability values were measured within 0.5–10 GHz using a network analyzer (Agilent 8720ET). The measured samples were prepared by mixing the Fe-Si-Al particles and wax homogeneously (weight ratio: alloy particles/wax = 4:1). The measured sample has an annular shape with inner and outer diameters of 3 and 7 mm, respectively.

The dependence of high-frequency complex permeability on particle shape is shown in **Figure 2**. Obviously, flaky particles have much larger values in both real and imaginary parts of permeability compared to the irregularly shaped particles. This finding is named "enhanced permeability." The enhanced permeability is more apparent at the lower frequency range. Besides, with increasing the milling hours, the enhanced permeability is stronger. For example, the μ' value of flakes after being milled for 30 hours is found to be 4.4 at 0.5 GHz. However, it is only 1.3 for the irregular shaped particle. Within 0.5–7 GHz, the μ' values of flakes after being milled for 30 hours are evidently larger than those of particles with irregular shapes. With regard to the imaginary part values (μ''), the flakes of Fe-Si-Al exhibit larger

Figure 1.
SEM pictures: (a) preliminarily pulverized particles and (b) particles with flaky shapes prepared by 30 hours of ball milling.

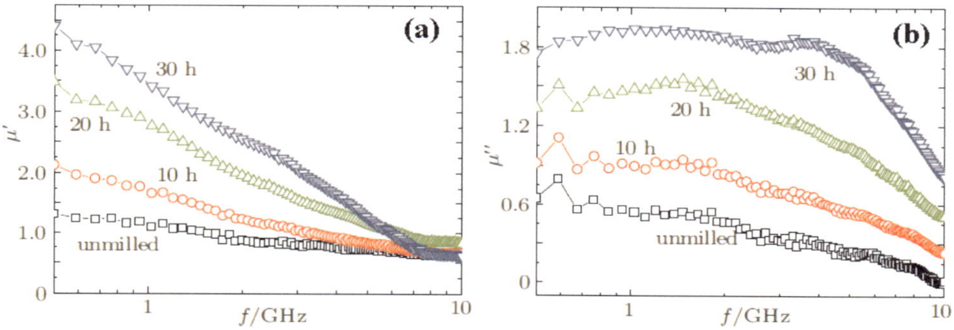

Figure 2.
Dependence of high-frequency permeability on particle shapes (copyright, 2013, IOP).

values within the whole frequency range studied than those of particles with irregular shapes, as shown in **Figure 2b**. Our explanations for the enhanced permeability are as follows: Fe-Si-Al alloy is a metallic alloy with a conducting feature. The eddy current effect is stronger for irregular shape but can be greatly suppressed by milling the irregular particles into flakes that have lower thickness. Besides, Snoek's law governing the relationship between the permeability and shape factor is given as [3]

$$(\mu_s - 1) f_r^2 = \left(\frac{\gamma}{2\pi}\right)^2 \times 4\pi M_s \times (H_k + 4\pi M_s D_z) \tag{1}$$

where D_z is a factor related to the shape, also known as the demagnetization factor for the normal direction of the particle plane. D_z is about $4\pi/3$ for a sphere and is about 4π for a flake. Accordingly, the enhanced permeability value can be observed by increasing D_z from $(4\pi/3)$ to (4π) by controlling particle shapes. That is also the reason why people often fabricate ferromagnetic thin films to obtain enhanced high-frequency permeability. Thin films can be viewed as an extreme case of "all flakes well aligned," which therefore are found to have much larger permeability values resulting from the "flakes" and in-plane induced uniaxial anisotropy. Furthermore, the well-known Snoek's law for bulk materials describes

$$(\mu_s - 1) f_r = \frac{2}{3} \gamma' 4\pi M_s \tag{2}$$

No shape-related demagnetization factor is found in this equation.

3. Particle size distribution

As discussed before, the high-frequency permeability can be greatly enhanced by controlling the shapes of particles and alignment of flakes (i.e., a thin film can be treated as a large quantity of well-aligned flakes in a plane). In this section, we will focus on the effect of size distribution of flakes on the dynamic permeability. We choose the composites containing Fe-Cu-Nb-Si-B alloy particles (a.k.a. "FINEMET" alloy) for this purpose. The reason to choose this Fe-based nanostructured material for its special nanoscale structure is the nano-grains are well dispersed in an amorphous matrix (it can be thought as a core-shell structure). The amorphous matrix has the larger resistivity which can reduce the detrimental effects of eddy current on high-frequency permeability. Besides, both amorphous phase and nano-grains are ferromagnetic, which avoids the effect non-magnetic coatings have on ferromagnetic particles (it will be discussed in the next section), and as a result both phases contribute to the permeability.

The phase transition temperatures of a "FINEMET" alloy can be identified by taking a differential scanning calorimeter (DSC) curve. As shown in **Figure 3**, two exothermal peaks are found in the DSC curve: one at 530°C (called T_{x1}) and the other at 672°C (called T_{x2}). T_{x1} is known as the primary crystallization temperature, and T_{x2} is the secondary crystallization temperature. In order to form nanoscale grains, as-prepared ribbons may be annealed above T_{x1} (see zone I) or T_{x2} (see zone II). It was previously reported that nanoscale Fe(Si) grains can be formed when the alloys are annealed above T_{x1}. Annealing at different temperatures and with different time will give rise to different volume fractions of ferromagnetic phases (amorphous phase and nanocrystallized phase), size of nano-grains and different kinds of magnetic phases ($T > T_{x2}$, zone II), such as Fe-B phases. More details on the phase transitions can be found in literature [4]. We believe that controlling the annealing (in other words, phase transition) will give us abundant ways to tailor the high frequency of Fe-based nanostructured materials to meet the specific requirement of an EM wave absorbing application. Here, we propose that without preparing materials with new compositions, just annealing the nanocrystalline soft magnetic alloys (such as FINEMET, NANOPERM or HITPERM) to tailor the high-frequency permeability is an economic approach from the perspective of mass production.

Preparation of FINEMENT amorphous ribbon, annealing treatments and milling processes can be found in our published paper [5]. Nanostructures of sample heat

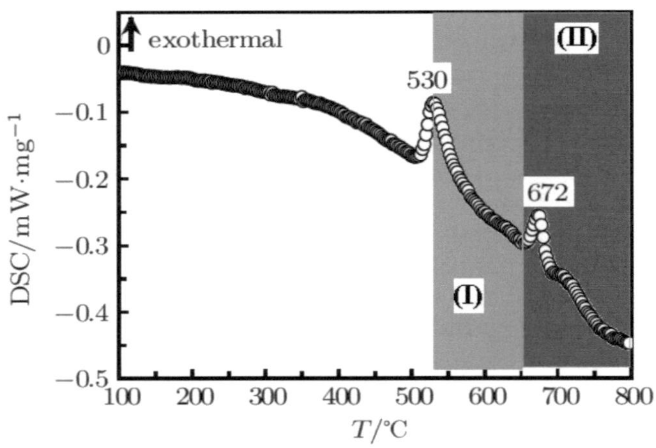

Figure 3.
DSC curve of FINEMET alloy.

treated at 540°C are characterized by a transmission electron microscopy (TEM) and shown in **Figure 4**. **Figure 4a** is the bright-field TEM image, whereas **Figure 4b** shows the typical features of Fe-based nanostructured alloys, where the amorphous matrix is surrounding the nanocrystalline grains. The average grain size is about 14.38 nm. XRD results show the nanocrystalline grains to be α-Fe(Si). It should be noted that the effect of annealing and ball milling on phase transitions can be investigated using Mössbauer technique and XRD measurements. Especially, Mössbauer technique can detect fine distinctions in the transformations of crystal structures due to the high-energy ball milling [5]. Using the milling procedure described in the previous section, we can effectively transform the annealed ribbons into flakes, which are shown in **Figure 5**. These flakes are divided into two categories by a shaking sieve: the large flakes (**Figure 5a**) and the small flakes (**Figure 5b**). The typical thickness of flakes of both categories is also shown in **Figure 5c** and **d**. For large flakes, the width of flakes is about 23–111 μm, average width is estimated to be about 81.1 μm and average thickness was found to be 4.5 μm. For small flakes, the width of flakes is about 3–21 μm, average width is estimated about 9.4 μm and thickness is averaged to be 1.3 μm.

Frequency dependence of permeability of these two kinds of flakes is shown in **Figure 6**. At the beginning of measurement frequency (i.e., at 0.5 GHz), the real part of permeability, which can be named "initial permeability" (μ_s'), is found to be about 4.6 for large flakes and about 3.9 for small flakes. Initial imaginary part of permeability (μ_s'') is about 1.3 for large flakes and about 0.9 for small flakes. Interestingly, it should be pointed out that wide magnetic loss spectra ($\mu \sim f$) can be found for both kinds of flakes, as indicated in **Figure 6b**. Wide magnetic loss peak is advantageous for electromagnetic noise attenuation composites with a wide working frequency band. The loss peak (μ'' is maximum) is found to be 3.0 GHz for large flakes and 5.5 GHz for small flakes, whereas μ'' (max) is 2.1 for the composite with large flakes, and for composites with small flakes, μ'' (max) is about 1.6.

For materials with high resistivity, such as ferrites magnetic loss above 1 GHz is often ascribed to the natural resonance mechanism. The frequency of natural resonance is closely associated with the magnetocrystalline anisotropy as per Snoek's law. One of typical magnetic materials in this case is M-type hexagonal ferrites. In our case, however, the sources of the observed broad magnetic loss peaks are believed to be due to the distribution of localized magnetic anisotropy field, which is the resultant of distribution of shapes (shape anisotropy fields), and distribution of interaction fields among particles. Moreover, eddy current effect is another cause

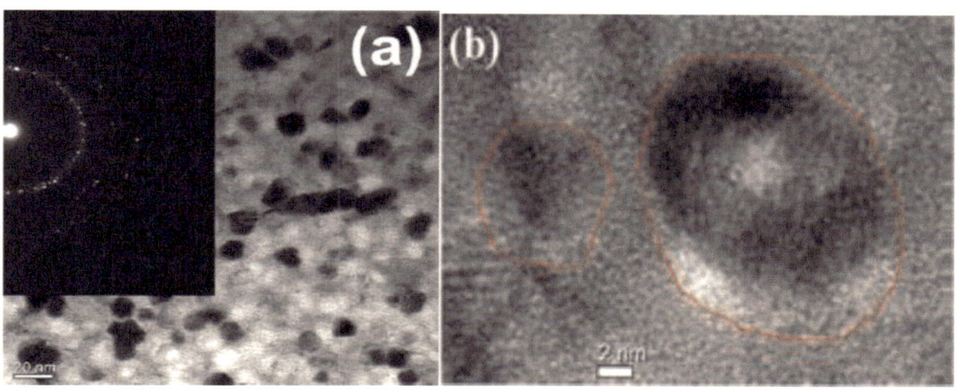

Figure 4.
TEM photographs of Fe-Cu-Nb-Si-B nanostructured alloys. (a) Image showing nano-grains. The inset showing the select area electron diffraction. (b) Nano-grains circled in the high-resolution image (copyright, 2014, IOP).

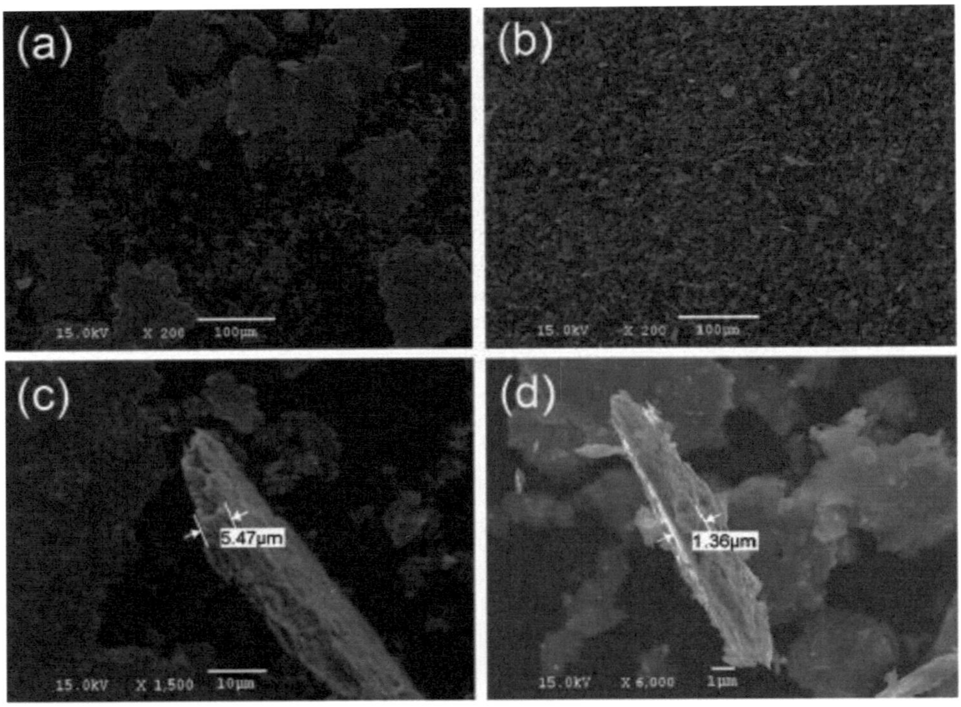

Figure 5.
SEM images of two categories of flakes: (a) large flakes; (b) small flakes; (c) typical thickness of large flakes; (d) typical thickness of small flakes (copyright, 2015, AIP).

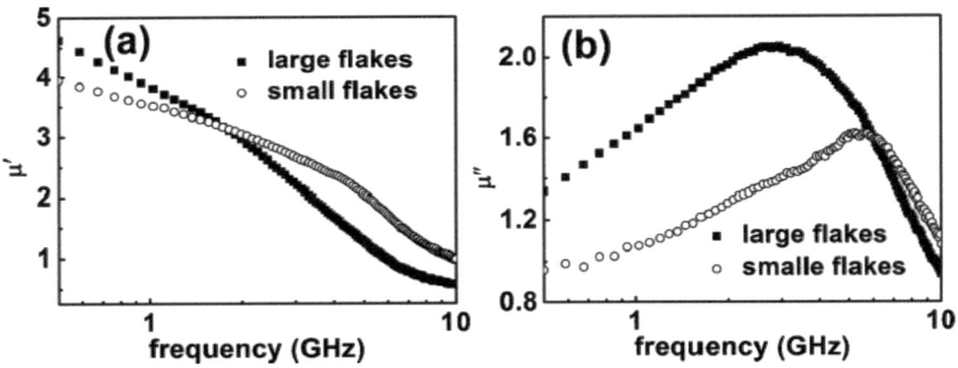

Figure 6.
High-frequency permeability of composites contained different sizes of Fe-Cu-Nb-Si-B flakes. (a) $\mu' \sim f$ spectra and (b) $\mu \sim f$ spectra (copyright, 2015, AIP).

for broadening the spectra of permeability. In order to interpret the observed dissimilarities in the high-frequency permeability of composites with these two categories of flakes, Snoek's law with shape factors included is employed and given as below:

$$\mu_s = 1 + \frac{4\pi M_s}{(H_k + 4\pi M_s N_h)} \tag{3}$$

$$f_r = \frac{\gamma}{2\pi}\sqrt{H_k^2 + 4\pi M_s H_k (N_\perp + N_h) + (4\pi M_s)^2 N_\perp N_h} \tag{4}$$

$$N_{\perp} = \frac{\alpha_r^2}{\alpha_r^2 - 1} \times \left(1 - \sqrt{\frac{1}{\alpha_r^2 - 1}} \times ars\sin\frac{\sqrt{\alpha_r^2 - 1}}{\alpha_r}\right) \tag{5}$$

$$N_h = \frac{1 - N_{\perp}}{2} \tag{6}$$

where α_r is the width/thickness ratio (often called "aspect ratio") of a flake. The demagnetization factor along the direction of thickness and width is $N_{\perp}$ and N_h, respectively. The saturation magnetization of material under studied is denoted as M_s. Magnetocrystalline anisotropy field is denoted as H_k. The total magnetic anisotropy field is given in the denominator of Eq. (3). For our samples, these two kinds of flakes are obtained under the same milling process and made from the same material. Therefore, large flakes and small flakes have same M_s and H_k values. The main differences among them are size distribution, aspect ratio and thickness. The aspect ratio is calculated as 7.23 and 18.02, respectively, for small flakes and large flakes according to the measured geometrical parameters. Subsequently, demagnetization factors ($N_{\perp}$ and N_h) have been calculated as per Eqs. (5) and (6) and are shown in **Table 1**. It can be seen that larger flakes having the larger initial permeability values are because they have smaller N_h values, as indicated by Eq. (3). Moreover, the finding that their magnetic loss peaks are found at lower frequencies can be understood according to Snoek's law: the inversely proportional relationship between the initial permeability and the loss peak. The $\mu' \sim f$ spectrum of large flakes drops more rapidly than small flakes. When f > 1.8 GHz, the real part (μ'(z, the real partflakes drops more rapidly ty and the loss peak. Theower frequencies, with increasing frequency, eddy current becomes a more serious issue in the large flakes. As we pointed out, FINEMET alloys are metallic and well-conducting materials. When they work above 1 GHz, the eddy current effect will be unavoidable. The electromagnetic wave will interact with the part of magnetic materials which are within the so-called "skin depth." As a result, high-frequency permeability spectrum depends on this eddy current effect. Stronger eddy current effect will give rise to the smaller volume fraction of magnetic materials interacting with the EM wave. Consequently, a faster dropping of $\mu' \sim f$ spectrum is observed.

For electromagnetic wave absorbing application, the simplest example is that composites containing the flakes work as single layer on a perfectly conducting substrate (such as the surface of aircrafts). The absorbing properties of a normal incident EM wave can be assessed by the reflection loss (RL, in dB) based on the equations as follows:

$$Z_{in} = Z_0\sqrt{\frac{\mu}{\varepsilon}}\tanh\left(j\frac{2\pi fd}{c}\sqrt{\mu\varepsilon}\right) \tag{7}$$

$$R.L. = 20\log\left|\frac{Z_{in} - Z_0}{Z_{in} + Z_0}\right| \tag{8}$$

where "*d*" is the thickness of composite layer, "*c*" is the velocity of light, Z_0 is the impedance of free space and Z_{in} is the characteristic impedance at the free

	$N_{\perp}$	N_h	$(N_{\perp} + N_h)$	$N_{\perp}N_h$
Small flakes	0.816	0.092	0.908	0.075
Large flakes	0.919	0.041	0.960	0.038

Table 1.
Demagnetization factors of two kinds of flakes.

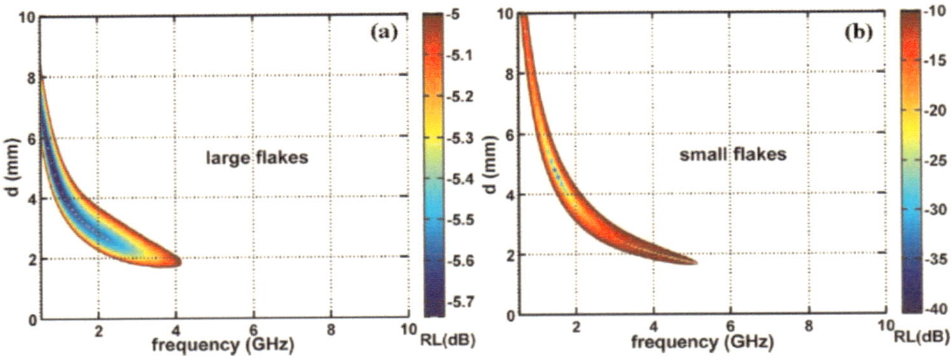

Figure 7.
Contour maps showing the absorbing properties of single layer composites with different flakes: (a) large flakes and (b) small flakes (copyright, 2015, AIP).

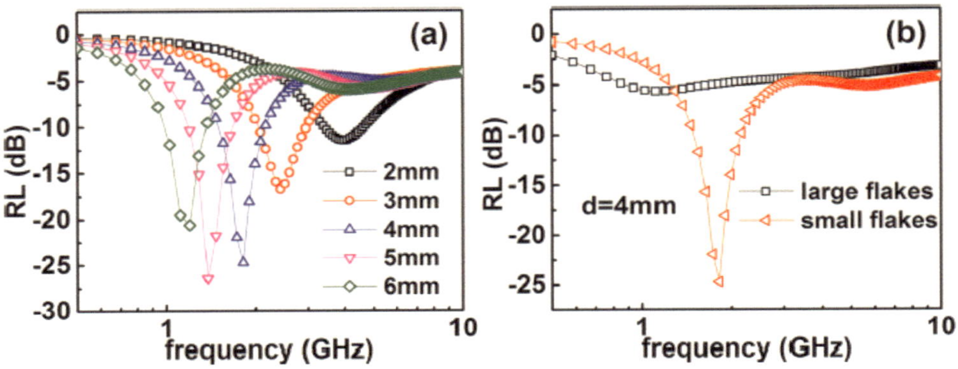

Figure 8.
(a) Composites filled with smaller flakes and with different thickness and (b) composite filled with different flakes but with same thickness (4 mm) (copyright, 2015, AIP).

space/absorber interface. In this chapter, all "t" values are in mm unit. "μ" and "ε" are the measured relative complex permeability and permittivity, respectively. The measured permittivity values can be found and have been studied in our paper [6]. The potential absorbing performances of composites with different thickness values are illustrated in **Figure** 7. Clearly, composites containing smaller flakes will have excellent absorbing performances in terms of RL as well as reduced absorber thickness. The superior absorbing properties are also shown in **Figure 8** by selecting a few thicknesses of single layer of composites filled with smaller flakes.

4. Coating treatments

Since high-frequency permittivity of metallic flakes is much larger than their permeability, this difference will cause a serious mismatch of impedance ($Z_{in} >> Z_0$), which will deteriorate the absorbing properties of flakes. Common means of reducing impedance mismatch is to coat the metallic particles with a layer of oxides with high resistivity (such as SiO_2, TiO_2, Al_2O_3, etc.). Although these layers are effective in decreasing the permittivity, they are not ferro- (or ferri-) magnetic and cannot take part in the absorbing of EM wave via magnetic losses. Therefore, we propose to coat the FINEMET flakes with ferrimagnetic layer (such as $NiFe_2O_4$ ferrite) so as to realize two objectives: to reduce the permittivity and to absorb the EM wave by the high-resistivity layer.

$NiFe_2O_4$ layers were fabricated on the Fe-Cu-Nb-Si-B flakes using a low-temperature chemical plating route. In a simplified description, Fe-Cu-Nb-Si-B nanocrystalline flakes were added into a flask containing a bath solution at 333 K. As for the bath solution, the well-designed molar ratios of $NiCl_2$, $FeCl_2$ and KOH solutions were prepared in the flask. Meanwhile, oxygen gas was introduced into the solution until the chemical reactions were completed. Subsequently, the flask was heated at 333 K for 40 min. When the chemical reaction was finished, the deionized flakes were collected and dried at 333 K for 12 h. Elaborative experimental descriptions can be found in our published paper [7]. The morphologies of coated flakes are presented in **Figure 9a** and **b**, respectively. TEM images in **Figure 9c** show that the thickness of coating layer is estimated to be about 17.73–55.61 nm. The energy dispersion spectrum (EDS) and XRD measurements of uncoated and coated flakes confirm the formation of $NiFe_2O_4$ spinel ferrite. XRD patterns are given in **Figure 10**, which indicate the formation of spinel ferrite phase. The magnetic hysteresis loops of coated and uncoated flakes are given in Ref. [7] and show that the saturation magnetization of coated flakes drops from 129.33 to 96.54 emu/g.

The impacts of spinel ferrite coating layer on the high-frequency permittivity and permeability are shown in **Figure 11**. When the flakes are coated, ε' drops from 61.49 to 33.02 at 0.5 GHz, whereas ε'' also drops from 21.39 to 1.16 at 0.5 GHz. As shown, the complex permittivity values are significantly decreased within the lower frequency band and are believed to result from increased resistivity. The permeability values are also found to be a little reduced, due to the fact that Ms value of

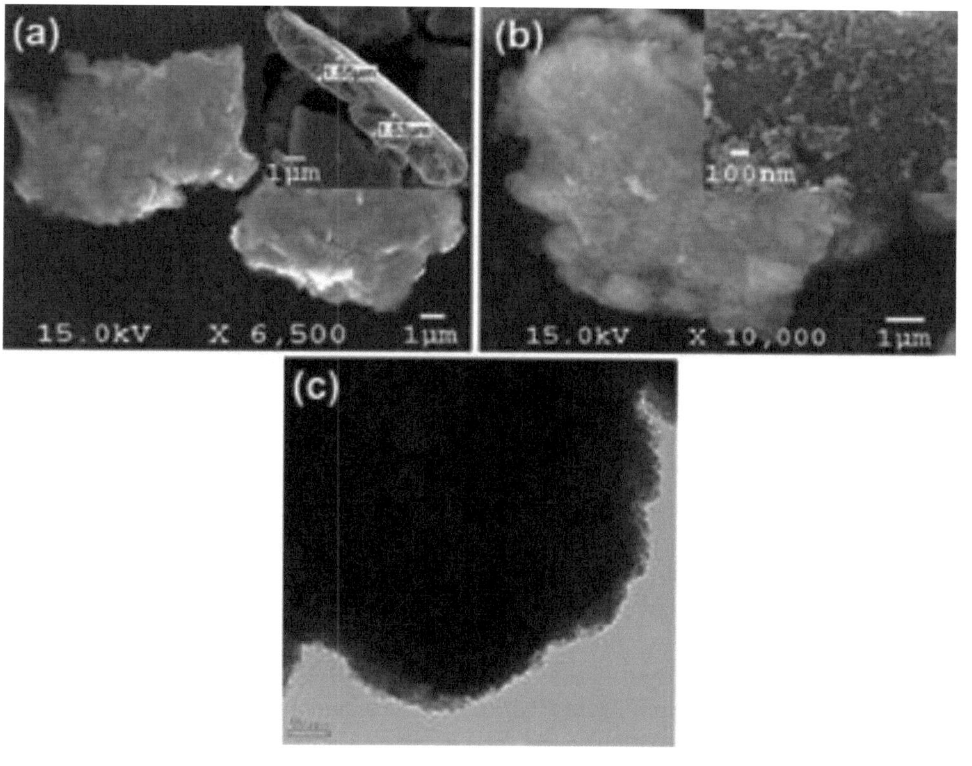

Figure 9.
SEM images of (a) uncoated flakes and (b) coated flakes. (c) TEM image of a coated flake. Inset in (a): typical thicknesses of the sample. Inset in (b): surface roughness of a coated flake (copyright, 2015, IEEE).

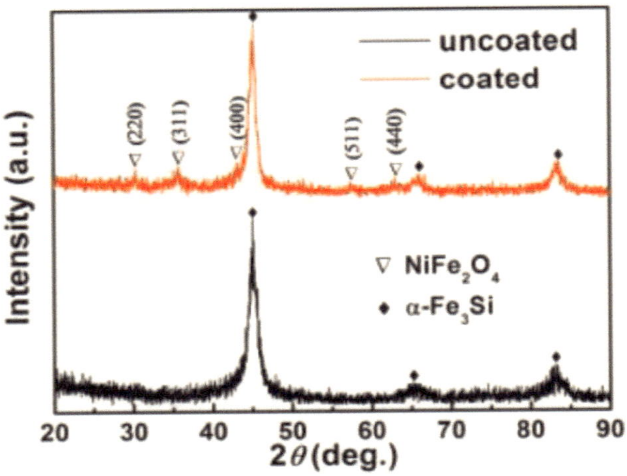

Figure 10.
XRD patterns of uncoated and coated flakes (copyright, 2015, IEEE).

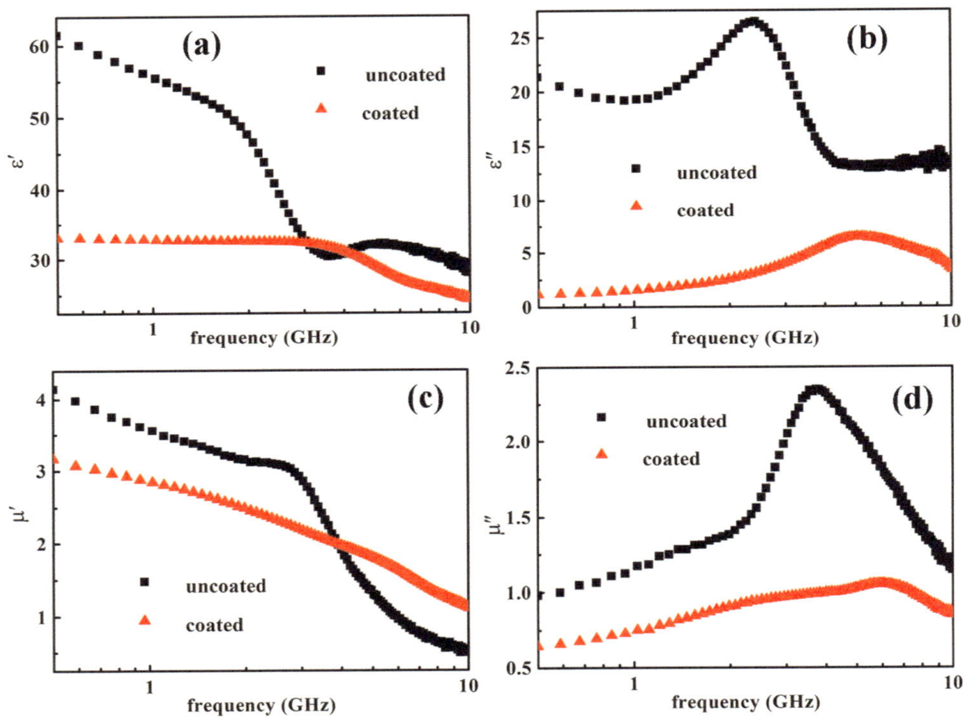

Figure 11.
The impact of ferrite coating on the permittivity and permeability of flakes. (a) $\varepsilon' \sim f$ and (b) $\varepsilon \sim f$. (c) $\mu' \sim f$ and (d) $\mu \sim f$ (copyright, 2015, IEEE).

coated flakes is less than that of uncoated flakes; see **Figure 11c** and **d**. Since the spinel ferrite layer has a smaller Ms value than the FINEMET alloy, the reduced permeability can be understood as per Snoek's law. It is interesting to point out that $\mu \sim f$ spectra of coated flakes are not significantly fluctuated as much as the uncoated flakes. The previous section showed that eddy current effect in uncoated flakes is strong and results in a large reduction of μ values. Such a large reduction is

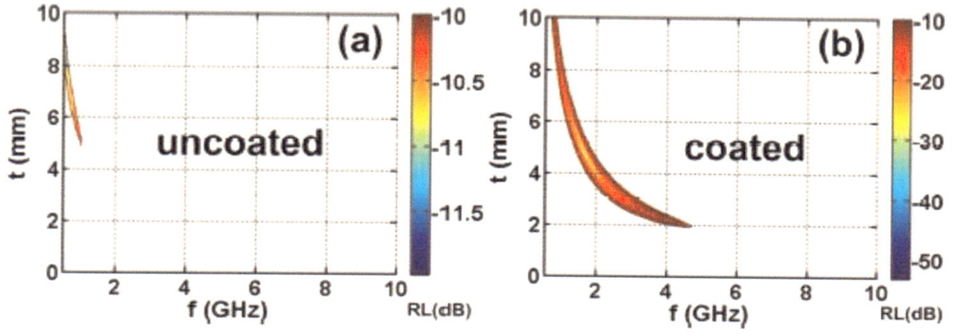

Figure 12.
Contour maps showing the absorbing properties (in terms of RL) of composites with different thicknesses. (a) Uncoated flakes and (b) coated flakes (copyright, 2015, IEEE).

greatly suppressed in coated particles, which means that the high-resistivity coating ferrite effectively reduces the impact of eddy currents on μ values.

The impact of coating with a high-resistivity ferrite on EM absorbing performances is shown in **Figure 12**. We use a contour map to illustrate the reflection loss of absorbers filled with flakes which are coated and uncoated. For composites containing uncoated flakes, the complex ε values are much larger than the complex μ values. Due to this large difference in ε and μ, the impedance mismatch (Z_{in} and Z_0) according to Eq. 7 is significant, and consequently it results in the worsening of absorbing properties compared to absorbers filled with coated flakes. Obviously, the high-resistivity coating can effectively lessen the impedance mismatch and improve absorbing performance. Moreover, the thickness of absorbing composite is greatly decreased if coated flakes are used as absorbent composite.

5. Origins of multi-peaks in the intrinsic permeability spectra

As discussed before, broad permeability spectra are commonly observed in the composites filled with magnetic particles. The debate on the causes is intense. We believe that a broad intrinsic permeability spectrum results from superposition of many natural resonance peaks. Intrinsic permeability can be retrieved from measured permeability using one of the mixing rules [8]. However, origins of multi-peaks in the intrinsic permeability spectra have not been well answered, and it is essential for the designing of electromagnetic devices or materials. In order to exclude impact of non-intrinsic factors (e.g., inhomogeneous microstructure, eddy current, size distribution and particle shape) on the understanding of the origin of multi-peaks, we design several "1 $\times$ 3" iron nanowire (Fe-NW) arrays. Each array has a different interwire spacing. Each nanowire in the array is identical in geometry. Each one has a cuboid shape: the length is 100 nm (set as the "z"axis). The cross section is 10 nm ("y" axis) $\times$ 20 nm ("x" axis).

In this study, we will discuss the impact of interwire distance on the intrinsic permeability for only two interwire distances: 2.5 and 60 nm, as depicted in **Figure 13**. The static magnetic properties and dynamic responses of Fe nanowire arrays are simulated using micromagnetic simulation. First-order-reversal-curve (FORC) technique is used to simulate and analyse the impact of interwire distance on the static magnetic properties. The dynamic response of magnetization under excitation of pulse magnetic field can be described by the Landau-Lifshitz-Gilbert equation:

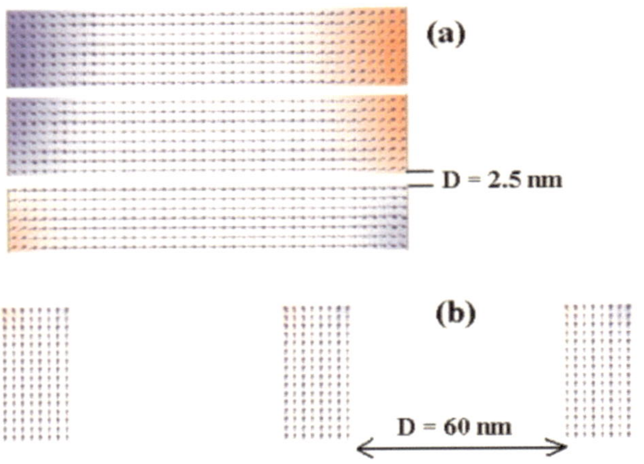

Figure 13.
The equilibrium states of magnetization for two kinds of interwire distance: (a) D = 2.5 nm and (b) D = 60 nm (only partial sections of nanowire are presented for saving space).

$$\frac{\partial \vec{M}(r_i,t)}{\partial t} = -\gamma \vec{M}(r_i,t) \times \vec{H}_{eff}(r_i,t) - \frac{\gamma\alpha}{M_s}\vec{M}(r_i,t) \times \left[\vec{M}(r_i,t) \times \vec{H}_{eff}(r_i,t)\right] \quad (9)$$

More details on simulation procedures and setting parameters can be found in our published paper [9]. FORC approach is based on the Preisach hysteresis theory and is helpful in investigating factors that determine local magnetic properties of materials. According to FORC measurement procedure, the array is at first saturated at a field value (H_s) in one direction; next, the magnetic field is decreased to a field (called "reversal field," H_a); and then the array is again saturated from H_a to H_s, which will trace out one partial magnetization curve (i.e., a FORC curve). Following the same procedure, a series of FORC curves can be obtained starting from different H_a values to the H_s value, which will fill the interior of the major hysteresis loop. Each data point of a FORC curve is denoted as M(H_a, H_b), where H_b is the applied field. According to the Preisach model, a major hysteresis loop is consisted to be a set of hysterons, and the probability density function ρ (H_a, H_b) of hysteron ensemble can be calculated by a mixed second derivative as follows, where ρ wh$_a$, Hb) is often called the FORC distribution, and is expressed as

$$\rho(H_a, H_b) = -\frac{1}{2}\frac{\partial^2 M}{\partial H_a \partial H_b} \quad (10)$$

Usually, the FORC distribution is illustrated in the diagram of (H_u vs. H_c), as shown in **Figure 14a**. H_u is the local interaction field and H_c the local coercivity. Relationship between (H_a, H_b) coordinate system and (H_u, H_c) coordinate system is given as $H_c = (H_b - H_a)/2$ and $H_u = (H_a + H_b)/2$.

Clearly, there are no domain walls existing in our simulations, as shown in **Figure 13**. Therefore, the impact of domain wall on the permeability can be excluded. In addition, coercivity is therefore only decided by localized effective magnetic field (no domain wall movements involved). Localized magnetic fields decide the precession of local magnetizations, which will precess around these effective local fields. Each precession has an eigenfrequency, which is also called "natural resonance frequency." Under the excitation of a pulse magnetic field ((h(t) = 100 exp(-10^9t), t in second, hour in A/m) is) perpendicular to "z" axis, the simulated intrinsic permeability spectra are shown in **Figures 14b** and **15b**.

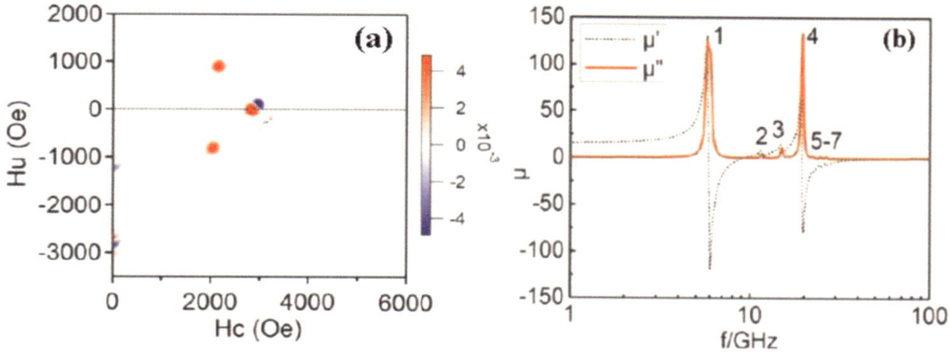

Figure 14.
FORC distribution and intrinsic permeability of nanowire array when interwire distance (D) is 2.5 nm (copyright, 2018, IOP).

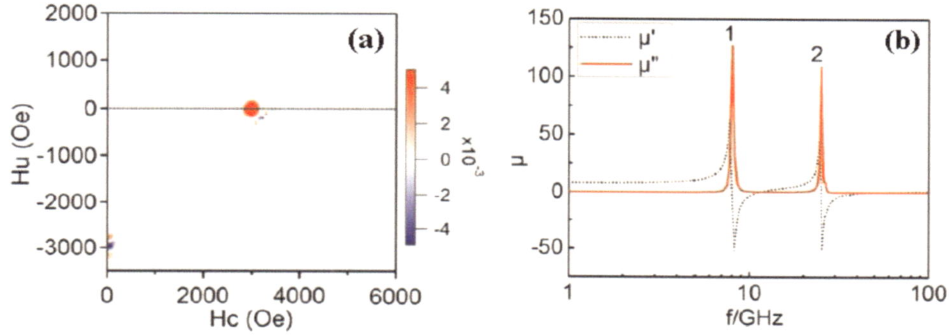

Figure 15.
FORC distribution and intrinsic permeability of nanowire array when D = 60 nm (copyright, 2018, IOP).

Corresponding FORC distributions are shown in **Figures 14a** and **15a**. Obviously as seen in **Figure 14**, two stronger resonance peaks at f = 5.75 and 19.5 GHz and several weak resonance peaks at f = 11.75, 15, 22.25, 24.25 and 26.5 GHz are found for the nanowire array with D as 2.5 nm. The previous studies of both ours and by others show that the resonance peaks found at lower frequency of 5.75 GHz are often named "edge mode" [10], which are resulting from precessions at the ends of nanowires. The frequencies of "edge mode" found by us are very close to those reported by others [10, 11]. The peak found at 19.5 GHz can be identified as "bulk mode." The eigenfrequency of bulk mode for an isolated nanowire can be calculated by the following equation:

$$f = \gamma' \sqrt{\{H + (Nx - Nz)M_s\} \times \{H + (Ny - Nz)M_s\}} \tag{11}$$

where H (the external DC magnetic field) is along the easy axis, demagnetization factor is N and γ' is the gyromagnetic ratio (for Fe: 2.8×10^6 Hz/Oe). For our nanowire arrays, Nx + Ny + Nz = 1, Nz is 0, Nx is 1/3 and Ny is 2/3. Then, the eigenfrequency frequency (natural resonance frequency) of bulk mode is calculated as 28.2 GHz. This calculated value is larger than those for bulk mode simulated in the arrays with different D values. This difference is a result from the fact that the calculated value is based on Eq. (11) which is for uniform precession without considering interactions among nanowires. However, the simulated values are obtained under the circumstance of interaction among NW. It was found that "the calculated f_r" and "the simulated f_r" are in good agreement in an isolated nanowire [10]. Our simulations show that when the interwire distance increases, FORC diagrams and the intrinsic $\mu \sim f$ spectra vary differently. In addition, the reversal process of nanowire array is found to be different. Please refer to our published paper for more details [9]. When interwire distance increases, magnetization reversal behaviour progressively changes from "the sequential mode" to "the independent mode." Moreover, the finding that the amount of weak resonance peaks decreases with increasing D is worth noting. With regard to the origin of weak resonance peaks in **Figure 14b**, we presume that it is the consequence of superposition of interaction field, magnetocrystalline field and shape anisotropy field. As depicted in FORC diagrams (see **Figures 14** and **15**), the interaction field can be either negative or positive. Effective field acting on some magnetization can be smaller (or larger) than the effective field related to "bulk mode" in an isolated nanowire. Accordingly, some resonance frequency is smaller (or larger) than that of bulk mode, which can be clearly observed in **Figure 14**. With increasing interwire distance, interaction among nanowires gradually disappears; therefore, these smaller peaks gradually vanish. From the stand point of electromagnetic (EM) attenuation application, it means that volume fraction of magnetic particles in a composite should not be diluted in order to have many resonance peaks for expanding the attenuation bandwidth. As shown in **Figure 15**, there are only two strong resonance peaks found when D is 60 nm, which may be explained using the same physical mechanism. The "calculated f_r" (bulk mode) is about 28.2 GHz and is close to the "simulated f_r" (about 25 GHz).

Impact of interwire distance on interaction among NWs is shown in **Figures 14a** and **15a**. Obviously, with increasing interwire distance (D), data points approach together (results of other D values were shown in Ref. 9), and H_u becomes "zero" when D is equal or greater than 60 nm. Here, when reversal resistance comes only from localized effective field (H_{eff}), then effective field can be approximated by coercivity field (i.e., $H_c \approx H_{eff}$). Hence, when D increases, localized H_c is always larger than zero. The "scattered" characteristic of H_c value vanishes when D is 60 nm, which means that reversal behaviour of M changes into the "independent mode," as shown in Ref. 9. In addition, peak magnitude of "edge mode" in an isolated "cylindrical" nanowire is usually smaller than "bulk mode" [10, 11]. However, all peak magnitudes of "edge mode" in our "cuboid" shape nanowires are comparable with bulk mode, which suggests that high-frequency magnetic loss due to the "edge mode" cannot be neglected in our case. Moreover, smaller fr value of "edge mode" means that localized effective magnetic field governing its precession is weaker. It means that reversal of magnetization commonly starts from the ends of nanowires, which has been observed in our study.

To better understand the changes of intrinsic $\mu \sim f$ spectra, studies on difference in magnetic moment orientations in equilibrium states are necessary, as illustrated in **Figure 13**. Apparently, magnetic moment orientations around the ends of nanowires at equilibrium states are distinct. Such orientation pattern of magnetic moments strongly relies on the local effective magnetic field. The first-order--reversal-curve diagram is useful to know the distribution of local effective magnetic fields. Hence, we believe any factor affecting the equilibrium state of magnetic moments will change the intrinsic $\mu \sim f$ spectra. If the equilibrium state is rebuilt, then the permeability spectra under same excitation can be recovered too (we name it as "***memory effect***"). Finally, we want to point out that such traits of "scattered" values of H_c and H_u in these simulations differ from the continuous distribution measured in most nanowires deposited in nanoporous templates (such as AAO templates). The reason is that continuous distribution of H_c and H_u in a FORC diagram results from inhomogeneity in nanopore sizes of template, nanoscale electrochemically deposited polycrystals and multimagnetic domains, which will give rise to plentiful irreversible magnetization processes.

6. Negative imaginary parts of permeability ($\mu'' < 0$)

As stated before, high-frequency permeability of Fe-based conducting nanostructured materials can be tailored by shapes, phase transitions, coating and size distributions. Most importantly, all imaginary parts of permeability are positive ($\mu'' > 0$), which are accepted as an unalterable natural principle. This is correct when there are no other excitations except AC magnetic field based on the fact that positive μ'' manifests energy dissipation of magnetization precession. What happens to the imaginary parts of permeability if there is another excitation, which can compensate the energy loss during the precession? Furthermore, is it possible to tailor the high-frequency permeability by electric current? If possible, it will be possible for us to design intelligent electromagnetic devices. In this section, we propose for the first time an approach to accomplish such a goal via spin transfer torque (STT) effect. According to STT effect, when polarized electrons flow through zone of metallic ferromagnetic material with nonuniformly oriented magnetic moments (e.g., magnetic domain walls), the spins of electrons will exert torques on them, which will change the dynamic responses of precession of magnetic moments [13]. When STT is strong enough, it can even switch the direction of magnetic moments. Here we only demonstrate an example using micromagnetic simulation; our detailed results can be found in Ref. [12]. The object of micromagnetic simulations is an isolated Fe nanowire with a length of 400 nm (set as "x" axis) and the diameter of 10 nm, as shown in **Figure 16**. The dynamic response of magnetization (a group of magnetic moments in a direction, also can be called "macro spin") can be simulated under two external excitations: AC magnetic field and electrically polarized current. The AC magnetic field is applied along the "z" axis, and polarized electrons are flowing along the "x" axis. The software of micromagnetic simulation is a three-dimensional object-oriented micromagnetic framework (OOMMF). The dynamic responses of precession are simulated by solving the Landau-Lifshitz-Gilbert (LLG) equation as a function of time. When the STT effect is incorporated, the modified LLG equation is expressed as [13, 14]:

$$\frac{d\vec{M}}{dt} = \gamma\vec{H}_{eff} \times \vec{M} + \frac{\alpha}{Ms}\left(\vec{M}_s \times \frac{d\vec{M}}{dt}\right) - \left(\vec{u}\cdot\vec{\nabla}\right)\vec{M} + \frac{\beta}{Ms}\vec{M} \times \left[\left(\vec{u}\cdot\vec{\nabla}\right)\vec{M}\right] \quad (12)$$

where α (damping constant) is set as 0.01 for the simulations; γ is the Gilbert gyromagnetic ratio (2.21×10^5 mA^{-1} S^{-1}); $\boldsymbol{M}$ is the magnetization; $\boldsymbol{H_{eff}}$ is the

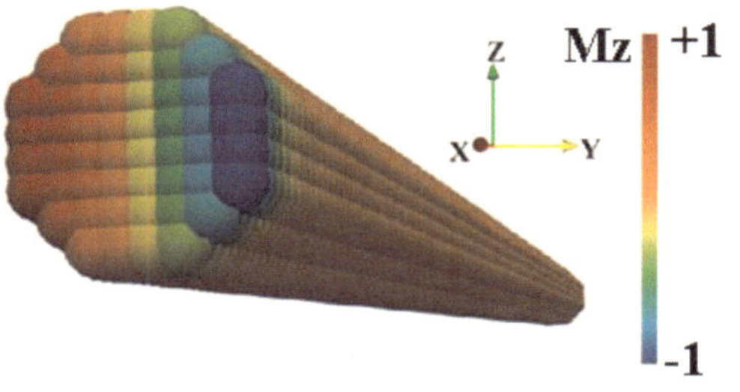

Figure 16.
Inhomogeneous orientation of magnetic moments of an isolated nanowire (copyright, 2018, IOP).

effective magnetic field which consists of demagnetization field, exchange interaction, anisotropic field and external applied field; and finally, β (nonadiabatic spin transfer parameter) is set as 0.02 and is used to consider the impact of temperature on the dynamics of precession. The vector $\boldsymbol{u}$ is defined as

$$\vec{u} = -\frac{g\mu_B P}{2eM_s}\vec{J} \tag{13}$$

where $\boldsymbol{g}$ is the Landé factor, $\boldsymbol{e}$ is the electron charge, μ_B is the Bohr magneton, $\boldsymbol{J}$ is the current density and $\boldsymbol{P}$ is the polarization ratio of current and set as 0.5. The simulation parameters for iron (Fe) nanowire are $\boldsymbol{Ms}$ = 17 × 10^5 A/m, magnetocrystalline anisotropy constant, $\boldsymbol{K_1}$ = 4.8 × 10^4 J/m^3, exchange stiffness constant and $\boldsymbol{A}$ = 21 × 10^{-12} J/m. Also, the nanowire is discretized into many tetrahedron cells (cell size: 5 × 1.25 × 1.25 nm). The cell is smaller than the critical exchange length. To obtain high-frequency permeability spectra, these steps are followed: Firstly, the remanent magnetization state should be acquired after the external field is removed. Secondly, a pulse magnetic field is applied perpendicular to the long axis (x-axis) of the wire. The pulse magnetic field has the form $\boldsymbol{h(t)}$ = 1000 exp(-10^9t) (***t*** in second, ***hour*** in A/m). A polarized current (density J is 3.0 × 10^{12} A/m^2) is flowing along x-axis. Thirdly, the dynamic responses of magnetization in time-domain are recorded under these two excitations. Using the Fast Fourier Transform (FFT) technique, the dynamic responses in frequency domain are obtained. The high-frequency permeability spectra are calculated based on the definition of permeability:

$$\mu(f) = 1 + \frac{m(f)}{h(f)} = 1 + \chi'(f) - i\chi''(f) \tag{14}$$

where $\boldsymbol{m(f)}$ is magnetization under both excitations and the pulse magnetic field ($\boldsymbol{h}$) works as the perturbing field. The following relations exist: $\mu' = \boldsymbol{1 + \chi'(f)}$, $\mu'' = \boldsymbol{\chi(f)}$. The differences in the high-frequency permeability under different excitations have been investigated.

When single nanowire is under the excitation of only an AC magnetic field (h), two Lorentzian-type resonance peaks are found in the permeability spectrum, as shown in **Figure 17a**. One is located approximately at 18 GHz, and the other is located approximately at 31.5 GHz. According to our previous studies, the major resonance is called "bulk mode," which has larger magnetic loss and is manifested by the larger μ'' value, which is ascribed to the precession of perfectly aligned magnetic moments within the nanowire body excluding the misaligned magnetic moments at both ends of nanowire. The minor resonance peak located at 18 GHz is often called "edge mode," whose smaller magnitudes of real and imaginary parts are due to smaller volume fractions of magnetic moments at the ends of nanowire. The edge mode is due to the precession and resonance of misaligned magnetic moments. Physically, if external field is not applied, the magnetizations are aligned along the local effective field ($\boldsymbol{H_{eff}}$). Once the perturbation field ($\boldsymbol{h}$) is applied to excite the precession, $\boldsymbol{M}$ will move away from their equilibrium positions with small angles, and $\boldsymbol{M}$ will then precess around H_{eff}. When precession goes on, the y-component of M ($\boldsymbol{M_y}$) will have nonzero values, as shown in **Figure 17b**. During the period of precession, the energy absorbed from the AC magnetic field will be gradually dissipated via damping mechanisms. M will restore to its equilibrium positions, and M_y will be zero; see **Figure 17b**. The energy loss is manifested by a positive imaginary part of permeability ($\mu'' > 0$). When the nanowire is excited under both the pulse magnetic field and polarized current, the high-frequency permeability

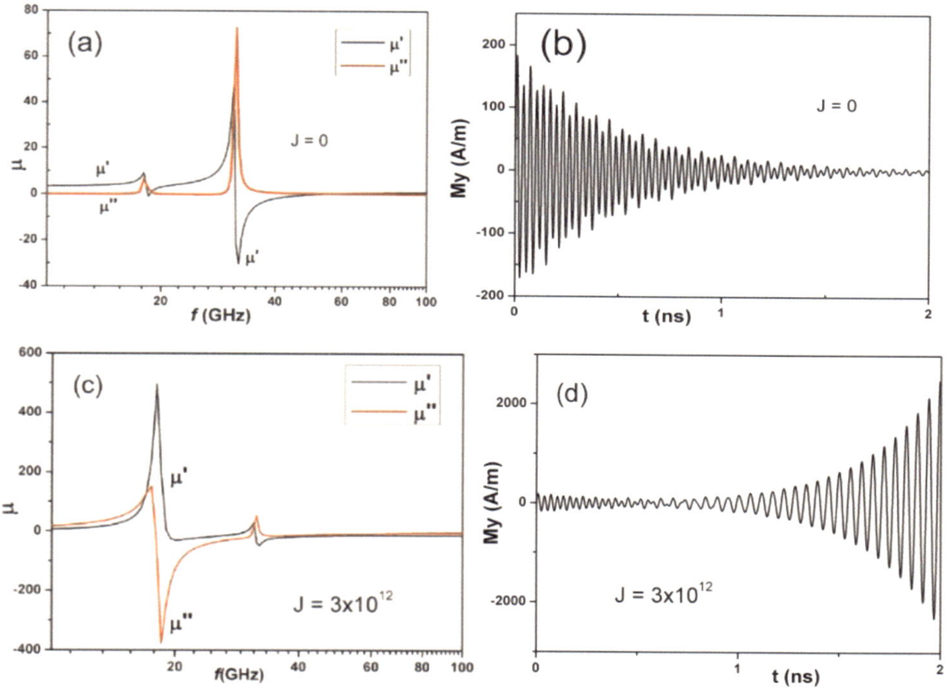

Figure 17.
Permeability spectra and responses of $\mathbf{M_y}$ *component without and with STT excitation (*$J \neq 0$*) (copyright, 2018, Elsevier).*

spectrum is naturally found to be different and is illustrated in **Figure 17c**. The previous minor resonance becomes the major resonance with negative μ'' values. The "M_y" component does not vanish gradually; see **Figure 17d**. According to the STT effect, a spin-polarized current flows through the nanowire; STTs only act on the magnetic moments at the ends of nanowire. These STTs will counteract the torques due to effective magnetic field, which will then bring the magnetizations back to their equilibrium positions. When the STT is strong enough, it can maintain the precession angle ($\mu'' = 0$) and even enlarge the precession angle ($\mu'' < 0$) or switch the direction of magnetization. Additionally, negative μ'' values are found only for "edge mode" at lower frequency. This is because the STT effect only acts on magnetic moments with a spatially nonuniform orientation, which are only found at both ends of nanowires. The cause for minor peak becoming major peak under STT effect is due to the fact that the spin transfer torques are consistently acting on

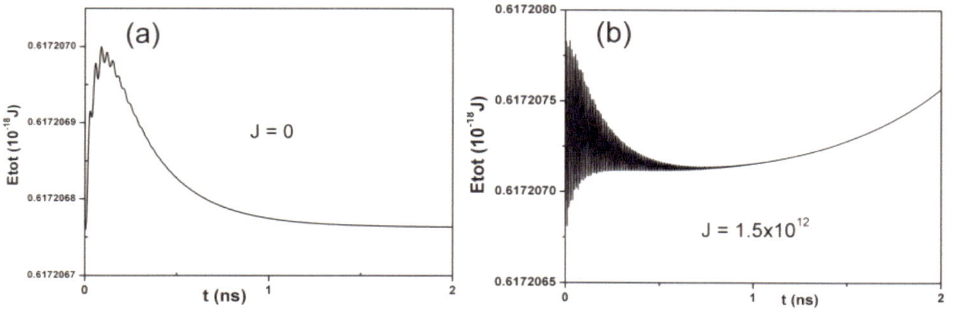

Figure 18.
The variation of total energy without and with STT excitation (copyright, 2018, Elsevier).

the magnetic moments; therefore, the precession angle is enlarging, and as a result, M_y component is also increased. From the perspective of total energy (E_{total}) of magnetic moment ensemble, when no STT effect is involved, E_{total} first increases due to the excitation of pulse magnetic field and then gradually attenuates due to energy loss via damping mechanisms. However, when STT effect exists, E_{total} is not attenuated to a constant; see **Figure 18**.

Finally, it should be pointed out that although others have also reported negative μ'' in other materials, there are no convincing physical mechanisms that have been provided to support their results ($\mu'' < 0$), and negative μ'' is mostly likely due to measurement errors.

7. Conclusions

In this chapter, we have shown several approaches to tailor the high-frequency permeability of ferromagnetic materials: shapes, particle size distribution, heat treatments, coating and spin transfer torque effect. Micromagnetic simulations are used to explain the origins of multi-peaks in permeability spectra of nanowire array. In addition, just like the simultaneous excitation method (pulse ***H*** field + ***STT*** effect) shown here, we want to propose the other possibilities (pulse ***H*** field + *the other excitation*) to tailor $\mu \sim f$ spectra and to realize negative imaginary parts of permeability ($\mu'' < 0$), such as femtosecond laser pulses or photomagnetic pulses.

Acknowledgements

This work is supported by the National Science Foundation of China (Grant No. 61271039) and the International Scientific Collaboration of Sichuan Province (Grant No. 2015HH0016).

Author details

Mangui Han
School of Materials and Energy, University of Electronic Science and Technology of China, Chengdu, China

*Address all correspondence to: han_mangui@yahoo.com

References

[1] Snoek JL. Dispersion and absorption in magnetic ferrites at frequencies above one mc/s. Physica. 1948;**14**:207. DOI: 10.1016/0031-8914(48)90038-X

[2] Han M, Deng L. Mössbauer studies on the shape effect of $Fe_{84.94}Si_{9.68}Al_{5.38}$ particles on their microwave permeability. Chinese Physics B. 2013; **22**(8):083303. DOI: 10.1088/1674-1056/22/8/083303

[3] Acher O, Adenot AL. Bounds on the dynamic properties of magnetic materials. Physical Review B. 2000;**62**: 11324. DOI: 10.1103/PhysRevB.62.11324

[4] McHenry ME, Willard MA, Laughlin DE. Amorphous and nanocrystalline materials for applications as soft magnets. Progress in Materials Science. 1999;**44**:291. DOI: 10.1016/S0079-6425(99)00002-X

[5] Han M, Guo W, Wu Y, et al. Electromagnetic wave absorbing properties and hyperfine interactions of Fe–Cu–Nb–Si–B nanocomposites. Chinese Physics B. 2014;**23**(8):083301. DOI: 10.1088/1674-1056/23/8/083301

[6] Wu Y, Han M, Liu T, Deng L. Studies on the microwave permittivity and electromagnetic wave absorption properties of Fe-based nano-composite flakes in different sizes. Journal of Applied Physics. 2015;**118**:023902. DOI: 10.1063/1.4926553

[7] Wu Y, Han M, Deng L. Enhancing the microwave absorption properties of Fe-Cu-Nb-Si-B nanocomposite flakes by coating with spinel ferrite $NiFe_2O_4$. IEEE Transactions on Magnetics. 2015; **51**(11):2802204. DOI: 10.1109/TMAG.2015.2444656

[8] Lagarkov AN, Rozanov KN. High-frequency behavior of magnetic composites. Journal of Magnetism and Magnetic Materials. 2009;**321**:2082. DOI: 10.1016/j.jmmm.2008.08.099

[9] Deng D, Han M. Discussing the high frequency intrinsic permeability of nanostructures using first order reversal curves. Nanotechnology. 2018;**29**: 445705. DOI: 10.1088/1361-6528/aad9c2

[10] Gerardin O, Le Gall H, Donahue MJ, Vukadinovic N. Micromagnetic calculation of the high frequency dynamics of nano-size rectangular ferromagnetic stripes. Journal of Applied Physics. 2001;**89**:7012. DOI: 10.1063/1.1360390

[11] Han XH, Liu RL, Liu QF, Wang JB, Wang T, Li FS. Micromagnetic simulation of the magnetic spectrum of two magnetostatic coupled ferromagnetic stripes. Physica B. 2010; **405**:1172-1175. DOI: 10.1016/j.physb.2009.11.030

[12] Han M, Zhou W. Unusual negative permeability of single magnetic nanowire excited by the spin transfer torque effect. Journal of Magnetism and Magnetic Materials. 2018;**457**:52-56. DOI: 10.1016/j.jmmm.2018.02.031

[13] Thiaville A, Nakatani Y, Miltat J, Suzuki Y. Micromagnetic understanding of current-driven domain wall motion in patterned nanowires. Europhysics Letters. 2005;**69**:990. DOI: 10.1209/epl/i2004-10452-6

[14] Yan M, Kakay A, Gliga S, Hertel R. Beating the walker limit with massless domain walls in cylindrical nanowires. Physical Review Letters. 2010;**104**: 057201. DOI: 10.1103/PhysRevLett.104.057201

Chapter 7

Ferrite Materials and Applications

Tsun-Hsu Chang

Abstract

This chapter starts from a generalized permeability and aims at providing a better understanding of the ferrites behavior in the microwave fields. The formula of the generalized permeability explains why the permeability of the ferrimagnetic or even the ferromagnetic materials strongly depends on the applied magnetic bias and the polarization of the wave. Right-hand circularly polarized (RHCP) wave may synchronize with the precession of the magnetic moment, resulting in a strong resonant effect. Characterizing the ferrites' properties, including the complex permittivity, the saturation magnetization, and the resonance linewidth, will be discussed. We then utilize these properties to design and fabricate various microwave devices, such as phase shifters, circulators, and isolators. Detailed analysis and simulation to demonstrate how these ferrite devices work will be shown. The mechanism will be discussed.

Keywords: ferrite, microwave, generalized permeability, circular polarization, phase shifter, circulator, isolator

1. Introduction

Ferrimagnetism is similar to ferromagnetism in many ways [1–4]. They all have hysteresis curves as the applied magnetic field changes, resulting in the saturation magnetization ($4\pi M_s$), the coercive field (H_c), and the remnant polarization (B_r). A ferri- or ferromagnetic material can be used to fabricate a hard or a soft magnet depending on its coercivity (H_c). Hard magnets are characterized by a high H_c, indicating that their magnetization is difficult to change and will retain their magnetization in the absence of an applied field as shown in **Figure 1(a)**. On the contrary, the soft magnets have low H_c values and, normally, have very weak, remnant magnetic field (low B_r) as in **Figure 1(b)**. Hard magnetism has been extensively used to make the permanent magnets and provides a strong DC magnetic field, while the soft magnetism can be used for the AC systems.

Soft magnetic materials include electrical steels and soft ferrites [3, 4]. Unlike the ferromagnetic metals which are conductors, soft ferrites have low electric conductivity, i.e., they are dielectric materials. The electrical steels have extensive applications in low-frequency systems, such as generators, motors, and transformers, while the soft ferrites are suitable for the high-frequency applications, such as circulators, isolators, phase shifters, and high-speed switches.

This chapter will focus on the properties of the soft ferrites and their applications at high-frequency systems. The ferrites are crystals having small electric conductivity compared to ferromagnetic materials. Thus they are useful in high-frequency devices because of the absence of significant eddy current losses. Ferrites

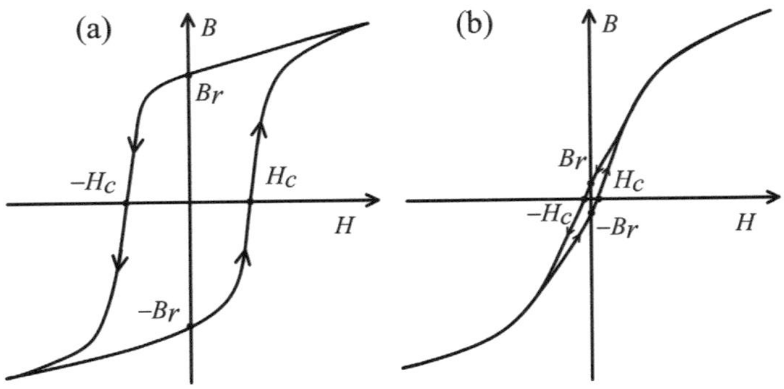

Figure 1.
The hysteresis (B-H) curves for (a) hard ferrimagnetism and (b) soft ferrimagnetism. The remnant polarization (B_r) and the coercive field (H_c) for the hard ferrites should be as large as possible. On the other hand, for the soft ferrites, the remnant polarization (B_r) and the coercive field (H_c) are very small or even close to zero.

are ceramic-like materials with specific resistivities that may be as much as 10^{14} greater than that of metals and with dielectric constants around 10 to 16 or greater. Ferrites are made by sintering a mixture of metal oxides and have the general chemical composition $MO{\cdot}Fe_2O_3$, where M is a divalent metal such as Mn, Mg, Fe, Zn, Ni, Cd, etc. Relative permeabilities of several thousands are common [5, 6]. The magnetic properties of ferrites arise mainly from the magnetic dipole moment associated with the electron spin [2].

The magnetic dipole moment precesses around the applied DC magnetic field by treating the spinning electron as a gyroscopic top, which is a classical picture of the magnetization process. This picture also explains the anisotropic magnetic properties of ferrites, where the permeability of the ferrite is not a single scalar quantity, but instead is a generally a second-rank tensor or can be represented as a matrix. The left and right circularly polarized waves have different propagation constant along the direction of the external magnetic field, resulting in the nonreciprocity of a propagating wave. Since the permeability should be treated as a tensor (matrix), not a scalar permeability, it is generally much difficult to understand and to have intuition, even for the researchers.

2. The susceptibility matrix

2.1 Magnetic dipole precesses around the applied magnetic field

The properties of ferrites are very intriguing. Without a DC bias magnetic field $\mathbf{H_0}$, the magnetic dipoles of ferrites are randomly orientated. They exhibit dielectric properties only. The dielectric loss can be high or low depending on the loss tangent of the soft ferrites. Interestingly, when the DC bias magnetic field is strong enough, and the ferrite is magnetically saturated, the magnetic dipole will precess around the DC bias field at a frequency called the Larmor frequency ($\omega_0 \propto H_0$). Waves with different circular polarizations will corotate or counter-rotate with the precession of the magnetic dipoles.

Figure 2 shows the Larmor precession with the circularly polarized fields [7]. Circular polarization may be referred to as right handed or left handed, depending on the direction in which the electric (magnetic) field vector rotates. For a

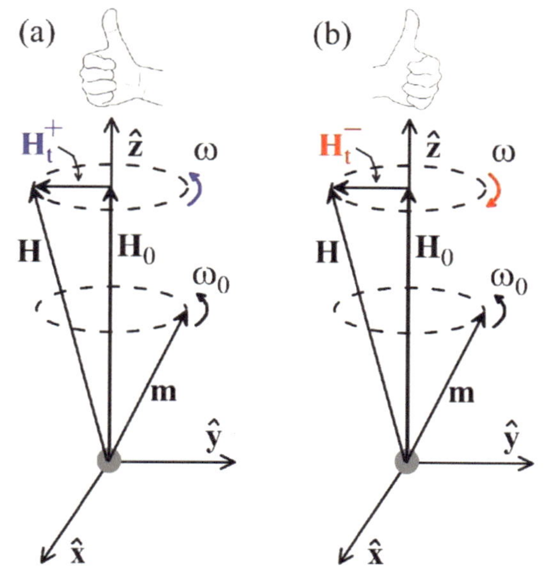

Figure 2.
Larmor precession of a magnetic moment **m** *around the applied DC bias field* $\mathbf{H}_o$ $(= H_0\hat{\mathbf{z}})$ *with (a) a right-hand circularly polarized (RHCP) wave and (b) a left-hand circularly polarized (LHCP) wave. The frequency of the Larmor precession in both cases are the same, i.e., the Larmor frequency* ω_0 $(= \mu_0\gamma H_0)$. H_t^+ *and* H_t^- *are the transverse components of the incident waves which rotate clockwise (RHCP) and counter-clockwise (LHCP) from the source viewpoint looking in the direction of propagation. The thumb points the direction of the wave propagation, and the fingers give the rotation of the transverse components [7].*

right-hand circularly polarized (RHCP) wave, the fields rotate clockwise at a given position from the source looking in the direction of propagation. The magnetic dipole moment **m** processes around the $\mathbf{H_0}$ field vector, like a top spinning precess around the z-axis at the Larmor frequency ω_0. The spinning property depends on the applied DC bias magnetic field. **Figures 2(a)** and **(b)** shows the RHCP and LHCP waves with the gyrating frequency of ω. When the RHCP wave is propagating along the direction of the DC bias field, it corotates with the precession of the magnetic dipole moments. On the other hand, the left-hand circularly polarized (LHCP) wave will counter-rotate with the precession of the dipole moments.

A linearly polarized incident wave can be decomposed into RHCP and LHCP waves of equal amplitude. The orientation of the linearly polarized wave changes after the wave propagates a certain distance because of the distinct propagation constants. The phenomenon is the famous Faraday's rotation [5, 6]. This unique property has various applications, such as phase shifters, isolators, and circulators. However, it is difficult to follow for students and even researchers in that the permeability is a tensor, not just a simple proportional constant.

Here we consider the simplest case for the pedagogic purpose—a circularly polarized plane wave is normally incident upon a semi-infinite medium. The wave characteristics such as the propagation constant k and the wave impedance Z are associated with the permeability μ, which is a tensor for the ferrite medium [5]. By finding the preferred eigenvalues, it will be shown that the properties of μ depend on the DC bias field H_0, the saturated magnetization M_s, and the operating frequency ω. By adjusting the frequency of the incident wave, i.e., ω, the permeability μ changes, especially close to the Larmor frequency (ω_0). Such an effect is called the ferrite magnetic resonance (FMR) or gyromagnetic resonance [7].

2.2 Derivation of the susceptibility matrix

The permeability $\mu_r(\omega)$ is a tensor. Note that the permittivity $\varepsilon_r(\omega)$ can be expressed in a tensor as well, but in the region of interest around 10 GHz, it can be treated as a complex proportional constant for many ceramics. Many microwave textbooks and literature have elaborated the derivation of the permeability tensor [5–7]. In this paper, a more accessible interpretation of the permeability tensor is provided. The magnetic properties of a material are due to the existence of magnetic dipole moment **m**, which arise primarily from its (spin) angular momentum **s**. The magnetic dipole moment and angular momentum have a simple relation, $\mathbf{m} = -\gamma\mathbf{s}$, where γ is the gyromagnetic ratio ($\gamma = e/m = 1.759 \times 10^{11}$ C/kg). When a DC bias magnetic field $\mathbf{B_0}$ ($= \mu_0\mathbf{H_0}$) is present, the torque $\boldsymbol{\tau}$ exerted on the magnetic dipole moment is

$$\boldsymbol{\tau} = \mathbf{m} \times \mathbf{B_0} = \mu_0\mathbf{m} \times \mathbf{H_0}. \tag{1}$$

Since the torque is equal to the time change rate of the angular momentum, we have

$$\boldsymbol{\tau} = \frac{d\mathbf{s}}{dt} = \frac{-1}{\gamma}\frac{d\mathbf{m}}{dt}. \tag{2}$$

By comparing Eqs. (1) and (2), we obtain

$$\frac{d\mathbf{m}}{dt} = -\mu_0\gamma\mathbf{m} \times \mathbf{H_0}. \tag{3}$$

A large number of the magnetic dipole moment **m** per unit volume give rise to an average macroscopic magnetic dipole moment density **M**. The torque exerting on the magnetization per unit volume **M** due to the magnetic flux **H** has the same form as Eq. (3):

$$\frac{d\mathbf{M}}{dt} = -\mu_0\gamma\mathbf{M} \times \mathbf{H}. \tag{4}$$

M and **H** in Eq. (4) differ slightly from **m** and $\mathbf{H_0}$ in Eq. (3) in that **M** and **H** can further be divided into two parts: the DC term and the high-frequency AC term. The DC term is, in general, much larger than the AC term. The applied DC bias magnetic field H_0 is assumed to be in the z-direction. When H_0 is strong enough, the magnetization will be saturated, denoted as M_s which aligns with the direction of H_0. If the AC term just polarizes in the transverse direction (i.e., the xy plane), the external magnetic bias field and the magnetization can be written as,

$$\mathbf{H} = H_x\hat{\mathbf{x}} + H_y\hat{\mathbf{y}} + H_0\hat{\mathbf{z}}, \tag{5}$$

$$\mathbf{M} = M_x\hat{\mathbf{x}} + M_y\hat{\mathbf{y}} + M_s\hat{\mathbf{z}}. \tag{6}$$

Since the AC terms have an $\exp(-i\omega t)$ dependence, by substituting Eq. (5) and (6) into Eq. (4) the transverse component terms read

$$(\omega_0^2 - \omega^2)M_x = \omega_0\omega_m H_x + j\omega\omega_m H_y; \tag{7}$$

$$(\omega_0^2 - \omega^2)M_y = -j\omega\omega_m H_x + \omega_0\omega_m H_y, \tag{8}$$

where $\omega_0 = \mu_0\gamma H_0$ and $\omega_m = \mu_0\gamma M_s$. In matrix representation, Eq. (7) and (8) can be rewritten as

$$\mathbf{M} = [\chi]\mathbf{H} = \begin{bmatrix} \chi_{xx} & \chi_{xy} & 0 \\ \chi_{yx} & \chi_{yy} & 0 \\ 0 & 0 & 0 \end{bmatrix} \mathbf{H}, \tag{9}$$

where $\chi_{xx} = \chi_{yy} = \omega_0\omega_m/(\omega_0^2 - \omega^2)$ and $\chi_{xy} = -\chi_{yx} = j\omega\omega_m/(\omega_0^2 - \omega^2)$. The eigenvalues of the susceptibility matrix are

$$\chi^{+} = \frac{\omega_m}{\omega_0 - \omega}, \tag{10}$$

$$\chi^{-} = \frac{\omega_m}{\omega_0 + \omega}. \tag{11}$$

The eigenvectors corresponding to these two eigenvalues are the right-hand circularly polarized wave (RHCP, denoted as +) and the left-hand circularly polarized wave (LHCP, denoted as -), respectively. The symbols, + and -, represent positive helicity and negative helicity, respectively. The LHCP wave has a relatively mild response over the entire frequency range. On the contrary, the RHCP wave has a much more dramatic response.

The permeabilities of the RHCP and LHCP waves are

$$\mu^{+} = \mu_0\left(1 + \frac{\omega_m}{\omega_0 - \omega}\right), \tag{12}$$

$$\mu^{-} = \mu_0\left(1 + \frac{\omega_m}{\omega_0 + \omega}\right). \tag{13}$$

Eq. (12) has a singularity when the wave frequency ω is equal to the Larmor frequency ω_0. This phenomenon is called the ferri-magnetic resonance (FMR, or called gyromagnetic resonance) [6]. On the contrary, Eq. (13) is relatively mild. Therefore, this study will mainly focus on the RHCP wave.

For a resonant cavity with a quality factor (Q), the loss effect can be introduced by using the complex resonant frequency $\omega_0(1 - i/2Q)$. By analogy with the resonant cavity, the loss part can be calculated by using the complex frequency $\omega_0 \rightarrow \omega_0(1 - i\Delta H\mu_0\gamma/2\omega_0)$, where ΔH is the ferrimagnetic resonance linewidth [2, 3]. Since

$$\frac{\Delta H\mu_0\gamma}{2\omega_0} = \frac{\Delta H\mu_0\gamma}{2H_0\mu_0\gamma} = \frac{\Delta H}{2H_0}, \tag{14}$$

the permeability for the RHCP wave now reads

$$\mu_r^{+} = \left(1 + \frac{\frac{\omega_m}{\omega_0}}{\left(1 - \frac{\omega}{\omega_0}\right) + i\frac{\Delta H}{2H_0}}\right). \tag{15}$$

To conduct a complete simulation of a ferrite device, we need to know its complex permittivity, the saturation magnetization, and the resonance linewidth. We will discuss how to characterize the ferrite's properties in the next section.

3. Characterization of ferrite materials

Here we will discuss the measurement of the most important properties of ferrites, including the dielectric properties $(\varepsilon_r + i\varepsilon_i)$, the saturation magnetization

(M_s or $4\pi M_s$), and the resonance linewidth (ΔH). The behavior of the spin wave linewidth should be considered when the field strength of an electromagnetic wave exceeds a threshold value, that is, the high-power condition. For the general purpose, only the first three properties will be used in the ferrite simulation.

3.1 Dielectric properties

Ferrites are ceramic-like materials with relative dielectric constants around 10 to 16 or greater. The resistivities of ferrites may be as high as 10^{14} greater than that of metals. Since ferrites are dielectric materials. The dielectric properties ($\varepsilon_r + i\varepsilon_i$) always play an important role with or without the influence of the magnetic field. The perturbation method is the most commonly employed resonant technique [8, 9]. The perturbation method is very good for small-size and low-dielectric samples. When measuring the high-dielectric samples, however, the fields and the resonant frequency change drastically. The perturbation technique may lead to a reduced accuracy. Recently, the field enhancement method was proposed [10, 11]. The field enhancement method operates at a condition opposite to the perturbation method. The resonant frequency and quality factor alter significantly and depend on not only the geometry of the cavity but the sample's size and complex permittivity as well. Luckily, both the perturbation method and the field enhancement method agree well for samples with the dielectric constant below 50, which is suitable for most of the ferrites.

Figure 3 shows the ideal of the field enhancement method. **Figure 3(a)** shows the resonant frequency as functions of the dielectric constant (ε_r) using a simulation setup as in **Figure 3(b)**. The ingot-shaped sample has a diameter of 16.00 mm and thickness of 5.00 mm. The solid blue line depicts the simulation result for the field enhancement method. From the measured resonant frequency, we can derive the corresponding dielectric constant. On the other hand, the dashed black line is obtained using a sample with the same diameter but much thinner in thickness of 1.0 mm. The response of the 1-mm-thick sample quite resembles the perturbation method. The imaginary part of the permittivity (ε_i) or the loss tangent ($\tan\delta = \varepsilon_i/\varepsilon_r$) can then be determined from the measured resonant frequency and the quality factor. The field enhancement method has very wide measuring range from unity to high-κ dielectrics and from lossless to lossy materials [10, 11].

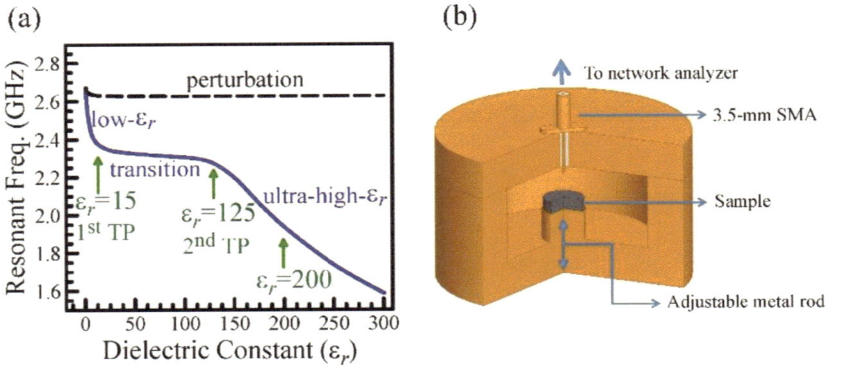

Figure 3.
(a) Resonant frequency versus dielectric constant based on full-wave simulations. The solid curve can be divided into three regions: low, transition, and ultrahigh. The dashed line is simulated with a much thinner sample of 1.00 mm in thickness, which exhibits the properties similar to those of perturbation. (b) Schematic diagram of the field enhancement method. It consists of a cylindrical resonant cavity and a metal rod. The sample is placed on the top of the metal rod. The metal rod focuses and enhances the electric field significantly. An SubMiniature version A (SMA) 3.5-mm adapter couples the wave from the top of the cavity [11].

3.2 Saturation magnetization

Ferrites have a strong response to the applied magnetic field. The magnetic properties of ferrites arise mainly from the magnetic dipole moment associated with the electron spin. Relative permeabilities of several thousands are common. The saturation magnetization (M_s or $4\pi M_s$) of a ferrite plays a key role as shown in Section 2. Researchers or engineers use the saturation magnetization as a design parameter that enters into the initial selection of a ferrimagnetic material for microwave device applications. Typical ferrimagnets exhibit values of $4\pi M_s$ between 300 gauss (G) and 5000 G. Static or low-frequency methods are generally used to measure $4\pi M_s$ [12]. From the measured hysteresis loop as shown in **Figure 4**, one can determine the saturation magnetization M_s.

Note that the saturation magnetization is denoted as M_s in the SI unit, but since the values are generally displayed in Gaussian unit (gauss, G), $4\pi M_s$ is commonly used. Also, the internal bias H_0 is different from the applied H-field (H_a). Demagnetization factor should be considered [5, 6]. The demagnetization factor allows us to calculate the H-field inside the sample denoted as H_0. In all, measurement of the saturation magnetization from the dynamic hysteresis loop characteristics can be used for the design and simulation of ferrite devices.

3.3 Resonance linewidth

The loss of ferrite material is related to the linewidth, ΔH, of the susceptibility curve near resonance. Consider the imaginary part of the susceptibility χ''_{xx} versus the bias field H_0. The linewidth ΔH is defined as the width of the curve of χ''_{xx} versus H_0, where χ''_{xx} has decreased to half of its peak value. For a fixed microwave frequency ω, resonance occurs when $\omega_0 = \mu_0 \gamma H_r$, such that $\omega = \omega_0 (= \mu_0 \gamma H_r)$. The linewidth, ΔH, is defined as the width of the curve of χ''_{xx} versus H_0, where χ''_{xx} has decreased to half its peak value. This is the idea that is introduced in [5]. However, obtaining the relation of χ''_{xx} versus H_0 is not easy. Here we adopt another commonly used technique [9, 12].

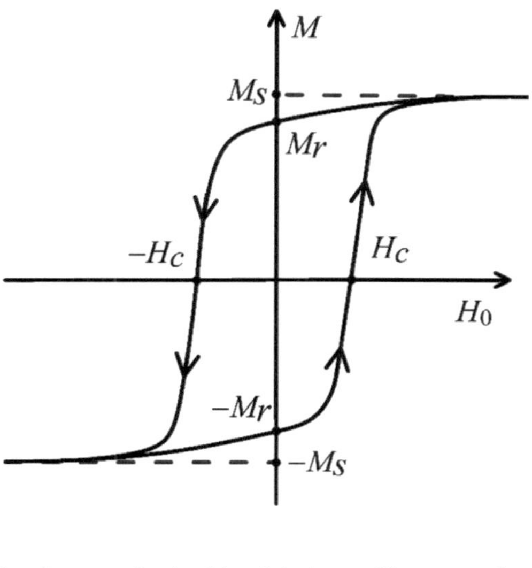

Figure 4.
The hysteresis curve regarding the magnetization M and the internal bias H_0. When the applied internal magnetic field H_0 is large enough, the magnetization will be saturated, denoted as (M_s or $4\pi M_s$). When H_0 decreases to zero, the remnant polarization is denoted as M_r. The polarization will change sign (from positive to negative) when H_0 is greater than $-H_c$ which is called the coercive field.

The idea is normally implemented using a TE_{10n} (n even) cavity in the X-band region [9, 12]. The test sample is placed at the *H*-field maximum. The sample is spherical with a diameter of approximately 0.040 inches, which is much easier to estimate than the internal bias field H_0. A cross-guide coupler is used with the coupling iris. The loaded Q (Q_L) of the empty cavity should be 2000 or greater. The sample, mounted on a fused silica or equivalent rod, is positioned away from the cavity wall at a point of minimum microwave electric field and maximum microwave magnetic field. A power meter can be used to read off the half-power points by adjusting the DC magnetic field and measuring the difference in H_0-field directly [12]. **Figure 5** shows how the resonant linewidth is determined.

The three key parameters are obtained in three experimental setups under different sizes and geometries of the samples. If the samples' properties are slightly different or the machining error is not negligible, the error will be large or even unacceptable. The ultimate goal is to integrate the measurements and to extract the parameters using one experimental setup. The three key parameters will be used in the design of the microwave ferrite devices in the next session.

4. Applications of ferrite materials

The ferrites are crystals having small electric conductivity compared to ferromagnetic materials. Thus they are useful in high-frequency situations because of the absence of significant eddy current losses. Three commonly used ferrite devices are discussed below. These are phase shifters, circulators, and isolators [13–16].

4.1 Phase shifter

The phase shifters are important applications of ferrite materials, which are two-port components that provide variable phase shift by changing the bias magnetic field. Phase shifters find application in test and measurement systems, but the most significant use is in phase array antenna where the antenna beam can be steered in space by electronically controlled phase shifters. Because of the demand, many different types of phase shifters have been provided. One of the most useful designs is the latching nonreciprocal phase shifter using a ferrite toroid in the rectangular waveguide. We can analyze this geometry with a reasonable degree of approximation using the double ferrite slab geometry.

Figure 6 shows the full-wave simulation for a two-port phase shifter using high-frequency structure simulator (HFSS, ANSYS). A standard waveguide WR-90 is employed with a width of 22.86 mm and height of 10.16 mm. The field patterns are

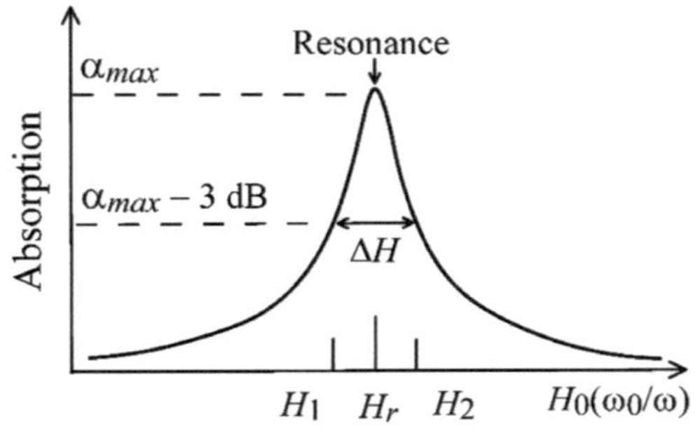

Figure 5.
The idea is to place the sample at the maximum of the H-field. It exhibits resonant absorption when the internal bias field is changed to H_r. By changing the magnetic field, we will obtain the absorption. H_1 and H_2 are associated with the 3-dB absorption. The difference between these two values is the resonance linewidth ΔH.

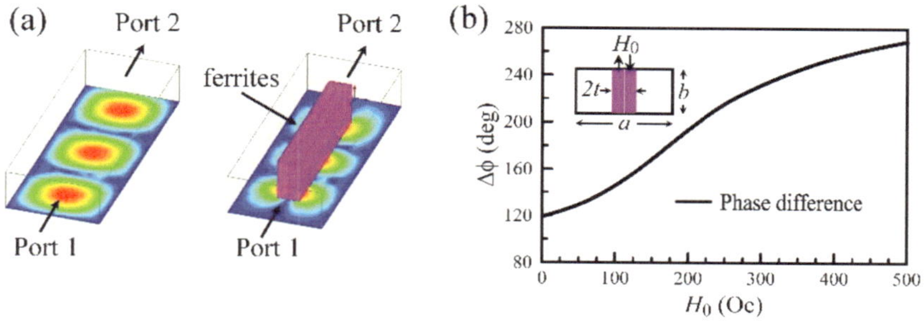

Figure 6.
Simulation results. (a) The field pattern to the left is for the empty waveguide. The right figure shows field strength with ferrites. (b) The phase difference $\Delta\phi$ as a function of the internal bias field H_0.

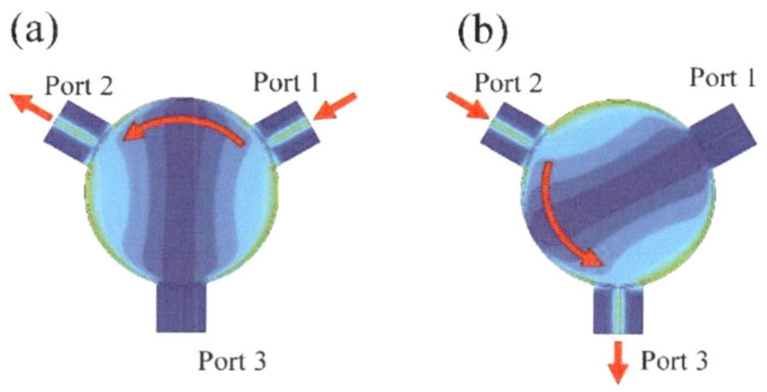

Figure 7.
Schematic diagrams of the operation of a stripline circulator. (a) The incident wave is injected from Port 1 and the transmitted wave ideally goes to Port 2. (b) The incident wave from Port 2 will go to Port 3. The color spectrum is the electric field pattern inside the ferrite disks. It is a three-port, nonreciprocal device. A full-wave solver, high-frequency structure simulator (HFSS), is used [18].

displayed in **Figure 6(a)** for an empty waveguide and a waveguide with two ferrite slabs. The phase difference $\Delta\phi$, shown in **Figure 6(b)**, is calculated under two conditions: with and without ferrites. The simulation parameters are the same as example 9.4 of Pozar's textbook. The saturation magnetization ($4\pi M_s$) is 1786 G, the dielectric constantε' is 13.0, and the resonance linewidth ΔH is 20 Oe. The length and thickness of the ferrite are L = 37.50 mm and t = 2.74 mm.

4.2 Circulator

Circulator, a nonreciprocal device, has been widely used in various microwave systems. **Figure 7** schematically shows the function of a stripline circulator. The circulator is, in general, a three-port device. If the incident wave is injected from Port 1, then the wave will ideally go to Port 2, while Port 3 will be isolated as shown in **Figure 7(a)**. On the other hand, if the wave is injected from Port 2, it will go to Port 3 and isolate from Port 1 as shown in **Figure 7(b)**. There are three figures of merit for a circulator: transmission, reflection, and isolation. The transmission from Port 1 to Port 2 should be as high as possible, i.e., the insertion loss should be as small as possible. The reflection received at Port 1 due to the incident wave of Port 1 (S_{11}) and the isolation from Port 1 to Port 3 (S_{13}) should be as small as possible. The nonreciprocity of the circulator can be used to protect the oscillators from the damage of the reflected power in plasma or material processing systems. It can also be used to separate the transmitted and the received waves in radar or communication systems [13–17].

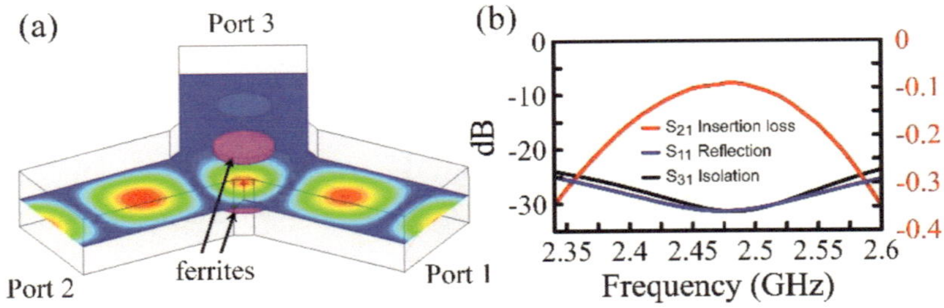

Figure 8.
(a) Schematic diagrams of the operation of a waveguide circulator. A full-wave solver, HFSS, is used with the saturation magnetization ($4\pi M_s$) is 1600 G, the dielectric constant ε' is 13.0, and the resonance linewidth ΔH is 10 Oe. The radius and thickness of the ferrite disks in rust red are R = 21.0 *mm and* t = 5 *mm, respectively. The waveguide is a standard WR 340 with 86.36 × 43.18 mm². The electric field pattern is displayed in color. (b) Simulation results of the waveguide circulator like the one in Part (a). The solid red curve is the transmission or insertion loss; the blue curve represents the reflection loss, and the black is the isolation.*

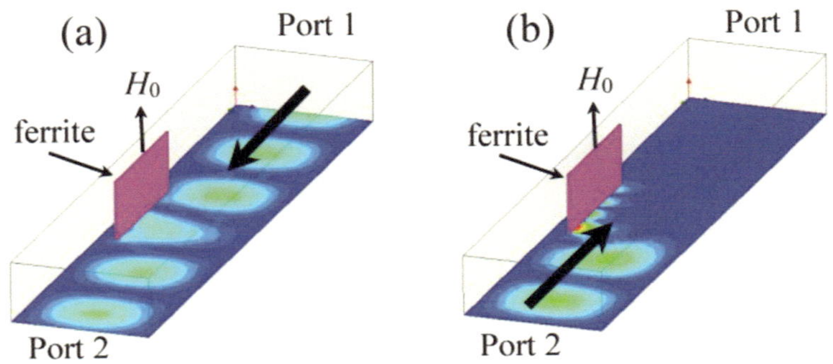

Figure 9.
The simulated field strength for a two-port isolator using the full-wave solver. (a) The high forward transmission (S_{21}) and (b) the low reverse transmission (S_{12}). The saturation magnetization ($4\pi M_s$) is 1700 G, the dielectric constant ε' is 13.0, and the resonance linewidth ΔH is 200 Oe. The length and thickness of the ferrite are L = 24.0 *mm and* t = 0.5 *mm.*

In addition to the stripline circulator, there are other types such as the microstrip circulator and the waveguide circulator. The microstrip circulator is similar to the stripline circulator in many ways. Here we show a waveguide circulator which is capable of high-power operation. **Figure 8(a)** shows the structure of the nonreciprocal device and the simulated electric field strength. The simulation parameters are described in the caption. The circulator is, in general, a three-port device. If the incident wave is injected from Port 1, then the wave will ideally go to Port 2, while Port 3 will be isolated as shown in **Figure 8(b)**. On the other hand, if the wave is injected from Port 2, it will go to Port 3 and be isolated from Port 1 as shown in **Figure 8(b)**.

4.3 Isolator

The isolator is one of the useful microwave ferrite components. As shown in **Figure 9**, the isolator is generally a two-port device having unidirectional transmission characteristics (nonreciprocity). From Port 1 to Port 2 (S_{21}), the forward transmission is high (i.e., low insertion loss in **Figure 9(a)**). However, from Port 2 to Port 1 (S_{12}), the reverse transmission is low (i.e., high isolation in **Figure 9(b)**). Besides, the reflection (S_{11} and S_{22}) should be as low as possible. The simulation parameters in **Figure 9** are the same as Ex. 9.2 of Pozar's textbook [5]. The simulation parameters and the sample's geometry are described in the caption.

An isolator is commonly used to prevent the high reflected power from damaging the precious and expansive microwave source. For example, the impedance of a plasma system changes a lot when the plasma is ignited. The radical change of the impedance will result in impedance mismatch and cause serious reflection which might kill the source instantly. An isolator can be used in place of a matching or tuning network. However, it should be realized that the reflected power will be absorbed by the ferrite of the isolator, as shown in **Figure 9(b)**. When the ferrite absorbs the reflected energy, the temperature will rise and the performance will be compromised. Therefore, a simple isolator can be implemented by using a circulator with one port well terminated [19]. For example, if Port 3 in **Figure 8(a)** is matched with water load, the power injected from Port 2 will go to Port 3 and will be isolated from Port 1. The power handling capability can be improved.

Circulator and isolator can be implemented using the self-bias [20–22], just like the latch phase shifter ($H_0 = 0$) shown in **Figure 6** using the remnant field B_r or M_r only. The self-biased ferrite devices will simplify the design and fabrication, but the overall performance is still not good enough. Further theoretical and experimental study is needed.

5. Conclusions

Ferrimagnetism and ferromagnetism share many magnetic properties in common, such as hard and soft magnets, but the conductivity differentiates these two materials. Ferrites are ceramic materials and suitable for the high-frequency operation. The electromagnetic properties of ferrite materials are difficult to understand in that the magnetic susceptibility is a tensor and depends on the saturated magnetization M_s, the internal bias field H_0, and the resonance linewidth ΔH. The magnetic susceptibility also depends on the frequency of the microwave ω as well as the polarization of the wave. The first two sessions explain the basic properties.

The complex permittivity $\varepsilon_r + i\varepsilon_i$, the saturation magnetization M_s, and the resonance linewidth ΔH are the most important electromagnetic properties of ferrites. How to measure the ferrite's properties are discussed in Section 3. The full-wave simulation is conducted to demonstrate how the phase shifter, circulator, and isolator work, which are shown in Section 4. Although the examples are discussed in many textbooks, Section 4 offers in-depth simulation results for the first time.

At high-power operation, the ferrite devices will be heated. The spin wave linewidth may be taken into account. Besides, the ferrites will become paramagnetism when the temperature exceeded the Curie temperature. These two factors are important for high-power operation, which are not considered in this chapter.

Acknowledgements

This chapter was supported in part by the Ministry of Science and Technology of Taiwan and in part by China Steel Company/HIMAG Magnetic Corporation, Taiwan. The author is grateful to the Taiwan Branch of ANSYS Inc. for technical assistance and to Dr. Hsein-Wen Chao and Mr. Wei-Chien Kao for their assistance in the full-wave simulation. Dr. Hsin-Yu Yao and Mr. Shih-Chieh Su are appreciated for the discussion of the ferrites' characterization.

Author details

Tsun-Hsu Chang
Department of Physics, National Tsing Hua University, Hsinchu, Taiwan

*Address all correspondence to: thschang@phys.nthu.edu.tw

References

[1] Okamoto A. The invention of ferrites and their contribution to the miniaturization of radios. IEEE Globecom Workshops. 2009:1-42. DOI: 10.1109/GLOCOMW.2009.5360693. ISBN 978-1-4244-5626-0

[2] Néel L. Magnetic Properties of Ferrites: Ferrimagnetism and Antiferromagnetism. Annales de Physique. 1948;**3**:137-198

[3] Chen CW. Magnetism and Metallurgy of Soft Magnetic Materials. Elsevier; 1977. 15–60 p. DOI: 10.1016/B978-0-7204-0706-8.X5001-1. ISBN: 9780444601193

[4] McGrayne SB, Suckling EE, Kashy E, Robinson FNH, Bleaney B. Magnetism Physics; 2018. Available from: https://www.britannica.com/science/magnetism/ [Accessed: 2018-12-20]

[5] Pozar DM. Chap. 9. In: Microwave Engineering. 4th ed. Hoboken, New Jersey: John Wiley & Sons, Inc.; 2011

[6] Collin RE. Chap. 6. In: Foundations for Microwave Engineering. 2nd ed. New York: McGraw Hill; 1992

[7] Chang TH. Gyromagnetically-induced transparency for ferrites. American Journal of Physics. 2016;**84**:279-283

[8] Cohn SB, Kelly KC. Microwave measurement of high-dielectric-constant materials. IEEE Transactions on Microwave Theory and Techniques. 1966;**14**:406-410

[9] Chen LF, Ong CK, Neo CP, Varadan VV, Varadan VK. Microwave Electronics: Measurement and Materials Characterization. Hoboken, New Jersey: John Wiley & Sons Inc.; 2004

[10] Chao HW, Wong WS, Chang TH. Characterizing the complex permittivity of high-κ dielectrics using enhanced field method. The Review of Scientific Instruments. 2015;**86**:114701

[11] Chao HW, Chang TH. Wide-range permittivity measurement with a parametric-dependent cavity. IEEE Transactions on Microwave Theory and Techniques. 2018;**66**:4641-4648

[12] Skyworks. Test for Line Width and Gyromagnetic Ratio [Internet]. 1999. Available from: http://www.skyworksinc.com/uploads/documents/Test_for_Line_Width_Gyromagnetic_Ratio_202837B.pdf [Accessed: 2018-12-20]

[13] Fuller AJB. Ferrites at Microwave Frequencies. London: Peter Peregrinus; 1987

[14] Pardavi-Horvath M. Microwave applications of soft ferrites. Journal of Magnetism and Magnetic Materials. 2000;**215**:171-183

[15] Schloemann E. Advances in ferrite microwave materials and devices. Journal of Magnetism and Magnetic Materials. 2000;**209**:15-20

[16] Harris VG, Geiler A, Chen Y, Yoon SD, Wu M, Yang A, et al. Recent advances in processing and applications of microwave ferrites. Journal of Magnetism and Magnetic Materials. 2009;**321**:2035-2047

[17] Helszajn J. The Stripline Circulators: Theory and Practice. Hoboken, New Jersey: John Wiley & Sons; 2008

[18] Chao HW, Wu SY, Chang TH. Bandwidth broadening for stripline circulator. The Review of Scientific Instruments. 2017;**88**:024706

[19] Linkhart DK. Microwave Circulator Design. 2nd ed. Norwood, MA: Artech House Microwave Library; 2014

[20] Wang J, Yang A, Chen Y, Chen Z, Geiler A, Gillette SM, et al. Self-biased Y-junction circulator at K_u-band. IEEE Microwave and Wireless Components Letters. 2011;**21**:292-294

[21] Peng B, Xu H, Li H, Zhang W, Wang Zhang YW. Self-biased microstrip junction circulator based on barium ferrite thin films for monolithic microwave integrated circuits. IEEE Transactions on Magnetics. 2011;**47**: 1674-1677

[22] Zuo X, How H, Somu S, Vittoria C. Self-biased circulator/isolator at millimeter wavelengths using magnetically oriented polycrystalline strontium M-type hexaferrite. IEEE Transactions on Magnetics. 2003;**39**: 3160-3162

Chapter 8

Electromagnetic Wave Absorption Properties of Core-Shell Ni-Based Composites

Biao Zhao and Rui Zhang

Abstract

Currently, high efficiency electromagnetic wave absorption plays an important role to keep away from the detection of aircraft by radar and reduce information leakage in various electronic equipment. Among the candidates of electromagnetic absorbers, ferromagnetic Ni materials possess high saturation magnetization and high permeability at high frequency (1–18 GHz), which is widely used to prepare thinner absorbing materials along with strong electromagnetic absorption properties. However, the metallic materials usually have relatively high electrical conductivity, and their permeability decreases rapidly at high frequency thanks to the eddy current losses, which is generally named as skin-depth effect. To address this issue, one effective way is to design core-shell structured Ni based composites combining magnetic cores with dielectric shells. This chapter focuses on the state-of-the-art of the microwave absorption properties of Ni-based core-shell composites, and the related electromagnetic attenuation theory about how to enhance absorption properties is also discussed in detail.

Keywords: ferromagnetic Ni, core-shell structure, dielectric loss, magnetic loss, impedance match

1. Introduction

The discovery of electromagnetic (EM) waves boosted the development of transmission technology. It is well accepted that the various electromagnetic waves are widely applied in numerous areas and make our daily life convenient [1]. Extensive electronics devices, such as mobile phones, WiFi, Near Field Communication (NFC), and wireless charging, are developing and play an important role in modern life [2, 3]. With the extensive practical applications of electronic devices and densely packed systems, electromagnetic interference (EMI) becomes a more and more serious problem, which would lead to pernicious impacts on equipment performance, human health, as well as surrounding environment [4]. Furthermore, our individual mini device produces unwanted EM waves, which would influence other nearby devices [5]. Moreover, the global need for some EM waves, such as for military radar stealth, is also boosting, which produces plentiful concerns for human health. Therefore, the protection of electromagnetic radiation has been widely concerned by the whole society. Recently, development of high performance microwave absorbing materials (MAMs) have attracted great interests to eliminate the electromagnetic pollution [6, 7].

Recently, considerable research attention has been focused on core-shell structure for innovative electromagnetic absorption due to the potential to combine the individual properties of each component or achieve enhanced performances through cooperation between the components [8–10]. Liu and co-workers have fabricated core-shell structured Fe_3O_4@TiO_2 with Fe_3O_4 as cores and hierarchical TiO_2 as shells and the Fe_3O_4@TiO_2 core-shell composites displayed the enhanced microwave absorption properties than pure Fe_3O_4 [11]. Chen and co-workers have successfully synthesized core/shell Fe_3O_4/TiO_2 composite nanotubes with superior microwave attenuation properties [12]. Combining Fe_3O_4 and TiO_2 can take advantage of both the unique magnetic properties of Fe_3O_4 and strong dielectric characteristics of TiO_2 as well as core-shell structure, and therefore offer an avenue to achieve excellent microwave-absorption performance. In this kind of core/shell configurations, the magnetic materials regarding as cores, which could improve the permeability of the composites, is conductive to the enhancement of the magnetic loss. The dielectric materials considering as shells, which are supposed to play the roles not only as a center of polarization but also as an insulating medium between the magnetic particles, lead to the increased dielectric loss and good impedance match. The high-efficiency microwave absorption properties resulted from the enhanced magnetic loss, dielectric loss, reduced eddy current loss and impedance match [8]. Thus, the traditional microwave absorbing materials holding a core-shell structure may improve their microwave absorption capabilities.

It is well known Ni is regarded as a typical magnetic metal material, which are supposed to have numerous applications in many fields such as magnetic recording devices, clinical medicine, catalysis and so on [13, 14]. It is worth pointing that Ni was also proved to be as a competitive candidate for high-efficiency electromagnetic absorption materials to address the electromagnetic interference and pollution problems because Ni can provide more beneficial features, such as high saturation magnetization, distinguishable permeability, and compatible dielectric loss ability in the gigahertz range when compared with those nonmagnetic EM absorption materials. However, single-component electromagnetic absorption materials easily suffer from mismatched characteristic impedance and poor microwave absorption performance. Moreover, Ni would generate eddy current induced by microwave in GHz range because of high conductivity. The eddy current effect may cause impedance mismatching between the absorbing materials and air space, which would make microwave reflection rather than absorption. This issue is a challenge to handle for scientists. Thus, for the sake of getting superior microwave absorption ability, a promising pathway is to compound the Ni products with an inorganic or nonmagnetic constituent to produce a core@shell configuration. Numerous literatures have been carried out to cover the magnetic Ni with inorganic or nonmagnetic shells. For example, Ni/SnO_2 core-shell composite [15], carbon-coated Ni [16], Ni/ZnO [17], Al/AlOx-coated Ni [18], Ni@Ni_2O_3 core-shell particles [19], Ni/polyaniline [20], and CuO/Cu_2O-coated Ni [21] show the better microwave absorption performance than the pure core or shell materials. Thus, the EM wave absorption abilities of Ni particles can be obviously enhanced after coating inorganic and nonmagnetic shells.

Herein, we report the microwave absorption properties of core-shell structured Ni based composited and discuss how does core-shell ameliorate the electromagnetic wave absorption properties and also investigate the related electromagnetic attenuation theory in detail.

2. Core-shell Ni@oxide composite as microwave absorbers

2.1 Core-shell Ni@ZnO composites as microwave absorbers

For the ZnO nanostructural materials, due to the features of lightweight semiconductive properties and its easily mass synthesis, they are expected to be the potential applications in EM wave absorbing materials [22]. Therefore, when the Ni particles were covered by ZnO shell, the electromagnetic properties of Ni would be boosted, correspondingly. Moreover, it is well accepted that the EM absorption properties are closely related with their morphologies. Herein, the various morphologies of Ni/ZnO composites were synthesized by control of the amounts of $NH_3{\cdot}H_2O$, and the microwave absorption properties of these Ni/ZnO composites have been investigated based on the complex permittivity and permeability.

2.1.1 Preparation of core-shell Ni@ZnO composites

Ni microspheres were prepared based on our previous publication [15]. Ni/ZnO composites were synthesized through a facile hydrothermal method [23]. Typically, 0.05 g of the as-obtained Ni microspheres was distributed in 60 mL distilled water. Then 0.45 g of $Zn(CH_3COO)_2{\cdot}2H_2O$ and a certain amounts of ammonia solution were added into the mixture solution. The mixture was transferred into a Teflon-lined stainless steel autoclave, and maintained at 120°C for 15 h. The precipitates were collected by centrifugation, washed several times with distilled water and absolute ethanol, respectively. For the convenience of discussion, the Ni/ZnO prepared at 1 mL $NH_3{\cdot}H_2O$, 2 mL $NH_3{\cdot}H_2O$ and 3 mL $NH_3{\cdot}H_2O$ were denoted as SA, SB and SC, respectively.

Figure 1a displays the representative FESEM image of the Ni particles, which possesses uniformly spherical shape and the diameter is about 1.0–1.2 μm. The SEM images of the Ni/ZnO obtained at different contents of $NH_3{\cdot}H_2O$ are displayed in **Figure 1b–d**. **Figure 1b** exhibits that the as-prepared Ni/ZnO product is composed of plentiful ZnO polyhedrons with the diameter of 0.2–0.5 μm covered on the surface of Ni particles to generate special core-shell structure if small amount of $NH_3{\cdot}H_2O$ (1 mL) was added. If the content of $NH_3{\cdot}H_2O$ is lifted to 2.0 mL, the football-like Ni/ZnO samples with the size of 4–5 μm could be observed (**Figure 1c**). One can infer that the thickness of ZnO polyhedron is about 2-3 μm, which is larger than that of Ni microsphere. Therefore, the Ni microspheres are completely coated by ZnO, thus, we could not see the existence if individual Ni microspheres. When the content of $NH_3{\cdot}H_2O$ is further improved to 3 mL, ZnO rods and Ni microsphere coexist in the final products (**Figure 1d**), separately. These results indicate that the morphology of Ni/ZnO composite can be effectively adjusted by controlling the $NH_3{\cdot}H_2O$ content.

Figure 2 depicts the schematic diagram of the generation for various shapes of Ni/ZnO composite. First, the distributed Ni microspheres are fabricated through a chemical reduction method. Following, the different shapes of Ni/ZnO composites are fabricated by the addition of various $NH_3{\cdot}H_2O$ contents. The ZnO nuclei is prone to plant along special crystal planes and finally generate polyhedron-like or rod-like ZnO products. It is accepted that ZnO is a polar crystal with a polar c-axis ([0001] direction) [24]. In the solution system, the $NH_3{\cdot}H_2O$ consists of the positive hydrophilic group (NH_4^+) and negative hydrophobic group (OH^-). The positive hydrophilic groups would link with the basic cells of crystalline growth $[Zn(OH)_4]^{2-}$ easily by Coulomb force, which means that the positive hydrophilic groups turn into the carriers of $[Zn(OH)_4]^{2-}$; on the other hand, due to existence

Figure 1.
(a) SEM image of pure Ni microspheres and (b–d) SEM images of as-prepared Ni/ZnO samples prepared at various concentration of $NH_3 \cdot H_2O$: (b) 1 mL, (c) 2 mL, and (d) 3 mL [23] (permission from Elsevier).

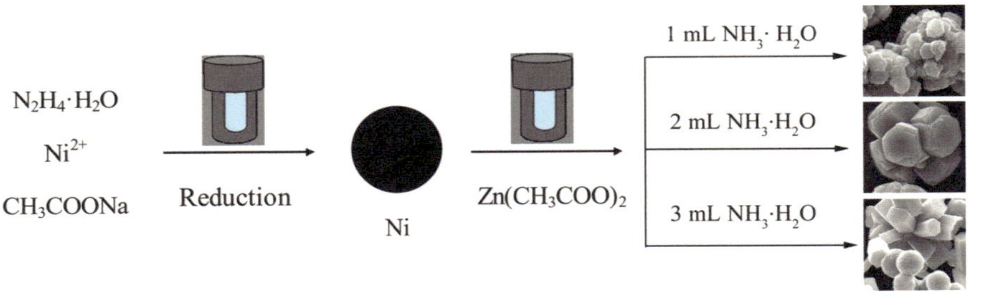

Figure 2.
Schematic illustration of formation of various morphologies of Ni/ZnO composites [23] (permission from Elsevier).

of van der Waals force, the connection between the negative hydrophobic groups and the non-polar lateral surfaces of ZnO would take place, indicating that the occurrence of hydrophobic films thanks to the negative hydrophobic groups on the non-polar lateral surfaces [25]. The basic cells of growing $[Zn(OH)_4]^{2-}$ attracted by the positive hydrophilic groups would move to the polar axial surface (0001) easily to integrate together and remain at the suitable lattice locations but difficultly reach the non-polar lateral remains due to the presence of the hydrophobic films. It indicates that the positive polar surface (0001) grows quicker than that of non-polar lateral surfaces in a fixed content of $NH_3 \cdot H_2O$. As a result, more OH^- ions assimilate and hamper the growth on the positively charged (0001) surface, forcing a shape transition [26]. At the high content of $NH_3 \cdot H_2O$, long ZnO nanorods could be produced because of the fast growth rate along the [0001] direction [27].

2.1.2 EM properties of core-shell Ni@ZnO composites

To reveal the electromagnetic wave absorption properties of SA, SB and SC paraffin composites, the reflection loss (RL) values of the Ni/ZnO samples are calculated based on following equations [28]:

$$RL = 20\log_{10}|(Z_{in} - Z_0)/(Z_{in} + Z_0)| \quad (1)$$

$$Z_{in} = Z_0\sqrt{\frac{\mu_r}{\varepsilon_r}}\tanh\left(j\frac{2\pi fd\sqrt{\mu_r\varepsilon_r}}{c}\right) \quad (2)$$

Herein Z_0 is the impedance of free space, Z_{in} is the input impedance of the material, f is the frequency of the microwave, c is the velocity of microwave in free space, μ_r and ε_r are, respectively, the relative complex permeability and permittivity, and d is the thickness of the absorber. The RL values of the three samples with a thickness of 2.0 mm are displayed in **Figure 3a**. The SA sample holds the outstanding EM wave absorption performances. A strong peak (−48.6 dB) could be seen at 13.4 GHz. The RL less than −10 dB (90% absorption) reaches 6.0 GHz (10.5–16.5 GHz). Furthermore, the RL less than −20 dB (99% microwave dissipation) is also obtained in the range of 11.5–14.2 GHz. But, for the SB and SC samples, they present inferior microwave dissipation capabilities. **Figure 3b** depicts the simulated RL of SA paraffin-composite with various thicknesses in the frequency of 1–18 GHz. Clearly, one can notice that the optimal RL shifts into lower frequency range along with an increased thickness, indicating that we could adjust the absorption bandwidth by tuning absorber thickness. From above analysis, one can note that the minimal RL of −48.6 dB could be observed at 13.4 GHz with a layer thickness of 2.0 mm. The effective absorption (below −10 dB) bandwidth could be monitored in the frequency of 9.0–18.0 GHz by control of the absorber thickness between 1.5 mm and 2.5 mm. Furthermore, the frequency with RL below −20 dB could be observed at 11.1–16.2 GHz with thickness of 1.8–2.2 mm. For the SA sample, the enhanced microwave absorption properties are stemmed from the good impedance match, synergistic effect between dielectric loss and magnetic loss, and special core-shell microstructures, which could induce the interference of microwave multiple reflection [29]. In addition, the compact polyhedron ZnO coating brings the metal/dielectric interfaces, in which the interface polarization boosts the microwave dissipation. For the football-like Ni/ZnO (SB), the size of ZnO is so big that the Ni microspheres could not interact with incident microwave, which

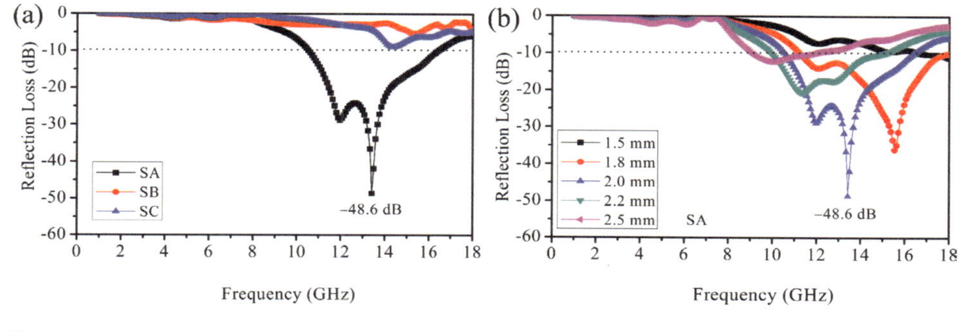

Figure 3.
(a) Frequency dependences of RL with the thickness of 2.0 mm for the three samples SA, SB, SC; (b) RL of Ni/ZnO (SA) paraffin composite of varying thicknesses [23] (permission from Elsevier).

gives rise to the mismatch between the magnetic loss and dielectric loss, leading to inferior microwave absorption. For the SC, due to the presence of uncover Ni, these uncoated Ni microspheres play a negative in the wave-absorption of materials thanks to the occurrence of a significant skin effect when its surface is irradiated by microwaves [30].

On the basis of transmission line theory, the suitable microwave absorption properties are determined by two key factors. One factor is the impedance match, which need the complex permittivity is close to the complex permeability, and the other one is the EM attenuation ability, which dissipates the microwave energy through dielectric loss or magnetic loss. The EM attenuation was determined by the attenuation constant α, which can be expressed as [31, 32]:

$$\alpha = \frac{\sqrt{2}\pi f}{c} \times \sqrt{\left(\mu''\varepsilon'' - \mu'\varepsilon'\right) + \sqrt{\left(\mu''\varepsilon'' - \mu'\varepsilon'\right)^2 + \left(\mu'\varepsilon'' + \mu''\varepsilon'\right)^2}} \quad (3)$$

where f is the frequency of the microwave and c is the velocity of light. **Figure 4** displays the frequency dependence of the attenuation constant. The SA possesses the biggest α in all measured frequency ranges, meaning the outstanding attenuation. Moreover, based on the above equation, one can notice that the attenuation constant is closely related to the values of ε″ and μ″. The highest ε″ and μ″ values are the responsible for the highest α in Ni@ZnO (SA) core/shell structures, which is related to interface polarization and relaxation. As a result, the enhancement of the microwave absorption properties for the dielectric coating originates from the improvement of dielectric loss and magnetic loss.

2.2 Core-shell Ni@CuO composites as microwave absorbers

Nowadays, CuO is well accepted as an important p-type semiconductor, which holds the unique features of narrow band gap (E_g = 1.2 eV), and has captured more and more interests. This material has proved to exhibit widespread potential applications in optical switches, anode materials, field emitters, catalyst, gas sensors, photoelectrode and high-temperature micro-conductors [33, 34]. Recently, CuO has been realized as an efficient material for the preparation of microwave absorbing materials [21, 35, 36]. Herein, we fabricated the core-shell structural composites with Ni cores and rice-like CuO shells via a facile method. The microwave absorption properties of Ni, CuO and Ni/CuO composites are studied in term of complex

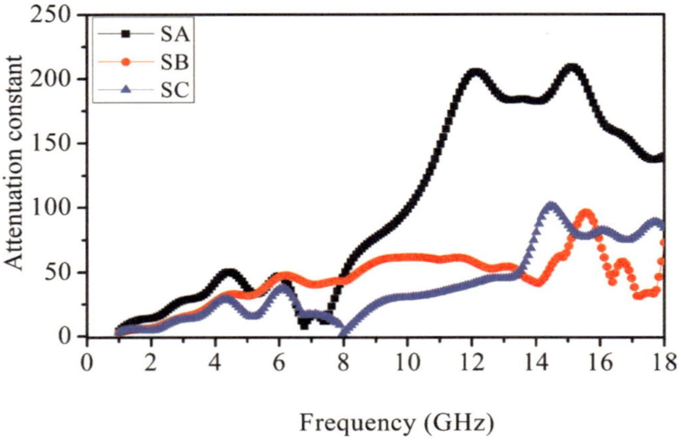

Figure 4.
Attenuation constant of Ni/ZnO samples-paraffin composites versus frequency [23] (permission from Elsevier).

permittivity and permeability. In comparison with pristine Ni and CuO, rice-like CuO-coated Ni composites displayed the enhanced microwave absorption properties. Furthermore, we also studied the effects of CuO amounts on microstructures and microwave absorption properties of Ni/CuO composites in detail.

2.2.1 Preparation of core-shell Ni@CuO composites

The Ni microspheres were prepared by a solvothermal method, which was described in our pervious literature [15]. Synthesis of CuO nanoflakes: $CuCl_2{\cdot}2H_2O$ (0.36 g) was dissolved in a mixture of distilled water (60 mL) and ammonia (2 mL) under continuously stirring (30 min); The final mixture was transferred into a Teflon-lined autoclave and heated hydrothermally at 150°C for 15 h.

Synthesis of CuO rice-coated Ni core/shell composites [37]: the as-prepared Ni microspheres (0.05 g) and $CuCl_2{\cdot}2H_2O$ (0.36 g) were both added in distilled water (60 mL). Then, the ammonia (2 mL) was introduced into the mixture. Finally, the prepared mixture was moved into a Teflon-lined autoclave. The Teflon-lined autoclave was sealed and kept at 150°C for 15 h. The Ni/CuO composites prepared at 0.18 g $CuCl_2{\cdot}2H_2O$, 0.36 g $CuCl_2{\cdot}2H_2O$ and 0.54 g $CuCl_2{\cdot}2H_2O$ were denoted as S-1, S-2 and S-2, respectively.

Figure 5c, d exhibits FESEM micrographs of Ni@CuO composites with different magnification after hydrothermal treatment at 150°C for 15 h. It can be observed that the products are composed of CuO rices-coated smooth Ni microspheres heterostructures with the diameter of 1.0–1.2 μm. One can notice that rice-like CuO/Ni composites hold rough surfaces, which results from compactly aggregated panicle-shape CuO nanostructures. In order to get more information about microstructure of Ni/CuO composite, TEM and HR-TEM images of Ni/CuO composites are carried out. The core-shell structure of Ni/CuO composite can be

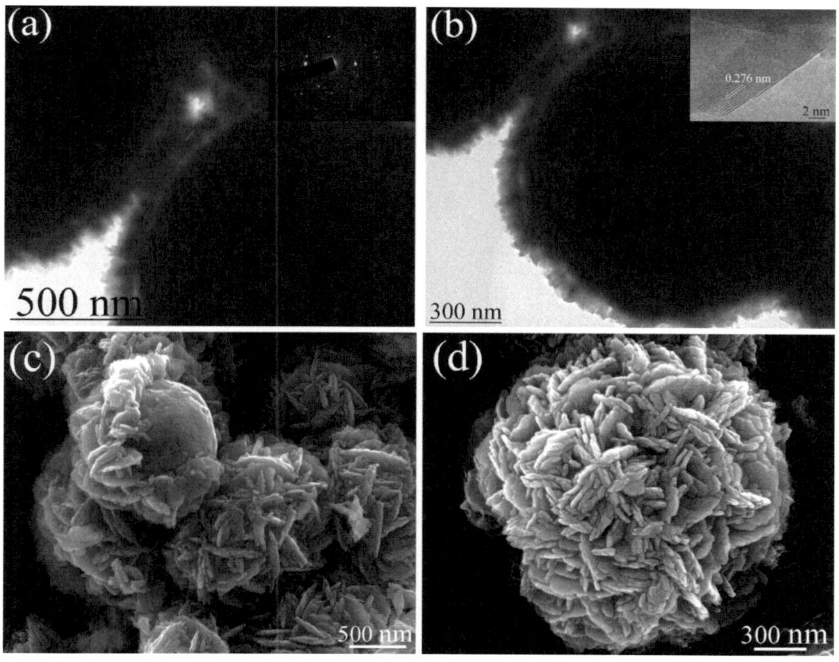

Figure 5.
Typical TEM and HRTEM images of the as-prepared Ni/CuO structures: (a) low magnification, inset of (a) shows the SAED pattern; (b) high magnification; inset of (b) shows the HRTEM. (c, d) FESEM images of the Ni microsphere-CuO rice core-shell structures [37] (permission from RSC).

clearly observed from **Figure 5a, b**. The inset SAED pattern of the CuO particles indicated that CuO particles are polycrystalline (**Figure 5a**). The HRTEM image (inset of **Figure 5b**) displays that the lattice spacing is 0.276 nm, which is in good agreement with the (110) lattice spacing of CuO. Based on the SEM and TEM results, it can be concluded that the CuO is deposited on the surface of Ni, the core-shell composites are obtained under this procedure.

Figure 6 exhibits the morphologies of the obtained products with different molar ratio of the $CuCl_2{\cdot}2H_2O$ to Ni microspheres. Noticeably, the surfaces of all samples turns coarser in comparison with the pure Ni microspheres, which indicates the successful coating of the CuO nanoparticles on the pristine Ni surfaces. Furthermore, the shape and coverage density of CuO materials could be controlled by tuning the content of precursor (Cu^{2+}). When the molar ratio of the $CuCl_2{\cdot}2H_2O$ to Ni microspheres in the precursor solution is 1: 0.85 (S-1), one can find (**Figure 6a, b**) that the Ni microspheres are coated by a large number of CuO nanorices. But, due to the low content of precursor (Cu^{2+}), we just could obtain thin CuO shell. If the

Figure 6.
FESEM images of hierarchical Ni/CuO core-shell heterostructures with different molar ratio: (a, b) S-1; (c, d) S-2; and (e, f) S-3 [37] (permission from RSC).

molar ratio of precursor is enhanced to 2: 0.85 (S-2), one can see that the aggregation occurs and CuO nanorices are compactly covered on the smooth surfaces of Ni microspheres to produce coarser thick CuO shells (**Figure 6c, d**). If the molar ratio is improved continuously to 3:0.85 (S-3), a thick layer of compact CuO nanoflakes coated on Ni microspheres could be observed (**Figure 6e, f**). Based on above results, the microstructures and coverage density of CuO shells can be effectively monitored by selecting a suitable content of Cu^{2+}.

2.2.2 EM properties of core-shell Ni@CuO composites

To compare and assess the EM wave absorption properties of Ni, Ni/CuO core-shell composites, and CuO nanoflakes, the paraffin (30 wt%, which is transparent to microwave) are mixed with as-obtained products, and pressed into a ring shape with an outer diameter of 7.00 mm and an inner diameter of 3.04 mm. The microwave absorption abilities of these as-fabricated products could be evaluated by the RL values, which could be simulated on the basis of the complex permeability and permittivity with the measured frequency and given layer thickness [38, 39]. As presented in **Figure 7a**, the three Ni/CuO composites show the superior microwave-absorption properties to those of the pure Ni microspheres and the CuO nanoflakes. Taking an example, when the thickness is 2 mm, the S-1 sample exhibits the enhanced EM-wave absorption with the minimal RL value of −15.6 dB at 11.9 GHz among the five samples. From Eqs. (1) and (2), one can find that the thickness of the absorber is one important factor, which would affect the position of minimal RL value and the absorption bandwidth. Therefore, the RL values of Ni/CuO samples with different thicknesses are also calculated. Compared with S-2 (**Figure 7c**) and S-3 (**Figure 7d**) samples, the S-1 (**Figure 7b**) displays the outstanding microwave absorption performances. The lowest RL of the S-1 sample is −62.2 dB at 13.8 GHz with the only thickness of 1.7 mm. The effective absorption (below −10 dB) bandwidth can be tuned between 6.4 GHz and 18.0 GHz by adjusting thickness in 1.3–3.0 mm.

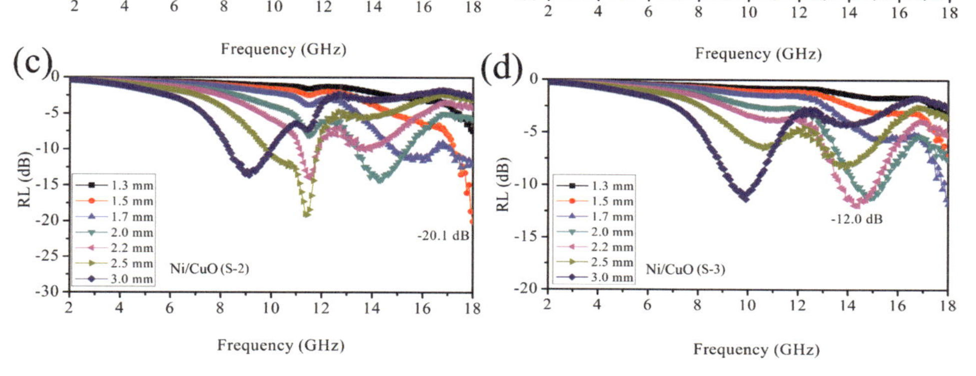

Figure 7.
(a) Comparison of RL of the five as-obtained samples with a thickness of 2.0 mm. The RL values of (b) S-1, (c) S-2, and (d) S-3 samples with various thicknesses [37] (permission from RSC).

Notably, the optimal RL peaks gradually shift toward lower frequencies with an increased absorber thickness, which can be described by quarter-wavelength cancelation model that the incident and reflected waves in the absorber are out of phase 180° and causing the reflected waves in the air-absorber interface are totally cancelled [40]. The enhanced microwave absorption property of core-shell Ni/CuO composites can be obtained by tuning the content of CuO. The rice-like CuO shell is expected to be helpful for the dissipation proprieties of the core/shell composites. The CuO shells are covered on the surface of Ni microspheres to produce the special core-shell structure, which brings metal-dielectric hetero-interfaces to cause interfacial polarization. It is supposed that the interfacial polarization taken place in heterostructures consisting of at least two constituents [17, 41, 42]. This type of polarization occurring at the interfaces results from the movement of charge carriers between different compositions, which accumulate the moving charge at these interfaces. When irradiated by alternating EM fields, the accumulated charge would redistribute periodically between Ni cores and CuO shells, which are favorable for the microwave dissipation. However, for the S-2 and S-3 samples, thanks to the high content of CuO, we cannot observe the synergistic effect between Ni cores and CuO shells, which gives rise to inferior microwave absorption.

3. Core-shell Ni@non-oxide composite as microwave absorbers

3.1 Ni@ZnS composites as microwave absorbers

It is well known that ZnS, a wide band-gap semiconductor with the band-gap energy (Eg) of 3.6 eV, has been widely applied in displays, sensors, and lasers for many years [43, 44]. As far as I am concerning, the publications about the microwave absorption properties of ZnS and core/shell structured Ni/ZnS are not reported. Herein, we synthesized the core-shell structured composites with Ni cores and ZnS nanowall shells through a facile method. The microwave absorption properties of Ni, ZnS and Ni@ZnS composites are detailedly investigated in the frequency of 2–18 GHz.

3.1.1 Preparation of core-shell Ni@ZnS composites

ZnS nanowall-coated Ni composites were fabricated via a two-stage method [45]. First, Ni microspheres were synthesized based on our previous paper [15]. Second, Ni microspheres are coated by ZnS nanowalls to generate the core-shell structural composites. In the modified procedure, the as-obtained Ni microspheres (0.05 g) and $Zn(CH_3COO)_2{\cdot}2H_2O$ (0.45 g) were added in a mixture solution of ethanol (30 mL) and distilled water (30 mL). Then, $Na_2S{\cdot}9H_2O$ (0.50 g) and ammonia solution (4 mL) were added into the mixture solution with intensely stirring for 20 min. Finally, the mixture was moved into a Teflon-lined stainless steel autoclave, and kept at 100°C for 15 h. In order to study the effect of core-shell structure on the microwave absorption properties of the Ni/ZnS composite, the pure ZnS particles were also prepared according to the above procedure without addition of Ni microspheres.

Inset of **Figure 8a** presents the XRD profiles of Ni microspheres, ZnS particles and Ni@ZnS composites. For the Ni microspheres, all the diffraction peaks can be well indexed to the face-centered cubic (fcc) structure of nickel (JCPDS No. 04-0850). For the ZnS particles, all diffraction peaks can be indexed to a typical zinc blende structured ZnS, which is consistent with the standard value for bulk ZnS (JCPDS Card No. 05-0566). The crystal structure of core/shell structured Ni/ZnS

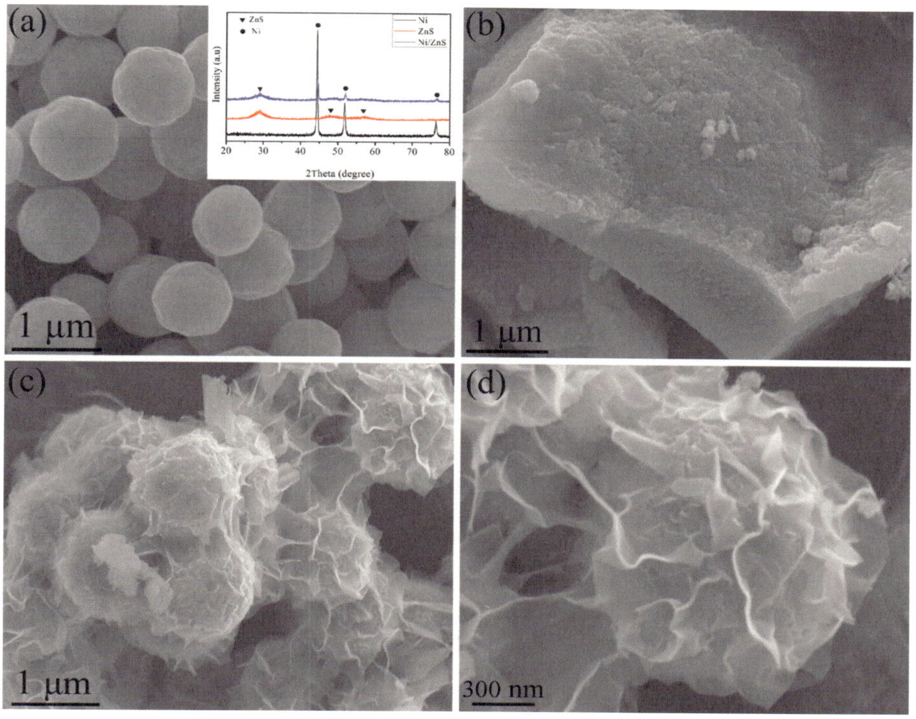

Figure 8.
SEM images of (a) Ni microspheres, (b) ZnS particles, and (c, d) the as-prepared Ni/ZnS composites. The inset in (b) is the magnified SEM image of ZnS particles. Inset in Figure 8a is the XRD patterns of Ni microspheres, ZnS particles and Ni/ZnS composites [45] (permission from RSC).

composites is also investigated by XRD measurements. Noticeably, we expectantly observed the diffraction peaks, which are in good accordance with Ni and ZnS, respectively. One can conclude that the as-obtained core/shell structural composites are made up of crystalline Ni and ZnS. **Figure 8a** presents the SEM image of the Ni microspheres. One can notice that the products have a relatively uniform spherical shape with the diameter of 0.7–1.0 μm. The pure ZnS products appear to have irregular shapes (**Figure 8b**). In raw ZnS particles, the formation of ZnS is via a two-step pathway. The fresh nanoparticles incline to aggregate for the sake of decreasing the surface energy. Therefore, we could obtain the irregular gathering ZnS particles. Whereas, as for the Ni/ZnS system, ZnS particles were generated via the template way (heterogeneous nucleation, raw Ni as template). Therefore, the variation of the shape and dimensions of ZnS particles in pure ZnS and core-shell Ni/ZnS composites could be seen. **Figure 8c, d** present the SEM images of core-shell Ni/ZnS. In comparison with pure Ni (**Figure 8a**), one significant distinction is clearly observed between the Ni/ZnS composites and the pure Ni particles. The distinction is that the Ni particles are absolutely wrapped by the ZnS nanowalls. The large-scale SEM image in **Figure 8d** suggests that the as-prepared Ni/ZnS composites show crinkled and rough textures, which are similar with the reduced graphene oxide sheets [46]. The thickness of ZnS nanowall is about 10 nm.

3.1.2 EM properties of core-shell Ni@ZnS composites

The relative complex permittivity (ε' and ε'') and permeability (μ' and μ'') of the Ni/paraffin, ZnS/paraffin and Ni@ZnS/paraffin composite samples are measured over a frequency of 2–18 GHz. **Figure 9a–c** manifest the real part (ε') and imaginary

part (ε″) of the complex permittivity as a function of frequency. The ε′ of Ni/paraffin composite shows a gradual decrease with frequency (**Figure 9a**).
However, the ε″ values are relative constant without significant change over the 2–18 GHz. The ε′ and ε″ of ZnS/paraffin composite presents constant value (4.5 and 0.5, respectively) in **Figure 9b**. The ε′ of the Ni/ZnS composite firstly reduces in the frequency of 2–15 GHz and then improves with increasing frequency (**Figure 9c**). Nevertheless, the ε″ exhibits the opposite tendency in the frequency of 2–18 GHz. One can note that the ε″ values of Ni/ZnS composite presents a peak in the 13–15 GHz range, which is originated from the natural resonance behavior of core-shell microstructure [21, 47]. Furthermore, it can be found that the ε″ values of Ni/paraffin composite are larger than those of ZnS/paraffin composite and Ni@ZnS/paraffin composite. Based on the free electron theory [17], $\varepsilon'' \approx 1/2\pi\varepsilon_0 \rho f$, where ρ is the resistivity. The lower ε″ values of ZnS/paraffin composite and Ni@ZnS/paraffin composite indicate the higher electric resistivity. In general, a high electrical resistivity is favorable for improving the microwave absorption abilities [48].

Figure 9 (d–f) present the curves of the real part (μ′) and imaginary part (μ″) of the complex permeability as a function of frequency for the Ni/paraffin, ZnS/paraffin and Ni@ZnS/paraffin composites. The μ′ and μ″ of Ni/paraffin composite are 0.81–1.59 and 0.05–0.51, respectively (**Figure 9d**). Compared with the complex permittivity (**Figure 9a**), the values of complex permeability is relatively small, which lead to mismatch impedance. The impedance match is required that complex

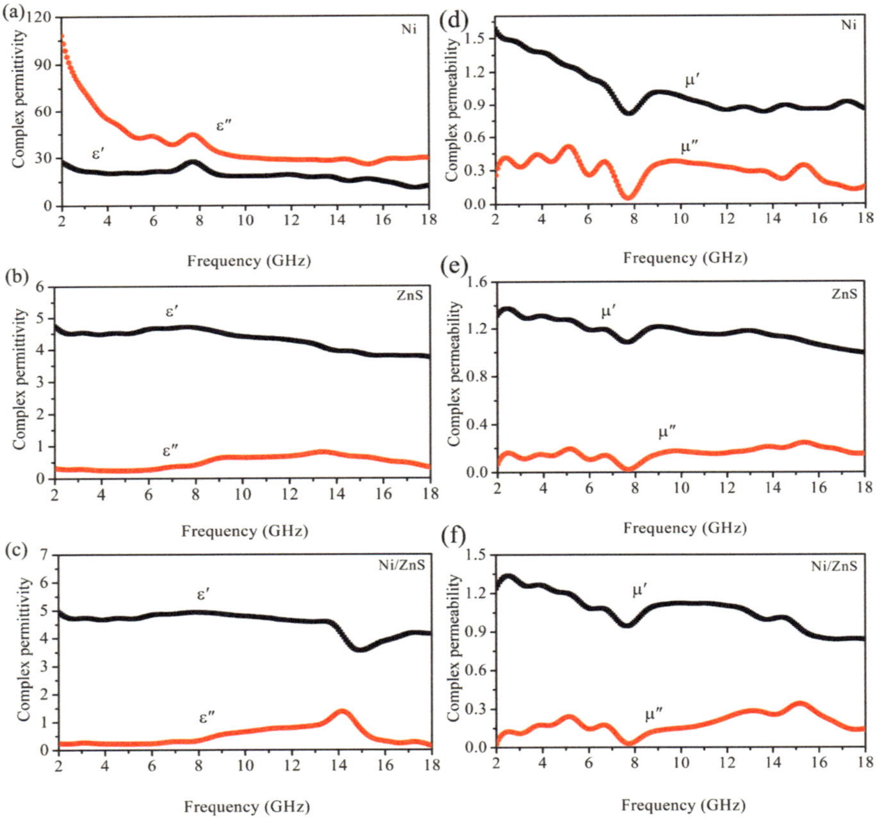

Figure 9.
Frequency dependence of the complex permittivity ($\varepsilon_r = \varepsilon' - j\varepsilon''$) of (a) Ni microspheres, (b) ZnS particles, and (c) Ni/ZnS composites; frequency dependence of the complex permeability ($\mu_r = \mu' - j\mu''$) of (d) Ni microspheres, (e) ZnS particles, and (f) Ni/ZnS composites [45] (permission from RSC).

permittivity is close to the permeability, which can make microwaves enter into the materials as much as possible [49]. The higher permittivity of the absorber plays a negative role in the impedance match [50], thus gives rise to inferior microwave absorption. From the **Figure 9e**, **f**, it can be found that the complex permeability of ZnS/paraffin and Ni@ZnS/paraffin composite exhibit the similar tendency with an increased frequency. The μ′ values of 0.99–1.38 and 0.84–1.34 could be observed in ZnS/paraffin and Ni@ZnS/paraffin composites, respectively. The μ″ values are in the range of 0.02–0.24 and 0.03–0.34 for the ZnS/paraffin and Ni@ZnS/paraffin composites, respectively. Given the complex permittivity (**Figure 9b, c**), it can be found that the relation between permittivity and permeability is prone to be close (good impedance match). The good impedance match is beneficial for the microwave absorption. On the basis of the above results, one can deduce that the impedance match of ZnS/paraffin and Ni@ZnS/paraffin is superior to that of Ni/paraffin composite. Thus, the ZnS/paraffin and Ni@ZnS/paraffin composites may possess better dissipation abilities of microwave energy.

It is widely accepted that the RL values could be utilized to evaluate the microwave absorption abilities of EM materials. **Figure 10a** exhibits the calculated RL values of Ni, ZnS and Ni/ZnS paraffin composites with 70 wt% amounts at the thickness of 2.5 mm in the frequency range of 2–18 GHz. Because a paraffin matrix is transparent to microwaves, these results are generally considered as the wave-absorption abilities of the filler itself. It is noting that the microwave absorption properties of Ni particles are weak and the optimal RL value is only −3.04 dB at 5.28 GHz, which is due to the mismatch impedance. Another possible factor is that the skin effect could be observed in Ni microspheres, which is harmful for microwave absorption [30]. Compared with Ni particles, ZnS particles and ZnS nanowall-coated Ni composite presents the superior microwave absorption abilities, which stems from good impedance match. It is worth pointing that, for Ni@ZnS composite, the minimal RL of −20.16 dB is observed at 13.92 GHz and RL values below −10 dB are seen in the 12–16.48 GHz rang. **Figure 10b** displays the relationship between RL and frequency for the paraffin wax composites with 70 wt% Ni/ZnS in various thicknesses. The optimal RL is −25.78 dB at 14.24 GHz with the corresponding thickness of 2.7 mm. The effective absorption (less than −10 dB) bandwidth reaches 4.72 GHz (11.52–16.24 GHz). Interestingly, with increasing the sample thickness, the location of minimal absorption peaks almost keeps the same at various thicknesses without moving to lower frequency, which has also been recorded by other' groups [51]. The location of absorption peaks is in accordance with the natural resonance, which means the resonance behavior in permittivity influences the microwave absorption.

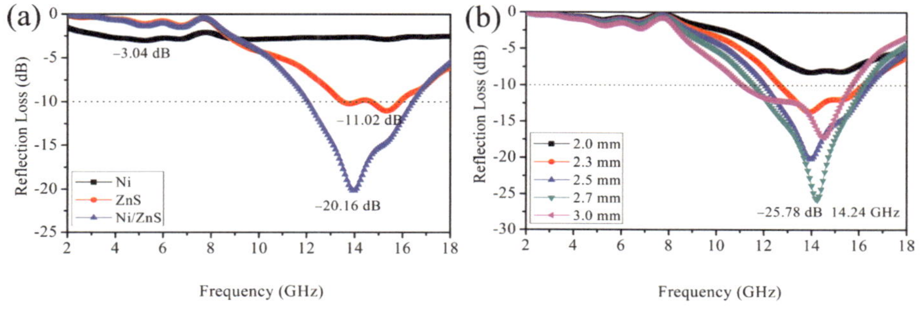

Figure 10.
(a) RL curves of Ni, ZnS and Ni/ZnS paraffin composite with 70 wt% loadings at the thickness of 2.5 mm; (b) RL curves of 70 wt% Ni/ZnS wax-composite at various thicknesses in the frequency of 2–18 GHz [45] (permission from RSC).

3.2 Urchin-like ZnS/Ni_3S_2@Ni composites as microwave absorbers

The EM absorption properties of nanomaterial are associated with the size, shape, and dimensionality. Metal sulfide nanomaterials have captured more and more attention due to their excellent properties and wide applications in electronic and optoelectronic devices. Nickel subsulfide and ZnS are the important categories in the metal sulfide family [52, 53] thanks to their various applications such as in lithium ion batteries, supercapacitors, dye-sensitized [54, 55] and charge transfer, anion exchange, electricity generation [56, 57], respectively. Due to the semiconductor properties of nickel subsulfide and ZnS, it can induce dipole and space charge polarizations when placed in the alternated electromagnetic field [16, 58]. Moreover, nickel subsulfide (Ni_3S_2) are highly metallic compared to insulating oxide compounds, which can cause conductive loss [59]. The core-shell Ni@ZnS composites with the improvement of electromagnetic properties were reported by our earlier literatures [45, 60]. To the best of our knowledge, the EM wave absorption of core-shell structured ternary ZnS/Ni_3S_2@Ni composite is hardly found in the published papers. Herein, a novel ZnS/Ni_3S_2@Ni composite with urchin-like core-shell structure is successfully synthesized, and it exhibits the excellent EM absorption and the absorption mechanism of such unique hierarchical microstructure is also discussed in detail.

3.2.1 Preparation of urchin-like ZnS/Ni_3S_2@Ni composites

The monodispersed Ni microspheres were prepared according to our previous literatures [37, 45, 62]. Synthesis of urchin-like core-shell structured ZnS/Ni_3S_2@Ni composite [61]: in brief, Ni (0.05 g) was dispersed in a mixture of aqueous solution of distilled water (30 mL) and ethanol (30 mL) containing 1.0 M NaOH. Then, 1 mmol $ZnCl_2$ and 2 mmol $Na_2S{\cdot}9H_2O$ were introduced into above mixture, respectively. The mixture was transferred to a Teflon-lined autoclave. The sealed autoclave was heated to 120°C for 15 h. To explore the possible generation mechanism of core-shell microstructure urchins and effects of temperatures on the shapes of target products, the temperature-control experiments (60°C, 80°C, 100°C and 120°C) were also conducted.

The phase constituent and structure of the as-prepared Ni microspheres and urchin-like ZnS/Ni_3S_2@Ni products are characterized by XRD. **Figure 11a** displays the XRD curve of uncoated Ni microspheres, which could be well assigned to face-centered cubic structure of Ni (JCPDS No. 040850). As presented in **Figure 11b**, except for the diffraction peaks of Ni, the other diffraction peaks could be attributed to the zinc-blende ZnS (JCPDS Card No. 05-0566) and Ni_3S_2 (JCPDS Card No. 76-1870). From these XRD patterns, it can be confirmed that the core-shell ZnS/Ni_3S_2@Ni composite is composed of nickel, nickel sulfide and zinc sulfide. It can be inferred that nickel has functioned as the template for in-situ generation of Ni_3S_2 and deposition of ZnS.

To investigate the morphology of the products, FESEM images are taken for Ni microspheres and ZnS/Ni_3S_2@Ni composite and the corresponding results were shown in **Figure 12**. From the **Figure 12a**, it can be seen that Ni products were composed of uniform distribution and smooth surface of microspheres. **Figure 12b, c** presents the different magnification FESEM images of core-shell ZnS/Ni_3S_2@Ni composite. Interestingly, from panoramic observation (**Figure 12b**), urchin-like products are optionally grown on the surfaces of Ni microspheres and the outline of Ni spheres cannot be clearly observed due to the formation of ZnS/Ni_3S_2. Noticeably, the decease size of Ni particles could be observed, which suggests the depletion of Ni products. The thorns possess about 1 μm and 50 nm in length and

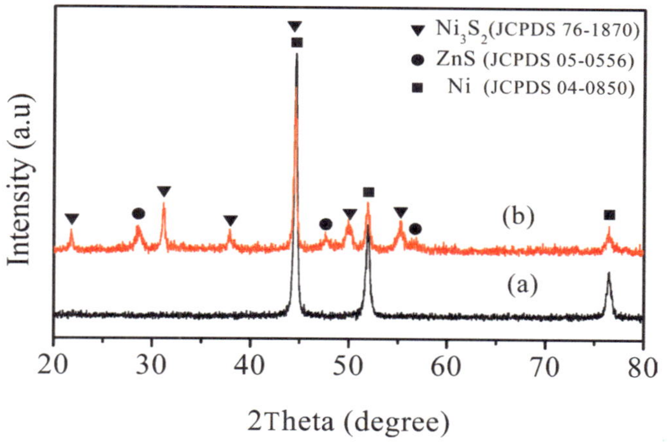

Figure 11.
XRD patterns of (a) pristine Ni microspheres and (b) as-prepared ZnS/Ni_3S_2@Ni [61] (permission from RSC).

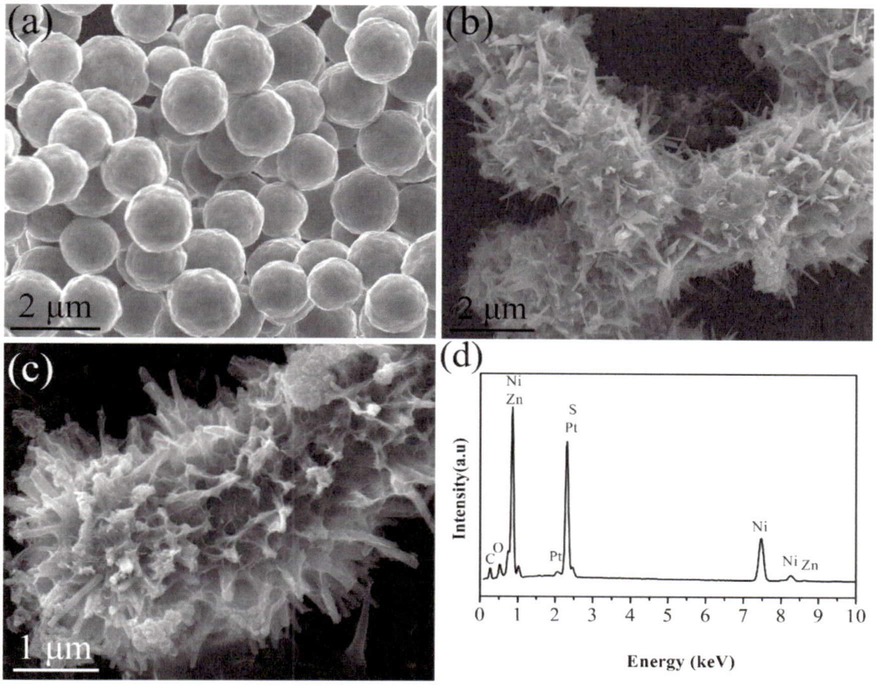

Figure 12.
FESEM images of (a) Ni microspheres, (b, c) urchin-like core-shell structured ZnS/Ni_3S_2@Ni composite, and (d) EDS profile of ZnS/Ni_3S_2@Ni composite [61] (permission from RSC).

diameter, respectively. Further observation from the high magnification FESEM image (**Figure 12c**), the existence of crumple products encircle Ni particles and the crinkled products are expected to be linked between Ni particles and thorns. **Figure 12d** presents the EDS of the ZnS/Ni_3S_2@Ni composite. The EDS reveals the presence of elements of S, Zn and Ni. Pt peaks are also seen in the EDS curve because the SEM sample is prepared by sputtering of platinum onto the sample.

It is assumed that the reaction temperature has an effect on the morphology of core-shell heterostructure. At low temperature (60°C), interestingly, there are plentiful waxberry-like products existed in **Figure 13a**. It is due to the fact that Ni

Figure 13.
FESEM images of core-shell $ZnS/Ni_3S_2@Ni$ composites prepared at various reaction temperatures: (a) 60°C, (b) 80°C, (c) 100°C, and (d) 120°C [61] (permission from RSC).

microspheres were coated by wrinkle ZnS products. Ni_3S_2 nanoparticles are in-situ formed by depletion of Ni particles and then covered on the surface of left Ni products. With elevating the reaction temperatures to high temperatures (80°C), one can notice that some protuberant prickles were grown on the surfaces of core-shell composites (**Figure 13b**). When the reaction temperature is increased to 100°C, the presence of more and strong protuberant stabs on the crumple surfaces of composite can be obviously observed (**Figure 13c**). With further enhancing the temperature to 120°C, the target urchin-like core-shell structural $ZnS/Ni_3S_2@Ni$ composites are formed with numerous of thorns or rods grown on the rugged surfaces (**Figure 13d**).

3.2.2 EM properties of urchin-like $ZnS/Ni_3S_2@Ni$ composites

With the purpose of revealing the electromagnetic wave absorption properties, the representation RL values of the core-shell $ZnS/Ni_3S_2@Ni$ paraffin-composites with different sample thicknesses are simulated. **Figure 14** reveals the RL values of the core-shell $ZnS/Ni_3S_2@Ni$ composites prepared at different temperatures with thickness varies from 0.8 to 2.5 mm in the frequency range of 1–18 GHz. It can be found that urchin-like $ZnS/Ni_3S_2@Ni$ composite exhibits outstanding electromagnetic absorption. The minimal RL is down to −27.6 dB at 5.2 GHz as the thickness is 2.5 mm. Notably, the reflection loss of urchin-like $ZnS/Ni_3S_2@Ni$ is −21.6 dB at 13.3 GHz with the absorber thickness of 1.0 mm, and the valuable bandwidth (RL below −10 dB) could reach 2.5 GHz (12.2–14.7 GHz), which is better than those of the literatures about microwave absorption performances of dielectric/magnetic composites, such as $Fe@SnO_2$ (>−10 dB) [63], and Co/CoO(−14.5 dB) [64]. Meanwhile, the absorption peaks would move to lower frequencies and dual RL peaks could be obtained with an increased absorber thickness above 2.5 mm. This phenomenon could be described by the quarter-wavelength cancelation model [65].

According to above results, we propose a possible electromagnetic wave absorption mechanism for the core-shell $ZnS/Ni_3S_2@Ni$ heterogeneous system When the

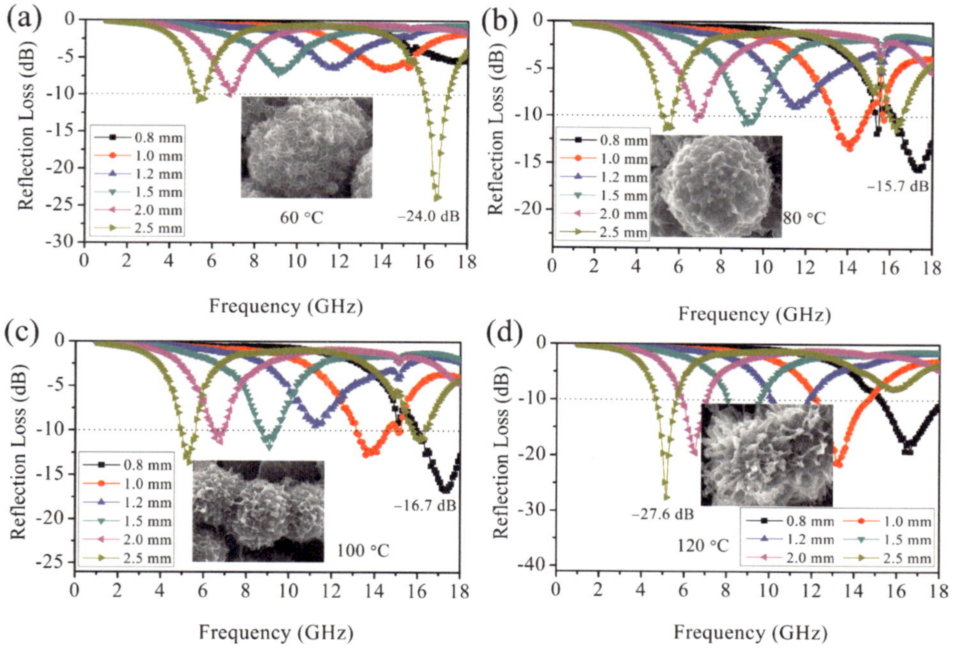

Figure 14.
The calculated RL of core-shell ZnS/Ni_3S_2@Ni composites prepared at various temperatures: (a) 60°C, (b) 80°C, (c) 100°C, and (d) 120°C with different thickness in the frequency range of 1–18 GHz [61] (permission from RSC).

ZnS/Ni_3S_2@Ni composite is subjected to EM wave radiation, the Ni_3S_2 thorns grown on the surfaces of Ni microspheres are expected as antenna receiver to allow electromagnetic waves penetrate into interior of absorber as much as possible, namely called good impedance match [66, 67]. Moreover, EM absorption is also understood from the viewpoint of Ohmic heating induced by an alternating magnetic field [68], in which ZnS/Ni_3S_2@Ni composite with relatively high electric conductivity. Thirdly, due to the heterogeneous systems of core-shell ZnS/Ni_3S_2@Ni composites, the multi-interfaces between the ZnS, Ni_3S_2, and Ni are favorable for improvement of electromagnetic wave absorption thanks to the interaction between electromagnetic radiation and charged multipoles at the interfaces [69]. The interfacial polarizations [42, 70] occurring at the interfaces are resulted from the migration of charge carriers through different conductivity properties of the composite material. During the irradiated by alternating electromagnetic field, an additional interfacial relaxation is produced, which is beneficial for the electromagnetic absorption [71]. On the other hand, the cooperation effect between dielectric loss at high frequency and magnetic loss at low frequency contribute to the improvement of electromagnetic absorption [72].

4. Conclusion

Various morphologies of Ni/ZnO composites are successfully fabricated by a hydrothermal method. The shapes of ZnO nanostructures could be effectively monitored by tuning the contents of $NH_3 \cdot H_2O$. One can find that the morphology of ZnO plays an important role on the microwave absorption capabilities. The core-shell structural Ni/polyhedron ZnO presents relatively high dielectric loss, magnetic loss and microwave dissipation abilities compared with the other Ni/ZnO samples.

The minimal RL could reach −48.6 dB at 13.4 GHz in the absorber thickness of 2.0 mm. The absorption band with RL below −10 dB could reach 6.0 GHz between 10.5 and 16.5 GHz. The enhanced microwave dissipation abilities are attributed to the synergetic effect between dielectric loss and magnetic loss, strong dissipation ability, as well as the multiple polarization of the core/shell interfaces.

The hierarchical Ni-CuO heterostructures have been successfully synthesized by a two-step process. The as-prepared Ni-CuO products display a rice-like coating composite. Moreover, by tuning the molar ratio of $CuCl_2 \cdot 2H_2O$ to the Ni microspheres, different shapes and coverage densities of CuO coating are obtained. The effects of CuO amounts on the Ni microspheres for microwave absorption properties have been investigated. The thin CuO-coated Ni composites (S-1) exhibit the enhanced electromagnetic absorption properties. The optimal RL is −62.2 dB at 13.8 GHz with only thickness of 1.7 mm. The outstanding microwave absorption properties result from the strongest attenuation constant, interfacial polarization of and the synergetic effect between the dielectric loss and magnetic loss.

ZnS nanowall-covered Ni composite is fabricated via a hydrothermal template method. The as-obtained Ni/ZnS composites display the crumble and rough features and the thickness of ZnS nanowall is about 10 nm. In comparison with raw Ni and ZnS particles, the Ni@ZnS composites show superior microwave dissipation abilities. The optimal RL of −25.78 dB could be obtained at 14.24 GHz and the valuable (less than −10 dB) band could reach 4.72 GHz (11.52–16.24 GHz) in the thickness of 2.7 mm. Moreover, the location of absorption peaks is almost similar at various thicknesses without moving to low frequency, which originates from natural resonance in permittivity.

Novel and interesting urchin-like ZnS/Ni_3S_2@Ni composites are synthesized through a two-step process including solution reduction and subsequently a template method. Crucially, the morphologies of the core-shell ZnS/Ni_3S_2@Ni composites are determined by the reaction temperature. Different ZnS/Ni_3S_2@Ni composites prepared at different temperatures show variable electromagnetic absorption responses, for which the urchin-like ZnS/Ni_3S_2@Ni obtained at 120°C show the enhanced EM absorption properties thanks to its promising dielectric loss and magnetic loss, good impedance match, as well as its unique urchin-like structure. Multiple dielectric resonances stemming from effective accumulation of different polarizations in the urchin-like structure are regarded to make a contribution to the enhancement of electromagnetic wave absorption. It is believed that *in situ* synthesis of core-shell ZnS/Ni_3S_2@Ni composites may open up a new avenue for the design and preparation of novel microwave absorbers with promising application potential.

In a word, the core-shell configuration is proved to be a promising pathway to design high-efficiency EM absorption properties of Ni based composites.

Conflict of interest

There are no conflicts to declare.

Author details

Biao Zhao[1,2]* and Rui Zhang[1,3]

1 Henan Key Laboratory of Aeronautical Materials and Application Technology, School of Material Science and Engineering, Zhengzhou University of Aeronautics, Zhengzhou, Henan, China

2 Department of Mechanical and Industrial Engineering, University of Toronto, Toronto, Ontario, Canada

3 School of Material Science and Engineering, Zhengzhou University, Zhengzhou, Henan, China

*Address all correspondence to: zhaobiao1813@163.com

References

[1] Jian X, Wu B, Wei Y, Dou SX, Wang X, He W, et al. Facile synthesis of Fe_3O_4/GCs composites and their enhanced microwave absorption properties. ACS Applied Materials & Interfaces. 2016;**8**(9):6101-6109. DOI: 10.1021/acsami.6b00388

[2] Lv H, Guo Y, Yang Z, Cheng Y, Wang LP, Zhang B, et al. A brief introduction to the fabrication and synthesis of graphene based composites for the realization of electromagnetic absorbing materials. Journal of Materials Chemistry C. 2017;**5**(3):491-512. DOI: 10.1039/C6TC03026B

[3] Zhao B, Deng J, Zhang R, liang L, Fan B, Bai Z, et al. Recent advances on the electromagnetic wave absorption properties of Ni based materials. Engineered Science. 2018;**3**:5-40. DOI: 10.30919/es8d735

[4] Wang G, Gao Z, Tang S, Chen C, Duan F, Zhao S, et al. Microwave absorption properties of carbon nanocoils coated with highly controlled magnetic materials by atomic layer deposition. ACS Nano. 2012;**6**(12):11009-11017. DOI: 10.1021/nn304630h

[5] Ye F, Song Q, Zhang Z, Li W, Zhang S, Yin X, et al. Direct growth of edge-rich graphene with tunable dielectric properties in porous Si3N4 ceramic for broadband high-performance microwave absorption. Advanced Functional Materials. 2018;**28**(17):1707205. DOI: 10.1002/adfm.201707205

[6] Liu Q, Cao Q, Bi H, Liang C, Yuan K, She W, et al. CoNi@SiO_2@TiO_2 and CoNi@Air@TiO_2 microspheres with strong wideband microwave absorption. Advanced Materials. 2016;**28**(3):486-490. DOI: 10.1002/adma.201503149

[7] Zhang Y, Huang Y, Zhang T, Chang H, Xiao P, Chen H, et al. Broadband and tunable high-performance microwave absorption of an ultralight and highly compressible graphene foam advanced materials. 2015;**27**(12):2049-2053. DOI: 10.1002/adma.201405788

[8] Xi L, Wang Z, Zuo Y, Shi X. The enhanced microwave absorption property of $CoFe_2O_4$ nanoparticles coated with a Co_3Fe_7–Co nanoshell by thermal reduction. Nanotechnology. 2011;**22**(4):045707. DOI: 10.1088/0957-4484/22/4/045707

[9] Zhou W, Hu X, Bai X, Zhou S, Sun C, Yan J, et al. Synthesis and electromagnetic, microwave absorbing properties of Core–Shell Fe_3O_4–Poly(3, 4-ethylenedioxythiophene) microspheres. ACS Applied Materials & Interfaces. 2011;**3**(10):3839-3845. DOI: 10.1021/am2004812

[10] Chen Y-J, Zhang F, G-g Z, X-y F, Jin H-B, Gao P, et al. Synthesis, multi-nonlinear dielectric resonance, and excellent electromagnetic absorption characteristics of Fe_3O_4/ZnO core/shell nanorods. The Journal of Physical Chemistry C. 2010;**114**(20):9239-9244. DOI: 10.1021/jp912178q

[11] Liu J, Che R, Chen H, Zhang F, Xia F, Wu Q, et al. Microwave absorption enhancement of multifunctional composite microspheres with spinel Fe_3O_4 cores and anatase TiO_2 shells. Small. 2012;**8**(8):1214-1221. DOI: 10.1002/smll.201102245

[12] Zhu C-L, Zhang M-L, Qiao Y-J, Xiao G, Zhang F, Chen Y-J. Fe_3O_4/TiO_2 core/shell nanotubes: synthesis and magnetic and electromagnetic wave absorption characteristics. The Journal of Physical Chemistry C. 2010;**114**(39):16229-16235. DOI: 10.1021/jp104445m

[13] Metin Ö, Mazumder V, Özkar S, Sun S. Monodisperse nickel nanoparticles and their catalysis in

hydrolytic dehydrogenation of ammonia borane. Journal of the American Chemical Society. 2010;**132**(5):1468-1469. DOI: 10.1021/ja909243z

[14] Xiong J, Shen H, Mao J, Qin X, Xiao P, Wang X, et al. Porous hierarchical nickel nanostructures and their application as a magnetically separable catalyst. Journal of Materials Chemistry. 2012;**22**(24):11927-11932. DOI: 10.1039/C2JM30361B

[15] Zhao B, Shao G, Fan B, Li W, Pian X, Zhang R. Enhanced electromagnetic wave absorption properties of Ni–SnO_2 core–shell composites synthesized by a simple hydrothermal method. Materials Letters. 2014;**121**:118-121. DOI: 10.1016/j.matlet.2014.01.081

[16] Wang H, Guo H, Dai Y, Geng D, Han Z, Li D, et al. Optimal electromagnetic-wave absorption by enhanced dipole polarization in Ni/C nanocapsules. Applied Physics Letters. 2012;**101**(8):083116. DOI: 10.1063/1.4747811

[17] Liu XG, Jiang JJ, Geng DY, Li BQ, Han Z, Liu W, et al. Dual nonlinear dielectric resonance and strong natural resonance in Ni/ZnO nanocapsules. Applied wPhysics Letters. 2009;**94**(5):053119. DOI: 10.1063/1.3079393

[18] Liu X, Feng C, Or SW, Jin C, Xiao F, Xia A, et al. Synthesis and electromagnetic properties of Al/AlOx-coated Ni nanocapsules. Materials Research Bulletin. 2013;**48**(10):3887-3891. DOI: 10.1016/j.materresbull.2013.05.110

[19] Wang B, Zhang J, Wang T, Qiao L, Li F. Synthesis and enhanced microwave absorption properties of Ni@Ni_2O_3 core–shell particles. Journal of Alloys and Compounds. 2013;**567**:21-25. DOI: 10.1016/j.jallcom.2013.03.028

[20] Dong XL, Zhang XF, Huang H, Zuo F. Enhanced microwave absorption in Ni/polyaniline nanocomposites by dual dielectric relaxations. Applied Physics Letters. 2008;**92**(1):013127. DOI: 10.1063/1.2830995

[21] Liu X, Feng C, Or SW, Sun Y, Jin C, Li W, et al. Investigation on microwave absorption properties of CuO/Cu_2O-coated Ni nanocapsules as wide-band microwave absorbers. RSC Advances. 2013;**3**(34):14590-14594. DOI: 10.1039/C3RA40937F

[22] Li H, Huang Y, Sun G, Yan X, Yang Y, Wang J, et al. Directed growth and microwave absorption property of crossed ZnO netlike micro-/nanostructures. The Journal of Physical Chemistry C. 2010;**114**(22):10088-10091. DOI: 10.1021/jp100341h

[23] Zhao B, Shao G, Fan B, Guo W, Xie Y, Zhang R. Facile synthesis of Ni/ZnO composite: Morphology control and microwave absorption properties. Journal of Magnetism and Magnetic Materials. 2015;**382**:78-83. DOI: 10.1016/j.jmmm.2015.01.053

[24] Wang ZL, Kong XY, Zuo JM. Induced growth of asymmetric nanocantilever arrays on polar surfaces. Physical Review Letters. 2003;**91**(18):185502. DOI: 10.1103/PhysRevLett.91.185502

[25] Yan J-F, Zhang Z-Y, You T-G, Zhao W, Yun J-N, Zhang F-C. Effect of polyacrylamide on morphology and electromagnetic properties of chrysanthemum-like ZnO particles. Chinese Physics B. 2009;**18**(10):4552. DOI: 10.1088/1674-1056/18/10/076

[26] Na J-S, Gong B, Scarel G, Parsons GN. Surface polarity shielding and hierarchical ZnO nano-architectures produced using sequential hydrothermal crystal synthesis and thin film atomic layer deposition. ACS Nano.

2009;**3**(10):3191-3199. DOI: 10.1021/nn900702e

[27] Jung S-H, Oh E, Lee K-H, Yang Y, Park CG, Park W, et al. Sonochemical preparation of shape-selective ZnO nanostructures. Crystal Growth & Design. 2007;**8**(1):265-269. DOI: 10.1021/cg070296l

[28] Flaifel MH, Ahmad SH, Abdullah MH, Rasid R, Shaari AH, El-Saleh AA, Appadu S. Preparation, thermal, magnetic and microwave absorption properties of thermoplastic natural rubber matrix impregnated with NiZn ferrite nanoparticles. Composites Science and Technology. 2014;**96**:103-108

[29] Li G, Xie T, Yang S, Jin J, Jiang J. Microwave absorption enhancement of porous carbon fibers compared with carbon nanofibers. The Journal of Physical Chemistry C. 2012;**116**(16):9196-9201. DOI: 10.1021/jp300050u

[30] Cooper ER, Andrews CD, Wheatley PS, Webb PB, Wormald P, Morris RE. Ionic liquids and eutectic mixtures as solvent and template in synthesis of zeolite analogues. Nature. 2004;**430**(7003):1012-1016. DOI: 10.1038/nature02860

[31] Zhang X, Dong X, Huang H, Lv B, Lei J, Choi C. Microstructure and microwave absorption properties of carbon-coated iron nanocapsules. Journal of Physics D: Applied Physics. 2007;**40**(17):5383. DOI: 10.1088/0022-3727/40/17/056

[32] Yan S, Zhen L, Xu C, Jiang J, Shao W. Microwave absorption properties of $FeNi_3$ submicrometre spheres and SiO_2@ $FeNi_3$ core–shell structures. Journal of Physics D: Applied Physics. 2010;**43**(24):245003. DOI: 10.1088/0022-3727/43/24/245003

[33] Pecquenard B, Le Cras F, Poinot D, Sicardy O, Manaud J-P. Thorough characterization of sputtered CuO thin films used as conversion material electrodes for lithium batteries. ACS Applied Materials & Interfaces. 2014;**6**(5):3413-3420. DOI: 10.1021/am4055386

[34] Kargar A, Jing Y, Kim SJ, Riley CT, Pan X, Wang D. ZnO/CuO Heterojunction branched nanowires for photoelectrochemical hydrogen generation. ACS Nano. 2013;**7**(12):11112-11120. DOI: 10.1021/nn404838n

[35] Zeng J, Xu J, Tao P, Hua W. Ferromagnetic and microwave absorption properties of copper oxide-carbon fiber composites. Journal of Alloys and Compounds. 2009;**487**(1-2):304-308. DOI: 10.1016/j.jallcom.2009.07.112

[36] Jun Z, Huiqing F, Yangli W, Shiquan Z, Jun X, Xinying C. Ferromagnetic and microwave absorption properties of copper oxide/cobalt/carbon fiber multilayer film composites. Thin Solid Films. 2012;**520**(15):5053-5059. DOI: 10.1016/j.tsf.2012.03.059

[37] Zhao B, Shao G, Fan B, Zhao W, Zhang R. Facile synthesis and enhanced microwave absorption properties of novel hierarchical heterostructures based on a Ni microsphere-CuO nano-rice core-shell composite. Physical Chemistry Chemical Physics. 2015;**17**(8):6044-6052. DOI: 10.1039/C4CP05229C

[38] Wang Z, Wu L, Zhou J, Cai W, Shen B, Jiang Z. Magnetite nanocrystals on multiwalled carbon nanotubes as a synergistic microwave absorber. The Journal of Physical Chemistry C. 2013;**117**(10):5446-5452. DOI: 10.1021/jp4000544

[39] Pan G, Zhu J, Ma S, Sun G, Yang X. Enhancing the electromagnetic performance of co through the phase-controlled synthesis of hexagonal

and cubic co nanocrystals grown on graphene. ACS Applied Materials & Interfaces. 2013;**5**(23):12716-12724. DOI: 10.1021/am404117v

[40] Wang C, Han X, Zhang X, Hu S, Zhang T, Wang J, et al. Controlled synthesis and morphology-dependent electromagnetic properties of hierarchical cobalt assemblies. The Journal of Physical Chemistry C. 2010;**114**(35):14826-14830. DOI: 10.1021/jp1050386

[41] Ortega N, Kumar A, Katiyar R, Rinaldi C. Dynamic magneto-electric multiferroics PZT/CFO multilayered nanostructure. Journal of Materials Science. 2009;**44**(19):5127-5142. DOI: 10.1007/s10853-009-3635-0

[42] Wen S, Liu Y, Zhao X, Cheng J, Li H. Synthesis, multi-nonlinear dielectric resonance and electromagnetic absorption properties of hcp-cobalt particles. Journal of Magnetism and Magnetic Materials. 2014;**354**:7-11. DOI: 10.1016/j.jmmm.2013.10.030

[43] Yan C, Xue D. Room temperature fabrication of hollow ZnS and ZnO architectures by a sacrificial template route. The Journal of Physical Chemistry B. 2006;**110**(114):7102-7106

[44] Gu F, Li CZ, Wang SF, Lü MK. Solution-phase synthesis of spherical zinc sulfide nanostructures. solution-phase synthesis of spherical zinc sulfide nanostructures. Langmuir. 2005;**22**(3):1329-1332

[45] Zhao B, Shao G, Fan B, Zhao W, Xie Y, Zhang R. ZnS nanowall coated Ni composites: facile preparation and enhanced electromagnetic wave absorption. RSC Advances. 2014;**4**(105):61219-61225. DOI: 10.1039/C4RA08095E

[46] Wang G-S, Wu Y, Wei Y-Z, Zhang X-J, Li Y, Li L-D, et al. Fabrication of Reduced Graphene Oxide (RGO)/Co3O4 nanohybrid particles and a RGO/Co3O4/Poly(vinylidene fluoride) composite with enhanced wave-absorption properties. ChemPlusChem. 2014;**79**(3):375-381. DOI: 10.1002/cplu.201300345

[47] Ren Y, Zhu C, Zhang S, Li C, Chen Y, Gao P, et al. Three-dimensional $SiO_2@Fe_3O_4$ core/shell nanorod array/graphene architecture: synthesis and electromagnetic absorption properties. Nanoscale. 2013;**5**(24):12296-12303. DOI: 10.1039/C3NR04058E

[48] Liu XG, Geng DY, Zhang ZD. Microwave-absorption properties of FeCo microspheres self-assembled by Al2O3-coated FeCo nanocapsules. Applied Physics Letters. 2008;**92**(24):243110. DOI: 10.1063/1.2945639

[49] Du Y, Liu T, Yu B, Gao H, Xu P, Wang J, et al. The electromagnetic properties and microwave absorption of mesoporous carbon. Materials Chemistry and Physics. 2012;**135**(2-3):884-891. DOI: 10.1016/j.matchemphys.2012.05.074

[50] He S, Wang G-S, Lu C, Liu J, Wen B, Liu H, et al. Enhanced wave absorption of nanocomposites based on the synthesized complex symmetrical CuS nanostructure and poly(vinylidene fluoride). Journal of Materials Chemistry A. 2013;**1**(15):4685-4692. DOI: 10.1039/C3TA00072A

[51] Li Y, Zhang J, Liu Z, Liu M, Lin H, Che R. Morphology-dominant microwave absorption enhancement and electron tomography characterization of CoO self-assembly 3D nano-flowers. Journal of Materials Chemistry C. 2014;**2**(26):5216-5222. DOI: 10.1039/C4TC00739E

[52] Lee CW, Seo S-D, Park HK, Park S, Song HJ, Kim D-W, et al. High-areal-capacity lithium storage of the kirkendall effect-driven hollow hierarchical NiSx nanoarchitecture.

Nanoscale. 2015;**7**(6):2790-2796. DOI: 10.1039/C4NR05942E

[53] Huang X, Willinger M-G, Fan H, Xie Z-l, Wang L, Klein-Hoffmann A, et al. Single crystalline wurtzite ZnO/zinc blende ZnS coaxial heterojunctions and hollow zinc blende ZnS nanotubes: synthesis, structural characterization and optical properties. Nanoscale. 2014;**6**(15):8787-8795. DOI: 10.1039/C4NR01575D

[54] Liao Y, Pan K, Pan Q, Wang G, Zhou W, Fu H. In situ synthesis of a NiS/Ni_3S_2 nanorod composite array on Ni foil as a FTO-free counter electrode for dye-sensitized solar cells. Nanoscale. 2015;**7**(5):1623-1626. DOI: 10.1039/C4NR06534D

[55] Wang Z, Li X, Yang Y, Cui Y, Pan H, Wang Z, et al. Highly dispersed [small beta]-NiS nanoparticles in porous carbon matrices by a template metal-organic framework method for lithium-ion cathode. Journal of Materials Chemistry A. 2014;**2**(21):7912-7916. DOI: 10.1039/C4TA00367E

[56] Xitao W, Rong L, Kang W. Synthesis of ZnO@ZnS-Bi_2S_3 core-shell nanorod grown on reduced graphene oxide sheets and its enhanced photocatalytic performance. Journal of Materials Chemistry A. 2014;**2**(22):8304-8313. DOI: 10.1039/C4TA00696H

[57] Zhu Y-P, Li J, Ma T-Y, Liu Y-P, Du G, Yuan Z-Y. Sonochemistry-assisted synthesis and optical properties of mesoporous ZnS nanomaterials. Journal of Materials Chemistry A. 2014;**2**(4):1093-1101. DOI: 10.1039/C3TA13636A

[58] Zhang Q, Li C, Chen Y, Han Z, Wang H, Wang Z, et al. Effect of metal grain size on multiple microwave resonances of Fe/TiO_2 metal-semiconductor composite. Applied Physics Letters. 2010;**97**(13):133115. DOI: 10.1063/1.3496393

[59] Watts PCP, Hsu WK, Barnes A, Chambers B. High permittivity from defective multiwalled carbon nanotubes in the x-band. Advanced Materials. 2003;**15**(7-8):600-603. DOI: 10.1002/adma.200304485

[60] Zhao B, Shao G, Fan B, Zhao W, Chen Y, Zhang R. Facile synthesis of crumpled ZnS net-wrapped Ni walnut spheres with enhanced microwave absorption properties. RSC Advances. 2015;**5**(13):9806-9814. DOI: 10.1039/C4RA15411H

[61] Zhao B, Shao G, Fan B, Zhao W, Zhang S, Guan K, et al. In situ synthesis of novel urchin-like ZnS/Ni_3S_2@Ni composite with a core-shell structure for efficient electromagnetic absorption. Journal of Materials Chemistry C. 2015;**3**(41):10862-10869. DOI: 10.1039/C5TC02063H

[62] Zhao B, Shao G, Fan B, Zhao W, Zhang R. Fabrication and enhanced microwave absorption properties of Al_2O_3 nanoflake-coated Ni core-shell composite microspheres. RSC Advances. 2014;**4**(101):57424-57429. DOI: 10.1039/C4RA10638E

[63] Liu X, Zhou G, Or SW, Sun Y. Fe/amorphous SnO_2 core-shell structured nanocapsules for microwave absorptive and electrochemical performance. RSC Advances. 2014;**4**(93):51389-51394. DOI: 10.1039/C4RA08998G

[64] Wang Z, Bi H, Wang P, Wang M, Liu Z, shen L, et al. Magnetic and microwave absorption properties of self-assemblies composed of core-shell cobalt-cobalt oxide nanocrystals. Physical Chemistry Chemical Physics. 2015;**17**(5):3796-3801. DOI: 10.1039/C4CP04985C

[65] Li G, Wang L, Li W, Ding R, Xu Y. $CoFe_2O_4$ and/or Co3Fe7 loaded porous activated carbon balls as a lightweight microwave absorbent. Physical Chemistry Chemical Physics.

2014;**16**(24):12385-12392. DOI: 10.1039/C4TA05718J

[66] Li X, Feng J, Du Y, Bai J, Fan H, Zhang H, et al. One-pot synthesis of $CoFe_2O_4$/graphene oxide hybrids and their conversion into FeCo/graphene hybrids for lightweight and highly efficient microwave absorber. Journal of Materials Chemistry A. 2015;**3**(10):5535-5546. DOI: 10.1039/C4TA05718J

[67] Tong G, Yuan J, Wu W, Hu Q, Qian H, Li L, et al. Flower-like Co superstructures: Morphology and phase evolution mechanism and novel microwave electromagnetic characteristics. CrystEngComm. 2012;**14**(6):2071-2079. DOI: 10.1039/C2CE05910J

[68] Jiang J, Li D, Li S, Wang Z, Wang Y, He J, et al. Electromagnetic wave absorption and dielectric-modulation of metallic perovskite lanthanum nickel oxide. RSC Advances. 2015;**19**(19):14584-14591. DOI: 10.1039/C5RA00139K

[69] Yang H-J, Cao W-Q, Zhang D-Q, Su T-J, Shi H-L, Wang W-Z, et al. Hierarchical nanorings on SiC: Enhancing relaxation to tune microwave absorption at elevated temperature. ACS Applied Materials & Interfaces. 2015;**7**(13):7073-7077. DOI: 10.1021/acsami.5b01122

[70] Liu Q, Zhang D, Fan T. Electromagnetic wave absorption properties of porous carbon/Co nanocomposites. Applied Physics Letters. 2008;**93**(1):013110. DOI: 10.1063/1.2957035

[71] Zhao B, Shao G, Fan B, Zhao W, Xie Y, Zhang R. Synthesis of flower-like CuS hollow microspheres based on nanoflakes self-assembly and their microwave absorption properties. Journal of Materials Chemistry A. 2015;**3**(19):10345-10352. DOI: 10.1039/C5TA00086F

[72] Wang Q, Lei Z, Chen Y, Ouyang Q, Gao P, Qi L, et al. Branched polyaniline/molybdenum oxide organic/inorganic heteronanostructures: Synthesis and electromagnetic absorption properties. Journal of Materials Chemistry A. 2013;**1**(38):11795-11801. DOI: 10.1039/C3TA11591G

Section 3

Applications of Electromagnetic Materials and Devices

Chapter 9

Electromagnetic Function Textiles

Hong Xiao, Meiwu Shi and Jianying Chen

Abstract

This chapter is about the electromagnetic (EM) functional textiles. There are three parts including the EM properties of the textiles, the EM functional textile materials, and the application of the EM functional textiles. In the first part, we outline the textile materials including the fibers, yarns, and fabrics and their EM properties. Generally, textiles are poor in EM properties such as low conductivity and low dielectric and magnetic properties. In the second part, EM functional textiles are defined and the manufacture method is stated. The EM functional fibers and yarns and fabrics can be got by various methods; the EM properties can be improved to a high level. In the third part, several typical EM functional textiles are introduced. These textiles are antistatic textiles, EM shielding textiles, EM scattering textiles, and two-dimensional and three-dimensional frequency selective textiles. The principle, the processing methods, the properties of the static electro or EM shielding or EM scattering or frequency selection, and so on are described one by one in detail.

Keywords: antistatic textiles, EM shielding textiles, EM scattering textiles, frequency selective textiles

1. Electromagnetic properties of textile materials

1.1 The outline of textile materials

The textile materials include various raw fiber materials which are used in textile and various products processed from textile fibers, such as the one-dimensional yarn, thread, rope, and so on; two-dimensional and shape-based fabrics, textile nets, flakes, and so on; and three-dimensional and form-based clothing, braids, utensils, and its reinforced composites. The basic textile processing process is shown in **Figure 1**.

As shown in **Table 1**, the textile materials are essentially different from traditional engineering materials; moreover, textile materials are flexible, easy to change their shape, and generally light weight; these characters can largely compensate the defects of engineering materials. According to the form, textile materials can be divided into fibers, yarns, flat fabrics, and three-dimensional fabrics. The fibers are spun to form the yarns, which are then weaved into the fabrics by weaving technology or knitting technology. In addition, the nonwoven fabrics are formed directly by winding the fibers.

1.1.1 Fiber

The shape of fibers is flexible and elongate, with length (**Figure 2**) and diameter ratio (**Figure 3**) of more than 10^3. Theoretically speaking, the fibers have round and

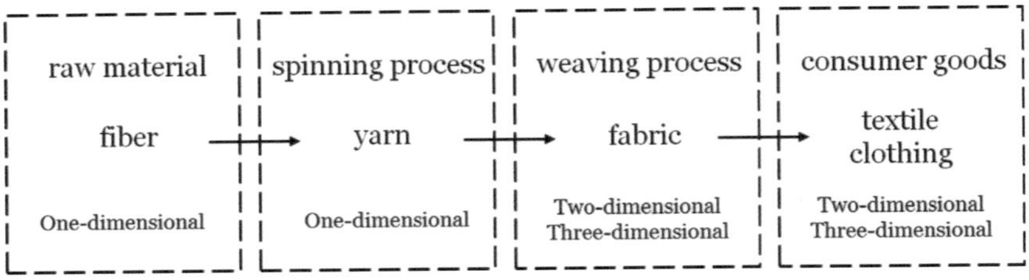

Figure 1.
Basic process of textile processing.

Traditional engineering materials	Transitional material	Textile materials
Rigidity	Flexible	Flexible
Homogeneous state	Solid state	Discontinuous state
Dense, non-permeable	Dense or porous	Porous
Smooth surface	Soft, fillable loose structure	Surface texture
No buckling state	Flexion or non-buckling	Multiple buckling state

Table 1.
The difference between textile materials and traditional engineering materials.

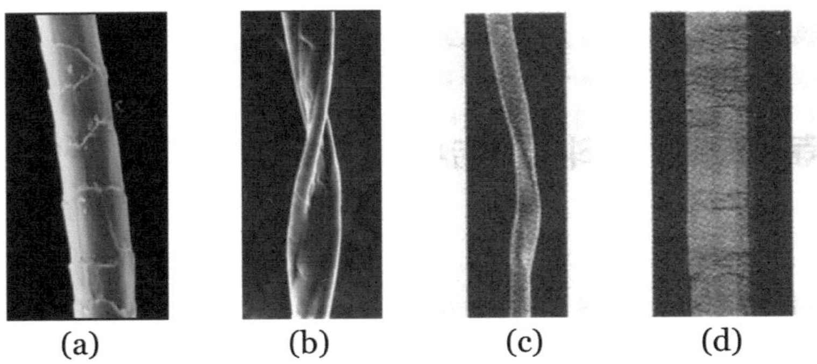

Figure 2.
The longitudinal morphology of natural fibers. (a) Wool fiber; (b) Cotton fiber; (c) Silk fiber; (d) Hemp fiber.

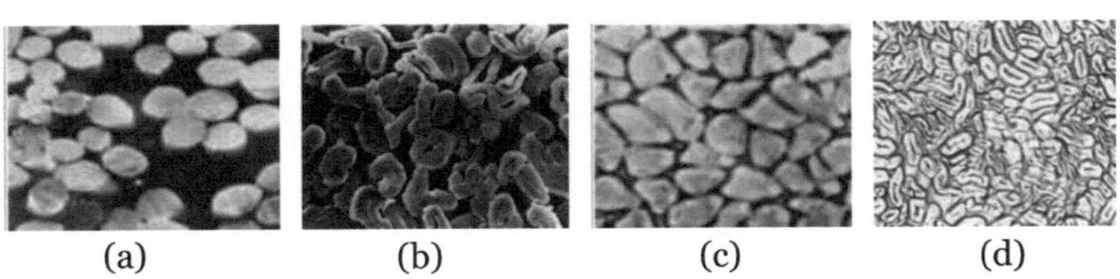

Figure 3.
The cross-sectional morphology of natural fibers. (a) Wool fiber; (b) Cotton fiber; (c) Silk fiber; (d) Hemp fiber.

slender bodies with the continuous homogeneous internal structure. But actually, they have a wide variety of cross-sectional shapes, and section shape changing along the length, heterogeneous internal structure, with the porosity form. According to the source of the fibers, they can be divided into natural fibers and chemical fibers. Fibers such as cotton, hemp, silk, and wool are the natural fibers with the longest history.

1.1.2 Yarn

The yarn is an elongated body having a certain strength and toughness, in which the fibers are arranged in parallel and are cohered or entangled by twisting or other methods. The yarn is an intermediate product of textile processing. A number of short fibers or filaments are arranged in an approximately parallel state and twisted in the axial direction to form an elongated object having a certain strength and linear density, which is called the "yarn." The strand of two or more single yarns is called the "thread" (**Figure 4**).

1.1.3 Fabric

A two-dimensional object having thin thickness, large length, and wide width formed by interweaving and interlacing textile fibers and yarns by a certain method is the flat fabric. There are a variety of fabrics (as shown in **Figure 5**); they could have various materials, forms, colors, structures, and formation methods.

According to the forming methods, the fabrics can be divided into woven fabric, knitted fabric, and nonwoven fabric. The woven fabric is composed by warp and weft yarns arranged perpendicularly to each other according to some organization rules. The knitted fabric is formed by the yarns bent into a loop. Nonwoven fabrics are reinforced by oriented or randomly arranged fiber webs.

1.2 Electromagnetic properties of textile materials

Traditional textile materials are mostly dielectric materials and are important electrical insulation materials. The electromagnetic properties of textile materials include electrical conductivity, dielectric properties, electrostatic and magnetic properties.

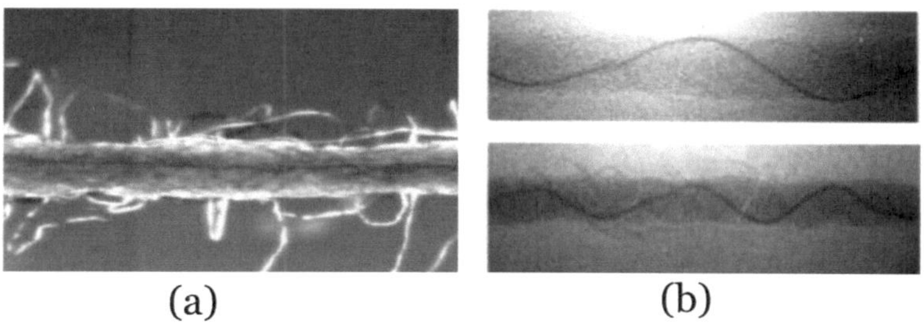

Figure 4.
The morphology of one kind of the yarn. (a) The appearance of the yarn and (b) The distribution of fibers in the yarn.

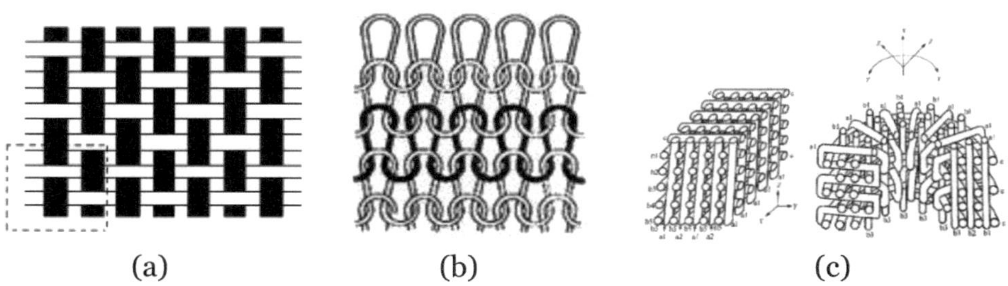

Figure 5.
The type of the fabric. (a) Woven fabric; (b) Knitted fabric and (c) Nonwoven fabric.

1.2.1 Conductive properties

The electrical conductivity of textile materials is expressed as specific resistance. There are usually three representations: volume specific resistance, mass specific resistance and surface specific resistance (**Table 2**).

According to the law of resistance, the resistance R of the conductor is proportional to the length L of the conductor, inversely proportional to the cross-sectional area S, and related to material properties. That is,

$$R = \rho_V \cdot \frac{L}{S} \tag{1}$$

where ρ_V is the resistivity or volume specific resistance and its unit is Ω cm, and it is the physical expression that indicates the electrical conductivity of a material.

For textile materials, the cross-sectional area or volume is not easy to measure; so, we usually use the mass specific resistance ρ_m rather than the volume specific resistance ρ_V to indicate the conductivity of textile materials, especially for the fibers and yarns.

$$\rho_m = d \cdot \rho_V \tag{2}$$

where ρ_m represents the mass specific resistance, and the unit is Ω cm/cm^2. d is the density of the material in 1 g/cm^3. In actual measurement, the moisture content of the fiber or the relative humidity of the air has a great influence on its electrical resistance. The dried textile fibers have extremely poor electrical conductivity, and their mass specific resistance is generally bigger than 10^{12} Ω cm/cm^2. For most textile materials, there is an approximate relationship between the moisture content M and the mass specific resistance ρ_m of the textile materials in the range of 30–90% relative humidity:

$$\lg\rho_m = -n\lg M + \lg K \tag{3}$$

where n and K are experimental constants.

Type of fiber	lg ρ_m	n	lg K
Cotton	6.8	11.4	16.6
Ramie	7.5	12.3	18.6
Silk	9.8	17.6	26.6
Wool	8.4	15.8	26.2
Washed wool	9.9	14.7	26.6
Viscose fiber	7.0	11.6	19.6
Acetate fiber	11.7	10.6	20.1
Acrylic	8.7	—	—
Acrylic (degreasing)	14	—	—
Polyester	8.0	—	—
Polyester (degreasing)	14	—	—

Table 2.
Mass specific resistance of textile materials.

1.2.2 Dielectric properties

1.2.2.1 Dielectric constant

The dielectric constant of dried fiber is 2–5 at the frequent of 50 or 60 Hz. The dielectric constant of the liquid water is 20 and the adsorbed water is 80. The dielectric constants of common textile fibers measured at the frequency of 1 kHz and the relative humidity of 65% are shown in **Table 3**.

As the relative dielectric constant of water is several tens of times larger than that of the dry textile material, the dielectric constant of the fiber is different when the moisture regain or moisture content of the textile material is different. The presence of frequency, temperature, and impurities also changes the dielectric constant of the materials.

1.2.2.2 Dielectric loss

A physical process in which a dielectric converts a portion of electrical energy into thermal energy under the action of an electric field is known as dielectric loss. The magnitude of the dielectric loss is related to the applied electric field frequency, electric field strength, fiber constant, and dielectric loss angle. In unit time, the heat energy P produced per unit volume of fiber is

$$P = 0.556f \cdot E^2 \cdot \varepsilon_r \cdot \tan\delta \cdot 10^{-12} \quad (4)$$

where P is the power consumed by the electric field (W/cm^3); f is the frequency of the applied electric field (Hz); E is the external electric field strength (V/cm); and $\tan\delta$ is the tangent of the dielectric loss angle δ.

The dielectric constant of dry textile material generally is 2–5, for which $\tan\delta$ is equal to 0.02–0.05. The dielectric constant of water is 20–80, for which $\tan\delta$ is 0.15–1.2. Therefore, the higher the moisture content of the textile material, the larger than $\tan\delta$.

1.2.3 Electrostatic performance

The specific resistance of textile materials with dielectric properties is generally high, especially for synthetic fibers with low hygroscopicity, such as polyester and acrylic fibers. Under normal atmospheric conditions, the mass specific resistance is

Fiber	Dielectric constant (ε)	Fiber	Dielectric constant (ε)
Cotton	18	Acetate	4.0
Wool	5.5	Nylon staple fiber	3.7
Viscose fiber	8.4	Nylon yarn	4.0
Viscose wire	15	Polyester staple fiber (deoiled)	2.3
Acetate staple fiber	3.5	Polyester staple fiber	4.2
Acrylic staple fiber (deoiled)	2.8	—	—

The high moisture regain of cotton and viscose leads to its high dielectric constant.

Table 3.
The dielectric constant of common textile fibers.

as high as 10^{13} Ω cm/cm^2 or more. In textile processing, the contact and friction between fibers or between fibers and machine parts tends to trigger charge transfer and static electricity generation. During the production process, static electricity will cause fiber hairiness, hairiness increase, filament winding mechanism, breakage, etc. In the course of the taking process, static electricity will cause clothes to stick and absorb dust.

Although the phenomenon of static electricity leads to many hazards during textile processing, the electrostatic properties of textile materials can also benefit to some processing technology, such as electrospinning and electrostatic flocking.

1.2.4 Magnetic properties

Ordinary textile materials are anti-magnets, which are negative. The magnetic susceptibility of some textile materials is shown in **Table 4**.

The magnetic properties of textile materials are not as much as those of electrical properties, but they are gradually being valued by people to develop various types of magnetic fibers and textiles. For example, magnetic powders such as iron, cobalt, nickel, and ferrite are added to a spinning solution, and fibers having magnetic properties are obtained by wet spinning.

2. Electromagnetic functional textile materials

Common textile materials have dielectric properties and electrostatic phenomena, but the electromagnetic parameter of them has not reached the order of magnitude of metals or semiconductors. Therefore, they generally do not own any electromagnetic function.

2.1 The definition of electromagnetic textile materials

Electromagnetic textile materials are a new type of functional textile materials obtained from fibers or yarns with good electrical and magnetic properties through textile processing technology or by applying the materials with metallic properties to common textile material. Meanwhile, electromagnetic textile materials have unique structure of textile materials and the electromagnetic properties of the metal materials [1, 2].

2.2 Preparation of electromagnetic materials

Ordinary fibers are generally made of nonconductive and non-magnetic polymer materials. To obtain the functionalization of textile materials, special materials must be introduced during the preparation process. Textile materials include fibers,

Material	Magnetic susceptibility (χ)	Material	Magnetic susceptibility (χ)
Ethylene	-10.3×10^{-6}	Polyester	-6.53×10^{-6}
Polypropylene	-10.1×10^{-6}	Nylon 66	-9.55×10^{-6}
Fluorine	-47.8×10^{-6}		

Table 4.
Magnetic susceptibility of some fibers.

yarns, and fabrics. Therefore, electromagnetic functionalization of fibers, yarns, and fabrics can be achieved by spinning, weaving, and finishing.

In the spinning process for fibers, metal fibers, carbon/graphite fibers, or intrinsically conductive polymer materials having intrinsic electromagnetic function may be used to take place of the ordinary fiber materials in whole or in part. It is possible to add the powder having electromagnetic properties to the spinning solution in the blending way during the spinning process.

In the spinning process for yarns, electromagnetic fibers such as metal fibers and magnetic fibers can be added to the ordinary fibers through different ways to combine, producing the electromagnetic yarn. Metal fibers have low elongation and poor toughness; so, they are not suitable to be used alone for weaving. They are often used to form the yarn containing metal fiber with ordinary textile fibers by blending, enveloping, etc.

In the weaving process, electromagnetically functionalized yarns can be directly woven. The common yarns can be interlaced into fabrics with the electromagnetically functionalized yarns.

The finishing process is suitable for fibers, yarns, and fabrics. For the fiber or the yarn that has been formed and does not have electromagnetic function, the surface of it may be coated with a metal coating or magnetic powder by electroless plating, electroplating, magnetron sputtering, or other ways. For ordinary fabrics without electromagnet properties, the surface can be treated by finishing, such as the electroplating, electroless plating, or embroidery to make it electromagnetic.

3. The application of electromagnetic textile materials

3.1 Antistatic textile materials

The electrostatic phenomenon of cellulose fibers in the processing process is not obvious; but the electrostatic interference of protein fibers is pretty serious. Although the wool fiber has high equilibrium moisture regain, its mass specific resistance is the highest in the natural fiber. The resistivity of synthetic fibers such as polyester, nylon, acrylic, and polypropylene, which are generally high in moisture regain, is as high as 10^{14} Ω cm, and the accumulation of electrostatic charge is obvious.

3.1.1 Electrostatic mechanism

The material is excited by various energies, causing the electrons to escape from the nucleus. The electrons overcome the binding of the nucleus, and the minimum energy required to escape from the surface of the material is called the work function. Different materials or the same material in different states have different work function. The generation and accumulation of electric charge causes the substance to carry static electricity, and the one that acquires the electron exhibits the negative electric property, and the one that loses the electron exhibits the positive electric property, which generate the electrostatic phenomenon.

The resistivity of conventional textile materials is up to 10^{10} Ω cm or more, and the generated charge is not easily dissipated, resulting in very serious electrostatic phenomenon. Therefore, the antistatic properties of textile materials have become an important property having a great influence on the processing of textile materials and the use of textiles.

3.1.2 Antistatic technology for textile materials

The antistatic technology of textile materials includes the preparation of antistatic fibers, the preparation of conductive yarns, and the conductive treatment of textiles.

3.1.2.1 The preparation of antistatic fibers

For textile materials with higher mass specific resistance, surfactants are often added to fibers in fiber factories, which absorb water molecules from the environment and reduce static interference in the yarns. The hydrophobic end of the surfactant molecule is adsorbed on the surface of the fiber; the hydrophilic group is pointed to the outer space [3]. Then, the fiber forms the polar surface and adsorbs water molecules in the air. The surface resistivity of the fiber is reduced, and the charge dissipation is accelerated. The method is simple and easy to make; however, the antistatic effect is poor in durability, and the surfactant is volatile and less resistant to washing.

In order to prepare relatively durable antistatic fiber, the methods are following: (1) Adding the surfactant to a fiber-forming polymer during blend spinning; (2) adding the hydrophilic group by block copolymerization; and (3) adding the hydrophilic group by graft modification in a fiber-forming polymer. These can make the fibers obtain durable hygroscopicity and antistatic properties.

In addition, there are also another methods, including fixing the surfactant to the surface of the fiber with a binder and crosslinking the surfactant on the surface of the fiber to form a film. The effect is similar to applying an antistatic varnish onto the surface of the plastic.

Antistatic fibers are usually blended with ordinary fibers, and a higher content of antistatic fibers is required to achieve a more feasible antistatic effect. The specific ratio between antistatic fibers and ordinary fibers should be based on the resistivity of the ordinary fibers used, the final use environment, and requirements of the products.

3.1.2.2 The preparation of conductive fibers

The electrical resistivity of the conductive fiber is smaller than that of the antistatic fiber, and it has a more significant antistatic effect. And during the blend fabrics with the same antistatic effects, the amount of conductive fiber added is much smaller than that of the antistatic fiber [4]. As long as a few thousandths to a few percent of the conductive yarn is added, the fabric can attain antistatic requirements. So with the widespread use of organic conductive fibers [5, 6], the field of application of antistatic fibers has been gradually reduced.

3.2 Electromagnetic shielding textile materials

3.2.1 Shielding effectiveness

Electromagnetic shielding is a technical measure to prevent or suppress the transmission of electromagnetic energy by using a shield. The shield used can weaken the electromagnetic field strength generated by the field source in the electromagnetic space protection zone. There are two main purposes for shielding: one is to limit the field source electromagnetic energy leaking out from the area that needs protection and the other is to prevent the external electromagnetic field energy entering into the area protected.

The shielding effectiveness equals to the ratio of electric field strength E_0 when a point in space is unshielded to the electric field strength E_S of the field after shielding, or the ratio of the magnetic field strength H_0 to the magnetic field strength H_S of the field after shielding, or the ratio of the power P_0 to the P_S of the field after shielding. The shielding effectiveness (SE) can be expressed as follows:

$$SE_E = 20\lg\frac{|E_0|}{|E_S|}, \; or \; SE_H = 20\lg\frac{|H_0|}{|H_S|}, \; or \; SE = 10\lg\frac{|P_0|}{|P_S|} \quad (5)$$

Arching method is usually used to measure shielding effectiveness. The schematic diagram is shown in **Figure 6**, which can characterize the shielding effectiveness by measuring the power of the receiving antennas and transmitting antennas.

3.2.2 Conductive grid structure of electromagnetic shielding fabrics

Woven fabric is composed of warp and weft yarns interlaced vertically with each other according to a certain regularity. It is well known that ordinary fiber yarns are transparent to the EM wave, while yarn containing metal fibers is conductive. As metal fiber yarns are closely spaced, continuous conductive paths can be established easily. For an electromagnetic shielding fabric composed of metal fiber yarns or coated with a metal layer or a functional layer, it has a remarkable and typical mesh structure [7], as shown in **Figure 7**. **Figure 7(a)** is a schematic view of the structure of the metal fiber-containing yarn fabric and **Figure 7(b)** is a structural model diagram of the metal fiber yarn extracted in **Figure 7(a)**. The geometric parameters in the figure correspond to the parameters in the fabric: h is the buckling wave height, similar to the thickness of the fabrics in the data; l is the length of the organization cycle; and a is the arrangement cycle spacing of the metal yarns in the fabric.

3.2.2.1 The metallic yarn grid model

The metal fiber yarn fabric is considered as a periodic grid structure model composed of conductive yarns [8], as shown in **Figure 8**. Assume that one parallel periodic array is composed of the warp yarn (as shown in **Figure 8(b)**) and the

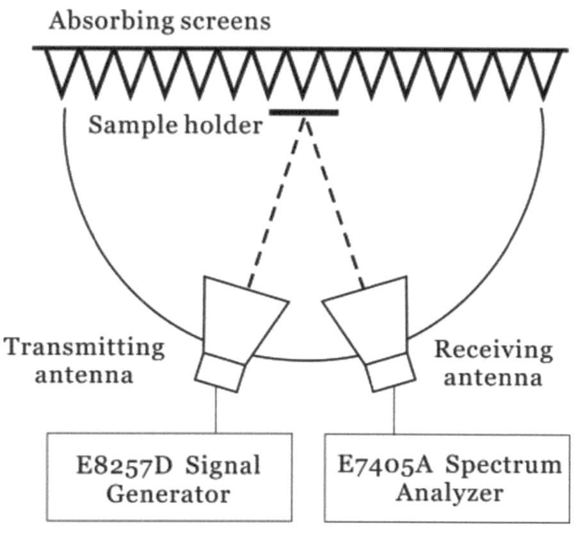

Figure 6.
Schematic diagram of arching measuring method for shielding effectiveness.

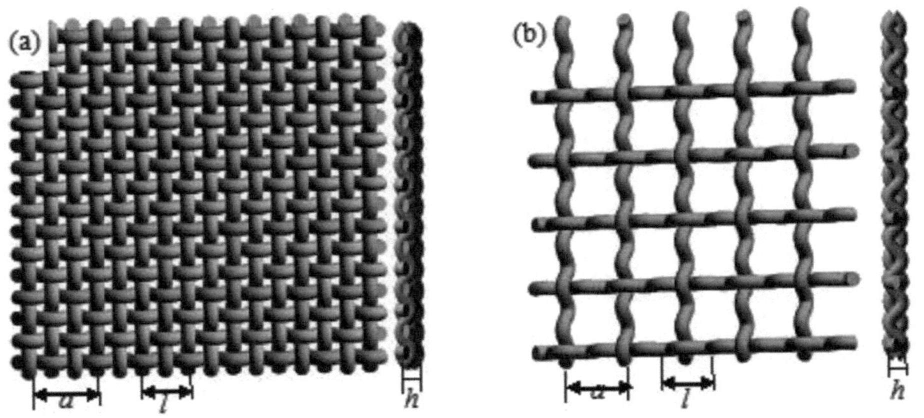

Figure 7.
The grid structure of electromagnetic shielding fabric. (a) Typical mesh structure (b) The structure of the metal fiber-containing yarn fabric.

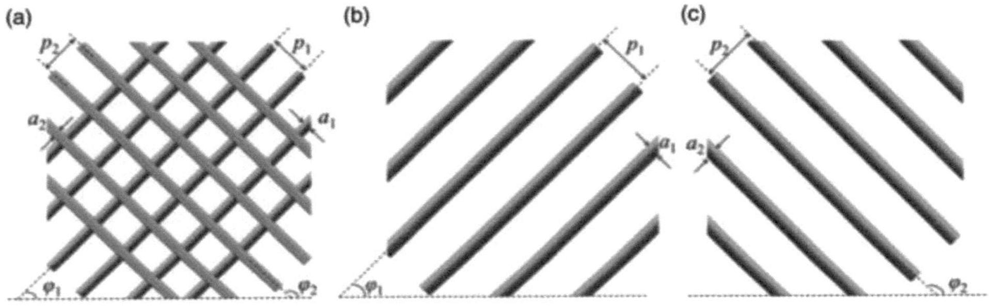

Figure 8.
Grid structure model. (a) Grid structure; (b) Period parallel array 1 and (c) Period parallel array 2.

other is composed of the weft yarn (as shown in **Figure 8(c)**). The two wire arrays are directly cascaded at a certain orientation angle. The contact impedances between the wires of the two arrays are assumed to be negligible, due to the *EM* coupling between the two grid arrays. In particular, the grid structure model is only composed of metal fiber yarns, that is, the effect of ordinary yarns in fabric is not considered.

The two parallel periodic arrays (as shown in **Figure 1**) have wire orientation ψ_1, ψ_2, and (°); spacing p_1 and p_2 (m); and wire diameter a_1 and a_2 (m), respectively. In most fabric structures, the grid is anisotropic and asymmetric, namely, $\psi_1 \neq \psi_2$, $p_1 \neq p_2$, and $a_1 \neq a_2$.

3.2.2.2 Shielding effectiveness of the grid model

The periodic grid is regarded as a stratified medium made of two periodic parallel arrays at a certain angle. The transmission matrix was established, and the *SE* for different polarization incident waves was calculated by analyzing the propagation of *EM* fields passing through the wire mesh. It can be used for the calculation of the *SE* of isotropic and anisotropic metal wire mesh structures for different polarization.

The grid is excited by a plane wave having normal incidence, and the incident *EM* wave can be decomposed into TM_Z (transverse magnetic: the magnetic field component only in the plane perpendicular to the propagation direction) and TE_Z (transverse electric: the electric field component only in the plane perpendicular to the propagation direction) polarized waves, that is, the vertical polarization wave and the horizontal polarization wave, respectively, as shown in **Figure 9**.

ε_0 and μ_0 are the dielectric constant and absolute permeability of free space, respectively. Assume that the metallic yarns are lossy and characterized by the per-unit-length impedance Z_w (unit: Ω/m). The skin depth is given by

$$\delta = \sqrt{\frac{1}{\pi f \mu_0 \mu_r \sigma}} \tag{6}$$

where δ is the skin depth (m); μ_0 is permeability of the vacuum (H/m); μ_r is relative permeability of the conductive yarns; σ is electrical conductivity (S/m); and f is frequency (Hz).

As shown in **Figure 10**, the global coordinate system (x, y, z) and the local one (ξ, ψ, ζ) are introduced. The ζ-axis is parallel to the wires of the array; the x-axis is parallel to the ξ-axis, and they are perpendicular to the plane of the periodic parallel array, which has yarn orientation ψ, spacing p, and yarn diameter a.

Assuming that the diameters of the metallic yarns and periodic spacing are small compared to the wavelength, then the parallel array can be modeled by a homogeneous thin anisotropic sheet with thickness a. Under the condition of the *EM* wave having normal incidence, the following relations can be written among the propagating field components expressed in the local coordinate system and averaged over the wire array period, on each side of the sheet.

$$E_\zeta(a) = E_\zeta(0) \tag{7}$$

$$E_\psi(a) = E_\psi(0) \tag{8}$$

$$H_\zeta(a) = H_\zeta(0) \tag{9}$$

$$H_\psi(a) = H_\psi(0) + J \tag{10}$$

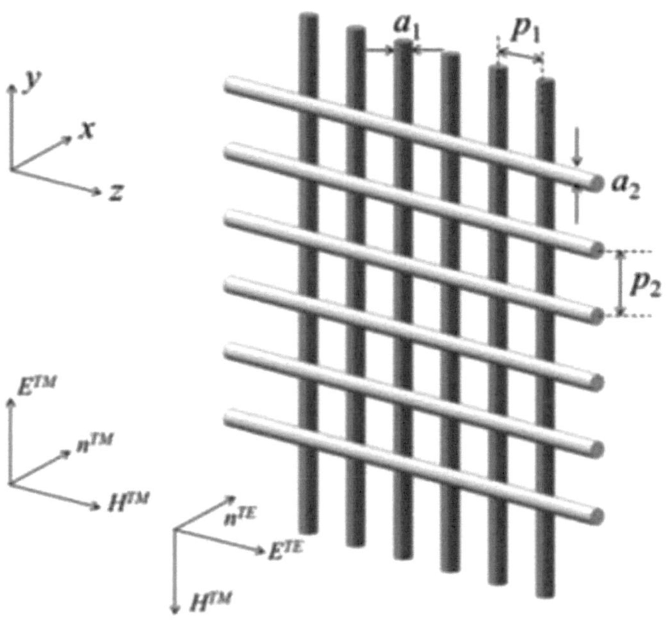

Figure 9.
Schematic wire mesh illuminated by a plane wave with normal incidence and transverse magnetic (TM) or transverse electric (TE) polarization.

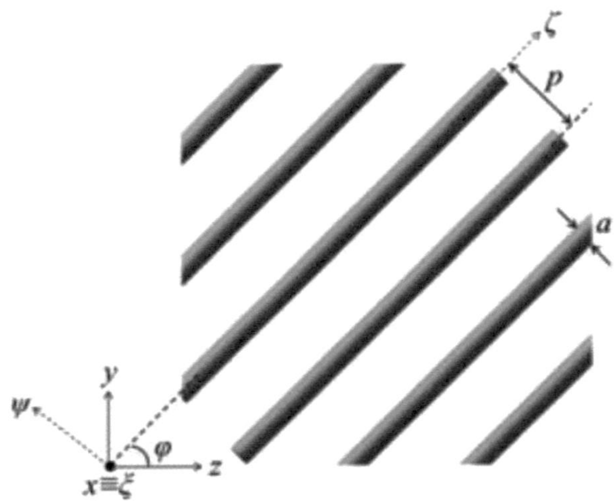

Figure 10.
The periodic array in the local coordinate system and the global coordinate system.

in which J is the current density (A/m^2) flowing inside the wire and all quantities are considered to be averaged over the wire array period. The current density J can be related to the tangential average electric field components along the wires. According to the following impedance condition,

$$E_\zeta(0) = Z_S J \tag{11}$$

in which the expressions of impedance Z_S for the periodic array are given by

$$Z_S = \left[Z_W p9 + j\alpha\sqrt{\mu/\varepsilon}\right] \tag{12}$$

where μ (H/m) and ε (F/m) are the magnetic permeability and dielectric constant of the substrate, respectively; α is a parameter that depends on the geometry structure of the periodic array and on the wavelength in the substrate λ_W (m).

$$\alpha = (p/\lambda_W)\log(p/(a\cdot\pi)) \tag{13}$$

Combining Eqs. (10) and (11) yield

$$H_\psi(a) = H_\psi(0) + E_\zeta(0)/Z_S \tag{14}$$

According to Eqs. (7)–(9) and (14), the boundary condition describing the relation among the *EM* field components tangential to the conductive grid is obtained, and its matrix expression is

$$\begin{bmatrix} E_\zeta(a) \\ E_\psi(a) \\ H_\zeta(a) \\ H_\psi(a) \end{bmatrix} = [\Phi_{loc}] \begin{bmatrix} E_\zeta(0) \\ E_\psi(0) \\ H_\zeta(0) \\ H_\psi(0) \end{bmatrix} \tag{15}$$

in which $[\Phi_{loc}]$ is the transformation matrix of the parallel array in the local coordinate system as follows:

$$[\Phi_{loc}] = \begin{bmatrix} 1 & 0 & 0 & 0 \\ 0 & 1 & 0 & 0 \\ 0 & 0 & 1 & 0 \\ 1/Z_S & 0 & 0 & 1 \end{bmatrix} \tag{16}$$

In the global coordinate system, the transformation matrix of the parallel array $[\Phi]$ is given by

$$[\Phi] = [R]^{-1}[\Phi_{loc}][R] \tag{17}$$

in which the transformation matrix $[R]$ from the global coordinate system to the local coordinate system is as follows:

$$[R] = \begin{bmatrix} \sin\varphi & \cos\varphi & 0 & 0 \\ -\cos\varphi & \sin\varphi & 0 & 0 \\ 0 & 0 & \sin\varphi & \cos\varphi \\ 0 & 0 & -\cos\varphi & \sin\varphi \end{bmatrix} \tag{18}$$

The grid array transmission matrix $[\Phi]$is

$$[\Phi] = \begin{bmatrix} [U] & [0] \\ [Y_T] & [U] \end{bmatrix} \tag{19}$$

where $[U]$ and $[0]$ are the bi-dimensional unit and null matrices, respectively, and $[Y_T]$ is the effective shunt admittance of the parallel array; its matrix expression is

$$[Y_T] = \frac{1}{Z_S}\begin{bmatrix} -\sin\varphi\cos\varphi & -\cos^2\varphi \\ \sin^2\varphi & \sin\varphi\cos\varphi \end{bmatrix} \tag{20}$$

The total thickness of the wire grid is $a_M = a_1 + a_2$; the orientations of two parallel arrays are ψ_1 and ψ_2, respectively, and the effective shunt admittances are $[Y_{T1}]$ and $[Y_{T2}]$, respectively. The transmission matrix of the grid is the product of two parallel array transmission matrixes $[\Phi_1]$ and $[\Phi_2]$ as shown below:

$$[\Phi_M] = [\Phi_2][\Phi_1] = \begin{bmatrix} [U] & [0] \\ [Y_M] & [U] \end{bmatrix} \tag{21}$$

in which the effective shunt admittance $[Y_M]$ of the grid is

$$[Y_M] = [Y_{T1}] + [Y_{T2}] \tag{22}$$

Therefore, the matrix of the boundary condition for the metal grid is given as

$$\begin{bmatrix} [E_z(a_M)] \\ [E_y(a_M)] \\ [H_z(a_M)] \\ [H_y(a_M)] \end{bmatrix} = [\Phi_M]\begin{bmatrix} [E_z(0)] \\ [E_y(0)] \\ [H_z(0)] \\ [H_y(0)] \end{bmatrix} \tag{23}$$

The shielding factors of the metal grid of against TM_z and TE_z polarized waves with normal incidence are F_{TM} and F_{TE}, respectively, shown as:

$$F_{TM} = \left|E^{TM,i}\right|^2 / \left|E^{TM}(a_M)\right|^2 \tag{24}$$

$$F_{TE} = \left|E^{TE,i}\right|^2 / \left|E^{TE}(a_M)\right|^2 \tag{25}$$

in which $\left|E^{TM,i}\right|$ and $\left|E^{TE,i}\right|$ are the amplitude of the incident electric field for the TM_z and TE_z polarized plane waves, respectively. $\left|E^{TM}(a_M)\right|$ and $\left|E^{TE}(a_M)\right|$ are the amplitude of the corresponding transmitted electrical field. $[\eta_0]$ is the free space wave impedance and $\left[E^i\right]$ is the vector of the incident electric field.

The electric field component on the back faces of the grid expressed in matrix form is given by

$$[E(a_M)] = 2\eta_0^{-1}\left[Z_{eq}\right]\left[E^i\right] \tag{26}$$

in which

$$\left[Z_{eq}\right] = \left\{[Y_M] + 2[\eta_0]^{-1}\right\}^{-1} \tag{27}$$

In the case where the incident wave is the TM_z polarized wave, set $\left[E^i\right] = [0E^y{}_i]_t$. Therefore, the amplitude of the transmitted field is

$$\left|E^{TM}(a_M)\right| = 2\eta_0^{-1}\left\{\left|Z_{eq}(1,2)\right|^2 + \left|Z_{eq}(2,2)\right|^2\right\}^{1/2}\left[E_y^i\right] \tag{28}$$

Similarly, for a TE_z polarized incident wave, set $\left[E^i\right] = [E^y{}_i, 0]_t$. Therefore, the amplitude of the transmitted field is

$$\left|E^{TE}(a_M)\right| = 2\eta_0^{-1}\left\{\left|Z_{eq}(1,1)\right|^2 + \left|Z_{eq}(2,1)\right|^2\right\}^{1/2}\left[E_z^i\right] \tag{29}$$

The shielding factors are, respectively

$$F_{TM} = \frac{\eta_0^2}{4}\left\{\left|Z_{eq}(1,2)\right|^2 + \left|Z_{eq}(2,2)\right|^2\right\}^{-1} \tag{30}$$

$$F_{TE} = \frac{\eta_0^2}{4}\left\{\left|Z_{eq}(1,1)\right|^2 + \left|Z_{eq}(2,1)\right|^2\right\}^{-1} \tag{31}$$

The *SE* against TM_z and TE_z polarized incident waves are, respectively

$$SE^{TM} = 10\lg F_{TM} \tag{32}$$

$$SE^{TE} = 10\lg F_{TE} \tag{33}$$

For isotropic material, the transmitted *EM* power due to incident TM_z or TE_z polarized plane waves is equal. Namely

$$\left|E^{TM}(a_M)\right| = \left|E^{TE}(a_M)\right| = |E(a_M)| \tag{34}$$

For anisotropic materials, the total incident and transmitted power are computed as the average of the two polarizations

$$P^i = \left[\left|E^{TM,i}\right|^2 + \left|E^{TE,i}\right|^2\right] / (4\eta_0) \tag{35}$$

$$P = |E(a_M)|^2/(2\eta_0) \tag{36}$$

From Eq. (6), we obtain

$$SE = 10\lg\left(\frac{F_{TM} + F_{TE}}{2}\right) \tag{37}$$

For the grid that is isotropic in the (y, z) plane, the *SE* against the two polarized waves is coincident. It results in

$$SE = SE^{TM} = SE^{TE} \tag{38}$$

3.2.3 Influencing factors for electromagnetic shielding effectiveness

In combination with the mesh structure of the electromagnetic shielding fabric, the key factors affecting the electromagnetic shielding effectiveness *SE* are the structural parameters and material parameters of fabric [9, 10]. Some key 1parameters are as following: the periodic spacing of metal yarns, the conductivity of the yarns (decided by the type of yarns, the type of metal fibers, the content of metal fibers), the diameters of the yarns, the arrangement method, the connection of intersections, the direction of incidence of electromagnetic field, and the frequency.

3.2.3.1 The influence of structural parameters on shielding effectiveness

3.2.3.1.1 The influence of periodic spacing on SE

Samples in which copper filaments are arranged in parallel at different intervals have different *SE*. As shown in **Figure 11**, the arrangement periodic intervals of copper filaments are 1, 2, 3, 4, and 5 mm, respectively. The *SE* of 10–14 GHz is 17–20, 7–12, 5–10, 2–5, and 2–4 dB, respectively. It can be seen that as the periodic spacing increases, the shielding effectiveness is significantly reduced.

Metal yarn arrangement interval has an important effect on *SE* of fabrics. The metal yarn periodic spacing is related to these parameters such as fabric density and

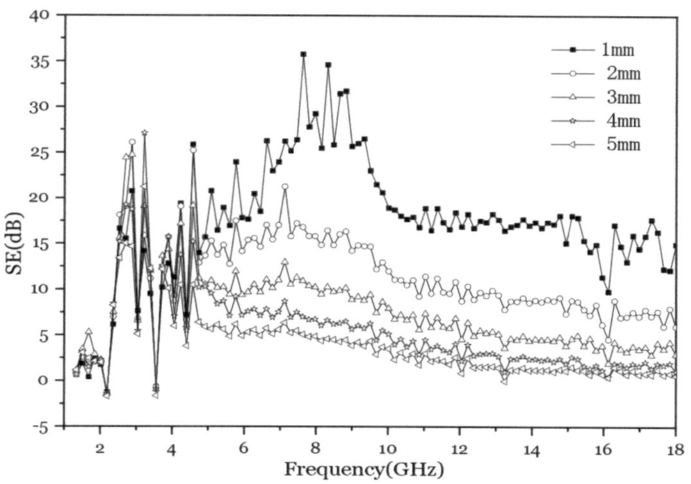

Figure 11.
The shielding effectiveness at different intervals.

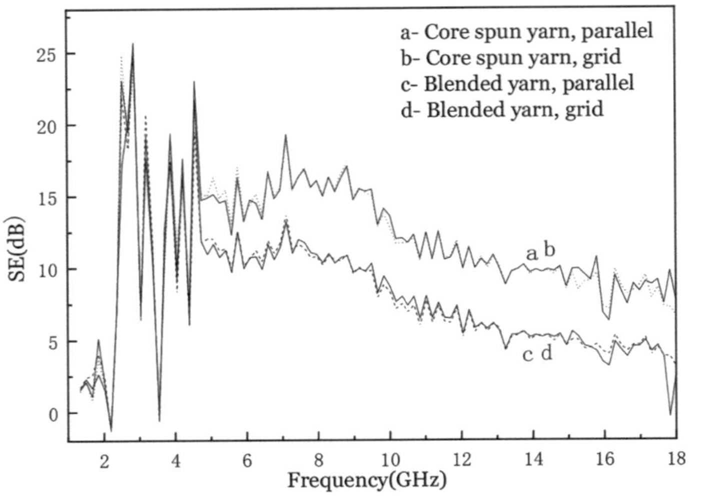

Figure 12.
The shielding effectiveness of metal fiber yarns in parallel and grid arrangement.

tightness. And as the fabric tightness and density increase, the spacing of the metal fibers reduces, the electromagnetic wave transmission decreases, and the *SE* increases.

3.2.3.1.2 The influence of the way of arrangement on SE

The woven fabric is interwoven from the yarns of two systems that are perpendicular to each other. Therefore, it is possible to introduce functional fibers into the parallel structure in only one system, or functional yarns are introduced to both systems to form a grid structure.

The stainless steel core spun yarn and the blended yarn were arranged in parallel as a sample with a spacing of 2 mm, and the distance of the grid structure sample is 2 mm in both vertical and horizontal directions. The shielding effectiveness is shown in **Figure 12**. It can be seen that the *SE* of the yarn model of the parallel arrangement structure is the same to the grid arrangement structure, and as the frequency increases, the *SE* gradually decreases.

3.2.3.1.3 The influence of intersection conduction on SE

Separating the yarns of the horizontal and vertical systems in the sample with a thin insulating plate is seen as a nonconducting state. The bare copper wire is arranged in two states with conduction and nonconduction at the intersection, and the periodic intervals of the grid samples are 1, 2, 3, 4, and 5 mm, respectively, the shielding effectiveness is shown in **Figure 13**. It can be seen that under the same periodic spacing, the shielding effectiveness curves of the copper wire mesh model samples almost coincide in the two states of conduction and nonconduction.

3.2.3.2 Effect of material parameters on shielding effectiveness

The material parameters mainly include the way of forming the metal yarns, the material of the metal fibers, and the content of the metal fibers.

3.2.3.2.1 The effect of the method of forming yarns on SE

Metal monofilaments can be used to make fabrics after they have been formed into yarns by a certain yarn forming method. For metal filaments, core yarns and twisted yarns are the common yarns. However, for metal staple fibers, blended

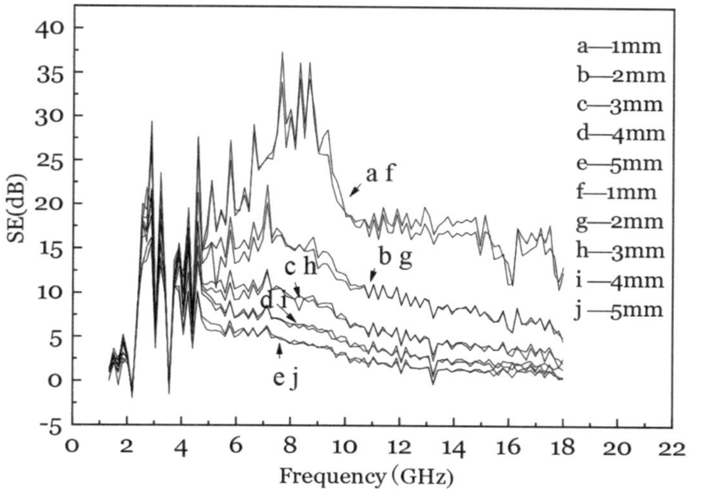

Figure 13.
The shielding effectiveness of metal fiber yarns in parallel and grid arrangement.

yarns are the commonly used yarn. The type of yarns affects the electromagnetic parameters of the yarns and fabrics, resulting in the difference in *SE*.

Stainless steel filaments, core-spun yarns, blended yarns, and twisted yarns composed of stainless steel/cotton with a stainless steel content of 30% are arranged in a grid sample with a periodic spacing of 2 mm. The *SE* is shown in **Figure 14**. At the same spacing, the blended yarns have the best shielding effectiveness, while the shielding effectiveness of stainless steel filaments, core spun yarns, and twisted yarns are equivalent. At the frequency of 8–16 GHz, the *SE* of the former is about 7 dB higher than that of the latter, and both decrease with increasing frequency.

3.2.3.2.2 The effect of the material of the metal fibers on SE

The metal fibers used in the fabric are different, and the different electrical conductivity of the metal may affect the electrical resistivity of the yarns and the

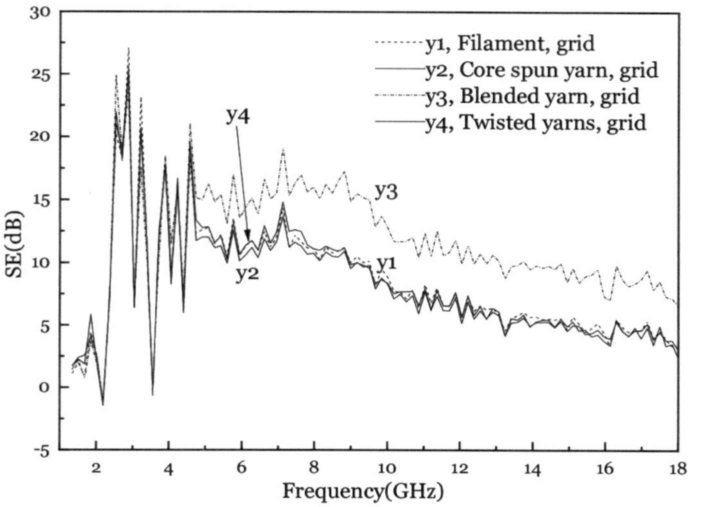

Figure 14.
The shielding effectiveness of different types of yarns.

fabrics. For example, the electrical conductivity of copper fiber is 5.8×10^7 S/m, and the electrical conductivity of aluminum is 3.54×10^7 S/m. When the metal fiber content, linear density, and fabric specification parameters are the same, the *SE* of copper fiber fabrics are better than that of aluminum fiber fabrics.

The five grid samples with completely nonconducting period of 2 mm is shown in **Figure 15**, for which each grid sample of five different materials consisting of stainless steel bare wire (the diameter is 35 μm), core spun yarn and blended yarn of stainless steel/cotton (containing the stainless steel content of 30%), silver-plated nylon filament (the diameter is 50 μm), and bare copper wire (the diameter is 80 μm). The *SE* of the samples made of, respectively, stainless steel blended yarns, silver-plated filaments, and bare copper wires are substantially equal and higher that of the samples of core spun yarn and stainless steel bare wire.

3.2.3.2.3 The effect of metal fiber content on SE

The two blended yarns with the stainless steel content of 20 and 30% are woven in both warp and weft directions, and the *SE* of the obtained sample in the range of 1–18 GHz are shown in **Figure 16**. It can be seen that except for the frequency range of 10 and 12–14 GHz, the *SE* of the fabric with 30% stainless steel content is 5 dB higher than that of the fabric with 20% stainless steel content, and the difference of *SE* in other frequency bands is unobvious. The content of stainless steel fibers has a certain effect on the shielding effectiveness of the fabric, but after reaching a certain level, the difference is not significant.

It is assumed that the fibers are evenly distributed in the yarn. As the content of the metal fibers increases, the shielding effectiveness of the fabrics will increase, but when it is increased to a certain extent, the bending stiffness and flexural modulus of the yarns will increase, and the porosity among fibers during the fabric will increase. So the *SE* of the fabrics becomes slow down or even lower. Considering the cost, the content of stainless steel is generally 20–30% for fabrics containing stainless steel fibers.

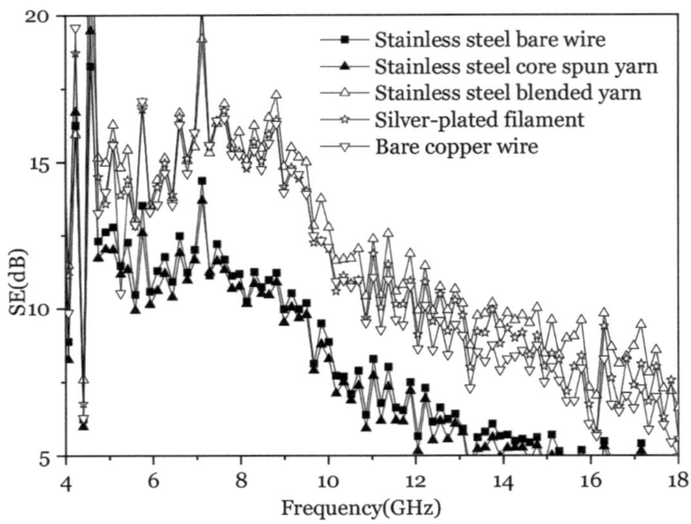

Figure 15.
Shielding effectiveness of different materials with grid period spacing of 2 mm in order.

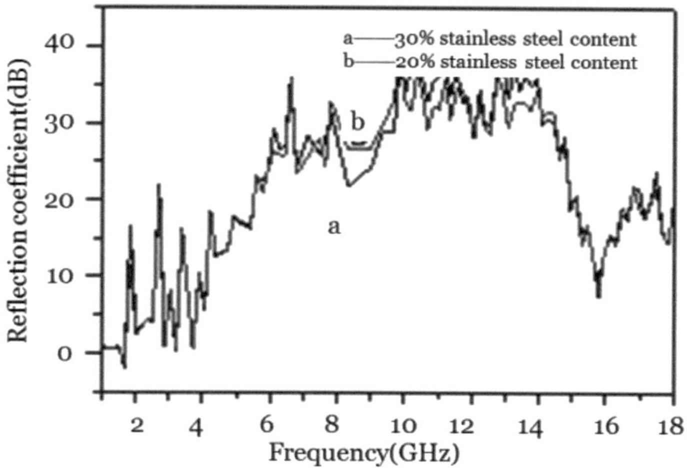

Figure 16.
The shielding effectiveness of the fabrics with different content of stainless steel.

3.3 Electromagnetic scattering fabrics

3.3.1 The mechanism of electromagnetic scattering fabrics

When electromagnetic waves radiate into the macroscopic object, which will causing induced electric charges and currents of the object, then the electromagnetic wave radiated into the object will be scattered into various directions. This process is called electromagnetic scattering [11, 12]. The electromagnetic scattering fabric is an electromagnetic functional material with the specific design structure, which makes the electromagnetic waves incident on the target are no longer reflected back along the way of the reflection of mirror, but radiated out into different directions. Thereby, it can reduce the radiated electromagnetic waves in the direction of propagation, and can make the human body and military targets invisible for certain direction Radar.

The textile technology is relatively mature in the preparation of three-dimensional structural fabrics. Thus, it is very feasible to design the three-dimensional structure of metallized fabrics which have good scattering properties for incident electromagnetic waves.

3.3.1.1 The electromagnetic wave scattering characteristics of linear column unit structures

As shown in **Figure 17**, the composite materials of the three-dimensional periodic structure can be simplified into two-phase dielectric materials when studying the transmission process of electromagnetic waves in the three-dimensional structure.

When a simple harmonic uniform plane wave is incident on the three-dimensional structure, the three-dimensional coordinate system XYZ shown in **Figure 18** is selected. The XOY plane of the coordinate system coincides with the lower surface of the object, and the Z axis is perpendicular to the interface of the upper and lower surfaces. Electromagnetic waves are incident from the upper surface with the incident angle θ, and reflected waves and transmitted waves are generated at the interface of the upper surface. The transmitted waves enter the

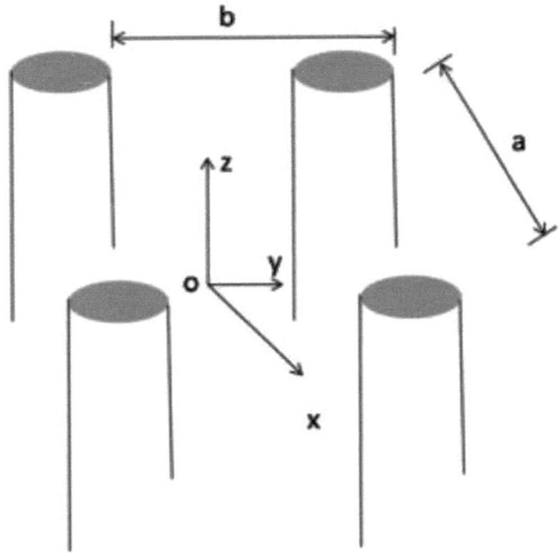

Figure 17.
Dielectric column periodic three-dimensional structure.

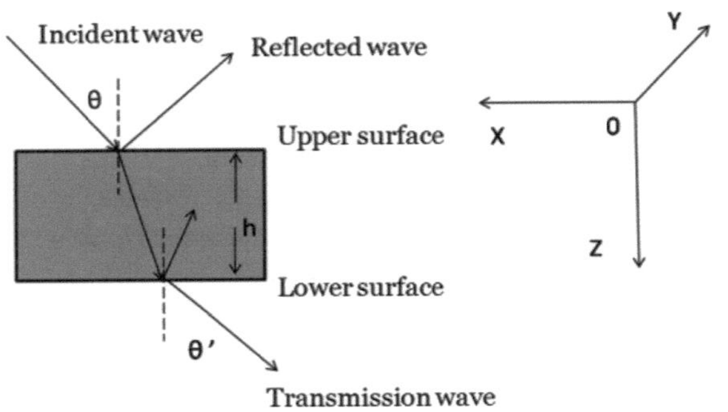

Figure 18.
The schematic diagram of electromagnetic wave incident three-dimensional structure.

periodic three-dimensional structure, and after being attenuated in the three-dimensional structure, the reflected waves and the transmitted waves are again generated at the lower surface interface.

3.3.1.2 The electromagnetic wave scattering characteristics of the concave-convex element structure

Electromagnetic waves are reflected at the interface of different medium, which conforms to the law of reflection, as shown in **Figure 19**. The concave-convex natural surfaces can be broken down into a series of planar elements with small-sized geometries, which is called roughness. The roughness of the scattering surface is very important in the surface scattering.

If the surface is smooth, the incident energy would form two plane waves after interacting with the surface. One is a surface-reflected wave whose angle with the normal is the same as the angle of incidence, and the direction is opposite, as shown in **Figure 20**. The other is refracted or transmitted waves with downward surface.

If the surface is rough, the incident energy interacts with the surface and then radiates and shoots in all directions, becoming a scattering field, as shown in **Figure 21**.

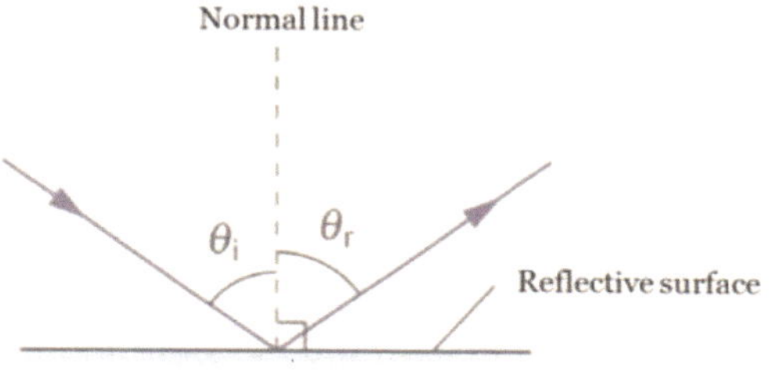

Figure 19.
The law of electromagnetic wave reflection.

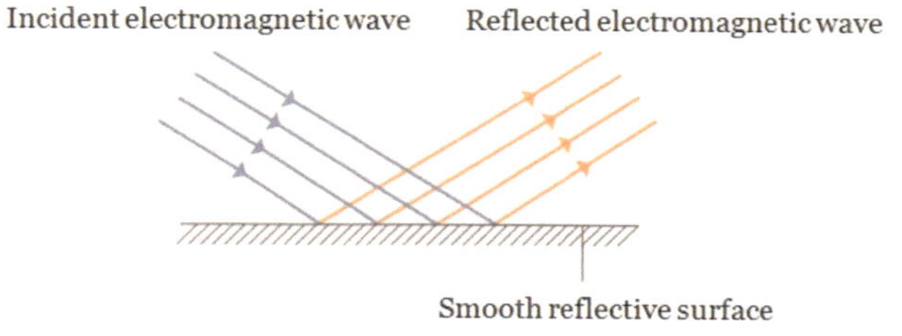

Figure 20.
The reflection of electromagnetic waves on smooth reflective surfaces.

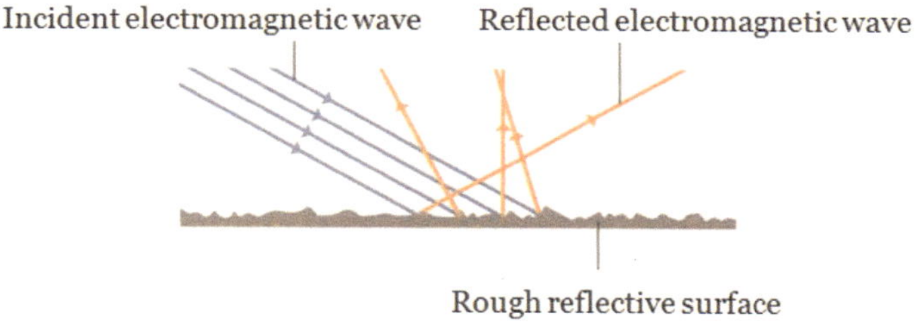

Figure 21.
The reflection of electromagnetic waves on rough reflective surfaces.

3.3.2 The scattering properties of electromagnetically functional textile materials

3.3.2.1 Velvet structure fabrics

The fibers having electromagnetic properties are scattered as the fluff of the fabric on the surface, or are consolidated as a U-shaped structural unit on the fabric to form the fluff, and thereby a velvet structure fabrics with good radar wave scattering property is obtained.

The woven velvet fabric for decoration and its reflection coefficient is displayed in **Figure 22**. It can be seen from the test results that the tested structural unit achieves attenuation of 5 dB in the bandwidth of 10 GHz, and the peak value reaches −30 dB. This is mainly due to the angle between the metal fluff of the structural unit and the plane of the sample. When electromagnetic waves are incident onto the sample, those metal fluffs with a certain angle in the plane have a certain scattering of the incident electromagnetic waves, which reduces the energy received by the receiving antenna, so that the reflection coefficient is reduced.

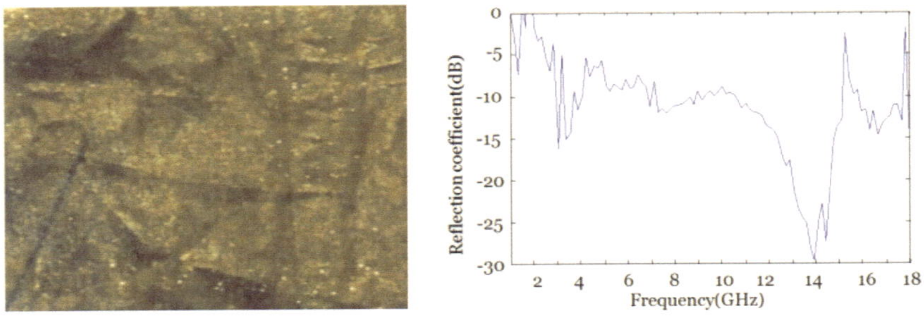

Figure 22.
The woven velvet fabric for decoration and its reflection coefficient.

3.3.2.2 The three-dimensional spacer fabrics

The silver fiber spacer fabric is prepared on the warp knitting machine, of which the silver-plated filaments with a fineness of 83dtex are used in the middle layer, and the upper and lower surfaces are all made of polyester fibers, as shown in **Figure 23**. The silver-plated fibers make the radar waves absorption, reflection, and multiple reflections happen in the intermediate layer.

The silver-plated fiber spacer fabric has a significant resonance peak, which should be related to the thickness of the intermediate layer of the fabric, and indicates that the shielding effect on the radar wave is not mainly due to the reflection radar wave mechanism. The reflectivity of silver-plated fiber spacer fabric is generally inferior to that of velvet fabrics, but its resonance peak can reach −30 dB, and when the reflection coefficient is below −5 dB, it has a wide bandwidth, even up to 18 GHz.

3.3.2.3 Cut flower structure fabrics

The cut fabrics are obtained by cutting fabrics containing metal fibers or metallized fibers into different shapes. The planar fabrics are formed into the three-dimensional structure through some support, and the cut flower units of fabrics become scattering units for radar waves, which are a kind of flexible, lightweight, wide-band radar stealth fabric.

The stainless steel/polyester/cotton blend fabric with a stainless steel content of 20% is cut as shown in **Figure 24**. The reflection coefficient of the fabric in the

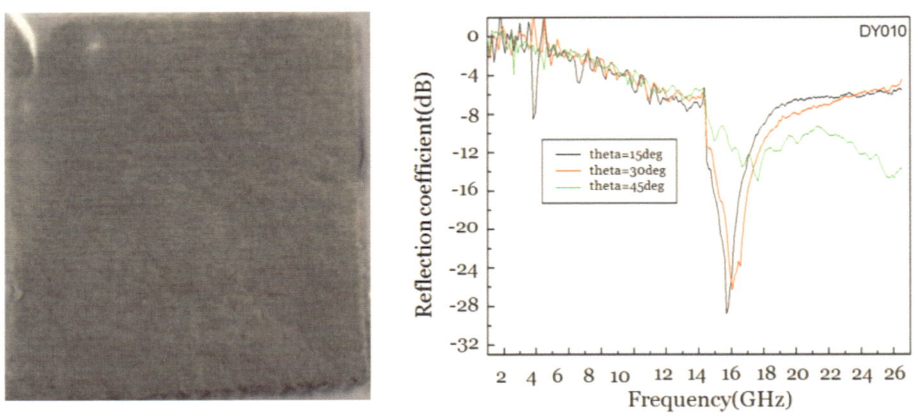

Figure 23.
The silver-plated fiber spacer fabric and its reflection coefficient.

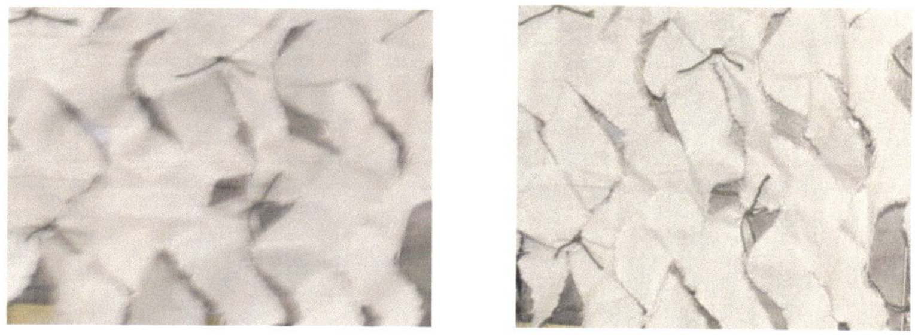

(a) (b)

Figure 24.
Cut flower structure fabric. (a) Three-dimensional cut flower fabric and (b) Flat cut flower fabric.

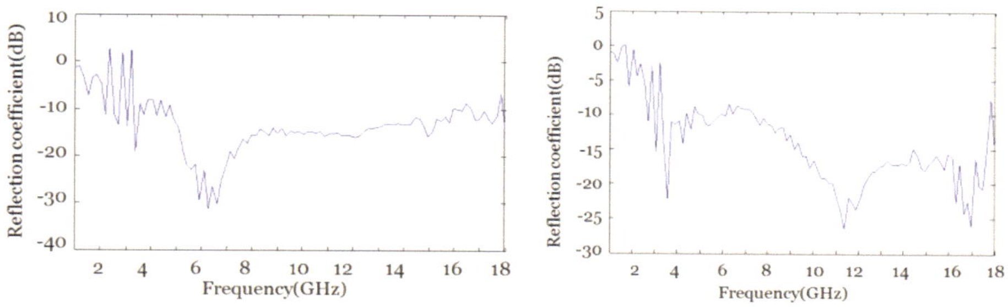

(a) (b)

Figure 25.
The reflection coefficient of cut flower structure fabric. (a) The reflection coefficient of three-dimensional cut flower fabric and (b) The reflection coefficient of flat cut flower fabric.

three-dimensional state and the state in which the fabric is flattened is as shown in **Figure 25**.

From the test results, it can be found that the reflection coefficient of the flat structural unit and the uneven structural unit are very large in a wide frequency range, and the coefficient of the uneven cut flower fabric can reach −10 dB at 2 GHz, and the flat cut flower fabric is −5 dB. In the test results, mainly because of the antenna used in the test, the results of the test in the frequency bands less than 3 GHz and greater than 17 GHz are not regular enough. The irregularity of the structural unit produces a strong scattering for electromagnetic waves, making the reflection coefficient smaller. However, the main difference in the structural unit of unevenness and flatness is the difference in the position of the resonance peak that appears. This is because when the mesh structure becomes flat, the size of the unit structure becomes small, so that the resonance peak shifts toward the higher frequency.

3.3.2.4 Uneven surface structure fabrics

By adopting the method of embedding the heat shrinkable yarns, the textured structure containing metal fibers or metallized fibers can be obtained, which imparts good electromagnetic wave scattering properties to the fabric.

The stainless steel fibers and the cotton fibers in a ratio of 40/60 were blended into a yarn of 116dtex linear density, and high heat-shrinkage polyester yarns with a shrinkage ratio of 53.7% in the boiling water and a linear density of 167dtex are embedded in the warp and weft directions. Mixed yarns and polyester yarns are

Figure 26.
The original plain weave.

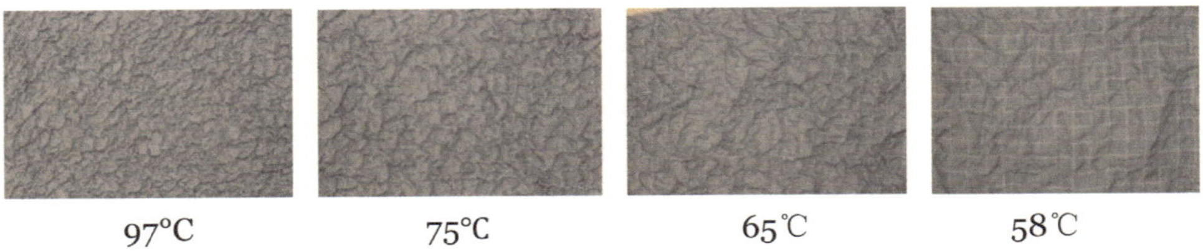

Figure 27.
Stainless steel fabrics with concave structure embedded with high heat shrinkage wires.

woven into a plain fabric with a square weight of 127 g/m^2 (as shown in **Figure 26**). The fabrics are treated at different temperatures to obtain fabrics with different concave and convex structures, as shown in **Figure 27**.

It can be seen from **Figure 27** that the fabrics have different degrees of unevenness at different heat processing temperatures. The higher the treatment temperature, the more obvious the uneven structure; the lower the treatment temperature, the smaller the uneven structure. At 58°C, the fabric has a smaller degree of shrinkage.

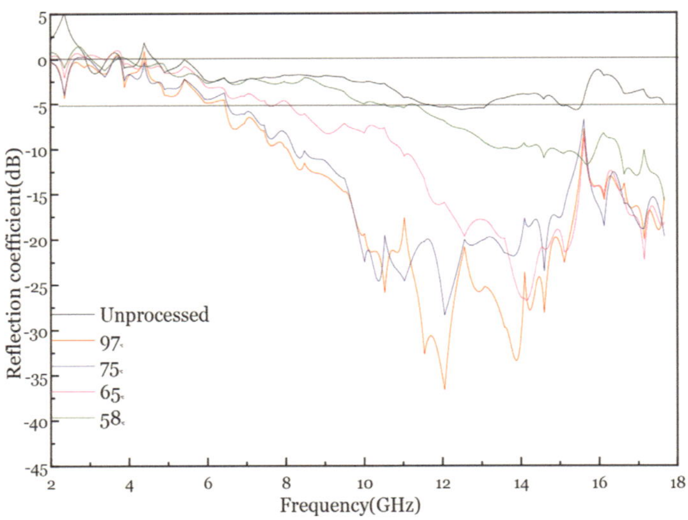

Figure 28.
Comparison of reflection coefficients under different conditions.

As **Figure 28** shows, in the range of 2–18 GHz, for the fabrics containing the heat shrinkage yarns in both directions, as the processing temperature is lowered, the degree of the uneven structure of the fabrics is reduced, the unit size of the concave and convex structure becomes larger, which make the fabrics have poor scattering performance for radar waves. Besides, the reflection coefficients are becoming more and more higher and the difference is obvious. At a frequency of 14 GHz, the reflection coefficients of fabrics having heat treatment temperatures of 97, 75, 65, and 58°C and untreated fabrics, respectively, are −39, −27, −24, −10, and −4 dB. It can be seen from the shrinkage structure at different temperatures that the degree of wrinkles of the fabrics after treatment at 97°C is significantly higher than that of the wrinkles treated at other temperatures. The unevenness of the fabric structure causes the electromagnetic waves to form the diffuse reflection in the structure; meanwhile, the electromagnetic wave scattering forms multiple absorptions on the adjacent two intersecting slopes.

3.4 Frequency selective surface textile materials

3.4.1 The structure and properties of frequency selective surface textile materials

Frequency selective surface (FSS) is an infinitely large periodic array structure that is one-dimensional, two-dimensional, etc. It is mainly divided into two types: patch type and aperture type, which have frequency selective characteristic for the propagation of electromagnetic wave in space. The patch type can totally reflect electromagnetic waves of a specific frequency, and the aperture type can transmit all electromagnetic waves of a specific frequency.

The textiles are light, soft, and flexible in processing. Relying on the media of textile materials, textile processing technology will qualify the textile products to attain the filtering characteristics and light, soft, and other characteristics, which can be applied in more fields [13]. Flexible periodic array structure prepared by textile processing technology is called frequency selective fabric (FSF). According to the filtering characteristics, the frequency selective fabric can be divided into four frequency response characteristics: high pass, low pass, band pass, and band stop, as shown in **Figure 29**.

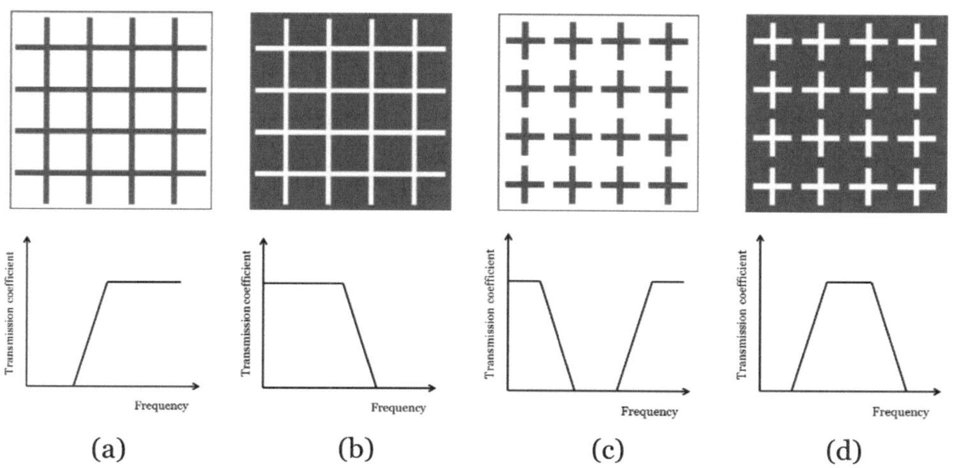

Figure 29.
The four frequency response characteristics of the frequency selective surface. (a) High pass; (b) low pass; (c) band stop; and (d) band pass.

3.4.2 The processing of frequency selective surface textiles

3.4.2.1 The method for preparing frequency selective surface textiles

The researches mainly focus on the preparation of frequency selective fabrics with high-precision two-dimensional periodic structure produced by different textile processing techniques, which can be roughly divided into four categories [14].

1. Continuous conductive yarns form a periodic structure in the fabric. The continuous carbon fibers are directly woven into a square or rectangular periodic structure, as shown in **Figure 30**. Since the conductive carbon fibers are continuously present in the fabric, the structure is actually a conductive grid formed by the conductive yarns in the woven fabric. This structure is more suitable for the preparation of isotropic electromagnetic shielding fabrics, not a true frequency selective periodic structure, which is confirmed by the absence of resonance peaks in the test curves reported in the article.

2. The cut commercialized conductive material unit is directly bonded to the nonconductive fabric substrate, as shown in **Figure 31**.

3. Depositing conductive materials on the surface of fabrics by screen printing, inkjet printing, and other textile finishing techniques can form the conductive structural unit, as shown in **Figure 32**.

4. The high conductive yarns are formed into a periodic structural unit by textile weaving processing techniques such as weaving, weft knitting, embroidery, and so on, as shown in **Figure 33**.

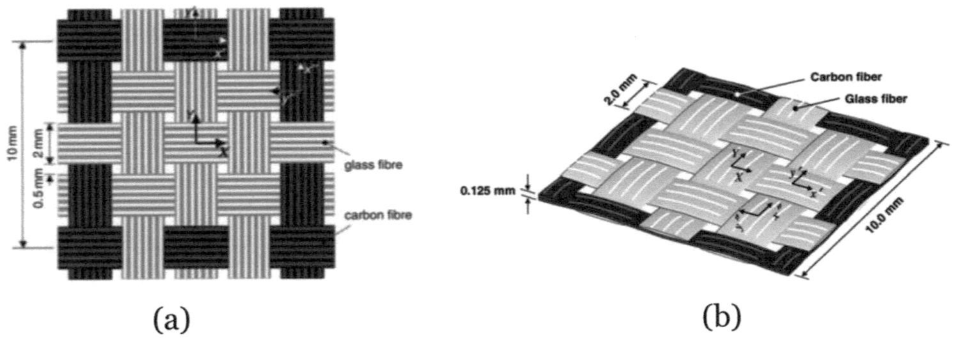

Figure 30.
Continuous conductive yarns form a periodic structure in the fabric. (a) Woven in a square structure and (b) Woven in a square structure rectangle.

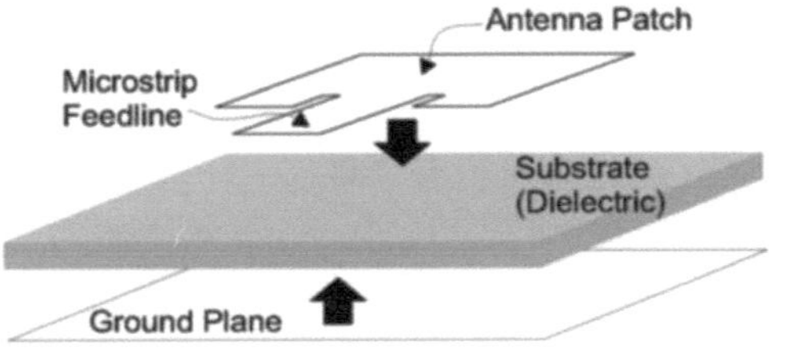

Figure 31.
The conductive material is adhered to the substrate to form a periodic structure.

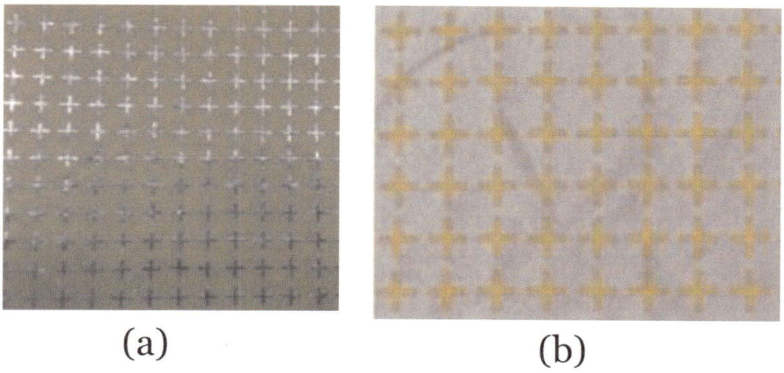

Figure 32.
Screen printing and inkjet printing form a periodic structure. (a) The screen printing and (b) The inkjet printing.

Figure 33.
The periodic structures produced by knitting and embroidery processes. (a) The samples of woven fabrics and (b) The samples of knitted fabrics.

In China, the team that studies the periodic structure of textile materials is mainly a joint research group composed of Professor Meiwu Shi in textile materials and Professor Qun Wang in electromagnetic materials. Based on preliminary sample preparation, theoretical simulation analysis, and the preliminary experimental results and research ideas of special electromagnetic functional textile materials, in the aspect of 2D FSF, various types of bandpass, band-stop filter fabrics, etc. have been prepared by weaving, electroless plating, embroidery, transfer printing, and so on. Through experiments, the effects of cell shape and dimensional changes, periodic spacing, and dielectric materials on transmission and reflection coefficients have been studied.

3.4.2.2 The comparison for frequency selective surface textile materials produced by different processing methods

The preparation of frequency selective surface for flexible materials mainly includes screen printing, laser processing, and computer embroidery [15].

The screen printing is to stretch and fix synthetic fibers, silk fabrics, or mental wire meshes on the frame, using the method of making the hand-painted film or photochemical plate to make the screen printing plate, and the metal ink is squeezed from the mesh of the pattern portion, which is a process for extruding onto a fabric to form a sample. **Figure 34** shows a ring-shaped frequency selective surface of a complementary structure prepared by the screen printing method.

The complementary cross-type frequency selective surface prepared by the laser processing method is directly produced by laser processing on the flexible medium

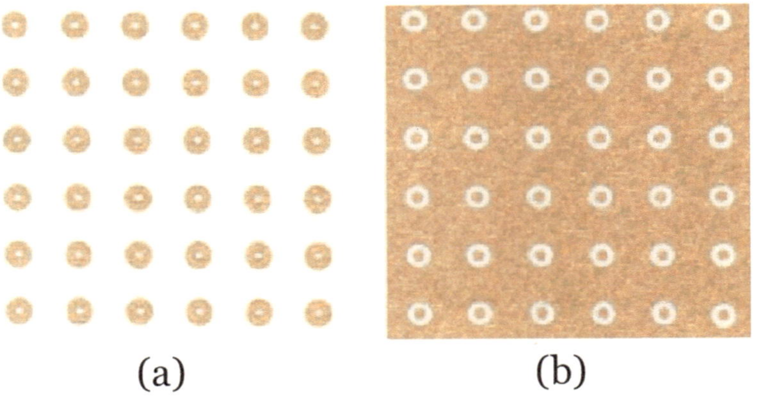

Figure 34.
The ways of screen printing processing. (a) The patch type and (b) The aperture type.

to which metal materials are pasted. The patch type (slit type) sample torn off the excess metal (patch metal) at the gap to obtain the final sample, thus ensuring the process precision of the patch type frequency selective surface (**Figure 35**).

The frequency selective surface with ring patch type is prepared by computer embroidery technology. According to the unit pattern and size of frequency selective surface period designed, the processing personnel uses the sample programming system to make the programming sample, and the needle position data of the design pattern is designed. Using these needle position data to control the computer embroidery machine, the silver-plated yarns are embroidered onto the fabric to produce the fabric material frequency selective surface (**Figure 36**).

The screen printing technology is more adaptable and can be applied to the flexible medium surface in printing; besides, the process is simple, the cost is low, and the quality is relatively stable. However, the screen printing processing has low production efficiency and is only suitable for small batch production, and the image accuracy produced is not high, which has a certain influence on the frequency response characteristics of the product.

The laser processing technology is characterized by high quality, high efficiency, and low cost. The laser processing is a kind of non-contact processing, and the frequency selection surface patterns at the sharp corners such as precise polygons can be obtained, and the products have high precision. Because the excess metal needs to be removed during the sample preparation processing to obtain the desired sample, the production efficiency of the product is affected.

The precision of computer embroidery technology when preparing flexible frequency selective surface is affected by the fineness of the needle, but the process

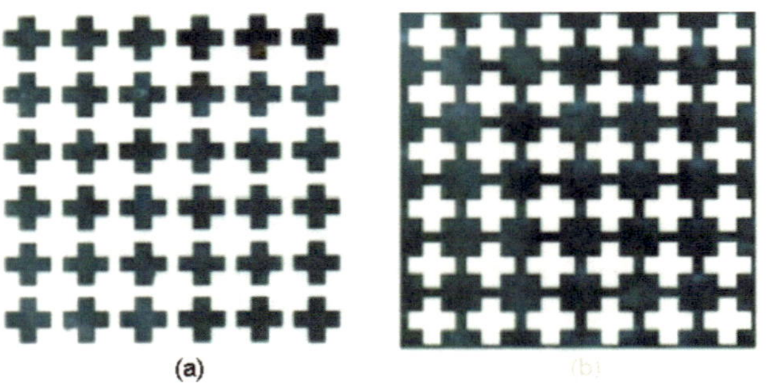

Figure 35.
The laser engraving. (a) The patch type and (b) The aperture type.

Figure 36.
The computer embroidery processing.

is simple and the production efficiency is the highest. It is suitable for mass production and exhibits better band resistance characteristics at the resonance frequency. The fabric-based frequency selective surface structure can be directly integrated into various textiles such as tents, clothing, and decorative products, and has the advantages of portability, maintenance-free, and low cost.

3.4.3 The frequency selective textile materials with three-dimensional periodic structure

The single-performance frequency selective surface can no longer meet complex electromagnetic wave environments. Incident angle stability, multi-band, wide passband, miniaturization, flexibility, and active frequency selective surfaces are the research hotspots of frequency selective surface in recent years. The use of various processing techniques to convert a two-dimensional FSF into a three-dimensional structure can bring more performance to the frequency selective surface [16].

The 3D FSF consists of the structural unit (white part in **Figure 37**), the dielectric unit (black part in **Figure 37**), and the base medium (gray part in **Figure 37**). In the z-axis direction, the conical stereoscopic periodic structure composed of the structural unit and the dielectric unit and a composite structure of the dipole plane periodically loaded with a base medium structure. At the same time, the 3D FSF has

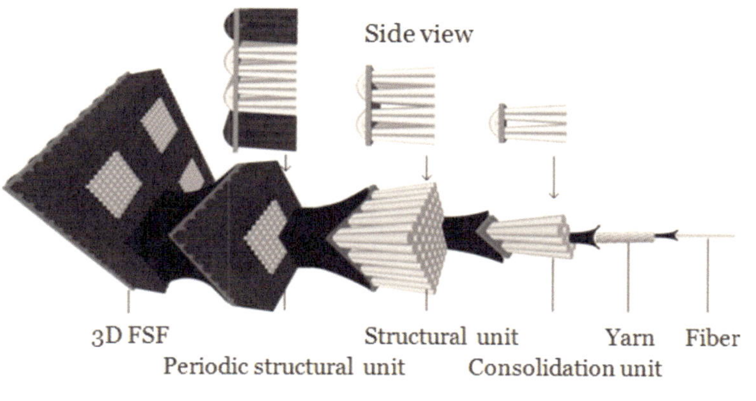

Figure 37.
The multi-scale structure of 3D FSF.

a multi-scale structure and easily deformable feature, and the electromagnetic parameters can be adjusted at multiple scales or by deformation.

Compared with metal periodic structural materials, 3D FSF is characterized by flexibility and can achieve deformation control. During use, the base medium will be the main bearer for external force, especially the elastic base medium with large deformation. The deformation of the substrate medium will cause changes in the size and other dimensions of the various scale structures fixed thereon, further leading to changes in electromagnetic response characteristics [17]. It can be seen that the electromagnetic response of the 3D FSF can be regulated by deformation.

Funding

The authors disclosed receipt of the following financial support for the research, authorship/or publication of this article: the authors acknowledge support from the National Natural Science Foundation of China (grant number 51673211).

Author details

Hong Xiao[1*], Meiwu Shi[1] and Jianying Chen[2]

1 Institute of Quartermaster Engineering and Technology, Institute of System Engineering, Academy of Military Science, PLA, Beijing, China

2 College of Textiles, Donghua University, Shanghai, China

*Address all correspondence to: 76echo@vip.sina.com

References

[1] Xiao H, Shi M. Research progress on electromagnetic textiles. Textile Research Journal. 2014;**35**(1):151-160. DOI: 10.3969/j.issn.0253-9721.2014.01.029

[2] Cheng KB, Lee ML, Ramakrishna S, et al. Electromagnetic shielding effectiveness of stainless steel/polyester woven fabrics. Textile Research Journal. 2001;**71**(1):42-49. DOI: 10.1177/004051750107100107

[3] Dai G, Xiao H, Shi M. Research progress and development direction of surface hydrophilic modification of polyester fiber. Textile Research Journal. 2015;**36**(8):156-164. DOI: 10.13475/j.fzxb.20140701609

[4] Gregory RV, Kimbrell WC, Kuhn HH. Conductive textiles. Synthetic Metals. 1989;**28**(1):823-835. DOI: 10.1016/0379-6779(89)90610-3

[5] Shi M, Nan Y. Study on the structure and properties of organic conductive fiber. Wool Textile Journal. 2001;(1):5-8. DOI: 10.3969/j.issn.1003-1456.2001.01.001

[6] Shi M, Nan Y, Liu J, et al. Antistatic effect of the organic conducting fiber. Wool Textile Journal. 2001;(3):9-11. DOI: 10.3969/j.issn.1003-1456.2001.03.002

[7] Xiao H, Tang Z, Wang Q, et al. Research on conductive grid structure and general influence factors to shielding effectiveness of electromagnetic shielding fabrics. Textile Research Journal. 2015;**36**(02):35-42. DOI: 10.13475/j.fzxb.20141002808

[8] Liang R, Cheng W, Xiao H, et al. A calculating method for the electromagnetic shielding effectiveness of metal fiber blended fabric. Textile Research Journal. 2018;**88**(9):973-986. DOI: 10.1177/0040517517693980

[9] Xiao H, Shi M, Chao S, et al. Influence of mental yarns arrangement mode on shielding effectiveness. Textile Research Journal. 2015;**36**(12):25-31. DOI: 10.13475/j.fzxb.20140701707

[10] Xiao H, Shi M, Chao S, et al. Influential factors of electromagnetic shielding effectiveness based on effective woven fabrics structure model. Journal of Industrial Textiles. 2015;**36**(07):43-49. DOI: 10.13475/j.fzxb.20140700908

[11] Kuester EF, Holloway CL. Comparison of approximations for effective parameters of artificial dielectrics. IEEE Transactions on Microwave Theory and Techniques. 1990;**38**(11):1752-1755. DOI: 10.1109/22.60028

[12] Regalado CM. A physical interpretation of logarithmic TDR calibration equations of volcanic soils and their solid fraction permittivity based on Lichtenecker's mixing formulae. Geoderma. 2004;**123**(1–2):41-50. DOI: 10.1016/j.geoderma.2004.01.026

[13] Guan F, Xiao H, Shi M, et al. The novel frequency selective fabrics and application research. Journal of Industrial Textiles. 2016;**46**(1):143-159. DOI: 10.1177/1528083715577934

[14] Guan F, Xiao H, Shi M, et al. Fabrication and application of fabric based frequency selective surface. Journal of Textile Research. 2016;**37**(2):141-148. DOI: 10.13475/j.fzxb.20151101308

[15] Lv Z, Tang Z, Xu X, et al. Flexible frequency selective surface preparation

and transmission performance. Safety & EMC. 2014;(4):79-81. DOI: 10.3969/j.issn.1005-9776.2014.04.016

[16] Xiao H, Cheng H, Shi M, et al. Exploration of 3D periodic structure frequency selective fabric based on U-taped velvet. Science & Technology Review. 2016;**86**(7):776-784. DOI: 10.3981/j.issn.1000-7857.2016.18.004

[17] Cheng H, Hong X, Shi M, et al. Research on 3D periodic structure velvet fabric and its frequency response characteristics. Textile Research Journal. 2016;**86**(7):776-784. DOI: 10.1177/0040517515596927

Chapter 10

Lightweight Electromagnetic Interference Shielding Materials and Their Mechanisms

Liying Zhang, Shuguang Bi and Ming Liu

Abstract

Motivated by the increase of stress over electromagnetic pollution issues arising from the fast-growing development and need for electronic and electrical devices, the demand for materials with high electromagnetic interference (EMI) shielding performance has become more urgently. Considering the energy consumption in real applications, lightweight EMI shielding materials has been attentive in this field of research. In this chapter, first of all, the EM theory will be briefly discussed. Secondly, the EMI shielding performance and corresponding mechanisms of three categories of lightweight materials, such as polymer-based composites, foams and aerogels, are reviewed. Finally, the summary and conclusions of this field will be addressed.

Keywords: lightweight, EMI shielding, polymer-based composite, foam, aerogel

1. Introduction

Electromagnetic (EM) waves are generated when an electric field comes in contact with a magnetic field. The oscillations of the electric field and the magnetic field are perpendicular to each other and they are also perpendicular to the direction of EM waves propagation. EM waves travel with a constant velocity of 3.0×10^8 m/s in vacuum. Unlike mechanical waves (sound waves) which need a medium to travel, EM waves can travel through anything, such as air, water, a solid material or vacuum. EM radiation refers to the EM waves, propagating through space–time, carrying EM radiant energy [1]. It is a form of energy that is all around us. Human activities like using global positioning system (GPS) device to navigate precise location, heating up a food in a microwave or using X-rays detection by a doctor would be impossible without EM radiation. **Figure 1** shows the EM spectrum used to describe different types of EM energy according to their frequencies (or wavelengths). The EM spectrum ranges from lower energy waves (longer wavelength), like radio waves and microwaves, to higher energy waves (shorter wavelength), like X-rays and gamma rays. As for the radiated emission which is focused on in this chapter, the frequency locates in the radio frequency spectrum (3 KHz–300 GHz).

Electromagnetic interference (EMI) is a disturbance generated by conduction or external radiation that affects an electrical circuit. The interference emission

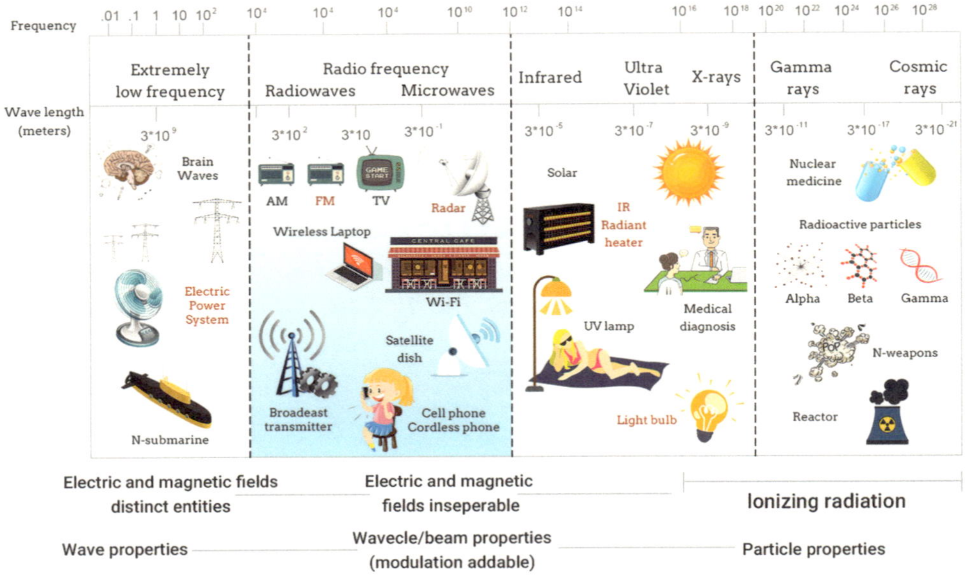

Figure 1.
A diagram of the EM spectrum showing various properties across the range of frequencies and wavelengths.

sources are from the conducted emission (several KHz–30 MHz) to the radiated emission (30 MHz–12 GHz) [2]. The conducted emission is the noise which is internally generated from the poor designed electrical circuit such as electrical cables and power wires. The radiated emission that is externally generated is in the form of transmitting EM waves such as the intended EM radiation from the radio broadcasting antenna and the unintended EM radiation from the high-speed transceivers. While detecting the EMI shielding of the device, it is usually relevant to the radiated emission lonely. The conducted emission is another subject especially for the noise prevention in system level.

EMI is encountered by all of us in our daily life and are expected to face exponential rise in future due to the growing numbers of wireless devices and standards, including cell phones, GPS, Bluetooth, Wi-Fi and near-field communication (NFC). Great effort has been dedicated for the development of EMI shielding materials. EMI shielding can be achieved by prevention of EM waves passing through an electric system either by reflection or by absorption of the incident radiation power. In the past, metals were conveniently used in many occasions. Among them, galvanized steel and aluminum are the most widely used. Copper, nickel, pre-tin plated steel, zinc and silver are also used for some purposes. When the trend in today's electronic devices become faster, smaller and lighter, metals are disadvantageous in weight consideration. Moreover, the EM pollution is not truly eliminated or mitigated since the EM signals are almost completely reflected at the surface of the metal protecting the environment only beyond the shield [3]. Hence, intensive research efforts have been focused on the development of EMI shielding materials that work by tunable reflection and absorption based on novel materials that possess lightness, corrosion resistance, flexibility, easy processing, etc.

This chapter is divided into two sections. In the next section, we will describe the EMI shielding theory in details and the parameters that influence the shielding by reflection and absorption. After that, we introduce three categories of lightweight EMI shielding materials, namely, polymer-based composites, foams and aerogels.

2. EMI shielding theory

The EMI capability of a material is called shielding effectiveness (SE). It is defined in terms of the ratio between the incoming power (P_i) and outgoing power (P_o) of an EM wave as [4]:

$$SE = 10 \log \left(\frac{P_i}{P_0} \right) \tag{1}$$

The unit of EMI SE is given in decibels (dB). According to Eq. (1), how much attenuation is blocked at given SE is given in **Table 1**.

2.1 Far field and near field

According to the distance r between the radiating source and the shield material, an EM wave can be divided into near field wave and far field wave relative to the total wavelength λ of the EM wave. As shown in **Figure 2**, the region within the distance $r > \lambda/2\pi$ is the far field while the distance $r < \lambda/2\pi$ is the near field.

In far field, the EM waves can be regarded as plane waves and EMI should consider both electric field (E) and magnetic field (H) effects. It fulfills the conditions as follows,

$$Z = \frac{|E|}{|H|} \tag{2}$$

$$E \perp H \tag{3}$$

SE (dB)	20	30	40	50	60	70
Attenuation %	99	99.9	99.99	99.999	99.9999	99.9999

Table 1.
Shielding effectiveness and attenuation %.

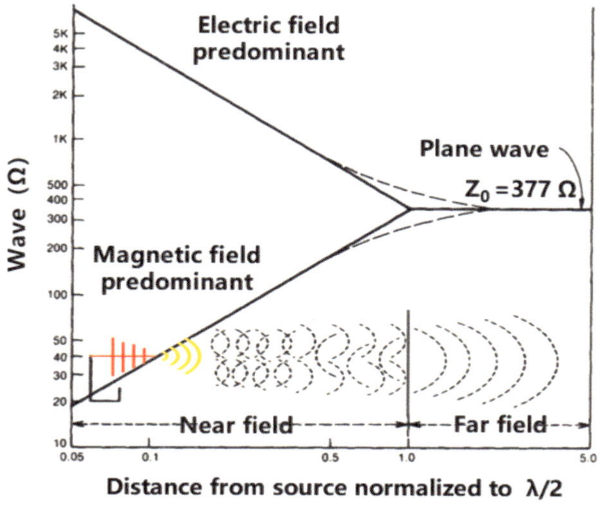

Figure 2.
Wave impedance in far field and near field [5].

where Z is the intrinsic impedance or what is sometimes called wave impedance. $|E|$ and $|H|$ are the electric and magnetic fields' amplitudes, respectively. For air ($\sigma = 0$, $\mu = \mu_0$, $\varepsilon = \varepsilon_0$), the wave impedance ($Z_0$) is always equal to 377 Ω and can be expressed as

$$z_0 = \sqrt{\frac{j\omega\mu}{\sigma + j\omega\varepsilon}} = \sqrt{\frac{j\omega\mu_0}{j\omega\varepsilon_0}} = \sqrt{\frac{\mu_0}{\varepsilon_0}} \approx 377\Omega \tag{4}$$

where σ is the electrical conductivity, μ_r is the relative permeability ($\mu = \mu_0\mu_r$), μ_0 is the permeability of air ($4\pi \times 10^{-7}$ H/m), ε_0 is the permittivity of air (8.85×10^{-12} F/m).

In near field, the wave front is curved instead of planar, so the wave front is not parallel to the surface of the shielding material. In this case, the wave impedance ($|E|/|H|$) is not constant and depends on the distance and the dominant field. For an electrical radiation source, the electrical field dominates. The wave impedance is higher than 377 Ω and decreases as the distance r increases. It can be expressed as [5].

$$z_0 = \frac{1}{2\pi f \varepsilon r} \tag{5}$$

For a magnetic radiation source, the near field is mainly magnetic. The wave impedance is lower than 377 Ω and increases as the distance r increases, it can be expressed as [5].

$$Z_0 = 2\pi f \mu r \tag{6}$$

In this chapter, all the formulations and results are taken based on far field condition because a distance of 48 cm associated with operating at a frequency of 100 MHz is already considered as far field.

2.2 EMI shielding mechanisms for homogeneous shield materials

Figure 3a illustrates the reflection and transmission of an EM wave when it strikes on a shield material. The uniform EM wave with the electric field E_i and magnetic field H_i is normal incident to the material. When the EM wave strikes the left boundary of the material, portions of the EM wave are reflected in the opposite

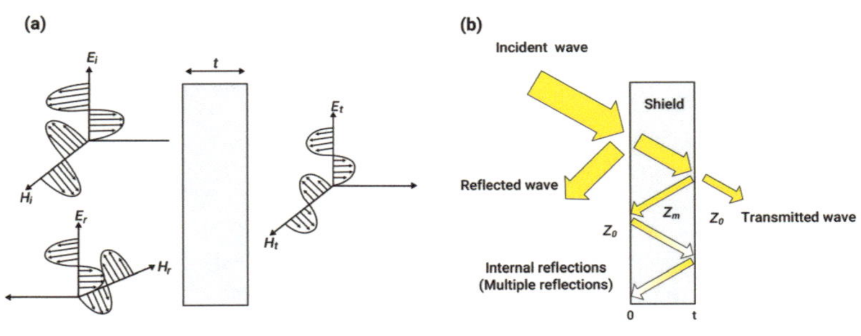

Figure 3.
(a) Schematic illustration of EM plane wave is normal incident to a material with thickness t *and (b) schematic illustration of attenuation of an incident EM wave by a shield material (thickness of shield material* = t*).*

direction with the electric field E_r and magnetic field H_r. Other portions of the EM wave are transmitted though the material with the electric field E_t and magnetic field H_t. The electric field SE can be expressed as:

$$SE = 20 \log \left| \frac{E_i}{E_t} \right| \tag{7}$$

The magnetic field SE can be expressed as:

$$SE = 20 \log \left| \frac{H_i}{H_t} \right| \tag{8}$$

Theoretically, the SE of a material is contributed from three mechanisms including reflection, absorption and multiple-reflections., the materials with mobile charge carriers (electrons or holes) can interact with the incoming EM wave to facilitate reflection. Absorption depends on the thickness of the shield materials. It increases with the increase of the thickness of the shield materials. For significant absorption, the shield materials possess electric and/or magnetic dipoles which could then interact with the EM fields. Multiple-reflections is the third shielding mechanism, which operates via the internal reflections within the shield material. Therefore, the overall SE is the sum of all the three terms:

$$SE_{overall} = SE_R + SE_A + SE_{MR} \text{ (dB)} \tag{9}$$

The EMI SE of the material depends on the distance between radiation source and the shielding material. When the radiation source is far from the shielding material, the SE is called as far field SE. In the case of the short distance between radiation source and the shielding material, the SE is called as near field SE.

Figure 3b illustrates three EMI shielding mechanisms in a conductive shield material. When an EM wave strikes the left boundary of the homogenous conductive material, a reflected wave and a transmitted wave will be created at the left external and right external surface, respectively. As the transmitted wave propagates within the shield material, the amplitude of the wave exponentially decreases as a result from absorption, and the energy loss due to the absorption will be dissipated as heat [6]. Once the transmitted wave reaches the internal right surface of the shield (t), a portion of wave continues to transmit from the shield material and a portion will be reflected into the shield material. The portion of internal reflected wave will be re-reflected within the shield material, which represents the multiple-reflections mechanism. The skin effect would affect the effect of multiple-reflections to the overall shielding to a great extent. The depth at which the electric field drops to ($1/e$) of the incident strength is call the skin depth (δ), which is given as follows [7]:

$$\delta = \frac{1}{\sqrt{\pi f \sigma \mu}} \tag{10}$$

where f is frequency (Hz). μ and σ are the magnetic permeability and the electrical conductivity of the shield material, respectively. If the shield is thicker than the skin depth, the multiple-reflections can be ignored. However, the effect of multiple-reflections will be significant as the shield is thinner than the skin depth.

As shown in **Figure 3b**, in case the shield material is a good conductor, $Z_m \ll Z_0$, then [8].

$$SE_R = 20\log\left|\frac{E_i}{E_t}\right| = 20\,log\left|\frac{(Z_0+Z_m)^2}{4Z_mZ_0}\right| \cong 20\,log\left|\frac{Z_0}{4Z_m}\right| \quad (11)$$

where $Z_m = \sqrt{\frac{j\omega\mu}{\sigma+j\omega\varepsilon}} \cong \sqrt{\frac{j\omega\mu}{\sigma}}$, $Z_0 = \sqrt{\frac{\mu_0}{\varepsilon_0}}$. So SE_{MR} can be expressed as [8].

$$SE_{MR} = 20\log\left(\frac{1}{4}\sqrt{\frac{\sigma}{\omega\mu_r\varepsilon_0}}\right) = 168 + 10\log\left(\frac{\sigma_r}{\mu_r f}\right) \quad (12)$$

where $\omega = 2\pi f$, $\sigma_r = \sigma/\sigma_{Cu}$ is the relative conductivity of the material, it is related to the electrical conductivity of the copper, the electrical conductivity of copper is $\sigma_{Cu} = 5.8 \times 10^7$ S/m. If the shield material possesses electric and/or magnetic dipoles, the attenuation of incident EM wave happens inside the shield material due to the absorption and multiple-reflections, the amplitude of the EM wave declines during wave traveling, and it can be expressed as [8].

$$SE_A = 131.4t\sqrt{f\mu_r\sigma_r} \quad (13)$$

$$SE_{MR} = 20\,log\left|1-\left(\frac{Z_0-Z_m}{Z_0+Z_m}\right)^2 e^{-2t/\delta}e^{-2j\beta t}\right| \cong 20\,log\left|1-e^{-2t/\delta}e^{-2jt/\delta}\right| \quad (14)$$

where t is the thickness of the shield material, δ is the skip depth under the operation frequency, β is the propagation constant.

The mechanism of multi-reflections is complicated. For a good conductor material, the multiple-reflection is usually insignificant because most of the incident EM waves are reflected from the external conductive surface of the shield material, and only few penetrated EM waves can be retained for multiple-reflections. The influence is more important for a material that has high permeability and low electrical conductivity. In this case, EM waves can easily penetrate through the external surface of the shield material and most penetrated EM waves are reflected from the second surface of the shield material. The influence is more important in low frequency and is reduced when the frequency gets higher because the ratio between material thickness and skin depth (t/δ) become larger as the frequency increases.

2.3 EMI shielding mechanisms for composites

Composites are made from fillers and matrices with significantly different physical or chemical properties. Hence, EMI shielding mechanisms are more complicated than those for homogeneous shield materials because of the huge surface area available for reflection and multiple-reflections. The EMI SE of composites can be measured experimentally, and it also can be calculated theoretically. The effective relative permittivity ε_{eff} of composites, which is one of the most important parameters in the calculation, can be approximately calculated from the Maxwell Garnett formula as [9]:

$$\varepsilon_{eff} = \varepsilon_e + 3f\varepsilon_e\frac{\varepsilon_i-\varepsilon_e}{\varepsilon_i+2\varepsilon_e-f(\varepsilon_i-\varepsilon_e)} \quad (15)$$

where ε_e is the relative permittivity of the matrix, ε_i is the relative permittivity of the filler and f is the volume fraction of the filler. If the filler are electrical conductive particles, the relative permittivity ε_i can be expressed as [10]:

$$\varepsilon_i = \varepsilon' - j\varepsilon'' = \varepsilon' - j\frac{\sigma}{\omega\epsilon_0} \tag{16}$$

where ε' and ε'' are the real and imaginary part of the complex relative permittivity of the filler, respectively. σ is the electrical conductivity of the filler. As shown in **Figure 3b**, the transmission coefficient T can be expressed as [10]:

$$T = \frac{T_1 T_2 e^{-\gamma_m D}}{1 + R_1 R_2 e^{-2\gamma_m D}} \tag{17}$$

where T_1 and T_2 are the transmission coefficients at the boundary 0 and t, respectively. R_1 and R_2 are the reflection coefficients at the boundary 0 and t, respectively. γ_m is the complex propagation constant. The T_1, T_2, R_1, and R_2 can further be expressed in terms of the impedance Z_0 and Z_m [10]:

$$T_1 = \frac{2Z_m}{Z_m + Z_0} \tag{18}$$

$$T_2 = \frac{2Z_0}{Z_m + Z_0} \tag{19}$$

$$R_1 = \frac{Z_m - Z_0}{Z_m + Z_0} \tag{20}$$

$$R_2 = \frac{Z_0 - Z_m}{Z_m + Z_0} \tag{21}$$

where Z_0 and Z_m are the impedance of the air and the composite material, respectively. Z_0 can be expressed in Eq. (4) and Z_m can further be expressed as:

$$Z_m = Z_o\sqrt{\frac{\mu_r}{\varepsilon_{eff}}} \tag{22}$$

The propagation constant γ_m can be expressed as [10]:

$$\gamma_m = j\omega\sqrt{\varepsilon_0\mu_0\left(\varepsilon'_{eff} - j\varepsilon''_{eff}\right)} \tag{23}$$

So, the SE can be calculated in terms of T,

$$SE = -20\ log(|T|) \tag{24}$$

3. Lightweight EMI shielding materials

When modern electronic devices are designed, high performance EMI shielding materials are highly demanded. In addition, lightweight is one additional important technical requirement for potential applications especially in the areas of automobile and aerospace. In the following section, we will briefly review state-of-the-art research work regarding polymer-based composite, foams and aerogels used for EMI shielding.

3.1 Polymer-based composites

Polymer/conductive fillers composites was seen as a promising advanced EMI shielding materials since the discovery that an insulating polymer would allow the flow of current through the conductive network stablished by conductive fillers above the percolation threshold. The conductive composite materials preserve the advantages of lightness of polymers, low cost, design flexibility and ease of processing, and the incorporation of conductive fillers circumvent intrinsic nature of polymers being transparent to EM waves through interaction between EM wave and the conductive fillers. Metallic fillers, intrinsically conductive polymers and carbon based electrically conductive fillers are discussed in this section with specific examples. Polymer/magnetic particles composites will also be briefly introduced as magnetic portion is an important component in EM waves that should not be ignored. This section aims to provide a general overview on the preparation of polymer-based EMI shielding materials and the advantages and challenges faced by each category and possible strategies towards enhancing the EMI shielding performances.

3.1.1 Polymer-based composites containing metallic fillers

Metals are typical wave-reflection materials used for EMI shielding purpose owing to their abundance in mobile charge carriers that can interact with the incident EM radiation. Metallic fillers of various physical forms, such as fibers or nanoparticles, were dispersed in the polymer matrix to increase the interaction with the incident EM radiation. Injection-molding provides a direct method to disperse metallic fillers into a polymer matrix. Stainless steel fibers (SSF) introduced into polycarbonate matrix through injection molding shown that EMI SE is heavily dependent on the molding parameters which would give an optimum electrical conductivity [11]. Blended textiles of polyester fibers with SSF showed that the EMI SE is more than 50 dB in the frequencies ranging from 30 MHz to 1.5 GHz [12] (see **Figure 4a**). As shown in **Figure 4b** and **c**, comparison of reflectance, absorbance and transmittance, (identified as reflectivity, absorptivity and transmissibility in **Figure 4**) for SSF and SSF/polyester fiber fabrics as a function of frequency revealed absorption as the dominant EMI shielding mechanism. In the case of SSF/polyester with 10 wt% SSF, EMI shielding by absorption increased from 30 MHz to maximum at 500 MHz and then decreased with the increase in frequency.

The challenges in achieving a good dispersion of metallic fillers and the weight increase make polymer/metallic fillers composites a less popular choice. Much attention was switched to intrinsically conductive polymers (including polyaniline, polyacetylene, and polypyrrole), carbon-based materials (including carbon fibers, carbon black, graphite, graphene, carbon nanotubes and mesoporous carbon), and magnetic materials like carbonyl iron and ferrites (including Fe_3O_4 and α-Fe_2O_3).

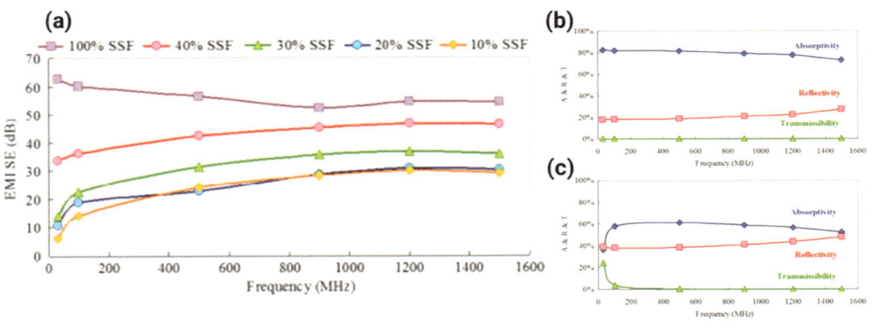

Figure 4.
(a) The EMI SE of the SSF/PET fabric as a function of frequency; (b) reflectivity/absorptivity/transmissibility of SSF fabric and (c) SSF/PET fabric with 10 wt% SSF as a function of frequency [12].

3.1.2 Intrinsically conductive polymers-based composites

Blends of a polymer with an intrinsically conductive polymer results in a composite combining the desired properties of the two components, that is, adequate mechanical properties of the polymer matrix for mechanical support and the electrically conducting component for interaction with the EM radiation. Conducting polymers are conjugated polymers, which on doping exhibit electronic conductivity. Distinctive to metallic fillers, the electrical conductivity of conducting polymers arises from the polymer molecular structure. Alteration of parameters such as chain size, doping level, dopant type and the synthesis route directly affect the molecular structure, hence the EMI shielding properties of the material.

Among the available conducting polymers, polypyrrole (PPY) and polyaniline (PANI) are the most widely used conductive fillers for EMI shielding purposes. PPY is known to possess high conductivity, easy synthesis, good environmental stability and less toxicological problem. Chemical and electrochemical polymerization of PPY on a polyethylene terephthalate (PET) fabric is given as an example for electrically conducting composite. Pyrrole was first dissolved in an aqueous solution containing 10 wt% polyvinyl alcohol (PVA) and sprayed on the PET fabric before subject to electrochemical polymerization at room temperature under a constant current density. The resultant PPY coated PET fabric was shown to exhibit EMI SE about 36 dB over a wide frequency range up to 1.5 GHz [13].

PANI was studied extensively for its various structures, unique doping mechanism, excellent physical and chemical properties, stability, and the readily obtainable raw materials. Lakshmi et al. [14] prepared PANI-PU composite film by adding aniline to polyurethane (PU) solution in tetrahydrofuran (THF). Doping of composites was done by adding camphor sulfonic acid to the composite solution. The EMI SE of the PU-PANI film was found to increase with thickness and the frequency specific material is ideal for shielding at 2.2 and 8.8 GHz.

Other intrinsically conducting polymers, such as poly(p-phenylene-vinylene) [15, 16] and poly(3-octylthiophene) [17], were also investigated for EMI shielding applications, but too much lesser extent, mainly due to the unsatisfactory performance and complex processing procedures involved.

In general, the EMI shielding performance arises by the addition of conductive polymer consequently dominated by reflection mechanism due to the increase of the level of impedance mismatch with air. One obvious advantage of such polymer-polymer system is the lightweight being preserved, also there is no issue on substrate flexibility as those associated with metallic or carbon-based fillers. However, the main drawbacks of such composites include (1) poor mechanical properties of the most of the intrinsically conducting polymers require a matrix material for structural support; (2) the insoluble and infusible characteristics caused conducting polymers to exhibit poor processability and (3) high filler (conducting polymer) level is usually needed for acceptable performances.

3.1.3 Polymer-based composites containing carbon-based fillers

Similar to metallic fillers, carbon-based fillers come in various shapes and aspect ratios. Carbon black (CB), including graphite and CB, is the generic name given to small particle size carbon pigments which are formed in the gas phase by thermal decomposition of hydrocarbons [18]. Carbon fibers (CFs) are 1D carbon structure of diameter generally lies between 50 and 200 nm and aspect ratios around 250 and 2000, largely produced by chemical vaporization of hydrocarbon [19, 20]. Carbon nanotubes (CNTs) can be considered as rolled-up hollow cylinders of graphene sheets of very high aspect ratio due to the small diameter, constituted of a single hollow cylinder, that is, single-walled carbon nanotubes (SWCNTs) or of a collection of graphene concentric cylinders, that is, multi-walled carbon nanotubes (MWCNTs) [21, 22]. Graphene sheet (GS), an atomically thick two-dimensional structure, exhibited excellent mechanical, thermal and electrical properties [23]. Both CNTs and graphene offer substantial advantages over conventional carbon fillers and the percolation threshold can be achieved by both at very low content if properly dispersed.

In general, carbon fillers with high aspect ratio are generally more effective in imparting electrical conductivities to a polymer matrix, hence it is no surprise to observe the highest SE from fillers with the highest aspect ratio, that is, SWCNTs > MWCNTs > CNFs > CB when the volume fraction of the fillers is the same. The different methods of fillers dispersion and various carbon filler surface modification methods were comprehensively reviewed in the published paper and will not be discussed in detail here [3, 24]. The EMI shielding performance of the polymer/carbon-fillers composites can also be found in Ref. [3, 7, 24, 25].

3.1.4 Polymer-based composites containing magnetic particles composites

A binary or even ternary component consists of two or more types of the fillers provide an effective way to bypass the inherent shortcomings of a single-filler composite. The incorporation of magnetic components will supplement the attenuation properties of a carbon-based EMI shielding material.

Physical blending or deposition of metallic particles within a polymer blend or structure is the most direct way of incorporation a third element, however, such method faces the problem of uniform dispersion and deposition at the bottom layers due to the higher density of metallic particles. Electroless plating of metals on carbon substrates provides a neat way of incorporating metal components uniformly into a system without excessive weight addition. Works by Kim et al. [26] and Yim et al. [27] dispersed nickel coated MWCNT through electroless plating in epoxy and high-density polyethylene, respectively. **Figure 5a** gives an illustration of

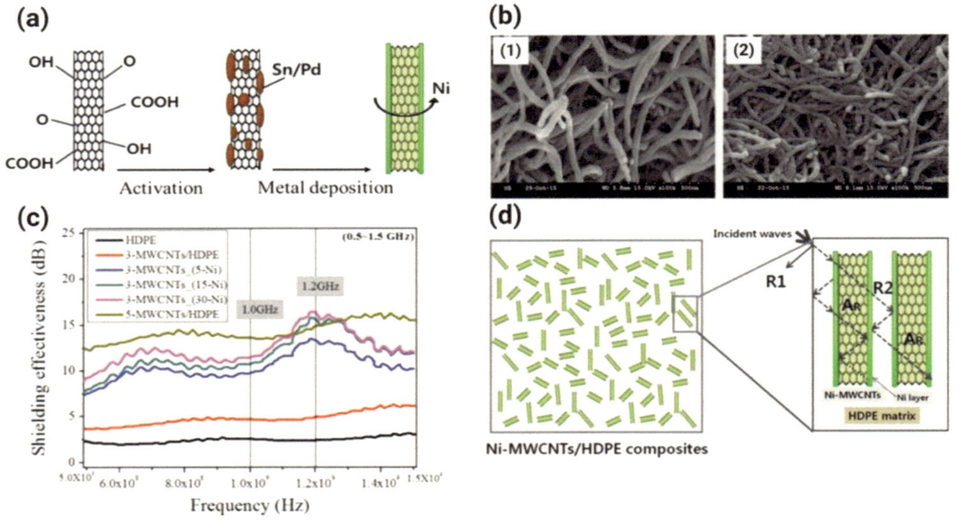

Figure 5.
(a) Schematic diagram of the electroless Ni-plating process; (b) SEM images of (1) pristine MWCNTs and (2) Ni-coated MWCNTs, respectively; (c) comparison of the EMI SE of MWCNTs/HDPE and Ni-MWCNTs/HDPE and (d) the proposed shielding mechanism of Ni-MWCNTs/HDPE [27].

the nickel coated MWCNTs. It is apparent that the nickel coated MWCNTs appeared rougher comparing to the pristine ones due to the presence of nickel particles as shown **Figure 5b**. Yim achieved 140% (at 1 GHz, **Figure 5c**) in enhancement of the EMI SE compared to the pristine MWCNT/polymer composites. The enhancement was attributed to the increased surface conductivity. **Figure 5d** shows the proposed shielding mechanism of Ni-MWCNTs/HDPE. EM wave was firstly reflected at the composite surfaces upon reaching the surface of the composite. When the penetrated EM wave meets the nickel layer on the MWCNTs, the metallic layer functioned as EM absorbable or reflective fillers. It is evident that the EMI absorbing nature of the metallic layer can be used as an effective additional shielding material despite the small amount present in the systems.

3.2 Foams and aerogels used in EMI shielding

In view of the rigid index of fuel-economy in the applications of automobile and aerospace, lightweight EMI shielding materials with the combination of reduced density and high EMI SE are much preferred. In this section, we aim to provide a general overview on the preparation of foam and aerogel materials used in EMI shielding and the advantages and challenges faced by each category and possible strategies towards enhancing their EMI shielding performances. The specific EMI SE, defined as the ratio of the EMI SE to the density (SSE) or both density and thickness (SSE/t), is a more appropriate criterion to compare the EMI shielding performance with those of other typical materials for the applications where lightweight is required.

3.2.1 Polymer-based composite foams

Conductive polymer-based composites foams offer significant reduction in weight, while the pores decrease the real part of the permittivity, accordingly reducing the reflection at the material surface. The porous structure enhances the energy absorption through wave scattering in the walls of the pores. Electrically conductive fillers, including CNFs, CNTs and graphene sheets, are commonly used to form a desirable conducting network within the inherently insulating polymer foam matrix. Yang et al. [28] first reported CNFs reinforced polystyrene (PS) composite foam as a conductive foam for EMI shielding application. The EMI SE of PS/CNFs foam containing 1 wt% CNFs was less than 1 dB, upon increasing CNFs content to 15 wt%, EMI SE increased to 19 dB. Following this work, the authors reported PS/CNTs composite foam with varying CNTs contents from 0 to 7 wt% [4]. The PS/CNTs composite foam achieved a higher EMI SE of above 10 dB compared to 3 dB for the PS/CNFs composite foam at the same filler content of 3 wt%. The difference in the results originated from the remarkable electrical and structural properties of CNTs, such as larger aspect ratio, smaller diameter, higher electrical conductivity and strength, compared to CNFs.

3.2.2 Syntactic foams

Syntactic foam, filling hollow spheres in a matrix, is a kind of lightweight composite materials. The approaches to enhance the EMI SE of syntactic foams include (i) hollow particles made of a conductive material; (ii) coating a conductive layer onto the surface of hollow particles and (iii) adding a second conductive filler in syntactic foam matrix.

Zhang et al. [29] added a second conductive filler, (CNFs, chopped carbon fiber (CCF), and long carbon fiber (LCF)), into syntactic foams containing conductive

Filler content (vol%)	CNF Aspect ratio: 500–1700	CCF Aspect ratio: 6–50	LCF Aspect ratio: 150–750
0.5	5.2	2.2	2.8
1.0	11.3	3.4	4.4
1.5	16.4	3.7	6.5
2.0	24.9	4.3	7.5

Table 2.
Comparison of the EMI SE (dB) of CNF, CCF, and LCF reinforced syntactic foam.

hollow carbon microspheres (HCMs). The EMI SE values of used syntactic foams at the same filler content were compared, as shown as **Table 2**. The results showed that CNFs is more effective in providing EMI shielding compared to CCF and LCF due to the larger aspect ratio of CNFs.

Zhang et al. [30] also demonstrated the effect of functionalization of HCMs on the EMI SE of the epoxy-HCMs syntactic foam. HCMs were coated with polydopamine (PDA) via the self-polymerization of dopamine. The PDA coating promotes dispersion and served as a reducing agent to deposit silver (Ag) particles on the surface of HCMs as illustrated in **Figure 6a**. The average EMI SE of the epoxy-HCMs syntactic foam containing Ag-PDA-HCMs with 28.5 and 30.5 wt% of silver in the X-band achieved 49.5 and 60.2 dB, respectively as shown in **Figure 6b**. The SSE reached up to 46.3 dB cm^3/g, demonstrating the prospect of epoxy/Ag-PDA-HCMs syntactic foam as a lightweight high-performance EMI shielding material. The corresponding EMI shielding mechanism of this syntactic foam was analyzed by comparing the values of reflectance (*R*), absorptance (*A*), and

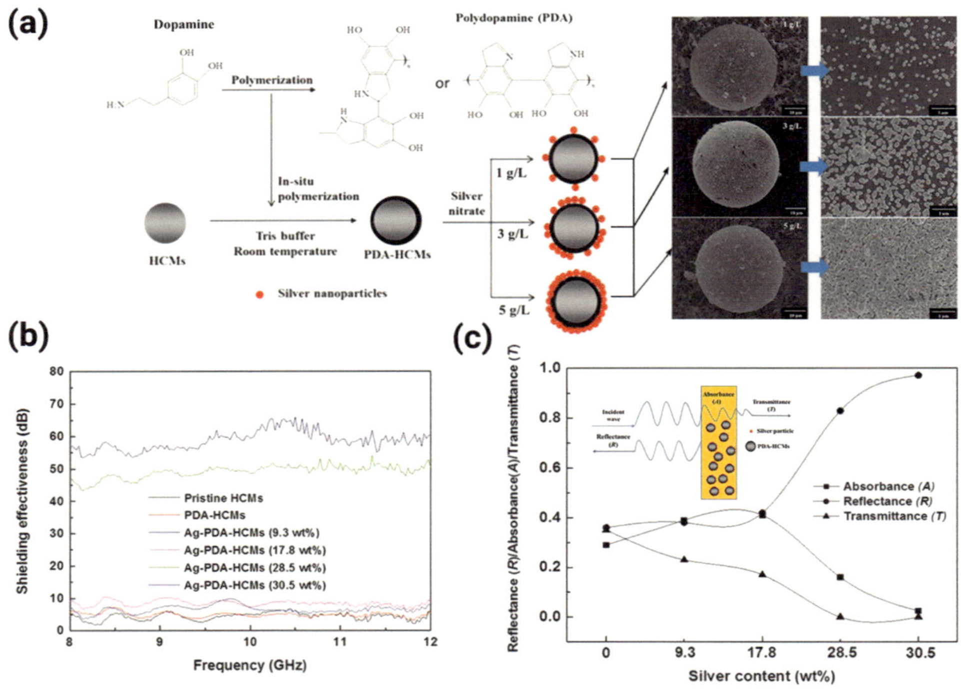

Figure 6.
(a) Schematic illustration of the procedure for preparation of PDA-HCMs and Ag-PDA-HCMs; (b) EMI SE in the frequency range from 8 to 12 GHz for syntactic foam containing pristine HCMs and Ag-PDA-HCMs with different silver contents; and (c) reflectance (R), absorbance (A), and transmittance (T) of EM radiation over syntactic foams containing Ag-PDA-HCMs with different silver content at 10 GHz [30].

transmittance (*T*) in **Figure 6c**. The specimens were both reflective and absorptive towards EM radiation at silver content less than 17.8 wt%. The contribution of reflection (0.83) towards EMI SE surpassed that from absorption (0.16) when silver content increased to 28.5%. The dense and thick electrically conductive silver formed due to further increasing the silver content to 30.5 wt% increased the *R* to 0.97 and resultant in reflection as the dominant shielding mechanism.

Xu et al. [31] fabricated syntactic foams ("hybridized epoxy composite foams" according to authors) through impregnating expandable epoxy/MWCNT/microsphere blends into a preformed, highly porous, and 3D silver-coated melamine foam (SF) sponge. The highly conductive SF resolved the problem of the foam reduction of high filled epoxy blends and provided channels for rapid electron transport. MWCNTs were used to offset the loss of conductive pathways due to the crystal defects in the silver layer and the insulating epoxy resin. As a result, the EMI SE of 68.1 dB was achieved with only 2 wt% of MWCNTs and 3.7 wt% of silver due to the synergy of the MWCNT and SF.

3.2.3 Carbon foams

Carbon foam is a class of three-dimensional (3D) architecture consisting of a sponge-like interconnected network of porous carbon. Carbon foams have been wildly used as candidates for realistic EMI shielding applications due to their excellent properties, such as low density, resistance to chemical corrosion, high thermal and electrical conductivity, and high temperature resistance.

Zhang et al. [32] prepared a novel ultralight (0.15 g/cm^3) carbon foam by direct carbonization of phthalonitrile (PN)-based polymer foam, as shown in **Figure 7a**. High EMI SE of ~ 51.2 dB (see **Figure 7b**, C1000 was labeled as the carbonization of 1000°C) was contributed by the high graphitic carbonaceous species and the intrinsic nitrogen-containing structure. The carbon foams showed the best SSE of 341.1 dB cm^3/g so far when mechanical property was considered. The carbon foam developed by Zhang provides an excellent low-density and high-performance EMI shielding material for use in areas where mechanical integrity is desired.

3.2.4 CNTs/graphene foams

The EMI SE of carbon foams was closely related to the char yield of polymer precursors and the demanding carbonization conditions. Therefore, a new kind of filler-free lightweight EMI shielding material, is in demand, which can be prepared without the stringent processing conditions. In view of the lightweight requirement, assembling one dimensional (1D) CNTs and two-dimensional (2D) graphene sheets into three dimensional (3D) macroscopic porous structures (e.g., sponges, foams and aerogels) emerged as an efficient approach.

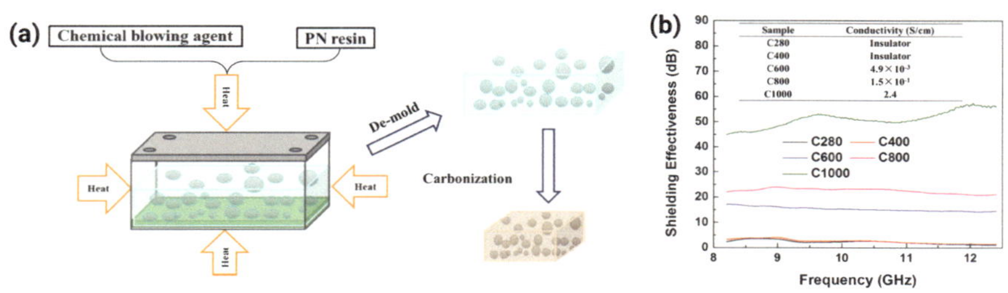

Figure 7.
(a) Schematic representation of the preparation of PN-based carbon foams and (b) EMI SE of carbon foams [32].

Lu et al. [33] synthesized a flexible CNTs sponge with a density of 10.0 mg/cm^3 via chemical vapor deposition (CVD) process, composed of self-assembled and interconnected CNT skeletons. The freestanding CNTs sponge showed the high EMI SE and SSE of 54.8 dB and 5480 dB cm^3/g in X-band, respectively. After composited with polydimethylsiloxane (PDMS) by directly infiltrating method, the CNT/PDMS composites still exhibited excellent EMI SE (46.3 dB) at the thickness of 2.0 mm, while the CNT loading content was less than 1.0 wt%.

Surface modification is employed to increase the EMI shielding ability of graphene foams. Zhang et al. [34] prepared surfaced modified 3D graphene foams via self-polymerization of dopamine with a subsequent foaming process, as shown in **Figure 8a**. The polydopamine (PDA) served as a nitrogen doping source and an enhancement tool to achieve higher extent of reduction of the graphene through providing wider pathways and larger accessible surface areas. The enhanced reduction of graphene sheets and the polarization effects introduced by PDA decoration compensated the negative effect of the barrier posed by PDA. As a result, the resultant EMI SE showed 15% improvement compared to PDA-free graphene foam as shown in **Figure 8b**. Wu et al. [35] also fabricated an ultralight, high performance EMI shielding graphene foam (GF)/poly(3,4-ethylenedioxythiophene):poly(styrene sulfonate) (PEDOT:PSS) composites by drop coating of PEDOT:PSS on the freestanding cellular-structured GFs, as illustrated in **Figure 8c**. The GF/PEDOT:PSS composites possess an enhanced electrical conductivity from 11.8 to 43.2 S/cm after the incorporation of PEDOT:PSS. The modified grapheme foam with a density of 18.2×10^{-3} g/cm^3 provide a remarkable EMI SE of 91.9 dB (identified as SE_T in **Figure 8d**).

3.2.5 Graphene aerogels

Aerogel is a synthetic porous ultralight material derived from a gel, in which the liquid component used in gel are replaced by air. In recent years, the great potential of graphene aerogel (GAs) in EMI shielding applications has been confirmed by several researchers. Song et al. [36] reported that the EMI SE of GA-carbon textile hybrid with a thickness of 2 mm was 27 dB. The 3D scaffold GA greatly enhances

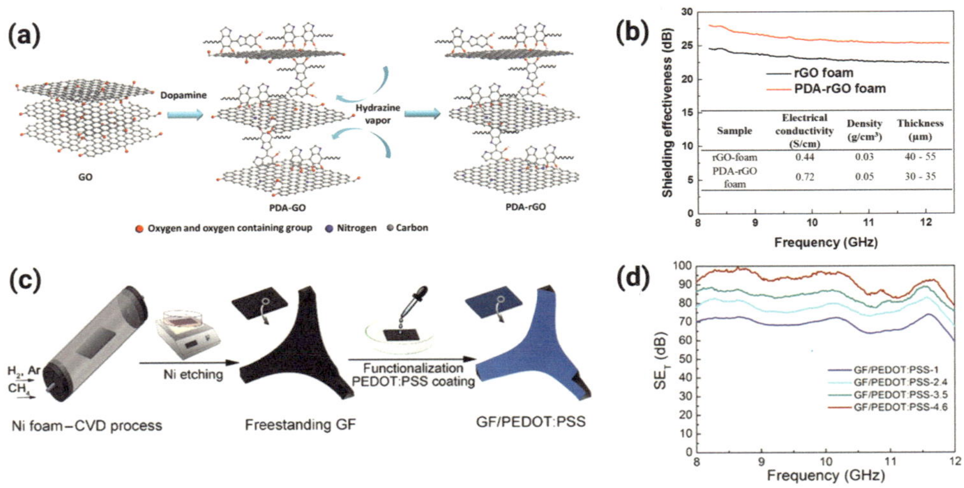

Figure 8.
(a) Schematic representation of the preparation of PDA-GO and PDA-rGO; (b) EMI SE of rGO foam and PDA-rGO foam [34]; (c) schematic procedure of the preparation of GF/PEDOT:PSS composites; (d) EMI SE of GF/PEDOT:PSS composites as a function frequency [35].

the conductive network while maintaining the advantage of light carbon textile. Singh et al. [37] studied the EMI SE of pure GA, which was 20 dB, with a density ~75 mg /cm^3 and a thickness of 2 mm. They discussed the EMI shielding mechanism by correlating the EM wave interaction with the 3D porous structure. Zeng et al. [38] fabricated an ultralight and highly elastic rGO/lignin-derived carbon (LDC) composite aerogel with aligned microspores and cell walls by directional freeze-drying and carbonization method. The EMI SE of rGO/LDC composite aerogels with a thickness of 2 mm could reach up to 49.2 and 21.3 dB under ultralow densities of 8.0 and 2.0 mg/cm^3, respectively.

The graphitization of GAs facilitates to improve its electrical conductivity, thus improving the EMI SE. Liu et al. [39] reported an effective method of manufacturing an integrated graphene aerogel (IGA) using a complete bridge between rGO sheets and polyimide macromolecules via graphitization at 2800°C, as shown in **Figure 9a**. The rGO sheets were efficiently reduced to graphene during graphitization, while the polyimide component was graphitized to turbostratic carbon to connect the graphene sheets, resulting in a high EMI SE of ~83 dB in X-band at a low density of 18 mg/cm^3, as shown in **Figure 9b**. The EMI shielding mechanism analysis for the porous IGA revealed that most of the incident EM wave was dissipated through absorption, thus forming an absorption-dominant EMI shielding mechanism.

Different reduction process of graphene oxide (GO), including chemical reduction and thermal reduction would affect the EMI shielding performance of GAs. Bi et al. [40, 41] carried out a comprehensive study of EMI shielding mechanisms of GAs solely consisted of graphene sheets to determine the main parameters of high EMI SE. As shown in **Figure 10a**, two types of ultralight (4.5–5.5 mg/cm^3) 3D GAs were prepared by chemical reduction and thermal reduction of GO aerogels. The EMI SE reached 27.6 and 40.2 dB for chemically reduced graphene aerogel (GAC) and thermally reduced graphene aerogel (GAT), respectively. The distinct graphene surface resulted from different processing pathway led to different EM wave

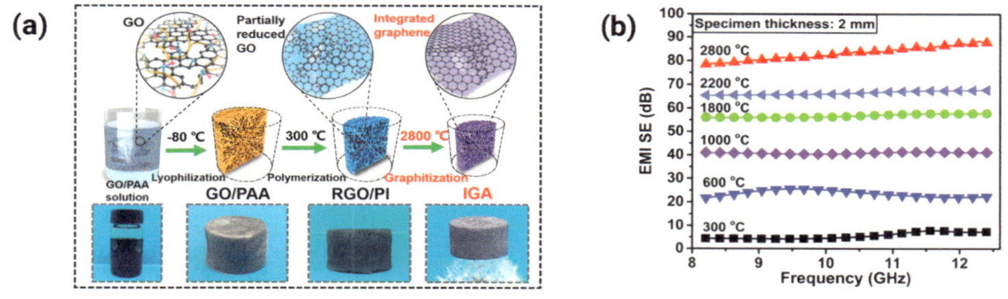

Figure 9.
(a) Schematic illustration for fabricating IGA and (b) effect of annealing temperature on EMI shielding performance of IGAs [39].

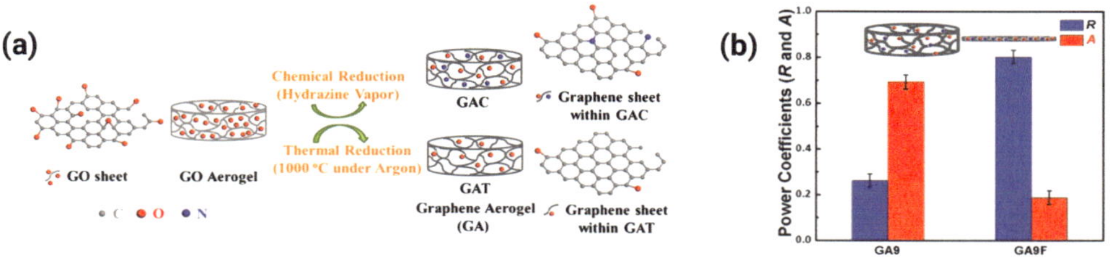

Figure 10.
(a) Schematic representation of the preparation process of GAC and GAT [42] and (b) R & A *of GA9 and GA9F [41].*

response upon striking the graphene/air interface. Nitrogen-doping and side polar groups induced strong polarization effects in GAC. Higher extent of reduction of the grapheme sheets in GAT left a smaller amount of side polar groups and formed more sp^2 graphitic lattice, both favored π-π stacking between the adjacent graphene sheets. The enhanced polarization effects and the increased electrical conductivity of GAT contributed to better EMI shielding performance. Bi further investigated the effect of porosity on EMI shielding mechanisms compressing the aerogel (GA9) into thin film (GA9F), as shown in **Figure 10b**. The highly connected conducting network resulted in a significant increase in the electrical conductivity of GA9F, while the EMI SE remained unchanged at constant rGO content. The observation was contradictory to the previous outcomes that higher electrical conductivity or better-connected network contributed to higher EMI SE. Hence, the fact can be believed that the EMI SE is highly dependent on the effective amounts of materials response to the EM waves. Despite the similar intrinsic properties of rGO, the amount of absorption of EM waves in GA9 was much higher than that in GA9F when the EM waves penetrated through the porous structure. The cavities within the highly porous GA absorbed the EM waves through multiple internal reflections and eventually depleted the energy. Hence, the tightly connected conducting network within GA9F changed the EMI shielding mechanism from absorption to reflection.

4. Conclusions

Generally, EMI shielding is defined as the prevention of the propagation of EM waves from one region to another by using shield materials. With the development of electronic industry, weight reduction is an additional technical requirement besides the good EMI shielding performance. Metal as a traditional EMI shielding material has been replacing with lighter materials, such as polymer-based composites, foams and aerogels. This chapter reviewed various types of lightweight materials with their EMI SEs corresponding to their EMI shielding mechanisms. To verify the benefits of using lightweight materials for EMI shielding applications, a comprehensive comparison was performed as shown in **Figure 11**. All the data in **Figure 11** were collected from the

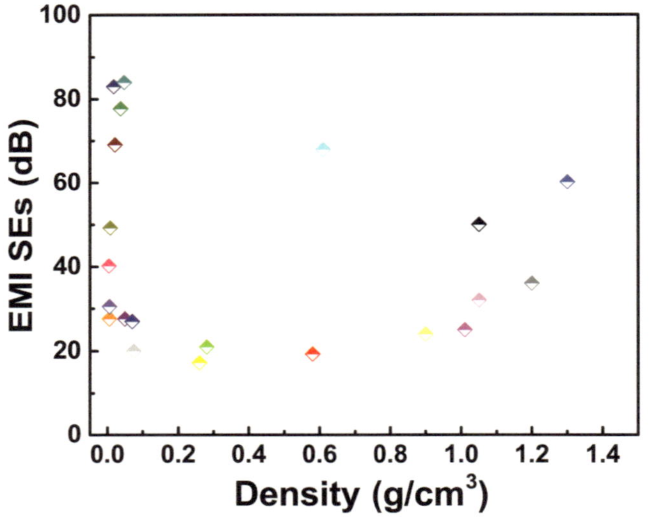

Figure 11.
Comparison of EMI SEs of lightweight materials as a function of density of materials.

reference papers listed in this chapter. Although the data are not involved all the published results, they are representative to the library of lightweight EMI shielding materials. The reported EMI SEs of polymer-based composites containing conductive fillers varied in the range of 20–60 dB corresponding to the densities higher than 0.8 g/cm^3. Polymer-based foams reinforced with additional conductive fillers and carbon foams outperform polymer-based composites in terms of EMI SE. They possessed comparable EMI SE of 20–80 dB with the lower density (<0.8 g/cm^3). Aerogels with ultralow densities (<100 mg/cm^3) exhibited high EMI SEs in the same range of polymer- and carbon-based foams, indicating they can be used as an ideal potential lightweight EMI shielding materials though the mechanical properties of aerogels still remain a big issue.

Acknowledgements

Liying Zhang would like to acknowledge the support by the initial research funds for young teachers of Donghua University. Shuguang Bi would like to acknowledge the financial support of Wuhan Engineering Center for Ecological Dyeing & Finishing and Functional Textiles, Key Laboratory of Textile Fiber & Product (Wuhan Textile University), Ministry of Education, Hubei Biomass Fibers and Eco-dyeing & Finishing Key Laboratory. Zhang and Bi would also thank the funding support by State Key Laboratory for Modification of Chemical Fibers and Polymer Materials, Donghua University (KF1827). Ming Liu would like to acknowledge the support from School of Materials Science and Engineering at Nanyang Technological University for this work.

Conflict of interest

No conflict of interest.

Author details

Liying Zhang[1*†], Shuguang Bi[2†] and Ming Liu[3]

1 Center for Civil Aviation Composites, Donghua University, Shanghai, China

2 Chemistry and Chemical Engineering College, Wuhan Textile University, Wuhan, China

3 Temasek Laboratories, Nanyang Technological University, Singapore

*Address all correspondence to: lyzhang@dhu.edu.cn

†These authors contributed equally to this work.

References

[1] Davidovits J. 30 years of successes and failures in geopolymer applications. Market trends and potential breakthroughs. In: Geopolymer Conference; 2002; Melbourne, Australia

[2] Perez R. Handbook of Electromagnetic Compatibility. London, United Kingdom: Academic Press; 1995

[3] Thomassin J-M, Jerome C, Pardoen T, Bailly C, Huynen I, Detrembleur C. Polymer/carbon based composites as electromagnetic interference (EMI) shielding materials. Materials Science & Engineering R-Reports. 2013;**74**:211-232

[4] Yang YL, Gupta MC. Novel carbon nanotube-polystyrene foam composites for electromagnetic interference shielding. Nano Letters. 2005;**5**:2131-2134

[5] Crooks LE. Noise reduction techniques in electronic systems (2nd ed.), Henry W. Ott. Wiley-Interscience: New York. 1988. Magnetic Resonance in Medicine. 1989;**10**:426-427

[6] Al-Saleh MH, Sundararaj U. Electromagnetic interference shielding mechanisms of CNT/polymer composites. Carbon. 2009;**47**:1738-1746

[7] Chung DDL. Electromagnetic interference shielding effectiveness of carbon materials. Carbon. 2001;**39**:279-285

[8] Paul CR. Introduction to Electromagnetic Compatibility. 2nd ed. New York, United States: John Wiley & Sons, Inc; 2006

[9] Sihvola A. Electromagnetic Mixing Formulas and Applications. London, United Kingdom: Institution of Engineering and Technology; 1999

[10] Koledintseva M, Ravva PC, DuBroff R, Drewniak J, Rozanov K, Archambeault B; Ieee. Engineering of composite media for shields at microwave frequencies. In: EMC 2005: IEEE International Symposium on Electromagnetic Compatibility. Proceedings. Vols. 1-3. 2005. pp. 169-174

[11] Chen C-S, Chen W-R, Chen S-C, Chien R-D. Optimum injection molding processing condition on EMI shielding effectiveness of stainless steel fiber filled polycarbonate composite. International Communications in Heat and Mass Transfer. 2008;**35**:744-749

[12] Shyr T-W, Shie J-W. Electromagnetic shielding mechanisms using soft magnetic stainless steel fiber enabled polyester textiles. Journal of Magnetism and Magnetic Materials. 2012;**324**:4127-4132

[13] Kim MS, Kim HK, Byun SW, Jeong SH, Hong YK, Joo JS, et al. PET fabric/polypyrrole composite with high electrical conductivity for EMI shielding. Synthetic Metals. 2002;**126**:233-239

[14] Lakshmi K, John H, Mathew KT, Joseph R, George KE. Microwave absorption, reflection and EMI shielding of PU-PANI composite. Acta Materialia. 2009;**57**:371-375

[15] Courric S, Tran VH. The electromagnetic properties of poly (p-phenylene-vinylene) derivatives. Polymer. 1998;**39**:2399-2408

[16] Courric S, Tran VH. The electromagnetic properties of blends of poly (p-phenylene-vinylene) derivatives. Polymers for Advanced Technologies. 2000;**11**:273-279

[17] Taka T. EMI shielding measurements on poly(3-octyl-thiophene) blends. Synthetic Metals. 1991;**41**:1177-1180

[18] Sanchez-Gonzalez J, Macias-Garcia A, Alexandre-Franco MF, Gomez-Serrano V. Electrical conductivity of carbon blacks under compression. Carbon. 2005;**43**:741-747

[19] Al-Saleh MH, Sundararaj U. A review of vapor grown carbon nanofiber/polymer conductive composites. Carbon. 2009;**47**:2-22

[20] Tibbetts GG, Lake ML, Strong KL, Rice BP. A review of the fabrication and properties of vapor-grown carbon nanofiber/polymer composites. Composites Science and Technology. 2007;**67**:1709-1718

[21] Iijima S, Ichihashi T. Single-shell carbon nanotubes of 1-nm diameter. Nature. 1993;**363**:603

[22] Iijima S. Helical microtubules of graphitic carbon. Nature. 1991;**354**:56

[23] Geim AK, Novoselov KS. The rise of graphene. Nature Materials. 2007;**6**:183

[24] Nazir A, Yu H, Wang L, Haroon M, Ullah RS, Fahad S, et al. Recent progress in the modification of carbon materials and their application in composites for electromagnetic interference shielding. Journal of Materials Science. 2018;**53**: 8699-8719

[25] Al-Saleh MH, Saadeh WH, Sundararaj U. EMI shielding effectiveness of carbon based nanostructured polymeric materials: A comparative study. Carbon. 2013;**60**: 146-156

[26] Kim B-J, Bae K-M, Lee YS, An K-H, Park S-J. EMI shielding behaviors of Ni-coated MWCNTs-filled epoxy matrix nanocomposites. Surface & Coatings Technology. 2014;**242**:125-131

[27] Yim Y-J, Rhee KY, Park S-J. Electromagnetic interference shielding effectiveness of nickel-plated MWCNTs/high-density polyethylene composites. Composites Part B-Engineering. 2016;**98**:120-125

[28] Yang YL, Gupta MC, Dudley KL, Lawrence RW. Conductive carbon nanoriber-polymer foam structures. Advanced Materials. 2005;**17**:1999

[29] Zhang L, Wang LB, See KY, Ma J. Effect of carbon nanofiber reinforcement on electromagnetic interference shielding effectiveness of syntactic foam. Journal of Materials Science. 2013;**48**:7757-7763

[30] Zhang L, Roy S, Chen Y, Chua EK, See KY, Hu X, et al. Mussel-inspired polydopamine coated hollow carbon microspheres, a novel versatile filler for fabrication of high performance syntactic foams. ACS Applied Materials & Interfaces. 2014;**6**:18644-18652

[31] Xu Y, Li Y, Hua W, Zhang A, Bao J. Light-weight silver plating foam and carbon nanotube hybridized epoxy composite foams with exceptional conductivity and electromagnetic shielding property. ACS Applied Materials & Interfaces. 2016;**8**: 24131-24142

[32] Zhang L, Liu M, Roy S, Chu EK, See KY, Hu X. Phthalonitrile-based carbon foam with high specific mechanical strength and superior electromagnetic interference shielding performance. ACS Applied Materials & Interfaces. 2016;**8**:7422-7430

[33] Lu D, Mo Z, Liang B, Yang L, He Z, Zhu H, et al. Flexible, lightweight carbon nanotube sponges and composites for high-performance electromagnetic interference shielding. Carbon. 2018;**133**:457-463

[34] Zhang L, Liu M, Bi S, Yang L, Roy S, Tang X-Z, et al. Polydopamine decoration on 3D graphene foam and its electromagnetic interference shielding properties. Journal of Colloid and Interface Science. 2017;**493**:327-333

[35] Wu Y, Wang Z, Liu X, Shen X, Zheng Q, Xue Q, et al. Ultralight graphene foam/conductive polymer composites for exceptional electromagnetic interference shielding. ACS Applied Materials & Interfaces. 2017;**9**:9059-9069

[36] Song W-L, Guan X-T, Fan L-Z, Cao W-Q, Wang C-Y, Cao M-S. Tuning three-dimensional textures with graphene aerogels for ultra-light flexible graphene/texture composites of effective electromagnetic shielding. Carbon. 2015;**93**:151-160

[37] Singh S, Tripathi P, Bhatnagar A, Patel CRP, Singh AP, Dhawan SK, et al. A highly porous, light weight 3D sponge like graphene aerogel for electromagnetic interference shielding applications. RSC Advances. 2015;**5**: 107083-107087

[38] Zeng Z, Wang C, Zhang Y, Wang P, Shahabadi SIS, Pei Y, et al. Ultralight and highly elastic graphene/lignin-derived carbon nanocomposite aerogels with ultrahigh electromagnetic interference shielding performance. ACS Applied Materials & Interfaces. 2018;**10**:8205-8213

[39] Liu J, Liu Y, Zhang H-B, Dai Y, Liu Z, Yu Z-Z. Superelastic and multifunctional graphene-based aerogels by interfacial reinforcement with graphitized carbon at high temperatures. Carbon. 2018;**132**:95-103

[40] Bi S, Zhang L, Mu C, Lee HY, Cheah JW, Chua EK, et al. A comparative study on electromagnetic interference shielding behaviors of chemically reduced and thermally reduced graphene aerogels. Journal of Colloid and Interface Science. 2017;**492**:112-118

[41] Bi S, Zhang L, Mu C, Liu M, Hu X. Electromagnetic interference shielding properties and mechanisms of chemically reduced graphene aerogels. Applied Surface Science. 2017;**412**: 529-536

[42] Bi S, Zhang L, Mu C, Lee HY, Cheah JW, Chua EK, et al. A comparative study on electromagnetic interference shielding behaviors of chemically reduced and thermally reduced graphene aerogels. Journal of Colloid and Interface Science. 2017;**492**:112-118

Chapter 11

Optical Propagation in Magneto-Optical Materials

Licinius Dimitri Sá de Alcantara

Abstract

Magneto-optical materials present anisotropy in the electrical permittivity controlled by a magnetic field, which affects the propagation characteristics of light and stands out in the design of nonreciprocal devices, such as optical isolators and circulators. Based on Maxwell's equations, this chapter focuses on the wave propagation in magneto-optical media. The following cases are covered: The propagation of a plane wave in an unbounded magneto-optical medium, where the phenomenon of Faraday rotation is discussed, and the guided propagation in planar magneto-optical waveguides with three and five layers, highlighting the phenomenon of nonreciprocal phase shift and its potential use on the design of nonreciprocal optical devices.

Keywords: magneto-optical media, light propagation, Faraday rotation, nonreciprocal phase shift, optical devices

1. Introduction

A material is classified as magneto-optical (MO) if it affects the propagation characteristics of light when an external magnetic field is applied on it. For ferromagnetic materials, which are composed by magnetically ordered domains, MO phenomena may also occur in the absence of an external magnetic field. A great number of magneto-optical phenomena are the direct or indirect outcome of the splitting of energy levels in an external or spontaneous magnetic field [1].

The MO effect depends on the polarization of the magnetic field. It also depends on the polarization of the light and on its propagation direction, so it is an anisotropic phenomenon, which has attracted great attention from researchers in optical devices. The MO materials can have their anisotropy controlled by a magnetostatic field (H_{DC}), and this behavior can be exploited on the design of nonreciprocal devices. By nonreciprocal devices or structures, it means that waves or guided modes supported by them have their propagation characteristics altered when the wave propagation sense is reversed. Optical isolators and circulators can be highlighted as examples of such devices. Isolators are designed to protect optical sources from reflected light and are present in optical amplification systems. The circulators are employed as signal routers and act in devices that extract wavelengths in WDM systems.

The design of optical devices with MO materials is addressed in several works such as [2–5]. The challenges for the design of such devices are the development of MO materials with high-induced anisotropy and high transparency at the optical

IntechOpen

spectrum. Therefore, research activities on the improvement of MO materials and structures have also great relevance and are covered in works such as [6–10]. Integration of MO materials and structures with other optical system components, with reduction of insertion losses, is also a target for researches in optical devices. Research of MO effects in optical structures such as photonic crystals has also been addressed [11–13].

This chapter presents analytical formalisms derived from Maxwell's and wave equations to analyze the propagation characteristics of transverse electromagnetic (TEM) waves in unbounded magneto-optical material. The guided propagation characteristics of transverse magnetic (TM) modes in three- and five-layered planar magneto-optical waveguides are also formalized and discussed. The analytical formalism is versatile so that each layer can be set as magneto-optical or isotropic in the mathematical model.

2. Wave propagation characteristics

This section focuses on the optical propagation analysis in magneto-optical media using Maxwell's equations as starting point. In a magnetized MO media, cyclotron resonances occur at optical frequencies, if the wave is properly polarized. This physical phenomenon induces a coupling between orthogonal electric field components in the plane perpendicular to the applied magnetostatic field H_{DC}, which affects the wave polarization. Depending on the orientation of the magnetostatic field, the configuration of the electric permittivity tensor changes. If H_{DC} is oriented along one of the Cartesian axes, the relative electric permittivity assumes the form

$$\overline{\overline{\varepsilon}}_r = \begin{bmatrix} n^2 & 0 & 0 \\ 0 & n^2 & j\delta \\ 0 & -j\delta & n^2 \end{bmatrix}, \text{ for } H_{DC} \parallel x-\text{axis}; \tag{1}$$

$$\overline{\overline{\varepsilon}}_r = \begin{bmatrix} n^2 & 0 & j\delta \\ 0 & n^2 & 0 \\ -j\delta & 0 & n^2 \end{bmatrix}, \text{ for } H_{DC} \parallel y-\text{axis}; \tag{2}$$

$$\overline{\overline{\varepsilon}}_r = \begin{bmatrix} n^2 & j\delta & 0 \\ -j\delta & n^2 & 0 \\ 0 & 0 & n^2 \end{bmatrix}, \text{ for } H_{DC} \parallel z-\text{axis}. \tag{3}$$

where n is the refractive index of the material and δ is the magneto-optical constant. The MO constant is proportional to H_{DC}. If the sense of H_{DC} is reversed, $\delta(-H_{DC}) = -\delta(H_{DC})$, and for $H_{DC} = 0$, the off-diagonal components of the electric permittivity tensor are zero [14, 15].

2.1 TEM wave in an unbounded magneto-optical medium

Let us consider a TM wave propagating in an unbounded MO medium, as shown in **Figure 1**.

From Maxwell's equations, the vectorial Helmholtz equation for anisotropic media and for the electric field $\overline{E}(x,y,z)$ can be written as

$$\omega^2\mu_0\varepsilon_0\overline{\overline{\varepsilon}}_r\overline{E} + \nabla^2\overline{E} - \nabla\left(\nabla\cdot\overline{E}\right) = \overline{0}, \tag{4}$$

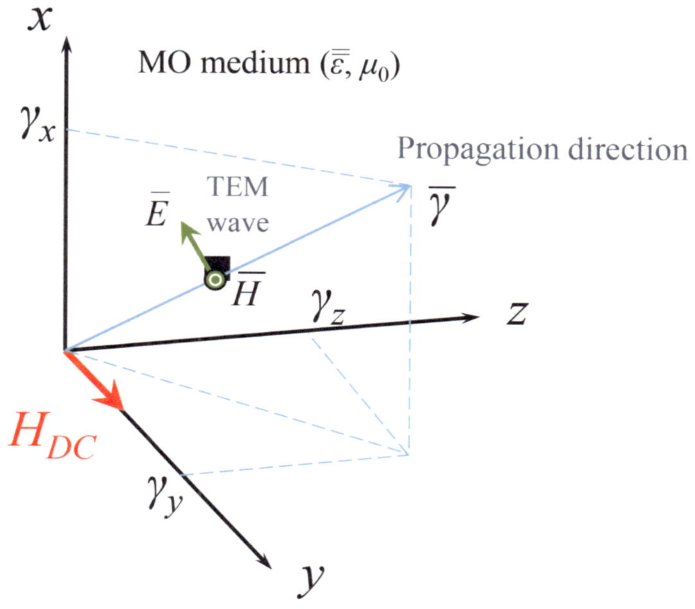

Figure 1.
TEM wave in an unbounded magneto-optical medium.

where ω is the angular frequency in rad/s, μ_0 is the magnetic permeability of the vacuum in H/m, and ε_0 is the electric permittivity of the vacuum in F/m.

To develop a plane wave solution for MO media, it is assumed that H_{DC} is parallel to the y-axis and $\bar{\bar{\varepsilon}}_r$ is given by Eq. (2) from now on. This assumption does not imply on lack of generality because it is assumed that the wave propagates at an arbitrary direction, with the electric field vector given by

$$\bar{E} = \bar{E}_0 \exp\left(j\omega t\right) \exp\left[-j\left(\gamma_x x + \gamma_y y + \gamma_z z\right)\right]. \tag{5}$$

where $\bar{\gamma} = \gamma_x \vec{\mathbf{i}} + \gamma_y \vec{\mathbf{j}} + \gamma_z \vec{\mathbf{k}}$ is the propagation constant vector.

From Gauss' law for a medium with equilibrium of charges, $\nabla.(\varepsilon_0 \bar{\bar{\varepsilon}}_r \bar{E}) = 0$, we obtain:

$$\nabla \cdot \bar{E} = j\frac{\delta}{n^2}\left(\frac{\partial E_x}{\partial z} - \frac{\partial E_z}{\partial x}\right). \tag{6}$$

Substituting Eq. (6) into Eq. (4) leads to

$$\omega^2 \mu_0 \varepsilon_0 \bar{\bar{\varepsilon}}_r \bar{E} + \nabla^2 \bar{E} - j\frac{\delta}{n^2}\nabla\left(\frac{\partial E_x}{\partial z} - \frac{\partial E_z}{\partial x}\right) = \bar{0}. \tag{7}$$

Expanding Eq. (7) in the Cartesian coordinates results in

$$\omega^2 \mu_0 \varepsilon_0 \left(n^2 E_x + j\delta E_z\right) + \frac{\partial^2 E_x}{\partial x^2} + \frac{\partial^2 E_x}{\partial y^2} + \frac{\partial^2 E_x}{\partial z^2} - j\frac{\delta}{n^2}\left(\frac{\partial^2 E_x}{\partial x \partial z} - \frac{\partial^2 E_z}{\partial x^2}\right) = 0, \tag{8}$$

$$\omega^2 \mu_0 \varepsilon_0 n^2 E_y + \frac{\partial^2 E_y}{\partial x^2} + \frac{\partial^2 E_y}{\partial y^2} + \frac{\partial^2 E_y}{\partial z^2} - j\frac{\delta}{n^2}\left(\frac{\partial^2 E_x}{\partial y \partial z} - \frac{\partial^2 E_z}{\partial x \partial y}\right) = 0, \tag{9}$$

$$\omega^2\mu_0\varepsilon_0\left(-j\delta E_x + n^2 E_z\right) + \frac{\partial^2 E_z}{\partial x^2} + \frac{\partial^2 E_z}{\partial y^2} + \frac{\partial^2 E_z}{\partial z^2} - j\frac{\delta}{n^2}\left(\frac{\partial^2 E_x}{\partial z^2} - \frac{\partial^2 E_z}{\partial x \partial z}\right) = 0. \tag{10}$$

The spatial derivatives in Eqs. (8)–(10) are now calculated by considering Eq. (5):

$$\left(\omega^2\mu_0\varepsilon_0 n^2 - |\gamma|^2 - j\frac{\delta}{n^2}\gamma_x\gamma_z\right)E_x + j\delta\left(\omega^2\mu_0\varepsilon_0 - \frac{1}{n^2}\gamma_x^2\right)E_z = 0 \tag{11}$$

$$\left(\omega^2\mu_0\varepsilon_0 n^2 - |\gamma|^2\right)E_y + j\frac{\delta}{n^2}\left(\gamma_y\gamma_z E_x - \gamma_x\gamma_y E_z\right) = 0, \tag{12}$$

$$-j\delta\left(\omega^2\mu_0\varepsilon_0 - \frac{1}{n^2}\gamma_z^2\right)E_x + \left(\omega^2\mu_0\varepsilon_0 n^2 - |\gamma|^2 - j\frac{\delta}{n^2}\gamma_x\gamma_z\right)E_z = 0, \tag{13}$$

where $|\gamma| = \sqrt{\gamma_x^2 + \gamma_y^2 + \gamma_z^2}$.

2.1.1 TEM wave with electric field vector parallel to H_{DC}

By observing Eqs. (11–13), we note that **when the electric field of the electromagnetic wave is polarized along the y-axis and is parallel to H_{DC}**, so that $E_x = E_z = 0$, the magneto-optical constant δ related to H_{DC} will have no effect on the propagation characteristics of the wave. In this case, from Eq. (12), the propagation constant modulus would be

$$|\gamma| = n\omega\sqrt{\mu_0\varepsilon_0}, \tag{14}$$

which is the same expression for a traveling wave in an isotropic material. Note that when the electric field is polarized along the y-axis, the wave is traveling in the plane xz, so that $\gamma_y = 0$.

2.1.2 The general expression for the propagation constant

In a general case, by solving the system formed by Eqs. (11) and (13), we obtain the following equation

$$\left(\omega^2\mu_0\varepsilon_0 n^2 - |\gamma|^2 - j\frac{\delta}{n^2}\gamma_x\gamma_z\right)^2 - \delta^2\left(\omega^2\mu_0\varepsilon_0 - \frac{1}{n^2}\gamma_x^2\right)\left(\omega^2\mu_0\varepsilon_0 - \frac{1}{n^2}\gamma_z^2\right) = 0. \tag{15}$$

Solving Eq. (15) for $|\gamma|$, we obtain:

$$|\gamma| = \sqrt{\omega^2\mu_0\varepsilon_0 n^2 - j\frac{\delta}{n^2}\gamma_x\gamma_z \pm \delta\sqrt{\left(\omega^2\mu_0\varepsilon_0 - \frac{1}{n^2}\gamma_x^2\right)\left(\omega^2\mu_0\varepsilon_0 - \frac{1}{n^2}\gamma_z^2\right)}}. \tag{16}$$

Note that when the MO constant $\delta = 0$, Eq. (16) reduces to Eq. (14).

The parameters γ_x and γ_z are projections of the propagation constant vector along the x and the y-axis, respectively.

2.1.3 TEM wave propagating parallel to H_{DC}

If the TEM wave is propagating along the H_{DC} direction (y-axis), so that $\gamma_x = \gamma_z = 0$, Eq. (16) assumes the simpler form:

$$|\gamma| = \omega\sqrt{\mu_0\varepsilon_0(n^2 \pm \delta)}, \tag{17}$$

and from Eq. (5), the electric field vector becomes

$$\overline{E} = \left(E_{0x}\vec{\mathbf{i}} + E_{0z}\vec{\mathbf{k}}\right)\exp\left[j\left(\omega t - y\omega\sqrt{\mu_0\varepsilon_0(n^2 \pm \delta)}\right)\right]. \tag{18}$$

From Eq. (11), we see that the electric field components are connected by

$$E_{0x} = -j\delta\frac{\omega^2\mu_0\varepsilon_0}{\left(\omega^2\mu_0\varepsilon_0 n^2 - |\gamma|^2\right)}E_{0z}. \tag{19}$$

Substituting Eq. (17) in Eq. (19), we obtain:

$$E_{0x} = \pm jE_{0z}. \tag{20}$$

Therefore, substituting Eq. (20) in Eq. (18), and given that $\pm j = \exp(\pm j\pi/2)$, the electric field components can be written as

$$E_x = E_{0z}\exp\left[j\left(\omega t - y\omega\sqrt{\mu_0\varepsilon_0(n^2 \pm \delta)} \pm \pi/2\right)\right], \tag{21}$$

$$E_z = E_{0z}\exp\left[j\left(\omega t - y\omega\sqrt{\mu_0\varepsilon_0(n^2 \pm \delta)}\right)\right]. \tag{22}$$

Eqs. (21) and (22) represent a circular polarized wave, which can be dismembered into two circular polarized eigenmodes propagating along the y-axis with different propagation constants. If the plus sign (in "±") is adopted for Eqs. (21) and (22), we obtain a counterclockwise (CCW) circular polarized eigenmode. Otherwise, if the minus sign is adopted, we obtain a clockwise (CW) circular polarized eigenmode, as shown in **Figure 2**. From Eq. (17), it is possible to associate an equivalent refractive index to each eigenmode:

$n^+ = \sqrt{n^2 + \delta}$, for the CCW circular polarized eigenmode;
$n^- = \sqrt{n^2 - \delta}$, for the CW circular polarized eigenmode.

A linear polarized wave propagating along the y-axis may be decomposed into two opposite circular polarized waves in the xz plane, as shown in **Figure 2**. Since these eigenmodes propagate with distinct propagation constants, the linear

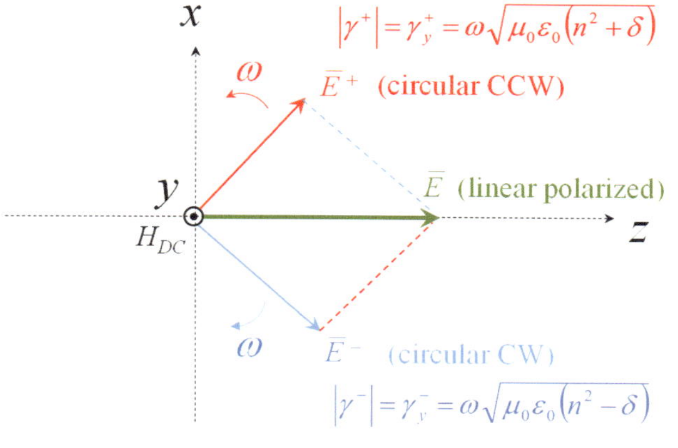

Figure 2.
Decomposition of a linear polarized TEM wave into two circular polarized components. The circular polarized components travel with distinct propagation constants in a MO medium.

polarization will rotate in the xz plane as the wave propagates along the y-axis, in a phenomenon known as **Faraday rotation**, which is depicted in **Figure 3**.

When the sense of the magnetostatic field H_{DC} is reversed, the magneto-optical constant δ changes its signal, and the values of n^+ and n^- are interchanged, and the sense of rotation of a linear polarized wave in the MO media will change.

The Faraday rotation angle (ϕ_F) may be calculated (in radians) as a function of the propagation distance y by

$$\phi_F = \frac{1}{2}(\phi^+ - \phi^-) = \frac{1}{2}\left(n^+\frac{2\pi}{\lambda_0}y - n^-\frac{2\pi}{\lambda_0}y\right) = \frac{\pi}{\lambda_0}\left(\sqrt{n^2+\delta} - \sqrt{n^2-\delta}\right)y, \quad (23)$$

where λ_0 is the optical wavelength in vacuum. The Faraday rotation effect is responsible for a periodic power transfer between the transverse components, in this case, E_x and E_z. This phenomenon in MO materials may be exploited for the design of optical isolators based on Faraday rotation.

When a MO waveguide, with H_{DC} applied along its longitudinal direction, supports degenerate orthogonal quasi TEM modes, the power transfer between these modes will be maximized. **Figure 4** shows a MO rib waveguide [16], where layers 1 and 2 are composed of bismuth yttrium iron garnet (Bi-YIG) grown on top of a gadolinium gallium garnet (GGG) substrate with n_{SR} = 1.94. For the Bi-YIG layers, the relative permittivity tensor has the form of Eq. (2), with δ = 2.4 × 10^{-4},

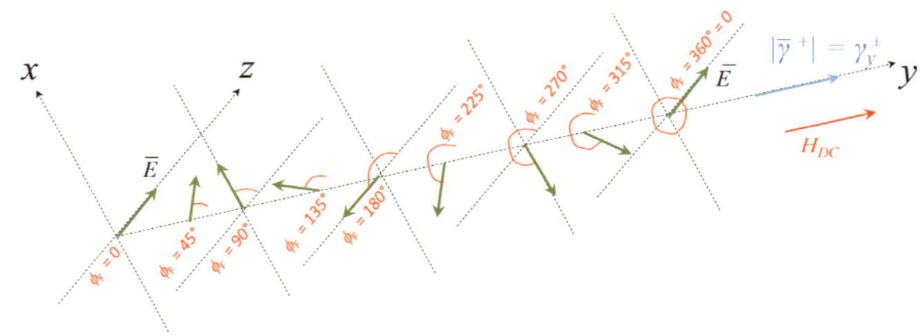

Figure 3.
Faraday rotation of a linear polarized TEM wave in a MO medium. The propagation direction is parallel to the magnetostatic field H_{DC}.

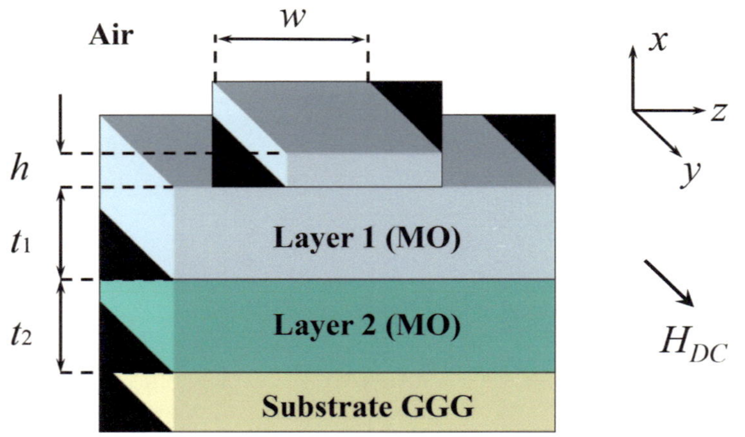

Figure 4.
Magneto-optical rib waveguide.

$n_1 = 2.19$, and $n_2 = 2.18$. The waveguide dimensions are $w = 8$ μm, $h = 0.5$ μm, $t_1 = 3.1$ μm, and $t_2 = 3.4$ μm. The optical wavelength is $\lambda_0 = 1.485$ μm.

Figure 5 shows numerical results for the power transfer between the transverse components along the propagation direction. These results were obtained using a finite difference vectorial beam propagation method (FD-VBPM) [17]. We observe that the length for maximum energy transfer is around 6800 μm. In practice, as observed in [16], the device length must be set at half that length (~3400 μm) so that a 45° rotation is achieved at the output port. Therefore, if a reflection occurs at this point, the reflected field will complete a 90° rotation at the input port, which can then be blocked with a polarizer without affecting the input field, so that an optical isolator is obtained.

In Eq. (23), by adopting $\delta = 2.4 \times 10^{-4}$, $n = n_1 = 2.19$, $\lambda_0 = 1.485$ μm, and $\phi_F = \pi/4$ (45°), we obtain $y = 3388$ μm, which is a propagation length that converges with the FD-VBPM result.

2.1.4 TEM wave propagating along the diagonal of an imaginary cube

Before finishing this section, let us consider another particular case of propagation direction—suppose, in **Figure 1**, that $\gamma_x = \gamma_y = \gamma_z = \gamma_u$, with $\gamma_u \neq 0$. **This case corresponds to a TEM wave propagating along the diagonal of an imaginary cube, adjacent to the Cartesian axes.** From Eq. (16), we obtain:

$$|\gamma| = \sqrt{\omega^2\mu_0\varepsilon_0 n^2 - j\frac{\delta}{n^2}\gamma_u^2 \pm \delta\left(\omega^2\mu_0\varepsilon_0 - \frac{1}{n^2}\gamma_u^2\right)}. \tag{24}$$

From the relation $|\gamma| = \sqrt{\gamma_x^2 + \gamma_y^2 + \gamma_z^2}$ we can also obtain:

$$|\gamma| = \gamma_u\sqrt{3}. \tag{25}$$

Equaling Eqs. (24)–(25) and solving for γ_u result in

$$\gamma_u = \omega\sqrt{\frac{\mu_0\varepsilon_0(n^2 \pm \delta)}{3 \pm \frac{\delta}{n^2} + j\frac{\delta}{n^2}}}. \tag{26}$$

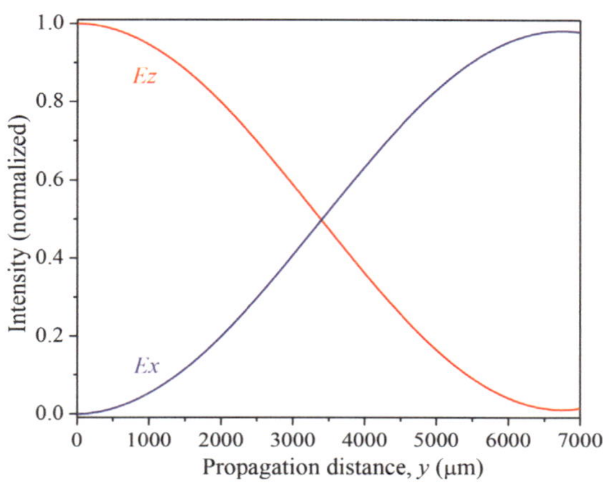

Figure 5.
Normalized intensity evolution of the transverse field components along the propagation direction (y-axis) of the MO waveguide.

Substituting Eq. (26) in Eq. (25), we obtain the propagation constant:

$$|\gamma| = \omega\sqrt{\frac{\mu_0\varepsilon_0(n^2 \pm \delta)}{1 \pm \frac{\delta}{3n^2} + j\frac{\delta}{3n^2}}}. \tag{27}$$

The corresponding electric field vector can be retrieved by substituting the results of Eqs. (26)–(27) in Eq. (11) to obtain

$$E_x = \pm jE_z. \tag{28}$$

However, for the considered propagation direction, the E_y component is not zero. From Eq. (12) we obtain:

$$E_y = -\frac{n^2 \pm \delta + j3(\delta \pm n^2)}{5n^2}E_z. \tag{29}$$

By using the results of Eqs. (26)–(29) in Eq. (5), we can express the electric field vector for this particular case by

$$\overline{E} = \left(\pm j\,\vec{\mathbf{i}} - \frac{n^2 \pm \delta + j3(\delta \pm n^2)}{5n^2}\,\vec{\mathbf{j}} + \vec{\mathbf{k}}\right)E_{0z}\exp\left[j\left(\omega t - \omega\sqrt{\frac{\mu_0\varepsilon_0(n^2 \pm \delta)}{3 \pm \frac{\delta}{n^2} + j\frac{\delta}{n^2}}}(x+y+z)\right)\right], \tag{30}$$

where **i**, **j**, and **k** are the unit vectors along the x-, y-, and z-axis, respectively.

As in the previous case of propagation, Eq. (30) provides two eigenmodes for TEM propagation. From Eq. (28) we can observe that, when projected in the *xz* plane, the electric field vector of each eigenmode is circular polarized. The combination of these eigenmodes will result in a wave with linear polarization progressively rotated as it propagates. The E_y component has the role of projecting the Faraday rotation to the plane perpendicular to the propagation direction (the diagonal of the cube), since the wave is TEM regarding this propagation direction. **Figure 6** shows a simulation of the TEM wave eigenmodes along the diagonal of an imaginary cube.

The simulations presented in **Figure 6** were performed for f = 193.4145 THz, n = 2, and δ = 0.2. Note that both eigenmodes present losses as they propagate. This is due the complex characteristic of the propagation constant expressed by Eq. (27), where the imaginary part depends on the magneto-optical constant δ. It was observed that increasing δ enhances the Faraday rotation but also increases the losses for diagonal propagation.

Equivalent refractive indexes for the circular polarized eigenmodes can be obtained from Eq. (27), which leads to the following equation to compute the Faraday rotation for diagonal propagation:

$$\phi_F = \frac{\pi}{\lambda_0}\mathrm{Re}\left(\sqrt{\frac{n^2 + \delta}{1 + \frac{\delta}{3n^2} + j\frac{\delta}{3n^2}}} - \sqrt{\frac{n^2 - \delta}{1 - \frac{\delta}{3n^2} + j\frac{\delta}{3n^2}}}\right)d, \tag{31}$$

where d is the propagation distance along the diagonal.

For n = 2, δ = 0.2, and λ_0 = 1.55 μm, we obtain ϕ_F/d = 0.27046 rads/μm. Comparing with the case for propagation along the y-axis (parallel to H_{DC}), by using Eq. (23), we obtain ϕ_F/y = 0.40549 rads/μm. These results show that we can obtain a better Faraday rotation when the propagation direction is aligned with the magnetostatic field, when considering TEM waves.

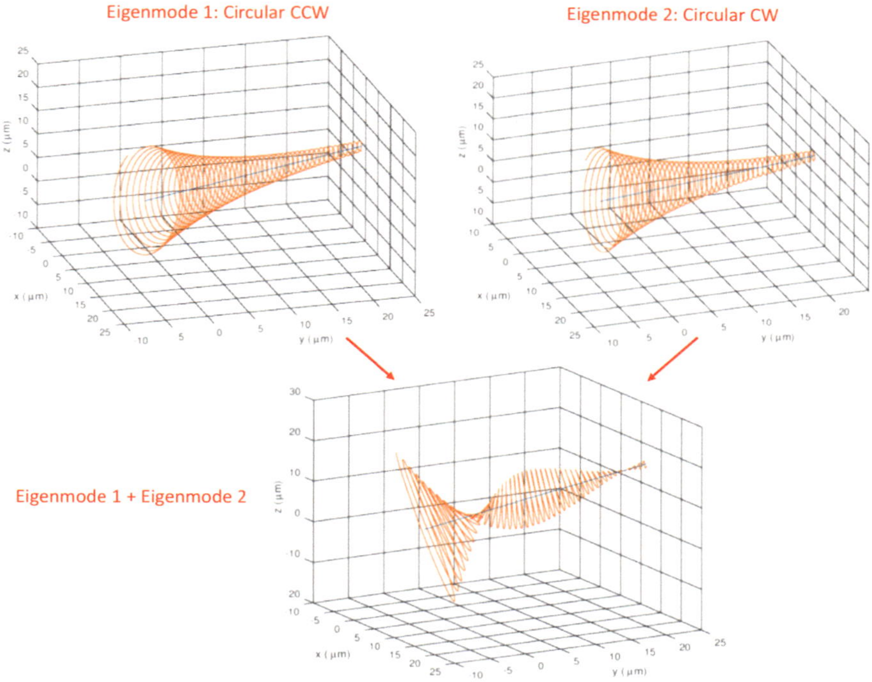

Figure 6.
TEM eigenmodes for diagonal propagation where $\gamma_x = \gamma_y = \gamma_z$. The trajectory of the electric field vector is represented by red lines.

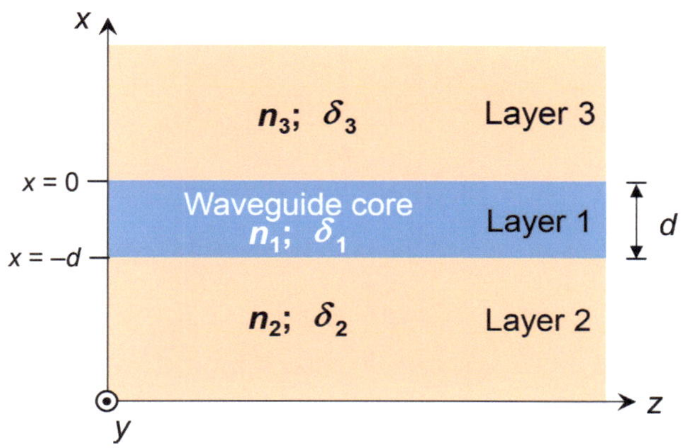

Figure 7.
Longitudinal section of a planar MO waveguide.

2.2 TM mode in a planar magneto-optical waveguide

Figure 7 presents a planar MO waveguide, which is composed by three MO layers. The magnetostatic field H_{DC} is applied along the y-axis. The propagation direction is now the z-axis. The planar waveguide supports transversal electric, TE, modes (H_x, E_y, H_z components) and transversal magnetic, TM, modes (E_x, H_y, E_z components). As discussed in Section 2.1.1, if H_{DC} is parallel to the electric field vector of the wave, then MO constant δ does not affect the propagation characteristics of the mode. Therefore, for the TE modes, no MO effect will be observed. For TM modes, however, the electric field components are perpendicular to H_{DC}, and nonreciprocal propagation characteristics will take place. In this section, mathematical expressions to calculate the propagation constants for TM modes in a MO planar waveguide will be derived. For the occurrence of guided modes in the structure shown in **Figure 7**, $n_1 > n_2$ and $n_1 > n_3$.

Defining $\overline{\overline{\xi}}$ as the inverse of the electric permittivity tensor of Eq. (2), we have:

$$\overline{\overline{\xi}} = \overline{\overline{\varepsilon}}_r^{-1} = \begin{bmatrix} \dfrac{n^2}{n^4-\delta^2} & 0 & -\dfrac{j\delta}{n^4-\delta^2} \\ 0 & \dfrac{n^2}{n^4-\delta^2} & 0 \\ \dfrac{j\delta}{n^4-\delta^2} & 0 & \dfrac{n^2}{n^4-\delta^2} \end{bmatrix} = \begin{bmatrix} \xi_{xx} & 0 & -\xi_{zx} \\ 0 & \xi_{yy} & 0 \\ \xi_{zx} & 0 & \xi_{zz} \end{bmatrix}. \tag{32}$$

From Maxwell's equations at the frequency domain, considering TM modes (E_x, H_y, E_z components) and no field spatial variations along the y-axis, we obtain:

$$j\omega\mu H_y = j\beta E_x + \frac{\partial E_z}{\partial x}, \tag{33}$$

$$E_x = \frac{1}{\varepsilon_0}\left(\xi_{xx}\frac{\beta}{\omega}H_y + j\frac{\xi_{zx}}{\omega}\frac{\partial H_y}{\partial x}\right), \tag{34}$$

$$E_z = \frac{1}{\varepsilon_0}\left(\xi_{zx}\frac{\beta}{\omega}H_y - j\frac{\xi_{zz}}{\omega}\frac{\partial H_y}{\partial x}\right), \tag{35}$$

where β is the propagation constant of the guided TM mode in radians per meter.

Substituting Eqs. (34)–(35) in Eq. (33), we obtain the following wave equation for nonreciprocal media in terms of the H_y component:

$$\frac{\partial^2 H_y}{\partial x^2} + \left(\frac{k_0^2 - \xi_{xx}\beta^2}{\xi_{zz}}\right)H_y = 0, \tag{36}$$

where $k_0 = \omega\sqrt{\mu\varepsilon_0}$.

The solution for H_y is expressed for each waveguide layer as.

$$H_y = C\exp(-\zeta x), \text{ for } x \geq 0. \tag{37}$$

$$H_y = C\cos(\kappa x) + D\,\text{sen}(\kappa x), \text{ for} - \text{d} \leq x \leq 0. \tag{38}$$

$$H_y = [C\cos(\kappa d) - D\,\text{sen}(\kappa d)]\exp[\gamma(x+d)], \text{ for } x \leq -\text{d}. \tag{39}$$

The solution for the component E_z at each layer is obtained by substituting the corresponding solution for H_y in Eq. (35), resulting in.

$$E_z = \frac{C}{\omega\varepsilon_0}\left(\xi_{zx}^{(3)}\beta + j\zeta\xi_{zz}^{(3)}\right)\exp(-\zeta x), \text{ for } x \geq 0. \tag{40}$$

$$E_z = \frac{1}{\omega\varepsilon_0}\left\{C\left[\xi_{zx}^{(1)}\beta\cos(\kappa x) + j\kappa\xi_{zz}^{(1)}\text{sen}(\kappa x)\right] + D\left[\xi_{zx}^{(1)}\beta\text{sen}(\kappa x) - j\kappa\xi_{zz}^{(1)}\cos(\kappa x)\right]\right\}, \text{ for} - \text{d} \leq x \leq 0. \tag{41}$$

$$E_z = \frac{C\cos(\kappa d) - D\text{sen}(\kappa d)}{\omega\varepsilon_0}\left(\xi_{zx}^{(2)}\beta - j\gamma\xi_{zz}^{(2)}\right)\exp[\gamma(x+d)], \text{ for } x \leq -\text{d}. \tag{42}$$

The superscripts between parentheses on the inverse permittivity tensor elements identify the corresponding waveguide layer, as specified in **Figure 7**. The continuity of E_z at $x = 0$ and at $x = -d$ leads to the following system:

$$C\left[\left(\xi_{zx}^{(3)}-\xi_{zx}^{(1)}\right)\beta+j\zeta\xi_{zz}^{(3)}\right]+D\left(j\kappa\xi_{zz}^{(1)}\right)=0, \tag{43}$$

$$\begin{aligned}&C\left\{\left[\left(\xi_{zx}^{(1)}-\xi_{zx}^{(2)}\right)\beta+j\gamma\xi_{zz}^{(2)}\right]\cos(\kappa d)-j\kappa\xi_{zz}^{(1)}\,sen(\kappa d)\right\}\\&+D\left\{\left[\left(\xi_{zx}^{(2)}-\xi_{zx}^{(1)}\right)\beta-j\gamma\xi_{zz}^{(2)}\right]\mathrm{sen}(\kappa d)-j\kappa\xi_{zz}^{(1)}\cos(\kappa d)\right\}=0.\end{aligned} \tag{44}$$

After solving this system formed by Eqs. (43)–(44), we obtain:

$$\tan(\kappa d)=\frac{\kappa\xi_{zz}^{(1)}\left[\zeta\xi_{zz}^{(3)}+\gamma\xi_{zz}^{(2)}-j\left(\xi_{zx}^{(3)}-\xi_{zx}^{(2)}\right)\beta\right]}{\left(\kappa\xi_{zz}^{(1)}\right)^2-\left[\left(\xi_{zx}^{(3)}-\xi_{zx}^{(1)}\right)\beta+j\zeta\xi_{zz}^{(3)}\right]\left[\left(\xi_{zx}^{(2)}-\xi_{zx}^{(1)}\right)\beta-j\gamma\xi_{zz}^{(2)}\right]}. \tag{45}$$

The constants ζ, κ, and γ can be determined by substituting Eq. (37), Eq. (38), or Eq. (39), respectively, in Eq. (36), resulting in

$$\zeta=\sqrt{\frac{\xi_{xx}^{(3)}\beta^2-k_0^2}{\xi_{zz}^{(3)}}}, \tag{46}$$

$$\kappa=\sqrt{\frac{k_0^2-\xi_{xx}^{(1)}\beta^2}{\xi_{zz}^{(1)}}}, \tag{47}$$

$$\gamma=\sqrt{\frac{\xi_{xx}^{(2)}\beta^2-k_0^2}{\xi_{zz}^{(2)}}}, \tag{48}$$

where $k_0 = 2\pi/\lambda_0$, and λ_0 is the optical wavelength.

From the roots of Eq. (45) for β, the dispersion curve for TM modes in MO waveguides can be retrieved. Assuming that $n_1 = 2.26$, $n_2 = 2.0$, $n_3 = 2.23$, $d = 1$ μm, and only the layer 3 is magneto-optical with $\delta = 0.019$, the dispersion curve for the fundamental and a superior TM mode is shown in **Figure 8**. We observe that the effective index profile changes when the propagation direction is reversed, which opens the possibility to the design of nonreciprocal devices. This phenomenon is known as *nonreciprocal phase shift*. If the magnetostatic field is not applied ($\delta = 0$), the effective index profile becomes reciprocal and converges to the dashed line

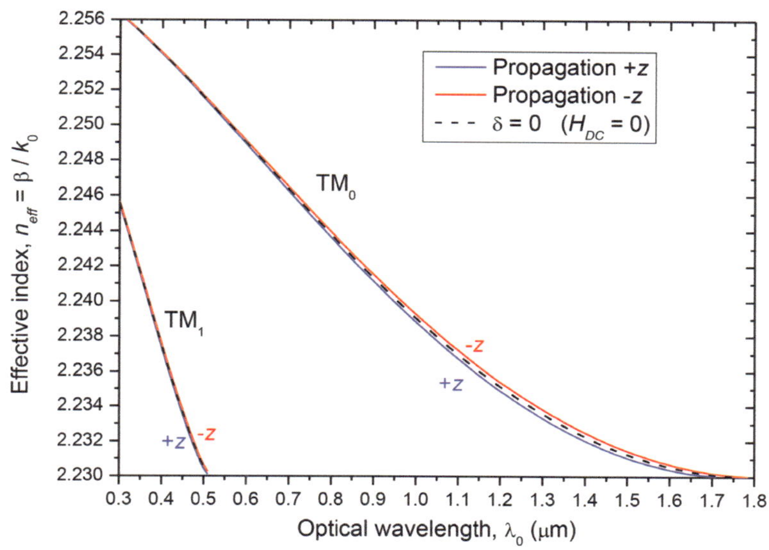

Figure 8.
Dispersion curves of the fundamental TM_0 mode and the superior TM_1 mode.

shown in **Figure 8**. The TM modes reach cutoff for optical wavelengths at which the effective index reaches the minimum value of 2.23. For greater optical wavelengths, the mode becomes irradiated and escapes through layer 3.

Figure 9 shows the transversal distributions of the H_y component at two distinct optical wavelengths. For this waveguide design, λ_0 = 1.55 μm is near cutoff, and the mode is highly distributed in the MO layer, which increases the nonreciprocal phase shift. Note from **Figure 8** that the difference between the effective indexes of the counter propagating TM modes are greater for optical wavelengths near cutoff, but as the wavelengths decreases, the mode becomes more confined at the waveguide core, and its interaction with the MO layer decreases, resulting in a decrease of the nonreciprocal phase shift effect, considering this waveguide configuration.

2.3 TM mode in a planar magneto-optical directional coupler

Now let us consider a five-layered MO planar structure as shown in **Figure 10**.

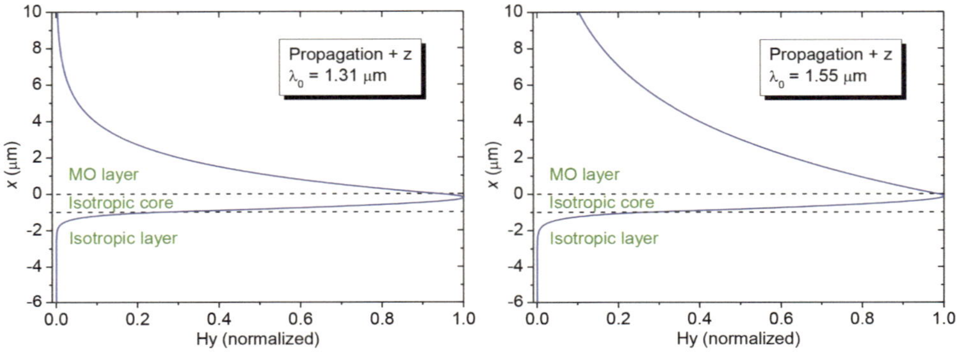

Figure 9.
Transversal distribution of the H_y *component of the fundamental* TM_o *mode at* λ_o = 1.31 μm *and at* λ_o = 1.55 μm.

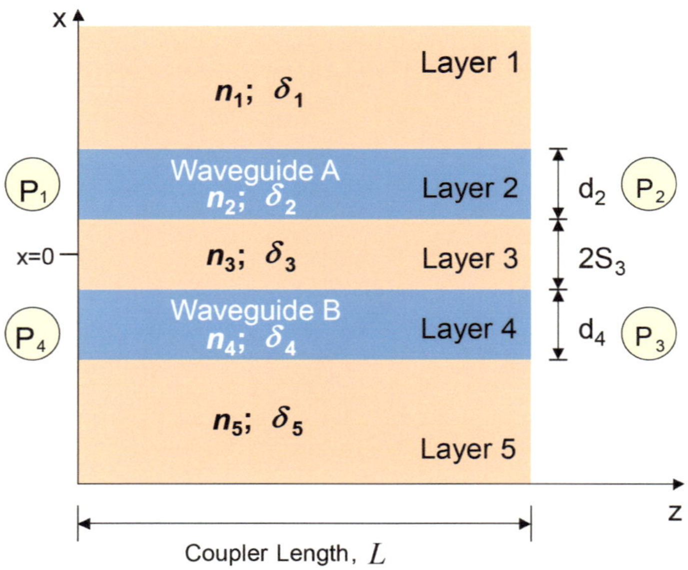

Figure 10.
Longitudinal section of the five-layered MO planar structure.

The solutions for Eq. (36) in each layer, making use of the proper radiation conditions, are [18]

$$
\begin{aligned}
H_y &= A_1 \exp\left[-\gamma_1(x - S_3 - d_2)\right], \text{ for } x \geq S_3 + d_2,\\
H_y &= A_2 \cos\left[\kappa_2(x - S_3 - d_2/2)\right] + A_3 \sin\left[\kappa_2(x - S_3 - d_2/2)\right] \text{ for } S_3 \leq x \leq S_3 + d_2,\\
H_y &= A_4 \exp(-\gamma_3 x) + A_5 \exp(\gamma_3 x), \text{ for } -S_3 \leq x \leq S_3,\\
H_y &= A_6 \cos\left[\kappa_4(x + S_3 + d_4/2)\right] + A_7 \sin\left[\kappa_4(x + S_3 + d_4/2)\right], \text{ for } -S_3-d_4 \leq x \leq -S_3,\\
H_y &= A_8 \exp\left[\gamma_5(x + S_3 + d_4)\right], \text{ for } x \leq -S_3-d_4,
\end{aligned}
$$

where A_1 through A_8 are constants to be determined, κ_i and γ_j are given by.

$$\kappa_i = \sqrt{\frac{k_0^2 - \xi_{xx}^{(i)}\beta^2}{\xi_{zz}^{(i)}}}, i = 2, 4, \tag{49}$$

$$\gamma_j = \sqrt{\frac{\xi_{xx}^{(j)}\beta^2 - k_0^2}{\xi_{zz}^{(j)}}}, j = 1, 3, 5, \tag{50}$$

where $k_0 = 2\pi/\lambda_0$, and λ_0 is the optical wavelength.

The electric field components E_x and E_z can be directly obtained with Eq. (34) and Eq. (35), respectively. Applying the boundary conditions for the tangential components H_y and E_z, one obtains a system of eight equations and eight unknowns, which can be conveniently written in matrix form as follows:

$$[M(\beta)]\mathbf{A} = \mathbf{0}. \tag{51}$$

Here, $[M(\beta)]$ is an 8×8 matrix that depends on the unknown longitudinal propagation constant β and $\mathbf{A} = [A_1\, A_2\, ...\, A_8]^{\mathrm{T}}$. The propagation constant can be easily found by solving the equation $Det([M(\beta)]) = 0$. The nonzero elements of the matrix $[M(\beta)]$ are listed below:

$$
\begin{aligned}
M_{11} &= 1; M_{12} = -\cos(\kappa_2 d_2/2); M_{13} = -\sin(\kappa_2 d_2/2);\\
M_{21} &= -j\xi_{zx}^{(1)}\beta + \gamma_1\xi_{zz}^{(1)}; M_{22} = j\xi_{zx}^{(2)}\beta\cos(\kappa_2 d_2/2) - \kappa_2\xi_{zz}^{(2)}\sin(\kappa_2 d_2/2);\\
M_{23} &= -j\xi_{zx}^{(2)}\beta\sin(k_2 d_2/2) + k_2\xi_{zz}^{(2)}\cos(k_2 d_2/2);\\
M_{32} &= \cos(\kappa_2 d_2/2); M_{33} = -\sin(\kappa_2 d_2/2); M_{34} = -\exp(-\gamma_3 S_3); M_{35} = -\exp(\gamma_3 S_3);\\
M_{42} &= -j\xi_{zx}^{(2)}\beta\cos(\kappa_2 d_2/2) - \kappa_2\xi_{zz}^{(2)}\sin(\kappa_2 d_2/2);\\
M_{43} &= j\xi_{zx}^{(2)}\beta\sin(\kappa_2 d_2/2) - \kappa_2\xi_{zz}^{(2)}\cos(\kappa_2 d_2/2);\\
M_{44} &= \left(j\xi_{zx}^{(3)}\beta - \xi_{zz}^{(3)}\gamma_3\right)\exp(-\gamma_3 S_3); M_{45} = \left(j\xi_{zx}^{(3)}\beta + \xi_{zz}^{(3)}\gamma_3\right)\exp(\gamma_3 S_3);\\
M_{54} &= \exp[\gamma_3 S_3]; M_{55} = \exp[-\gamma_3 S_3]; M_{56} = -\cos(\kappa_4 d_4/2); M_{57} = -\sin(\kappa_4 d_4/2);\\
M_{64} &= -\left(j\xi_{zx}^{(3)}\beta + \xi_{zz}^{(3)}\gamma_3\right)\exp[\gamma_3 S_3]; M_{65} = -\left(j\xi_{zx}^{(3)}\beta + \xi_{zz}^{(3)}\gamma_3\right)\exp[-\gamma_3 S_3];\\
M_{66} &= j\xi_{zx}^{(4)}\beta\cos(\kappa_4 d_4/2) - \xi_{zz}^{(4)}\kappa_4\sin(\kappa_4 d_4/2);\\
M_{67} &= j\xi_{zx}^{(4)}\beta\sin(\kappa_4 d_4/2) + \xi_{zz}^{(4)}\kappa_4\cos(\kappa_4 d_4/2);\\
M_{76} &= \cos(\kappa_4 d_4/2); M_{77} = -\sin(\kappa_4 d_4/2); M_{78} = -1;\\
M_{86} &= -j\xi_{zx}^{(4)}\beta\cos(\kappa_4 d_4/2) - \xi_{zz}^{(4)}\kappa_4\sin(\kappa_4 d_4/2);\\
M_{87} &= j\xi_{zx}^{(4)}\beta\sin(\kappa_4 d_4/2) - \xi_{zz}^{(4)}\kappa_4\cos(\kappa_4 d_4/2);\\
M_{88} &= j\xi_{zx}^{(5)}\beta + \xi_{zz}^{(5)}\gamma_5;
\end{aligned}
$$

As an example, **Table 1** shows the material parameters and layer thicknesses for each layer. Layers 1 and 5 are unbounded, and their thicknesses are theoretically infinite for the analytical model. The optical wavelength is λ_0 = 1.55 μm.

Figure 11 shows a plot of guided supermodes that occurs in the planar structure for forward propagation (along +*z*). The guided propagation along the five-layered structure, as well the periodical energy exchange of light between the two waveguides, can be expressed as a linear combination of these supermodes. The coupling length for the structure is given by $L_\pi = \pi/|\beta_1 - \beta_2|$, where β_1 and β_2 are the propagation constants of the supermodes obtained from the roots of $Det([M(\beta)]) = 0$. The computed coupling length, which refers to the propagation along the +*z* axis, is $L_\pi^+ = 1389.84\ \mu m$.

Figure 12 shows the plot of the supermodes, now considering backward propagation of the TM mode (along -*z*). The computed coupling length, which refers to the backward propagation along the *z*-axis, is $L_\pi^- = 689\,\mu m$.

Layer	Parameters		
	n	δ	Thickness (μm)
1	2.23	−0.019	∞
2	2.26	0	1.20
3	2.00	0	0.75
4	2.26	0	1.23
5	2.23	−0.019	∞

Table 1.
Material and geometric parameters of the MO directional coupler.

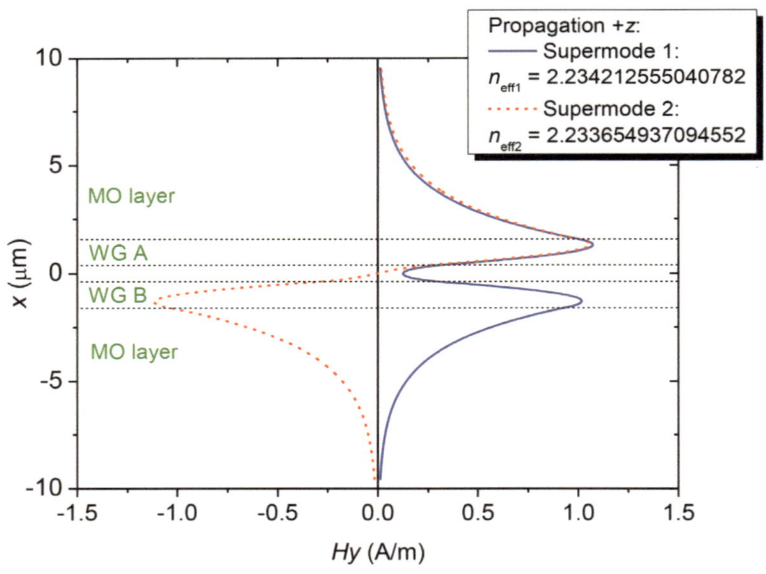

Figure 11.
Transversal distribution of the supermodes (H_y *component*) *for forward propagation (+z).*

Considering both propagation senses, when the condition $L = L_\pi^+ = 2L_\pi^-$ for the length of the directional coupler is achieved, we obtain an optical isolator calibrated for the given optical wavelength. The operation of the optical isolator is depicted in **Figure 13**. If an optical source is placed at the port 1 of the waveguide A, all optical power will be coupled into port 3 of the waveguide B, if the length of the directional coupler is $L = L_\pi^+$. If some light is reflected at port 3, since $L = 2L_\pi^-$, all optical power is directed to the port 4. Therefore, the optical source at port 1 becomes isolated from the reflected light. **Figures 14, 15** show simulations of the forward and backward optical propagation in the MO directional coupler via a propagation projection of a linear combination of the corresponding supermodes.

The MO directional coupler of **Figure 10** also acts as an optical circulator, considering the following sequence of input and output ports: 1 to 3; 3 to 4; 4 to 2; and 2 to 1.

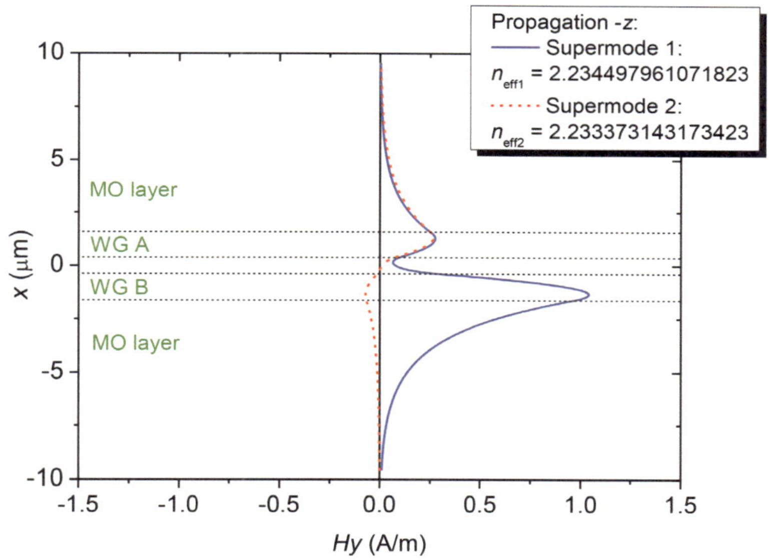

Figure 12.
Transversal distribution of the supermodes (H_y component) for backward propagation ($-z$).

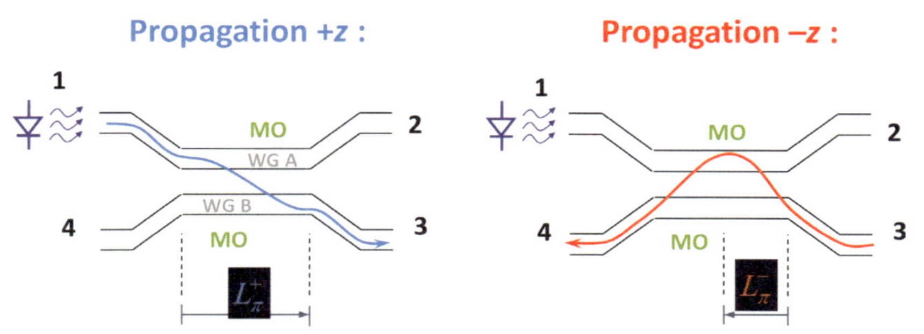

Figure 13.
Operation of an optical isolator based on nonreciprocal phase shift.

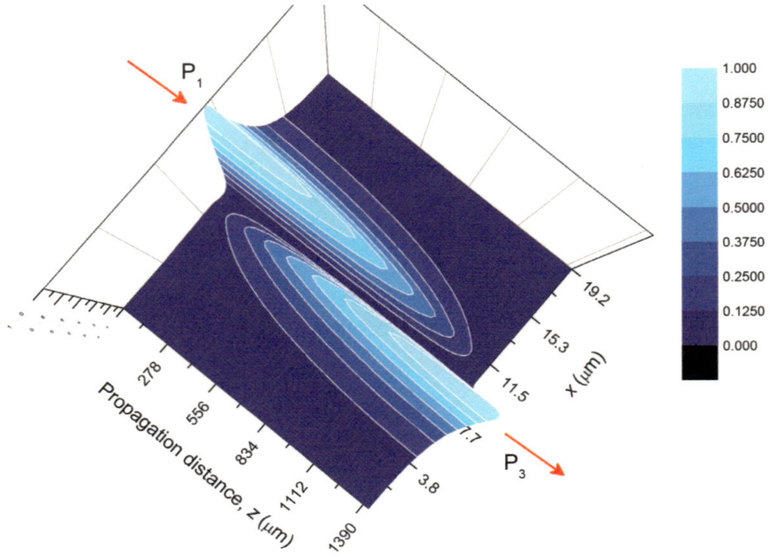

Figure 14.
Forward propagation simulation of the TM mode component H_y *excited at port 1 (P_1) of the five-layered structure. The light exits through port 3 (P_3). The starting transversal* H_y *field was supermode 1 plus supermode 2 of* ***Figure 11***.

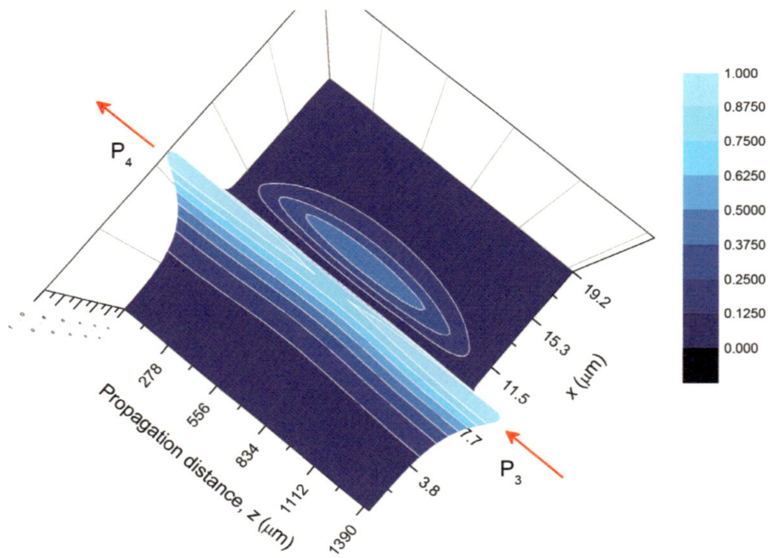

Figure 15.
Backward propagation simulation of the TM mode component H_y *excited at port 3 (P_3) of the five-layered structure. The light exits through port 4 (P_4). The starting transversal* H_y *field was supermode 1 minus supermode 2 of* ***Figure 12***.

3. Conclusions

The propagation characteristics of optical waves in magneto-optical media and in planar waveguides with three and five MO layers were exposed. The effects of Faraday rotation and nonreciprocal phase shift were discussed with mathematical background to support the analyses. The propagation of TEM waves in unbounded MO media was discussed, where it was shown that the Faraday rotation is maximized when the propagation occurs in the same direction of the applied magnetostatic field. It was also mathematically shown that if there is no such alignment, losses may be added to the wave propagation. A planar MO waveguide and a directional coupler were also analyzed in the context of their nonreciprocity. For these structures, nonreciprocity is observed for TM-guided modes. The theoretical analyses confirm that magneto-optical materials have great potential to be employed on the design of nonreciprocal optical devices, such as isolators and circulators.

Author details

Licinius Dimitri Sá de Alcantara
Cyberspatial Institute, Federal Rural University of Amazon, Belém, Brazil

*Address all correspondence to: licinius@ufra.edu.br

References

[1] Antonov V, Harmon B, Yaresko A. Electronic Structure and Magneto-Optical Properties of Solids. Dordrecht: Springer; 2004. 528 p. DOI: 10.1007/1-4020-1906-8

[2] Levy M. The on-chip integration of magnetooptic waveguide isolators. IEEE Journal of Selected Topics in Quantum Electronics. 2002;**8**(6):1300-1306. DOI: 10.1109/JSTQE.2002.806691

[3] Ando K. Waveguide optical isolator: A new design. Applied Optics. 1991; **30**(9):1080-1095. DOI: 10.1364/AO.30.001080

[4] Bahlmann N, Chandrasekhara V, Erdmann A, Gerhardt R, Hertel P, Lehmann R, et al. Improved design of magnetooptic rib waveguides for optical isolators. Journal of Lightwave Technology. 1998;**16**(5):818-823. DOI: 10.1109/50.669010

[5] Li TF, Guo TJ, Cui HX, Yang M, Kang M, Guo QH, et al. Guided modes in magneto-optical waveguides and the role in resonant transmission. Optics Express. 2013;**21**(8):9563-9572. DOI: 10.1364/OE.21.009563

[6] Bolduc M, Taussig AR, Rajamani A, Dionne GF, Ross CA. Magnetism and magnetooptical effects in Ce-Fe oxides. IEEE Transactions on Magnetics. 2006; **42**(10):3093-3095. DOI: 10.1109/TMAG.2006.880514

[7] Fratello VJ, Licht SJ, Brandle CD. Innovative improvements in bismuth doped rare-earth iron garnet Faraday rotators. IEEE Transactions on Magnetics. 1996;**32**(5):4102-4107. DOI: 10.1109/20.539312

[8] Pedroso CB, Munin E, Villaverde AG, Medeiros Neto JA, Aranha N, Barbosa LC. High Verdet constant Ga:S:La:O chalcogenide glasses for magneto-optical devices. Optical Engineering. 1999;**38**(2). DOI: 10.1117/1.602080

[9] Kalandadze L. Influence of implantation on the magneto-optical properties of garnet surface. IEEE Transactions on Magnetics. 2008; **44**(11):3293-3295. DOI: 10.1109/TMAG.2008.2001624

[10] Nomura T, Kishida M, Hayashi N, Ishibashi T. Evaluation of garnet film as magneto-optic transfer readout film. IEEE Transactions on Magnetics. 2011; **47**(8):2081-2086. DOI: 10.1109/TMAG.2011.2123103

[11] Inoue M, Arai K, Fujii T, Abe M. One-dimensional magnetophotonic crystals. Journal of Applied Physics. 1999;**85**(8):5768-5770. DOI: 10.1063/1.370120

[12] Koerdt C, Rikken GL, Petrov EP. Faraday effect of photonic crystals. Applied Physics Letters. 2003;**82**(10): 1538-1540. DOI: 10.1063/1.1558954

[13] Zvezdin AK, Belotelov VI. Magnetooptical properties of two-dimensional photonic crystals. The European Physical Journal B–Condensed Matter and Complex Systems. 2004;**37**: 479-487. DOI: 10.1140/epjb/e2004-00084-2

[14] Haider T. A review of magneto-optic effects and its application. International Journal of Electromagnetics and Applications. 2017;**7**(1):17-24. DOI: 10.5923/j.ijea.20170701.03

[15] Zak J, Moog ER, Liu C, Bader SD. Magneto-optics of multilayers with arbitrary magnetization directions. Physical Review B. 1991;**43**:6423-6429. DOI: 10.1103/PhysRevB.43.6423

[16] Wolfe R, Lieberman R, Fratello V, Scotti R, Kopylov N. Etch-tuned ridged waveguide magneto-optic isolator. Applied Physics Letter. 1990;**56**: 426-428. DOI: 10.1063/1.102778

[17] Alcantara LDS, Teixeira FL, Cesar AC, Borges BH. A new full-vectorial FD-BPM scheme: Application to the analysis of magnetooptic and nonlinear saturable media. Journal of Lightwave Technology. 2005;**23**(8):2579-2585. DOI: 10.1109/JLT.2005.850811

[18] Alcantara LDS, De Francisco CA, Borges BH. Analytical model for magnetooptic five-layered planar waveguides. In: SBMO/IEEE MTT-S International Microwave and Optoelectronics Conference (IMOC'17); 27–30 August 2017; Águas de Lindóia, Brazil. pp. 1-5. DOI: 10.1109/IMOC.2017.8121106

Chapter 12

Compact, Efficient, and Wideband Near-Field Resonant Parasitic Filtennas

Ming-Chun Tang, Yang Wang and Ting Shi

Abstract

As a hybrid component in RF front-end systems, filtennas possess the distinctive advantages of simultaneously combining filtering and radiating performance characteristics. Consequently, filtennas not only save space and costs but also reduce transmission losses. In this chapter, three sorts of filtennas have been proposed: the first sort is band-pass/band-stop filtennas, which are mainly realized by assembling band-pass/band-stop filters and antennas to achieve the combined functions; the second sort is multi-resonator-cascaded filtennas, which are obtained by altering the coupled-resonators in the last stage of the filters to act as the radiating elements; and the third sort is near-field resonant parasitic, bandwidth-enhanced filtennas, which are accomplished through organically combining radiator and filtering structures. For the second and third sorts, it is worth noting that the design methods witness significant electrical size reduction without degrading the radiation performance of the filtennas in general.

Keywords: antennas, filter, filtennas, radiating performance, impedance bandwidth, near-field resonant parasitic element

1. Introduction

A filtenna is a co-designed antenna which integrates a radiating element and filter to be a single device. Due to its self-contained filtering characteristic, filtenna possesses several main properties compared with other general antennas while receiving a signal. First of all, the interconnection losses could be decreased, which emerge while a common receiving antenna is assembled to a filter in the fabrication process. In addition, it restrains unwanted signals which occur out of the operational band. Finally, from the aspect of practice, it promoted a RF front-end system with more compact and lower cost features. Consequently, more attention has been seized to propose all kinds of filtennas into engineering practice.

In this chapter, three main sorts of filtennas are introduced to demonstrate their design methods and performance characteristics. For the first sort, band-pass or band-stop filtennas focus on introducing band-notch filters into ultra-wideband (UWB)/wideband antennas using a variety of high-Q band-stop structures or embedding band-pass filter structures into various types of basic antennas [1–4]. Two printed planar ultrawideband (UWB) antennas are designed and fabricated. To further improve its high frequency characteristics, a multimode-resonator filter consisting of a single-wing element is combined with the slot-modified UWB antenna. These filtennas would be depicted in Section 2 in detail. For the second sort, multi-resonator-cascaded filtennas are obtained by altering the coupled-resonators in the last stages of the filters to act as the radiating elements [5, 6]. In Section 3, two planar efficient wideband electrically small monopole filtennas are proposed. The first one is directly evolved from a common planar capacitively loaded loop (CLL)-based filter. The second filtenna consists of a driven element augmented with a CLL structure and with slots etched onto its ground plane. Both the filtennas are electrically small. For the third sort, near-field resonant parasitic (NFRP), bandwidth-enhanced filtennas are accomplished through organically combining radiator and filtering structures. In Section 4, a filtenna possessing compact geometry with bandwidth enhancement is developed by a novel design method. It expanded an impedance bandwidth which is over three times improvement compared to its component near-field resonant parasitic (NFRP) monopole antenna alone. Then, a set of compact filtennas with the NFRP element is simulated, fabricated, and analyzed to validate the filtennas' reliability.

2. Planar ultrawideband filtennas

The degradation of the radiation pattern at higher frequency of the UWB range reveals a serious drawback for the planar design. For the purpose of decreasing this defect, some design methods have been published, such as adding electromagnetic band gaps (EBGs) [7], varying the radiating patches [8], reconstructing the ground planes [9], and turning to a trident-shaped strip integrated with a tapered impedance transformer connected to the feedline [10].

Alternatively, through assembling an asymmetrical single-wing filter into a feedline section of a modified arc-slot UWB antenna, the broadside gain of the antenna in the upper portion of the UWB band is increased. For example, the simulated broadside gains at 10 GHz are increased from −3.89 to 4.16 dBi for the single-wing antenna. Moreover, integrating a filter element into the antenna strengthens the sharp cutoff performance at both edges of the frequency range for the UWB. Additionally, the developed co-design method makes the size compact for the whole system constituted by the filter and antenna effectively. Eventually, the experiment results in good agreement with simulations that could validate the proposed strategy.

2.1 UWB antenna with an arc-shaped slot

The geometries of a traditional patch UWB monopole antenna as a reference together with its arc-slot modified case are illustrated in **Figure 1**. The reference antenna is evolved from the reported printed planar UWB monopole antenna designs [11]. Its radiating patch is elliptical in shape, and its ground plane is designed with a rectangular slot at its upper edge for impedance matching. As its modified case, an arc-shaped slot is engineered to be symmetric within the radiating patch and to be close to the throat of the microstrip-feed strip. Both the reference antenna and the antenna with an arc-shaped slot have their comparison on *S*-parameters, and broadside realized gains are shown in **Figure 2**. It is shown that the reference antenna has a very wide −10 dB impedance bandwidth from 2.855 up to 14.0 GHz in the simulation. In contrast, the simulated (measured) bandwidth of the antenna with the arc-slot is shown to be from 2.615 (2.775 GHz) up to 14.0 GHz. Moreover, by etching the arc-shaped slot, the antenna with the arc-shaped slot achieves improved broadside realized gains, particularly at the high frequency side

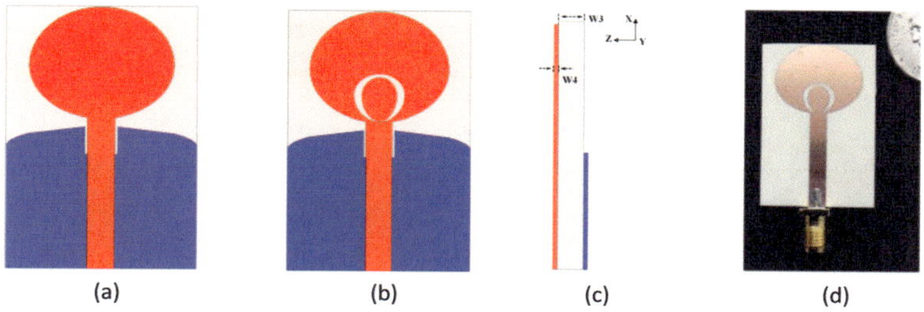

Figure 1.
The printed planar UWB antenna with an arc-shaped slot and its reference design. (a) Reference antenna, (b) top view, (c) side view, and (d) the fabricated prototype.

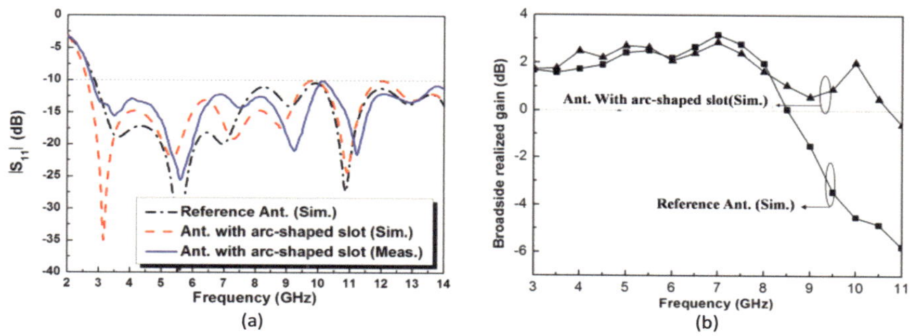

Figure 2.
Comparisons between the reference antenna and the antenna with the arc-shaped slot. (a) $|S_{11}|$ values and (b) broadside realized gain.

of the UWB band, e.g., a 6 dB increase in the realized gain near 10 GHz. The reason is that the arc-shaped slot produces a parasitic element to resonate at TM10 mode around 10 GHz to remedy the radiation performance. However, as shown in **Figure 3**, the arc-shaped slot modification could ensure the broadside gain improvement in the upper portion of the UWB band. But there is no corresponding improvement in the *H*-plane, i.e., it does not exhibit an omni-directional radiation pattern.

2.2 UWB multimode resonant filter

Numerous stub-loaded multimode-resonator-based UWB bandpass filters have been reported in recent years [12, 13]. We found one compact filter design that has several attractive features, including simple designs, compact sizes, low losses, flat group delays, enhanced out-of-band rejection, and easy integration with other microwave components in the UWB frequency range.

First, based on these advantages, a circular stub-loaded single-wing filter was designed, fabricated, and measured. In detail, its layout and the equivalent circuit network, together with the fabricated prototype and S-parameters, are shown in **Figure 4**. The filter is composed of a single-wing resonator and a pair of interdigital-coupled lines. The resonator creates and adjusts several sequential modes within the UWB passband [13]. The interdigital-coupled lines are equivalent to two pairs of single transmission lines attached in their middle to a J-inverter susceptance. The simulated (measured) results demonstrate that the single-wing filter provides 3 dB passband bandwidth from 2.806 (2.824) to 10.892 GHz (10.760 GHz), which covers the entire UWB band. Moreover, the +10 dB return loss bandwidth is from 3.025

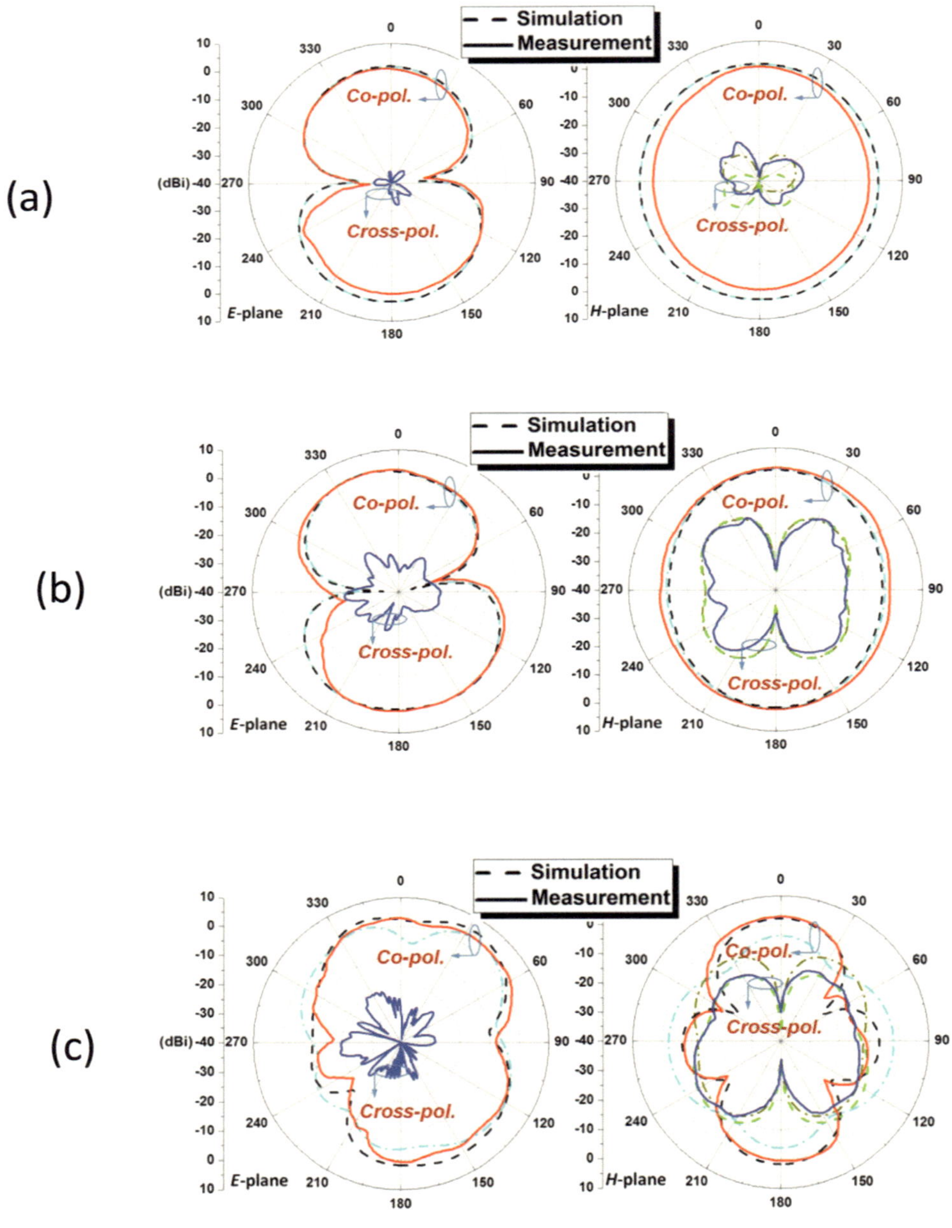

Figure 3.
Realized gain of the arc-slot modified UWB antenna at (a) 3.0, (b) 6.5, and (c) 10 GHz.

(2.989) to 11.010 GHz (10.842 GHz). Two transmission zeros are generated at 2.12 GHz (2.085 GHz) and at 11.5 GHz (11.449 GHz).

2.3 Integration of a UWB filter into an antenna with an arc-shaped slot

The single-wing filter was integrated into the arc-slot antenna as shown in **Figure 5**. The filter was connected directly to the microstrip feedline section. As shown in **Figure 5**, the UWB filter-antenna design was optimized, fabricated, and measured. As depicted, the simulated (measured) −10 dB impedance bandwidth of the antenna with the single-wing filter is from 2.995 (2.949) to 11.047 GHz

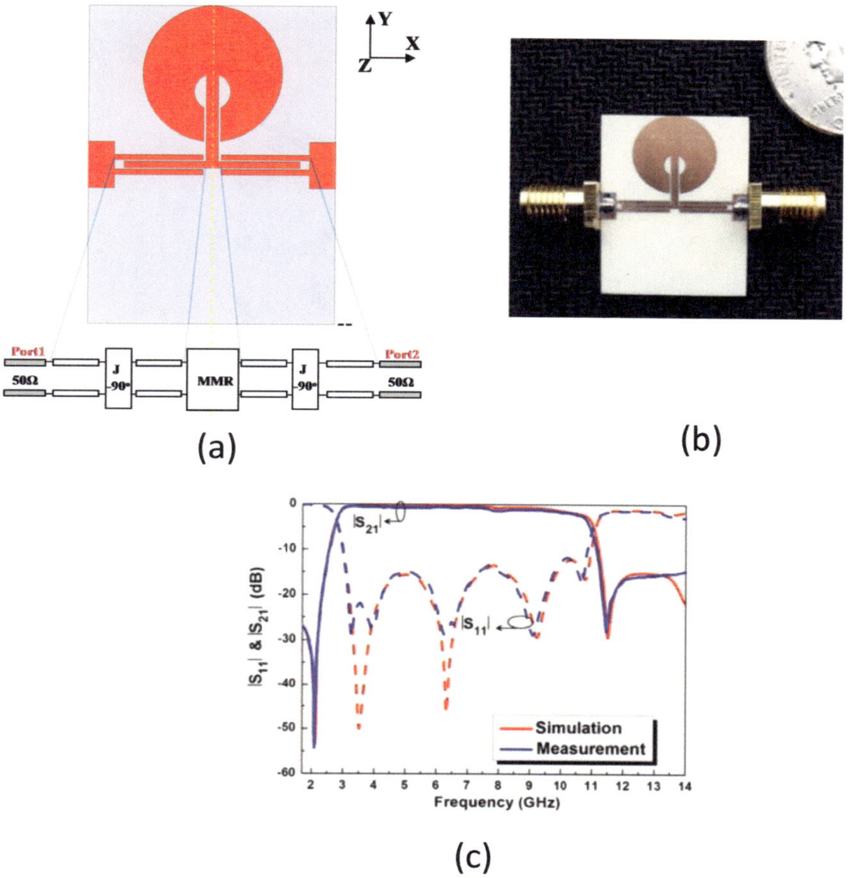

Figure 4.
The circular stub-based single-wing multimode-resonator filter. (a) Design layout of the filter and its equivalent circuit network, (b) fabricated prototype, and (c) its simulated and measured $|S_{11}|$.

(10.817 GHz). Clearly, the measured lower frequency bound is downshifted by ~46 MHz, and its upper frequency edge is downshifted by ~230 MHz.

The far-field realized gain patterns are presented in **Figures 6** and 7. By comparing the results in **Figure** 7, it is clear that the integration of the single-wing filter further increases the broadside gain values in the higher frequency range, while maintaining its original radiation patterns in the lower frequency range. The broadside realized gain values of the single-wing version increase to 4.16 dBi in simulation and to 4.25 dBi in experiment. It must be noted that the single-wing filter antenna has very good omnidirectional radiation performance in the *H*-plane and exhibits some improvements in the cross-polarization values.

3. Compact, planar, and wideband monopole filtennas

Two electrically small, efficient planar monopole filtennas based on capacitively loaded loop (CLL) resonators are presented. Taking advantage of the characteristics of filters that are based on a pair of electrically coupled CLL resonators, the filtenna is designed, fabricated, and measured. The experimental results demonstrate that this electrically small system had a 6.27% fractional impedance bandwidth, high out-of-band rejection, and stable omnidirectional radiation patterns. An additional

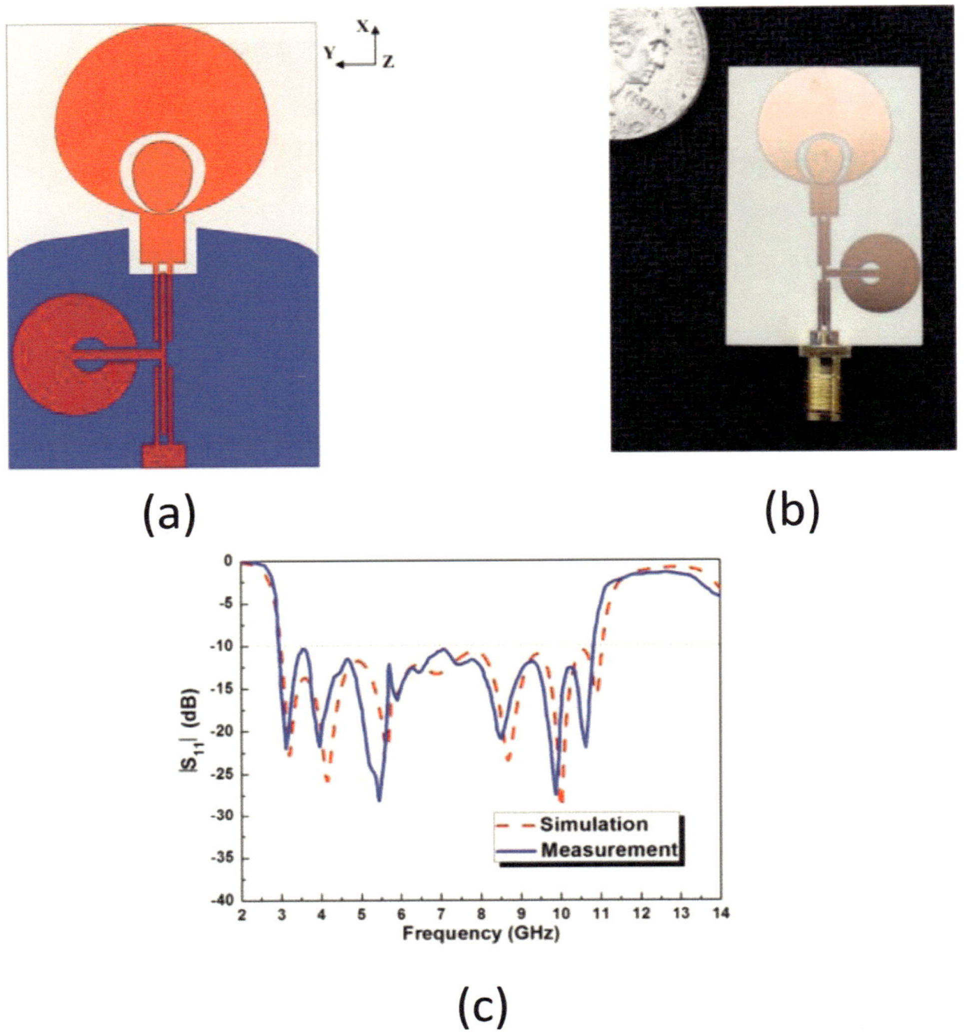

Figure 5.
The UWB antenna with both the arc-slot and the multimode resonator filter T. (a) Design model of the antenna, (b) fabricated prototype, and (c) its simulated and measured $|S_{11}|$.

CLL structure, as a near-field resonant parasitic (NFRP) element, is then integrated systematically into the system to achieve a wider operational bandwidth. The resulting filtenna owns a 7.9% fractional bandwidth, together with a flat gain response, stable omnidirectional radiation patterns, and high out-of-band rejection characteristics.

3.1 Square CLL-based bandpass filter design

A bandpass filter with a 0° feed structure based on rectangular microstrip CLL [14] is revealed in **Figure 8(a)**. The plane is symmetrical about the dashed lines O–O′ and T–T′ along the x- and y-axis, respectively. The Rogers substrate modeled 4350B, with permittivity ε_r = 3.48, dielectric loss tangent tan σ = 0.0037, and permeability μ_r = 1, was chosen to construct the filter. The total size of it is

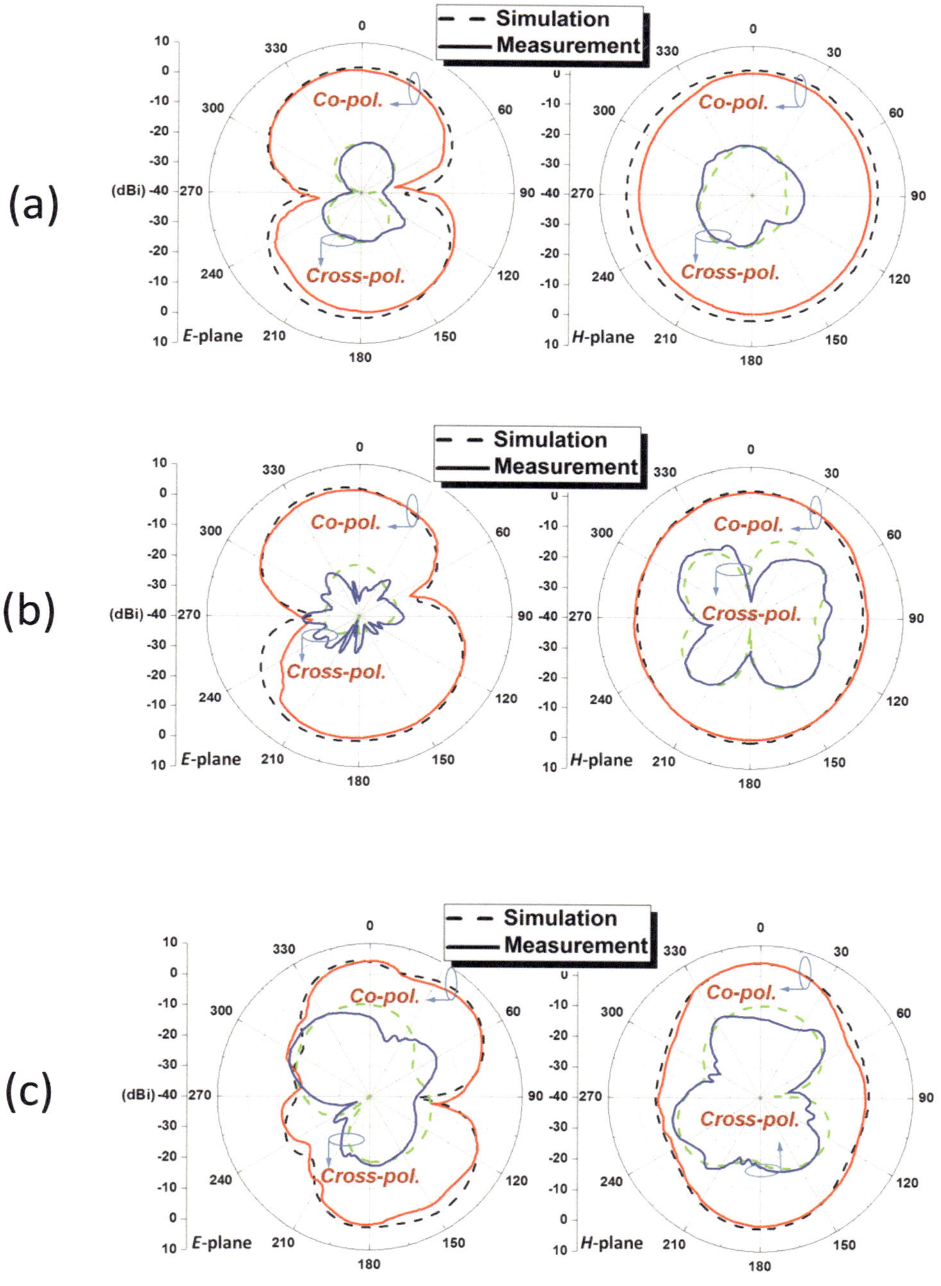

Figure 6.
Realized gain of the arc-slot modified UWB antenna with the single-wing filter at (a) 3.0, (b) 6.5, and (c) 10 GHz.

29 × 27 × 1 mm. For this design, two square CLLs oriented with each other gap to gap were etched on the substrate. This geometry introduced an electrical coupling between the two components [14–16], for instance, which has been exploited previously to efficiently improve the microwave field transmission by a metallic aperture with subwavelength [17]. As depicted in **Figure 8(a)**, the two feed ports of the

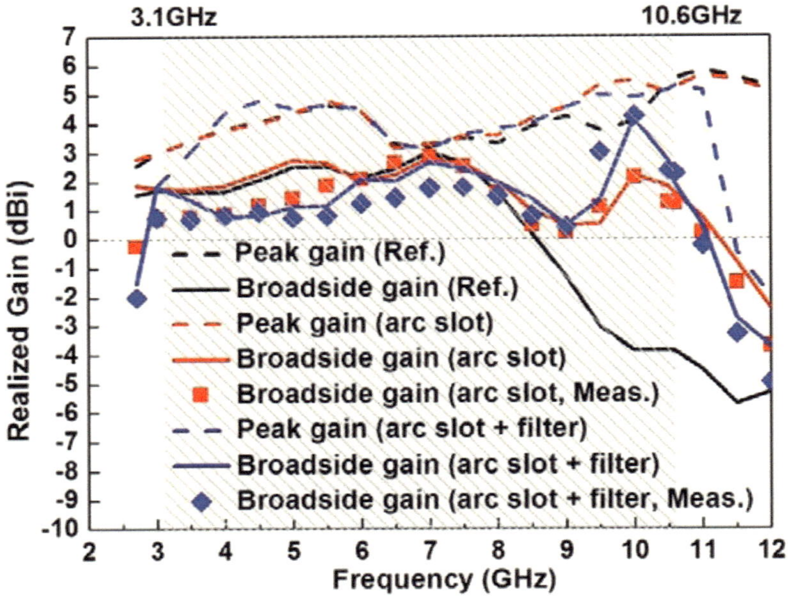

Figure 7.
The maximum simulated and measured realized gain values in the broadside direction and any direction for the three UWB antennas.

filter, which were connected to 50-Ω microstrip lines (with the width 2.2 mm), are placed to be centrally symmetric about the midpoint of line O–O′, so as to create another two transmission zeros in the stopband, and the passband response remains the same. As presented in [14], 0° feed geometry is superior to having two CLLs on one side.

Figure 8(b) correspondingly reveals the equivalent circuit with lumped elements. Its L and C values represent the natural self-inductance and self-capacitance of the uncoupled resonators alone. Notation C_m denotes the two resonators' mutual capacitance. When the symmetrical line T–T′ is substituted by an electric wall (a short circuit), the corresponding circuit has a lower resonant frequency $f_e = 1/\{2\pi [L(C + C_m)]^{1/2}\}$. In the same way, when it is replaced by a magnetic wall (an open circuit), the corresponding circuit has a higher resonant frequency $f_m = 1/\{2\pi [L(C - C_m)]^{1/2}\}$ [14].

Figure 9 demonstrates the simulated S-parameters of the filter, while varying the distance between the two resonators ($d1$) from 1.1 to 0.3 mm. The results show that when the distance $d1$ increases large enough (e.g., larger than 1.1 mm), the resonant frequency f_O remains unchanged, and also the resonant intensity ($|S_{11}|$ dip) at f_O presents slight variation. Because of the slight mutual coupling between the resonators, it hardly impacts the resonant frequency f_O, and the mutual capacitance approaches zero, i.e., $C_m \approx 0$. Therefore, in this scarcely coupled condition, $f_e \approx f_m \approx f_O = 1/[2\pi(LC)^{1/2}]$ is achieved. For comparison, when the distance $d1$ decreases more enough (e.g., smaller than 0.7 mm), the resonant frequency f_O results to be completely divided into two adjacent frequencies, i.e., f_e and f_m, which contribute to enhance the passband. It could be attained from the simulations that the surface currents of the two CLLs are in phase at the lower resonant frequency f_e. This conclusion agrees well with the selection of the electric wall substituting the plane labeled T–T′. For another, at higher resonance frequency f_m, the surface currents of the split rings are out of phase. The magnetic wall agrees well to replace

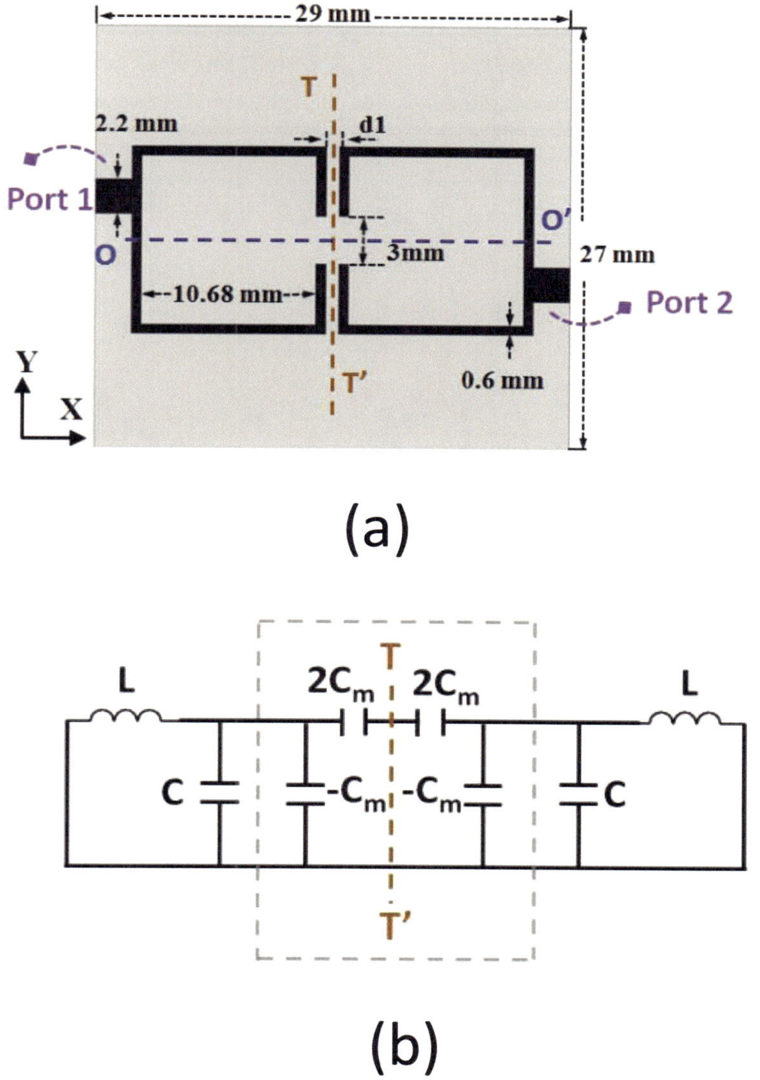

Figure 8.
Filter with electrically coupled resonators. (a) Design layout. (b) Equivalent circuit network [15].

the symmetrical line labeled T–T′. Furthermore, through decreasing the distance $d1$ continually, it produced even stronger mutual coupling, generating an even larger C_m value. As a result, both f_e and f_m departed each other away from f_0 as shown in **Figure 9**. Thus, with a narrower distance $d1$ for the filter configuration in **Figure 8(a)**, a much stronger coupling between the CLL resonators could be observed, and this case causes the corresponding two poles to depart from each other, leading to an enhanced bandwidth.

This can be verified by calculating the electric coupling coefficient (k_e) between two resonators. It is readily obtained using the expression [15].

$$k_e = \pm\frac{1}{2}\left(\frac{f_2}{f_1}+\frac{f_1}{f_2}\right)\sqrt{\left(\frac{f_m^2-f_e^2}{f_m^2+f_e^2}\right)^2-\left(\frac{f_2^2-f_1^2}{f_2^2+f_1^2}\right)^2} \tag{1}$$

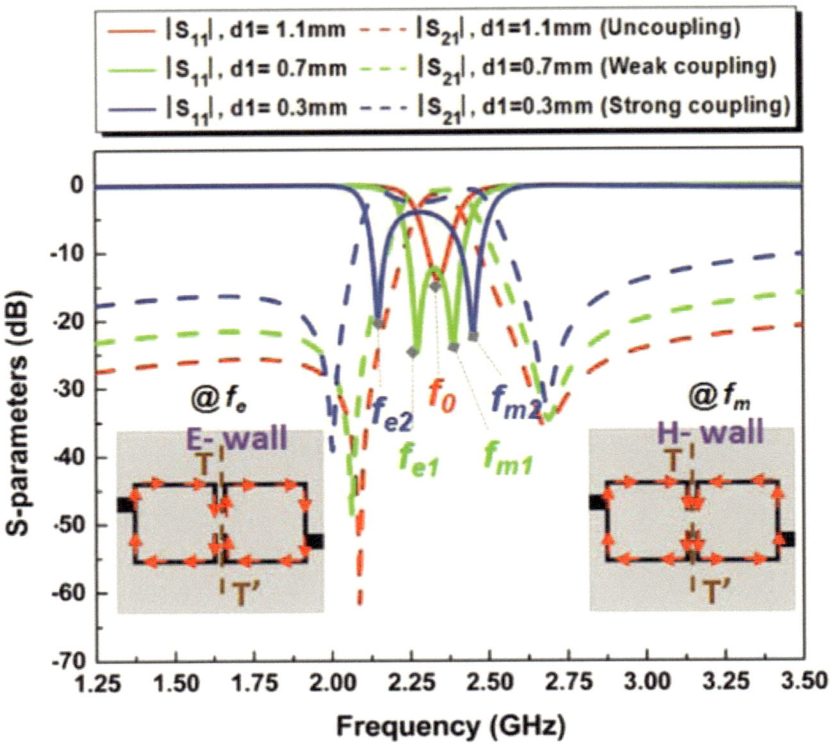

Figure 9.
S-parameters of the simulation for the electrically coupled filter while the separate distance d1 *varying from* 1.1 *to* 0.3 *mm, together with the surface current distribution behaviors on the resonance at* f_e *and* f_m.

where f_1 and f_2 indicate the resonance frequencies of each independent CLL resonator. Because the capacitance C_m in **Figure 8** is positive, the plus sign is selected. Furthermore, because the two CLL resonators are identical, one knows that $f_1 = f_2$. Thus, Eq. (1) reduces to the following:

$$k_e = \frac{f_m^2 - f_e^2}{f_m^2 + f_e^2}. \tag{2}$$

Consequently, as expected for the uncoupled, weak coupling, and strong coupling cases given in **Figure 9**, k_e is, respectively, 0, 0.14, and 0.49.

3.2 Electrically small filtenna design

A filtenna having a second-order filter was co-designed and optimized. It is shown in **Figure 10**. **Figure 10(a)** indicates that one CLL element acts as the directly driven element. A fan-shaped radiator with no ground plane on the back side of it acts as a NFRP element in the presence of the monopole (CLL-based) antenna [18]. The choice of this special fan-shaped radiator establishes an even smoother impedance transition over the desired wider bandwidth. Simply starting with the resonance frequencies near to each other facilitates a straightforward numerical approach to optimize and finalize the actual antenna design. Note that the fan-shaped part of the NFRP element is placed on the opposite side of the feed port. This arrangement facilitates the creation of dual transmission zeros on the two edges of the passband. This arrangement enhances the out-of-band rejection level.

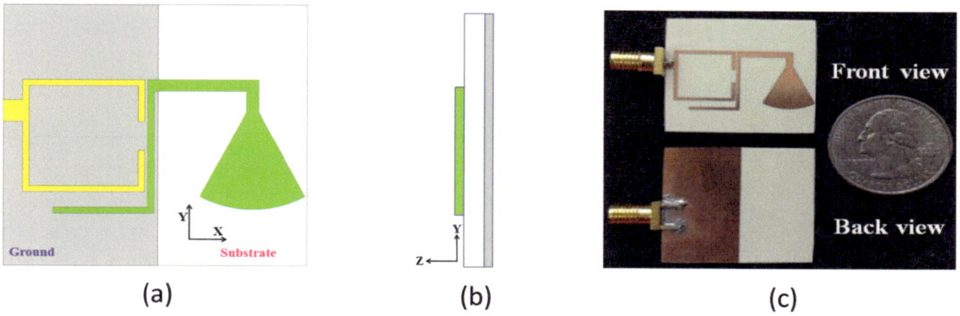

Figure 10.
First electrically small filtenna. (a) Top and (b) side views of the HFSS simulation model. (c) Front and back views of the fabricated prototype.

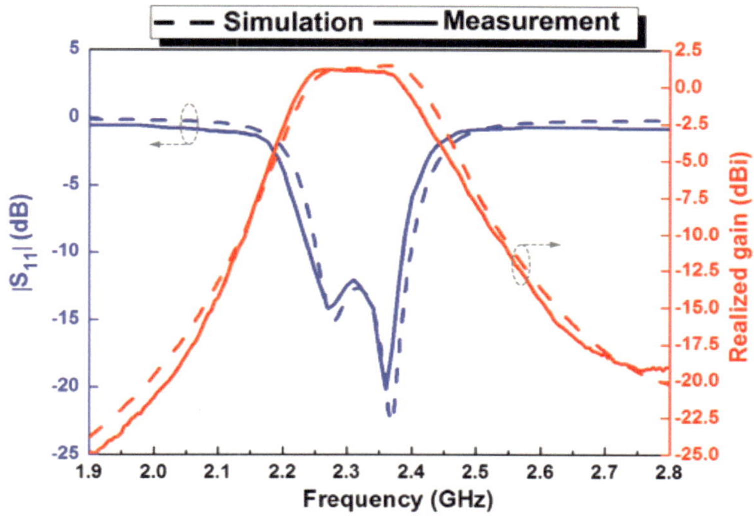

Figure 11.
Measured and simulated $|S_{11}|$ and realized gain values of the first filtenna as functions of the source frequency.

The measured and simulated $|S_{11}|$ values are presented in **Figure 11**. The measured values confirmed that the filtenna had a -10 dB impedance bandwidth from 2.24 to 2.385 GHz (6.27% fractional impedance bandwidth) in good agreement with the simulated values 2.252–2.398 GHz (6.28% fractional impedance bandwidth). The electrical size of the measured prototype is $ka \sim 0.93$, while its simulated value was $\sim$0.935. **Figure 11** demonstrates that the prototype filtenna has a flat realized gain response within its passband. The measured (simulated) peak value was 1.15 (1.41) dBi. The simulated radiation efficiency was higher than 80.93% throughout the operational band. This realized filtenna prototype clearly has very good band-edge selectivity and stopband suppression.

3.3 Filtenna with enhanced bandwidth

For many applications, it is desirable to have an even wider bandwidth. Consequently, the second design shown in **Figure 12** was considered. In order to improve the flatness of the transmission performance within the passband while maintaining its wideband operation and steep skirts, a third resonator was introduced without increasing the total overall dimension of the filtenna.

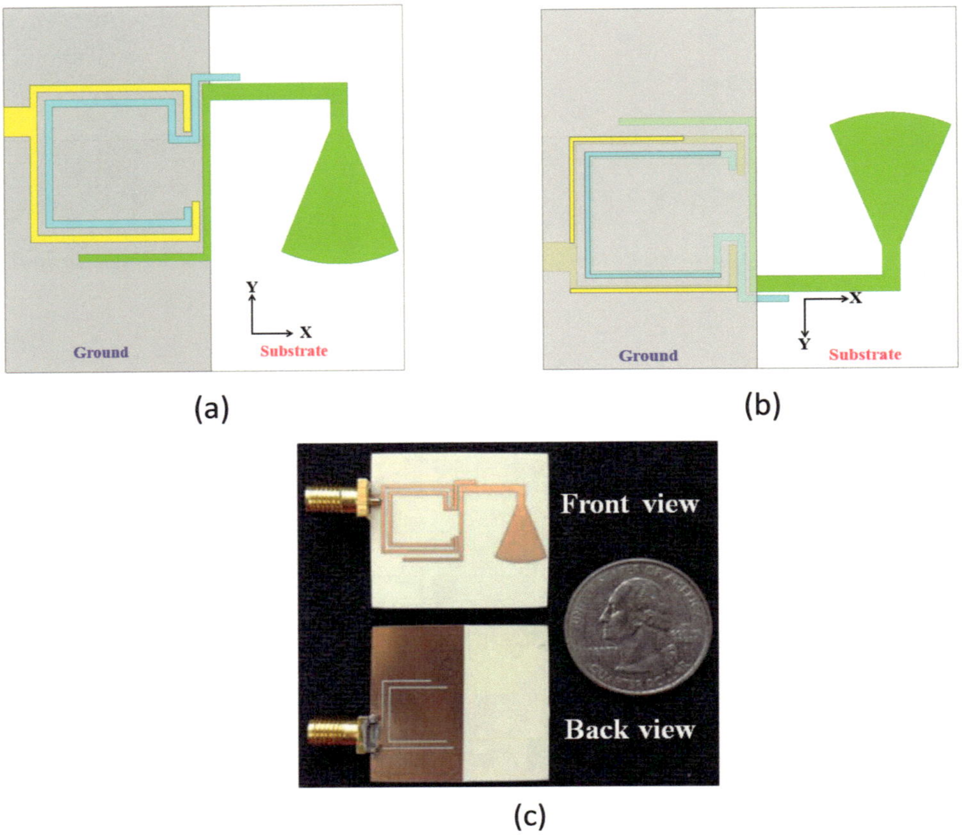

Figure 12.
Enhanced bandwidth filtenna with slots in its ground strip. (a) Top and (b) back views of the HFSS simulation model. (c) Front and back views of the fabricated prototype.

The third resonator is an additional CLL element, shown in blue in **Figure 12(a)**. Its gap position coincides with the driven CLL element, and it has an arm included to facilitate its coupling to the NFRP element. This collocated arrangement of the two CLLs provides a means to control the mutual coupling, further expanding the bandwidth without increasing the total overall dimensions of the filtenna. Three slots were etched in the ground strip directly beneath the two CLL elements to achieve a smoother realized gain curve. The length of the additional CLL element is set nearly equal to the driven CLL's size to make their resonance frequencies close to one another.

The simulated and measured $|S_{11}|$ and realized gain values of the second filtenna with the ground strip slots are given in **Figure 13**. The simulated (measured) realized gain values indicate that the simulated peak realized gain value is improved from 1.659 to 1.75 dBi. The corresponding measured value is 1.376 dBi, revealing more losses than expected in fabrication. For the simulated $|S_{11}|$ values exhibited in **Figure 13**, the impedance bandwidth ranges from 2.264 to 2.46 GHz (about 8.3% fractional bandwidth, i.e., a 32.2% improvement) and was from 2.261 to 2.447 GHz (7.9% fractional bandwidth, i.e., a 26% improvement) in the measurement. Similarly, the simulated $ka \sim 0.94$ and measured $ka \sim 0.938$ values verify that the filtenna is electrically small. Furthermore, the simulated radiation efficiency across the entire operational bandwidth is higher than 82.87%. Again, very good agreement between the simulated and measured performance characteristics was obtained.

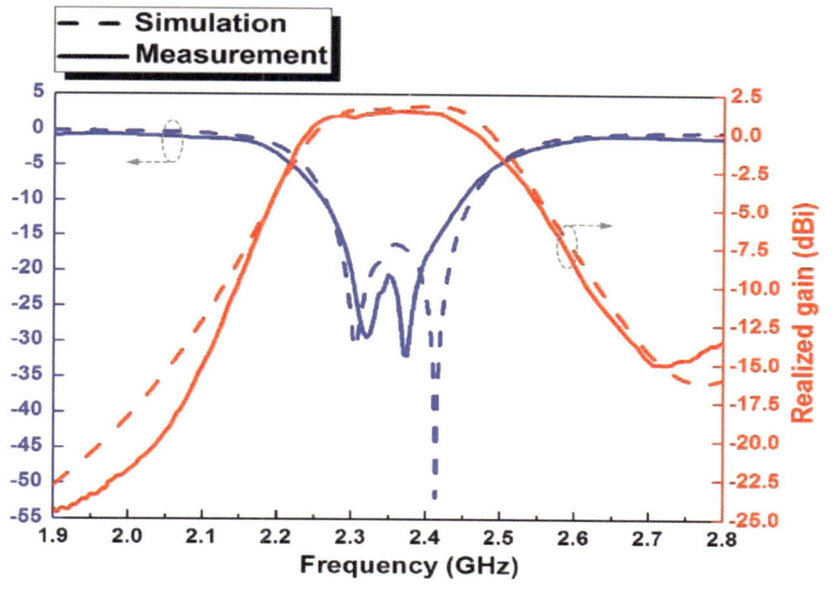

Figure 13.
Measured and simulated $|S_{11}|$ and realized gain values as functions of the source frequency for the enhanced bandwidth filtenna with slots in its ground strip.

4. Compact filtennas with enhanced bandwidth

Two filtennas are proposed by a design strategy with the merits of both a compact structure and enhanced bandwidth. The reliability of the filtennas is verified though simulations and analysis of a compact NFRP filtenna which is proposed and fabricated. The reported design employs a Rogers 4350B substrate with relative permittivity ε_r = 3.48, relative permeability μ_r = 1.0, and dielectric loss tangent tan σ = 0.0037.

4.1 Design of compact filtennas

A well-designed compact NFRP antenna is selected as the radiator [19, 20]. Then a compact NFRP antenna is designed, which consists of a traditional monopole and a rectangular microstrip capacitively loaded loop (CLL)-based band-pass filter [21, 22].

4.1.1 A CLL-based filtenna design

The elaborate geometry of the filtenna is shown in **Figure 14**. As depicted in **Figure 14(a)** and **(e)**, the compact electrically small antenna (ESA) with NFRP was chosen as the radiating element. The NFRP element is proposed to etch upon one side of the semi-circle board, while the monopole microstrip is located on the other side, with the design principle corresponding to the reported NFRP ESAs [23–25]. The composite structure of this radiator element and filtering element, which is based on CLL resonators, is well shown in **Figure 14(a)–(d)**. The enlarged filter is shown in **Figure 14(b)**. This filter structure is a typical band-pass design [21, 26, 27], and is set to be symmetric about the S–S′ line. One end is connected to the printed monopole and the other to the SMA.

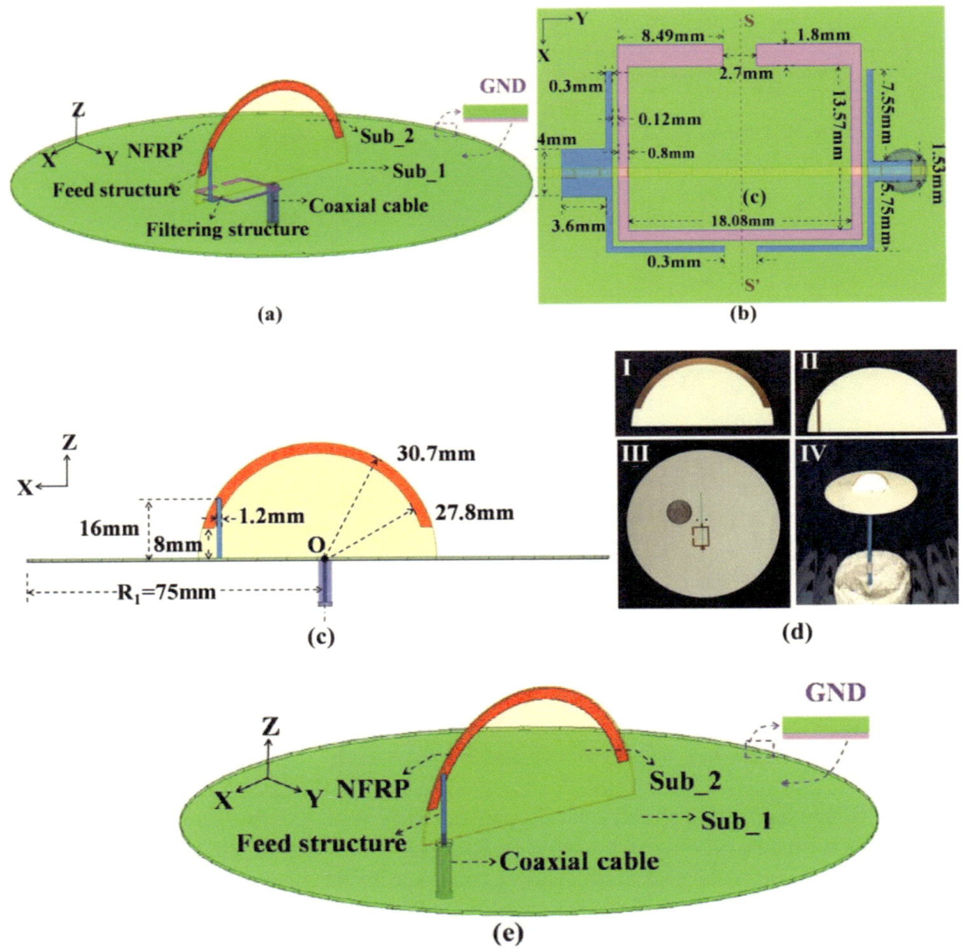

Figure 14.
Prototype of miniaturized filtenna with a NFRP structure. (a) 3D graphic of the NFRP filtenna. (b) Filtering structure. (c) Side views of the ESA and filtenna. (d) Fabricated module of the filtenna in various side views. (e) 3D graphic of the ESA with a NFRP structure.

4.1.2 Simulation and measured results

Figure 15(a) demonstrates the simulated and measured $|S_{11}|$ and peak realized gain values versus the source frequency of the optimized filtenna. As a reference, the simulated reflection coefficient of the optimized NFRP ESA alone (depicted in **Figure 14(e)**), is shown in **Figure 15(b)**. The Agilent E8361A PNA vector network analyzer (VNA) is exploited to quantify the impedance matching. With regard to the NFRP ESA, a 30.3 MHz −10 dB impedance bandwidth is realized corresponding to the center frequency which is located at 1.26 GHz (corresponding to FBW of 2.4%) and with ka_{rad} calculated to be 0.81 (while a_{rad} represents the smallest radius for the sphere which could entirely cover the radiating structure at the lowest operation frequency f_L, then $k = 2\pi/\lambda_L = 2\pi f_L/c$ represents the number of relevant waves in free space). It is worth noting that the ground size is R1 = 75 mm, i.e., $ka_{ground} \sim 1.96$. Although the ground size has a certain influence on the gain and front-to-back ratio, its effect on the impedance matching level and bandwidth is deemed slight.

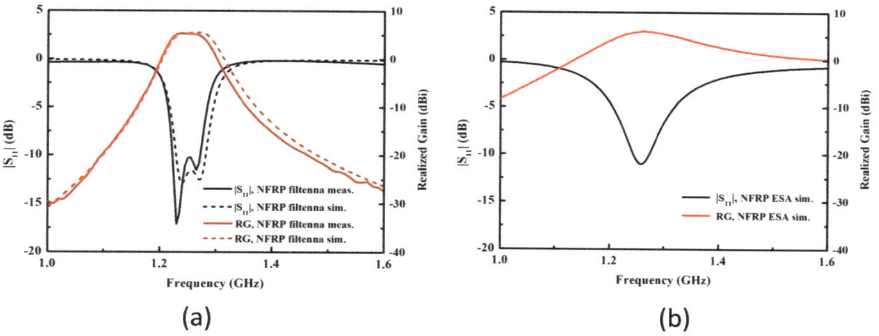

Figure 15.
*The simulated and measured $|S_{11}|$ and peak realized gain values versus the source frequency (a) for the filtenna design shown in **Figure 14(a)** and **(b)** for the CLL-based NFRP ESA alone in **Figure 14(e)**.*

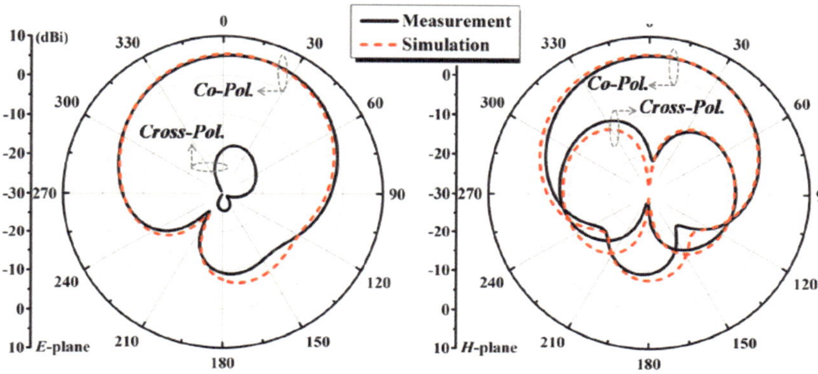

Figure 16.
The simulated (measured) peak realized gain patterns in the E- *and* H-*planes at the lower resonance frequency of the filtenna shown in **Figure 14**—1.24 (1.230) GHz.*

Within the operational band, the realized gain (along +*z* axis) and radiation efficiency ranges in 5.73–5.94 dBi and 94–95%, respectively. The expected two overlapping resonances are depicted in the result of the fabricated filtenna. As exhibited in **Figure 15(a)**, the measured (simulated) $|S_{11}|_{min}$ is respectively situated at 1.23 (1.24) and 1.265 (1.272) GHz. Then −10 dB bandwidth is expanded to 49 (50) MHz, ranging in 1.223 (1.23)–1.272 (1.28) GHz, i.e., the proposed filtenna processes a 3.93% (4.2%) FBW. It is comparably flat for the peak realized gain values in this operational band, which ranges from 4.73 (4.25) to 5.43 (5.23) dBi. This fairly flat realized gain curve indicates that an essentially stable response is obtained through the whole operational band. As observed, the measured ones shifted a little to the lower band.

Figure 16 indicates the measured (simulated) *E*- and *H*-plane peak realized gain patterns for the proposed filtenna at the lower resonance frequency of 1.230 (1.240) GHz. The measured (simulated) peak gain was 5.12 (5.36) dBi. On the whole, the measured results of our proposed filtenna are in good agreement with the simulated ones.

4.2 Variations of compact filtennas

As shown in **Figure 17**, by altering the orientation, position, and configuration of the filter element, certain advantages could be obtained. In contrast with the filtenna depicted in **Figure 14**, the configurations of the NFRP element, the printed monopole unit and the CLL resonator part shown in **Figure 17(a)–(c)** were all left the same. The results of the corresponding simulation are presented in **Figure 18**.

Figure 18 exhibits any of the three proposed filtennas that could introduce two adjacent resonance frequencies and thus reveals an expected, notably enhanced operation bandwidth. As is depicted, there is nearly no fluctuation for the values for peak realized gain traced with the +*z*-axis over the whole operation band. Furthermore, the improved suppression along the edges of the band remains unchanged as well. **Table 1** summarizes the performance properties of the various filtenna designs. In addition, **Table 1** reveals that it is electrically small (i.e., $ka < 1$) for all of the new simulated geometries which are composed of the radiating and filtering

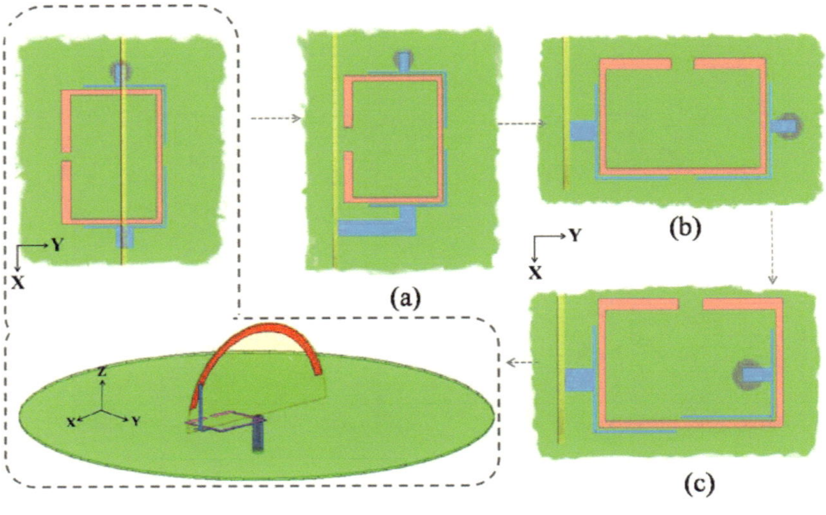

Figure 17.
Exploring variations of the filtenna design shown in ***Figure 17****. Change in the filter (a) position, (b) orientation, and (c) structure.*

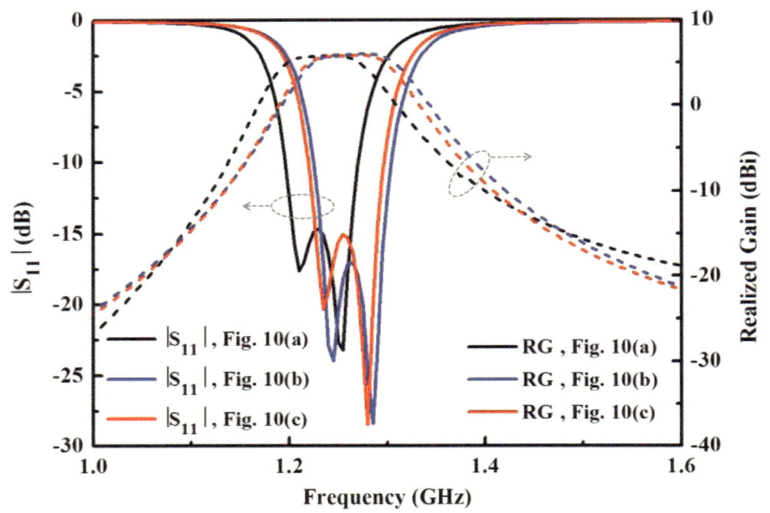

Figure 18.
The simulated $|S_{11}|$ and peak realized gain values as functions of the source frequency for the three cases shown in ***Figure 17****.*

Reported filtennas	FBW_{-10dB} (%)	*ka* Radiator & filter	Realized gain (dBi)	Radiation efficiency (%)
Figure 17(a)	6.07	0.79	4.85–5.83	76–86
Figure 17(b)	6.32	0.89	5.04–6.04	78–88
Figure 17(c)	6.13	0.90	4.92–5.89	77–86
ESA alone	2.4	0.81	5.73–5.94	94–95

Table 1.
Summary of the performance characteristics of the proposed one-stage filtennas.

elements, and also the fractional bandwidth remains two times broader than the electrically small CLL-based NFRP antenna alone.

4.3 Bandwidth enhancement of the filtennas

A wider impedance bandwidth could be obtained by adding more stages to the filter structure. As shown in **Figure 19**, this filtenna is evolved from the design in **Figure 14**. It is composed of the NFRP ESA and a two-stage filtering structure. The filter structure consists of two rectangular CLLs etched on the substrate with a gap-to-gap orientation. This arrangement produces a known electrical coupling between the two elements.

The details of the design parameters of the filtenna shown in **Figure 19** are listed in **Table 2**. Referring to the inset figure, the microstrip transmission line is placed on the right side of the upper CLL. It has a 50 Ω characteristic impedance and is connected directly to the center conductor of the coaxial feedline. A straight coupling line, which lies between the two CLLs along the y-axis, is utilized to further tune the coupling levels between the two CLLs. The impedance matching and

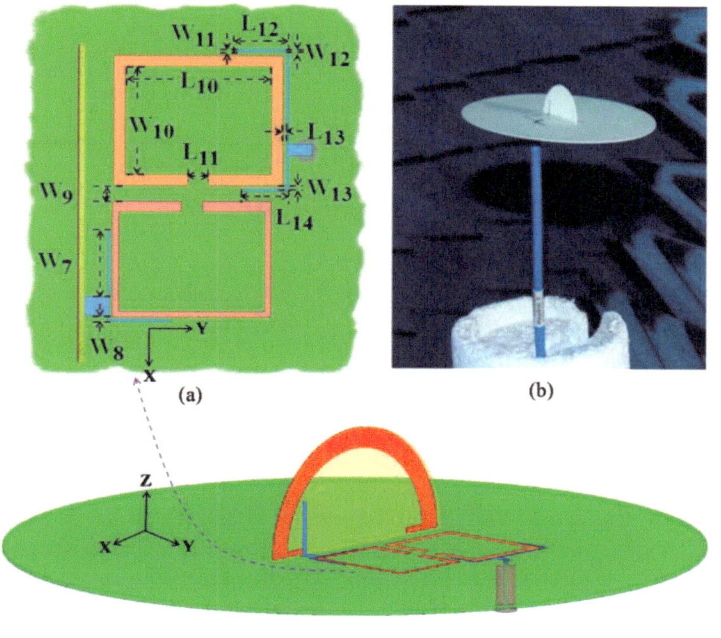

Figure 19.
The NFRP filtenna with two filter stages. (a) The geometry of the two filter stages and (b) fabricated prototype of the filtenna.

$R_1 = 75$	$R_2 = 20.7$	$R_3 = 16.8$	$W_1 = 2.5$	$W_2 = 1.4$
$W_3 = 12.73$	$W_4 = 12.3$	$W_5 = 1.7$	$W_6 = 4.3$	$W_7 = 9.01$
$W_8 = 0.49$	$W_9 = 1.0$	$W_{10} = 14.4$	$W_{11} = 0.2$	$W_{12} = 0.3$
$W_{13} = 0.3$	$L_1 = 0.9$	$L_2 = 7.7$	$L_3 = 3.0$	$L_4 = 8.7$
$L_5 = 0.3$	$L_6 = 0.13$	$L_7 = 0.76$	$L_8 = 7.95$	$L_{10} = 18.52$
$L_{11} = 2.52$	$L_{12} = 6.0$	$L_{13} = 0.2$	$L_{14} = 6.0$	$h_1 = 9.6$
$h_2 = 1.0$	$h_3 = 0.762$	$h_4 = 0.017$	$h_1 = 0.762$	Null

Table 2.
*Dimensions of the NFRP filtenna with two filter stages shown in **Figures 4–6** (all dimensions are in millimeters).*

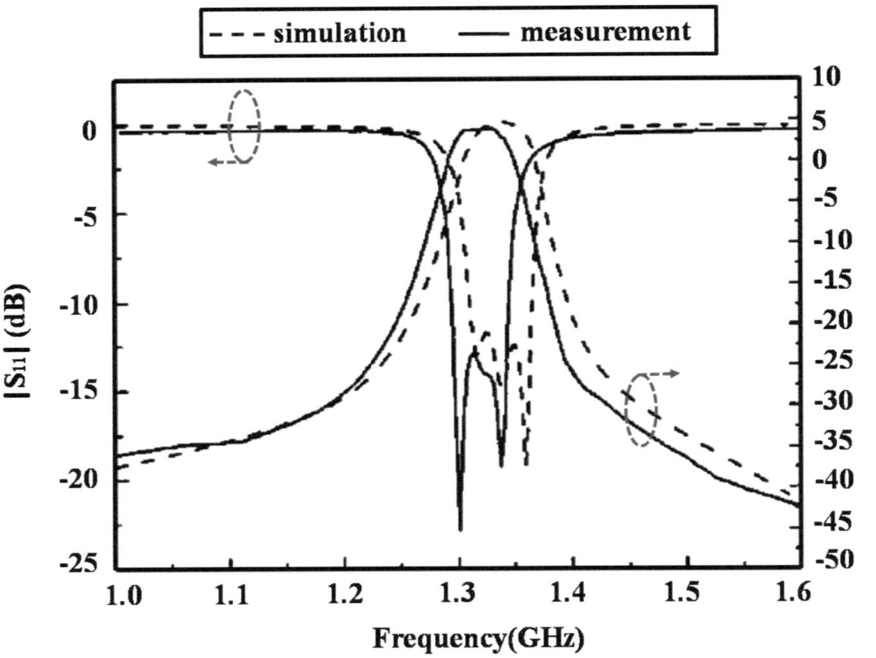

Figure 20.
*The simulated and measured $|S_{11}|$ and realized gain values as functions of the source frequency for the two-stage NFRP filtenna shown in **Figure 19**.*

far-field radiation performance characteristics of this two-stage NFRP filtenna were also studied experimentally. The simulated (measured) results shown in **Figure 20** demonstrate that the addition of the second CLL resonator introduces another resonance and produces a 55 (50) MHz impedance bandwidth, from 1.321 (1.29) to 1.376 (1.34) GHz, i.e., a 4.0% (3.8%) fractional bandwidth. The measured operational frequency range exhibits only a slight red shift from the simulated one. A flat realized gain response and excellent band-edge selectivity are again witnessed. The measured and simulated realized gain curves demonstrate that the two-stage NFRP filtenna also exhibits an essentially uniform and stable radiation performance over its entire operational bandwidth.

Author details

Ming-Chun Tang*, Yang Wang and Ting Shi
School of Microelectronics and Communication Engineering,
Chongqing University, Chongqing, China

*Address all correspondence to: tangmingchun@cqu.edu.cn

References

[1] Barbuto M, Trotta F, Bilotti F, Toscano A. Horn antennas with integrated notch filters. IEEE Transactions on Antennas and Propagation. 2015;**63**(2):781-785. DOI: 10.1109/TAP.2014.2378269

[2] Horestani AK, Shaterian Z, Naqui J, Martín F, Fumeaux C. Reconfigurable and tunable S-shaped split-ring resonators and application in band-notched UWB antennas. IEEE Transactions on Antennas and Propagation. 2016;**64**(9):3766-3776. DOI: 10.1109/TAP.2016.2585183

[3] Sun G-H, Wong S-W, Zhu L, Chu Q-X. A compact printed filtering antenna with good suppression of upper harmonic band. IEEE Antennas and Wireless Propagation Letters. 2016;**15**: 1349-1352. DOI: 10.1109/LAWP.2015.2508918

[4] Duan W, Zhang X-Y, Pan Y-M, Xu J-X, Xue Q. Dual-polarized filtering antenna with high selectivity and low cross polarization. IEEE Transactions on Antennas and Propagation. 2016; **64**(10):4188-4196. DOI: 10.1109/TAP.2016.2594818

[5] Chu H, Jin C, Chen J-X, Guo Y-X. A 3-D millimeter-wave filtering antenna with high selectivity and low cross-polarization. IEEE Transactions on Antennas and Propagation. 2015;**63**(5): 2375-2380. DOI: 10.1109/TAP.2015.2411282

[6] Mao C-X, Gao S, Wang Y, Luo Q, Chu Q-X. A shared-aperture dual-band dual-polarized filtering-antenna-array with improved frequency response. IEEE Transactions on Antennas and Propagation. 2017;**65**(4):1836-1844. DOI: 10.1109/TAP.2017.2670325

[7] Fereidoony F, Chamaani S, Mirtaheri SA. UWB monopole antenna with stable radiation pattern and low transient distortion. IEEE Antennas and Wireless Propagation Letters. 2011;**10**:302-305. DOI: 10.1109/LAWP.2011.2141106

[8] Zaker R, Abdipour A. A very compact ultrawideband printed omnidirectional monopole antenna. IEEE Antennas and Wireless Propagation Letters. 2010;**9**:471-473. DOI: 10.1109/LAWP.2010.2050852

[9] Qu S-W, Li J-L, Chen J-X, Xue Q. Ultrawideband strip-loaded circular slot antenna with improved radiation patterns. IEEE Transactions on Antennas and Propagation. 2007;**55**: 3348-3353. DOI: 10.1109/TAP.2007.908847

[10] Wu Q, Jin R, Geng J, Ding M. Printed Omni-directional UWB monopole antenna with very compact size. IEEE Transactions on Antennas and Propagation. 2008;**56**:896-899. DOI: 10.1109/TAP.2008.917018

[11] Tang M-C, Xiao S, Deng T, Wang D, Guan J, Wang B, et al. Compact UWB antenna with multiple band-notches for WiMAX and WLAN. IEEE Transactions on Antennas and Propagation. 2011; **59**(4):1372-1376. DOI: 10.1109/TAP.2011.2109684

[12] Wang J, Zhao J, Li J-L. Compact UWB bandpass filter with triple notched bands using parallel U-shaped defected microstrip structure. Electronics Letters. 2014;**50**(2):89-91. DOI: 10.1049/el.2013.3077

[13] Zhu L, Sun S, Menzel W. Ultra-wideband (UWB) bandpass filters using multiple-mode resonator. IEEE Microwave and Wireless Components Letters. 2005;**15**(11):796-798. DOI: 10.1109/LMWC.2005.859011

[14] Tsai C-M, Lee S-Y, Tsai C-C. Performance of a planar filter using a 0° feed structure. IEEE Transactions on

Microwave Theory and Techniques. 2002;**50**(10):2362-2367. DOI: 10.1109/TMTT.2002.803421

[15] Hong J-S, Lancaster MJ. Microstrip Filters for RF/Microwave Applications. New York, NY, USA: Wiley; 2001

[16] Cohn SB. Microwave bandpass filters containing high-Q dielectric resonators. IEEE Transactions on Microwave Theory and Techniques. 1968;**16**(4):218-227. DOI: 10.1109/TMTT.1968.1126654

[17] Guo Y, Zhou J. Total broadband transmission of microwaves through a subwavelength aperture by localized E-field coupling of split ring resonators. Optics Express. 2014;**22**(22):27136-27143. DOI: 10.1364/OE.22. 027136

[18] Ziolkowski RW, Jin P, Lin C-C. Metamaterial-inspired engineering of antennas. Proceedings of the IEEE. 2011;**99**(10):218-227. DOI: 10.1109/JPROC.2010.2091610

[19] Jin P, Ziolkowski RW. Multi-frequency, linear and circular polarized, metamaterial-inspired, near-field resonant parasitic antennas. IEEE Transactions on Antennas and Propagation. 2011;**59**(5):1446-1459. DOI: 10.1109/TAP.2011.2123053

[20] Tang M-C, Shi T, Ziolkowski RW. Electrically small, broadside radiating Huygens source antenna augmented with internal non-Foster elements to increase its bandwidth. IEEE Antennas and Wireless Propagation Letters. 2017;**16**:712-715. DOI: 10.1109/LAWP.2016.2600525

[21] Hong J-S, Shaman H, Chun Y-H. Dual-mode microstrip open-loop resonators and filters. IEEE Transactions on Microwave Theory and Techniques. 2007;**55**(8):1764-1770. DOI: 10.1109/TMTT.2007.901592

[22] Tang M–C, Chen Y, Shi T, Ziolkowski RW. Bandwidth-enhanced, compact, near-field resonant parasitic filtennas with sharp out-of-band suppression. IEEE Antennas and Wireless Propagation Letters. 2018;**17**(8):1483-1487. DOI: 10.1109/LAWP.2018.2850325

[23] Tang M–C, Shi T, Ziolkowski RW. Flexible efficient quasi-Yagi printed uniplanar antenna. IEEE Transactions on Antennas and Propagation. 2015;**63**(12):5343-5350. DOI: 10.1109/TAP.2015.2486807

[24] Tang M–C, Ziolkowski RW, Xiao S, Li M, Zhang J. Frequency-agile, efficient, near-field resonant parasitic monopole antenna. IEEE Transactions on Antennas and Propagation. 2014;**62**(3):1479-1483. DOI: 10.1109/TAP.2013.2295228

[25] Tang M–C, Ziolkowski RW, Xiao S, Li M. A high-directivity, wideband, efficient, electrically small antenna system. IEEE Transactions on Antennas and Propagation. 2014;**62**(12):6541-6547. DOI: 10.1109/TAP.2014.2361891

[26] Hong J-S, Lancaster MJ. Couplings of microstrip square open-loop resonators for cross-coupled planar microwave filters. IEEE Transactions on Microwave Theory and Techniques. 1996;**44**(12):2099-2109. DOI: S 001 8-9480(96)07906-9

[27] Zhang X-Y, Xue Q. Novel centrally loaded resonators and their applications to bandpass filters. IEEE Transactions on Microwave Theory and Techniques. 2008;**56**(4):913-921. DOI: 10.1109/TMTT.2008.919648

Chapter 13

Metal- and Dielectric-Loaded Waveguide: An Artificial Material for Tailoring the Waveguide Propagation Characteristics

Vishal Kesari

Abstract

In the present chapter a number of loaded structures are studied with circular cross-section to explore the deviation in their dispersion characteristics from their parent circular waveguide. The dielectric and/or metal loading to the waveguide tailors its dispersion characteristics. In general, the dielectric depresses and the metal elevates the dispersion characteristics from the characteristics of their parent circular waveguide. The axial periodicity results in periodic dispersion characteristics with a lower and an upper cut-off frequency (bandpass). However such a characteristic is not reported for the azimuthal periodic structures. The bandpass characteristic arises due to the shaping of the dispersion characteristics. Therefore the dispersion shaping is only possible with axial periodicity and not with the azimuthal periodicity. The sensitivity of the structure (geometry) parameters on the lower and upper cut-off frequencies, the extent of passband and the dispersion shaping are also included. In the axial periodic structures, the periodicity is found to be the most sensitive parameter for tailoring the dispersion characteristics and the disc-hole radius is the most sensitive parameter for shifting the dispersion characteristics over the frequency axis.

Keywords: periodically loaded waveguide, metal and dielectric loading, disc-loaded circular waveguide, dispersion characteristics, dispersion shaping

1. Introduction

Waveguides, due to their low insertion loss and high power handling capabilities at microwave and millimeter-wave frequencies, are the transmission line commonly used for transmitting or propagating or guiding the signals of these frequencies from one point to another. The propagation characteristics of a guiding structure are generally represented by its dispersion characteristics. The dispersion characteristics are the study of structure supported frequency for a given phase propagation constant, and most commonly being plotted as the supported angular frequency ω ($= 2\pi f, f$ being frequency) against the phase propagation constant β. Therefore, the dispersion characteristics are also known as $\omega - \beta$ characteristics. The dispersion ($\omega - \beta$) characteristics of a waveguide is a hyperbola that has a cutoff frequency, more specifically a lower cutoff frequency. All the signals having

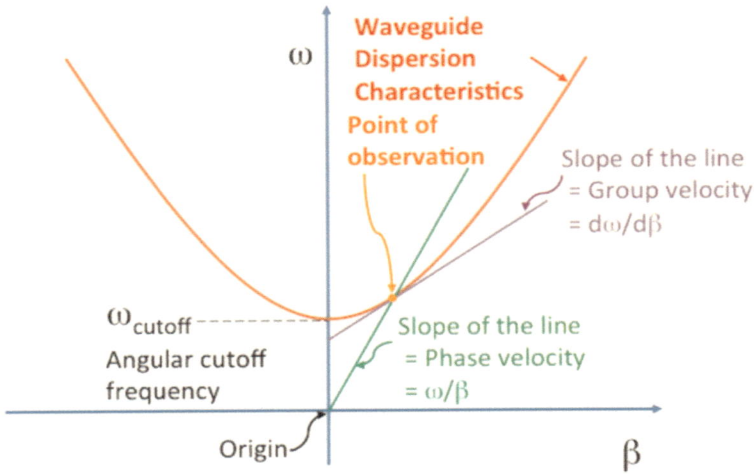

Figure 1.
$\omega - \beta$ dispersion characteristics of a circular waveguide showing the waveguide cutoff frequency and phase and group velocities.

frequencies above this lower cutoff frequency are allowed to propagate through the waveguide, and the signals having frequencies below this frequency will face a high reflection. Because of this characteristics a waveguide is inherently a high pass filter. The waveguide supports two kinds of velocities namely the phase velocity and the group velocity. The phase velocity at a chosen frequency is the one with which the signal of constant phase travels, which is represented by the slope of a line joining a chosen frequency point on $\omega - \beta$ dispersion characteristics to the origin (point representing zero frequency and zero phase propagation constant), i.e. mathematically given as ω/β. The group velocity at a chosen frequency point is the one with which the energy in signal travels, which is represented by the slope of the $\omega - \beta$ dispersion characteristics at the chosen frequency point, i.e. mathematically given as $d\omega/d\beta$ (**Figure 1**). Thus, one can control the supported phase and group velocities in a waveguide by tailoring its dispersion characteristics. Such tailoring can be achieved by loading the waveguide by metal and/ or dielectric in to the smooth wall waveguide [1–5]. The characteristics (propagation or dispersion) of the conventional (smooth wall) waveguide changes with the metal and/or dielectric loading, and the same cannot be generated naturally. Therefore, the metal- and/or dielectric-loaded waveguide may be considered as artificially created material or artificial material. In part of the chapter to follow, a number of circular waveguide models containing various metal and/or dielectric loading are considered (Section 2). The electromagnetic boundary conditions (Section 3) and the dispersion relations (Section 4) of these loaded waveguides are outlined. Further, the dispersion characteristics of all the considered loaded waveguides are discussed with their sensitivity against variation in structure (geometrical) parameters (Section 5). Finally, the conclusion is drawn (Section 6).

2. Structure models

Although, the considered structures being a single conductor structure support TE ($E_z = 0$) as well as TM ($H_z = 0$) modes, they are being analyzed for the TE modes. The structures excited in these modes are of the interest for a specific class of vacuum electronic fast-wave devices, specifically the gyro-devices. In the category of gyro-devices, for the broadband amplifier namely the gyro-traveling-wave tube (gyro-TWT) the growth rate of the TM-mode vanishes at higher frequencies [6]. The models are also restricted to circular waveguide and the analyses are carried out in cylindrical (r, θ, z) system of coordinates.

2.1 Dielectric-loaded circular waveguide

In this section, we will explore two variants of dielectric-loaded structure: (i) the circular waveguide with dielectric lining on metal wall (**Figure 2**), and (ii) the circular waveguide with dielectric coaxial insert (**Figure 3**) for their dispersion characteristics and the tailoring of these characteristics with change of the relative permittivity of the dielectric material.

2.1.1 Circular waveguide with dielectric lining on metal wall

This model (model-1) includes a metallic circular waveguide of inner radius r_W, inner wall of which is containing a dielectric lining of inner radius r_L and relative permittivity ε_r for the full length of the waveguide [7]. (Here, the waveguide is considered to be infinitely long and there is no reflection of the traveling signals

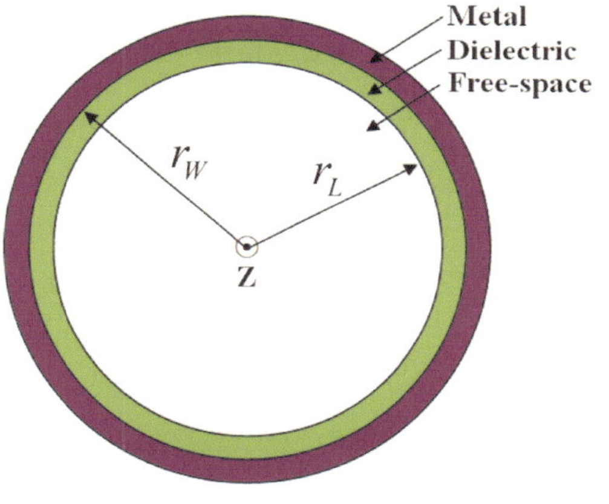

Figure 2.
Circular waveguide with dielectric lining on metal wall [7].

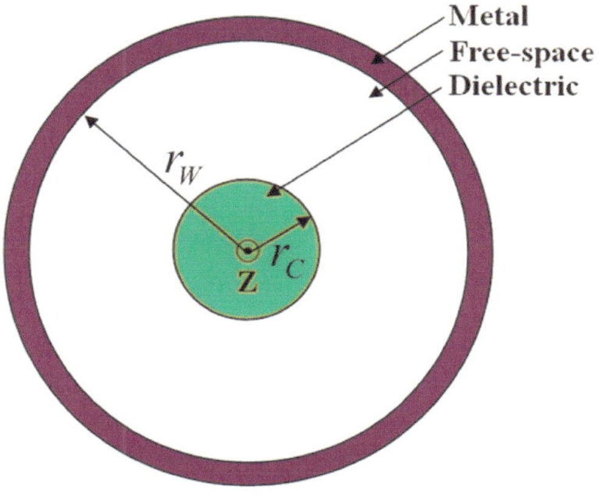

Figure 3.
Circular waveguide with dielectric coaxial insert [7].

from the waveguide extremes). Thus the radial thickness of the dielectric lining can be calculated as $r_W - r_L$. For the sake of analysis, the structure may be divided into two analytical regions, the central free-space (dielectric free) region I: $0 \le r < r_L$, $0 \le z < \infty$; and the dielectric filled region II: $r_L \le r < r_W$, $0 \le z < \infty$ (**Figure 2**). The relevant (axial magnetic and azimuthal electric) field intensity components may be written as [7]:

In region I:

$$H_z^I = \sum_{n=-\infty}^{+\infty} A_n^I J_0\{\gamma_n^I r\} \exp j(\omega t - \beta_n z) \tag{1}$$

$$E_\theta^I = j\omega\mu_0 \sum_{n=-\infty}^{+\infty} \frac{1}{\gamma_n^I} A_n^I J_0'\{\gamma_n^I r\} \exp j(\omega t - \beta_n z) \tag{2}$$

In region II:

$$H_z^{II} = \sum_{n=-\infty}^{+\infty} \left(A_n^{II} J_0\{\gamma_n^{II} r\} + B_n^{II} Y_0\{\gamma_n^{II} r\}\right) \exp j(\omega t - \beta_n z) \tag{3}$$

$$E_\theta^{II} = j\omega\mu_0 \sum_{n=-\infty}^{+\infty} \frac{1}{\gamma_n^{II}} \left(A_n^{II} J_0'\{\gamma_n^{II} r\} + B_n^{II} Y_0'\{\gamma_n^{II} r\}\right) \exp j(\omega t - \beta_n z) \tag{4}$$

where J_0 and Y_0 are the zeroth-order Bessel functions of the first and second kinds, respectively. Prime with a function represents the derivative with respect to its argument. A_n^I, A_n^{II} and B_n^{II} are the field constants, superscript identifying its value, in different analytical regions. $\gamma_n^I \left[= \left(k^2 - \beta_n^2\right)^{1/2}\right]$ and $\gamma_n^{II} \left[= \left(\varepsilon_r k^2 - \beta_n^2\right)^{1/2}\right]$ are the radial propagation constants in regions I and II, respectively. β_n and k are the phase and the free-space propagation constants, respectively [7].

2.1.2 Circular waveguide with dielectric coaxial insert

Similar to model-1, this model (model-2) also contains a metallic circular waveguide of inner radius r_W, inner wall of which is free from any dielectric. The model includes a coaxial dielectric insert of radius r_C and relative permittivity ε_r for the full length of the waveguide (**Figure 3**) [7].

For the sake of analysis, the structure may be divided into two analytical regions, the central dielectric filled region I: $0 \le r < r_C$, $0 \le z < \infty$; and the free-space (dielectric free) region II: $r_C \le r < r_W$, $0 \le z < \infty$. The relevant (axial magnetic and azimuthal electric) field intensity components may be written same as for model-1 (1)–(4), in which the radial propagation constants γ_n^I and γ_n^{II} are interpreted as: $\gamma_n^I = \left(\varepsilon_r k^2 - \beta_n^2\right)^{1/2}$ and $\gamma_n^{II} = \left(k^2 - \beta_n^2\right)^{1/2}$, respectively [7].

2.2 Metal-loaded circular waveguide

In Section 2.1, we have studied the dispersion characteristics of dielectric-loaded structures. In the present section, we will explore three variants of all metal-loaded structure: (i) the conventional annular metal disc-loaded circular waveguide (**Figure 4**), (ii) the interwoven-disc-loaded circular waveguide (**Figure 5**), and (iii) metal vane-loaded circular waveguide (**Figure 6**) for their dispersion characteristics and the effect of change of the geometry parameters on these characteristics.

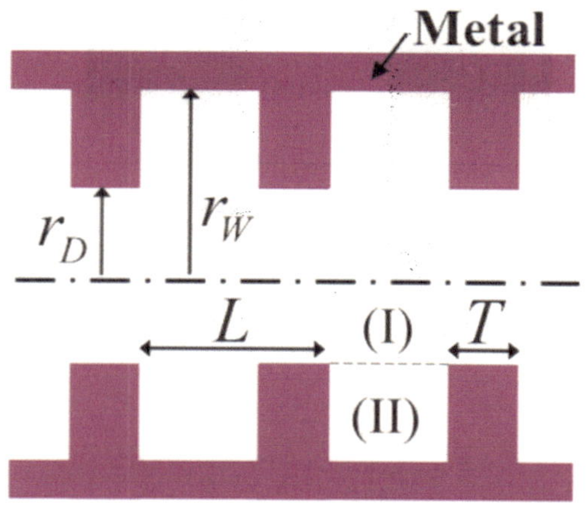

Figure 4.
Circular waveguide with annular metal discs [9–13].

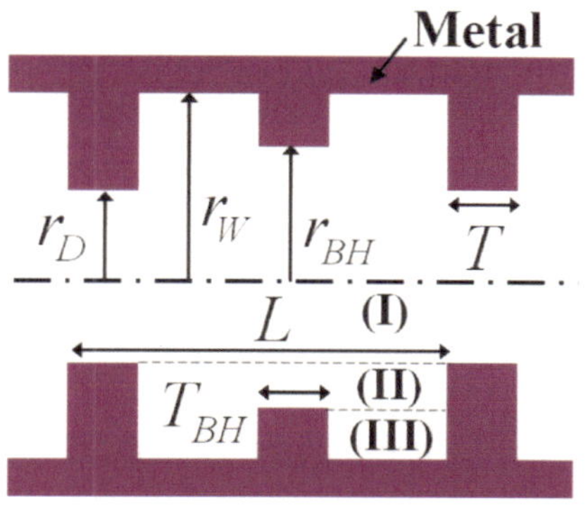

Figure 5.
Interwoven disc-loaded circular waveguides [2, 9, 14, 15].

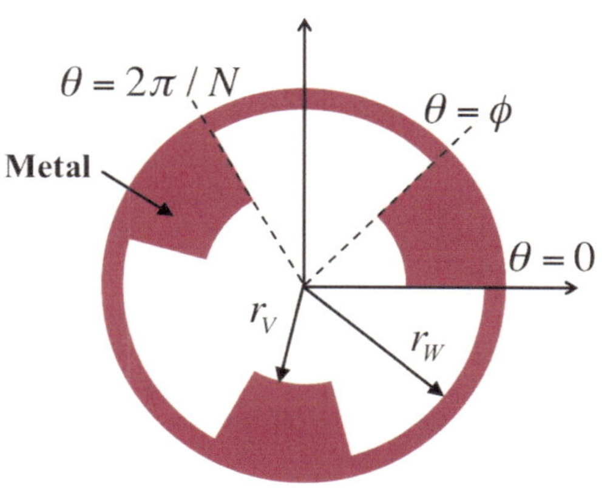

Figure 6.
Circular waveguide loaded with metal vanes [16–18].

2.2.1 Circular waveguide with annular metal discs

In this model (model-3) a circular metallic waveguide of inner radius r_W is considered in which annular disc of thickness T, inner radius r_D and outer radius r_W

are arranged periodically with periodicity L. The structure is commonly known as the disc-loaded circular waveguide (conventional) (**Figure 4**) [2, 5, 8–13]. As the structure is periodic, therefore one period of the structure coupled with Floquet's theorem is sufficient for the analysis of the infinitely long structure [1, 2, 5, 8, 9]. For the sake of analysis, the structure may be divided into two analytical regions, the central free-space (disc free) region I: $0 \le r < r_D$, $0 \le z < \infty$; and the disc occupied region II: $r_D \le r < r_W$, $0 \le z \le L - T$ (**Figure 4**). The disc free and disc occupied regions are assumed to support propagating (traveling) and stationary waves, respectively. The relevant (axial magnetic and azimuthal electric) field intensity components in the region I is given by (1) and (2) and in the region II may be written as [2, 5, 9–13]:

In region II:

$$H_z^{II} = \sum_{m=1}^{\infty} A_m^{II} Z_0\{\gamma_m^{II} r\} \exp(j\omega t) \sin(\beta_m z) \tag{5}$$

$$E_\theta^{II} = j\omega\mu_0 \sum_{m=1}^{\infty} \frac{1}{\gamma_m^{II}} A_m^{II} Z_0'\{\gamma_m^{II} r\} \exp(j\omega t) \sin(\beta_m z) \tag{6}$$

where $Z_0\{\gamma_m^{II} r\} = J_0\{\gamma_m^{II} r\} Y_0'\{\gamma_m^{II} r_W\} - J_0'\{\gamma_m^{II} r_W\} Y_0\{\gamma_m^{II} r\}$; A_m^{II} is the field constants, superscript identifying its value, in different analytical regions. $\gamma_m^{II}\left[= \left(k^2 - \beta_m^2\right)^{1/2}\right]$ is the radial propagation constant in region II. β_n, defined as $\beta_n = \beta_0 + 2\pi n/L$, is the axial phase propagation constant in region I with β_0 as the axial phase propagation constant for fundamental space harmonic, and n [= 0, ±1, ±2, ±3, ...] as space harmonic number. β_m, defined as $\beta_m = m\pi/(L - T)$, is the axial propagation constants in region II with m (= 1, 2, 3, ...) as the modal harmonic numbers in region II [2, 5, 9–13].

2.2.2 Interwoven-disc-loaded circular waveguide

This model (model-4) differs from the conventional disc-loaded circular waveguide due to different additional disc included in between two identical consecutive discs of conventional disc-loaded circular waveguide [2, 9, 14, 15]. Thus, this model is considered with a circular metallic waveguide of inner radius r_W in which annular disc of thickness T, inner radius r_D and outer radius r_W are arranged periodically with periodicity L. In addition, another annular disc of thickness T_{BH}, inner radius r_{BH} and outer radius r_W are also arranged periodically with periodicity L such that the disc of thickness T_{BH} is placed in middle of two identical consecutive discs of thickness T. The structure is known as the interwoven-disc-loaded circular waveguide. Similar to conventional disc-loaded circular waveguide, this structure is also periodic, therefore considering unit cell of the structure with Floquet's theorem suffices for the analysis of the infinitely long structure. The analytical regions of the model may be considered as: (i) region I: $0 \le r < r_D$, $0 \le z < \infty$; (ii) region II: $r_D \le r < r_{BH}$, $0 \le z \le L - T$; and (iii) region III: $r_{BH} \le r < r_W$, $0 \le z \le (L - T - T_{BH})/2$, where $L - T$ and $(L - T - T_{BH})/2$ represent the axial-gaps between two consecutive discs of smaller hole and between discs of bigger and smaller holes (**Figure 5**).

Similar to the conventional disc-loaded circular waveguide (model-3), it is assumed that the disc free (I) and disc occupied (II and III) regions, respectively, support propagating and standing waves. The relevant (axial magnetic and azimuthal electric) field intensity components in the region I is given by (1) and (2) and in the regions II and III may be written as [2, 9, 14, 15]:

In region II:

$$H_z^{II} = \sum_{m=1}^{\infty} \left[A_m^{II} J_0\{\gamma_m^{II} r\} + B_m^{II} Y_0\{\gamma_m^{II} r\}\right] \exp(j\omega t) \sin(\beta_m z) \tag{7}$$

$$E_\theta^{II} = j\omega\mu_0 \sum_{m=1}^{\infty} \frac{1}{\gamma_m^{II}} \left[A_m^{II} J_0'\{\gamma_m^{II} r\} + B_m^{II} Y_0'\{\gamma_m^{II} r\}\right] \exp(j\omega t) \sin(\beta_m z) \tag{8}$$

In region III:

$$H_z^{III} = \sum_{p=1}^{\infty} A_p^{III} Z_0\left\{\gamma_p^{III} r\right\} \exp(j\omega t) \sin\left(\beta_p z\right) \tag{9}$$

$$E_\theta^{III} = j\omega\mu_0 \sum_{p=1}^{\infty} \frac{1}{\gamma_p^{III}} A_p^{III} Z_0'\{\gamma_m^{III} r\} \exp(j\omega t) \sin\left(\beta_p z\right), \tag{10}$$

where $Z_0\left\{\gamma_p^{III} r\right\} = J_0\left\{\gamma_p^{III} r\right\} Y_0'\left\{\gamma_p^{III} r_W\right\} - J_0'\left\{\gamma_p^{III} r_W\right\} Y_0\left\{\gamma_p^{III} r\right\}$; A_m^{II}, B_m^{II} and A_p^{III} are the field constants, superscript identifying its value, in different analytical regions. $\gamma_m^{II}\left[= \left(k^2 - \beta_m^2\right)^{1/2}\right]$ and $\gamma_p^{III}\left[= \left(k^2 - \beta_p^2\right)^{1/2}\right]$ are the radial propagation constants in regions II and III, respectively. The axial phase propagation constants β_n in region I and β_m in region II are defined in the same manner as for model-3. β_p $[= 2p\pi/(L - T - T_{BH})]$ is the axial phase propagation constants in region III; here p is the modal harmonic number in region III [2, 9, 14, 15].

2.2.3 Circular waveguide loaded with metal vanes

This model (model-5) considers a circular waveguide of radius r_W and N number of metal vanes of vane-inner-tip radius r_V and vane angle ϕ extending axially over the length of the waveguide arranged on the waveguide wall to maintain the azimuthal periodicity (**Figure 6**) [16–18]. Clearly, the azimuthal periodicity is $2\pi/N$. For the analysis of the structure, it may be divided into two regions; (i) the central cylindrical vane-free free-space region I: $0 \le r < r_V$, $0 \le \theta < 2\pi$, and (ii) the free-space region II between two consecutive metal vanes $r_V \le r \le r_W$, $\phi < \theta < 2\pi/N$(**Figure 6**). The relevant (axial magnetic and azimuthal electric) field intensity components in the regions I and II may be written as [16–18]:

In region I:

$$H_z^I = \sum_{q=-\infty}^{+\infty} A_q^I J_q\{\gamma^I r\} \exp(-jq\theta) \tag{11}$$

$$E_\theta^I = \frac{j\omega\mu_0}{\gamma^I} \sum_{q=-\infty}^{+\infty} A_q^I J_q'\{\gamma^I r\} \exp(-jq\theta) \tag{12}$$

In region II:

$$H_z^{II} = \sum_{v=0}^{+\infty} \left[A_v^{II} J_v\{\gamma^{II} r\} + B_v^{II} Y_v\{\gamma^{II} r\}\right] \cos\left(\frac{v\pi(\theta - \phi)}{2\pi/N - \phi}\right) \tag{13}$$

$$E_\theta^{II} = \frac{j\omega\mu_0}{\gamma^{II}} \sum_{v=0}^{+\infty} \left[A_v^{II} J_v'\{\gamma^{II} r\} + B_v^{II} Y_v'\{\gamma^{II} r\}\right] \cos\left(\frac{v\pi(\theta - \phi)}{2\pi/N - \phi}\right) \tag{14}$$

where A_q^I, A_v^{II} and B_v^{II} are the field constants; J and Y are the ordinary Bessel function of first and second kinds, respectively, with their primes representing the derivatives with respect to their arguments. q is an integer; and v is a non-negative integer. $\gamma^{I} = \gamma^{II}\left(= \left(k^2 - \beta^2\right)^{1/2}\right)$ and β are the radial and the axial phase propagation constants, respectively. In order to include the effect of azimuthal harmonics due to angular periodicity of the structure, the azimuthal dependence is considered as $\exp(-jv\theta)$, such that $v = s + qN$, where s is also an integer [16–18].

2.3 Metal- and dielectric-loaded circular waveguide

In Sections 2.1 and 2.2, we have respectively explored the independent dielectric- and metal-loaded structures. However, in this section we will study the metal as well as dielectric loading in the circular waveguide.

2.3.1 Circular waveguide loaded with dielectric and metal discs

This model (model-6) is formed by alternatively stacking the metal and dielectric discs each of same disc hole radii r_D. This is similar to conventional disc-loaded circular waveguide in which the volume between two consecutive metal discs is filled with dielectric of relative permittivity ε_r. Similar to conventional disc-loaded circular waveguide for the sake of analysis, one may divide the structure into two regions: central disc free region I: $0 \le r < r_D$, $0 < z < \infty$, and disc occupied region II: $r_D \le r < r_W$, $0 < z < L - T$ (**Figure 7**). The relevant (axial magnetic and azimuthal electric) field intensity components in the regions I and II may be given by (1), (2), (5) and (6). In (5) and (6), the radial propagation constant in region II is interpreted as $\gamma_m^{II} = \left(\varepsilon_r k^2 - \beta_m^2\right)^{1/2}$ [19].

2.3.2 Circular waveguide loaded with dielectric and metal discs having different hole radius

This model (model-7) is similar to that of model-6, which has a circular waveguide consisting of alternate dielectric and metal discs, such that the hole-radius of metal discs is lesser than that of dielectric discs [20]. For the purpose of analysis, one may divide the structure into three regions: i) the central disc free region I: $0 \le r < r_D$, $0 < z < \infty$; ii) the disc occupied free space region II: $r_D \le r < r_{DD}$, $0 < z < L - T$; and iii) the dielectric filled disc occupied region III: $r_{DD} \le r < r_W$, $0 < z < L - T$; where r_{DD} is the hole radius of dielectric disc. It is considered that the region I (disc free region) supports propagating and regions II and III (disc occupied regions) support standing waves [20].

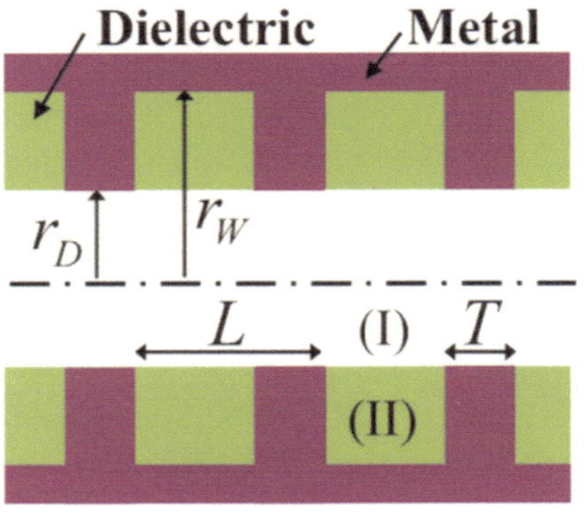

Figure 7.
Circular waveguide loaded with dielectric and metal discs [19].

The relevant (axial magnetic and azimuthal electric) field intensity components in the region I may be given by (1) and (2), and in regions II and III are written as [20]:

In region II:

$$H_z^{II} = \sum_{m=1}^{\infty} \left[A_m^{II} J_0\{\gamma_m^{II} r\} + B_m^{II} Y_0\{\gamma_m^{II} r\}\right] \exp(j\omega t)\ \sin(\beta_m z) \tag{15}$$

$$E_\theta^{II} = j\omega\mu_0 \sum_{m=1}^{\infty} \frac{1}{\gamma_m^{II}} \left[A_m^{II} J_0'\{\gamma_m^{II} r\} + B_m^{II} Y_0'\{\gamma_m^{II} r\}\right] \exp(j\omega t)\ \sin(\beta_m z) \tag{16}$$

In region III:

$$H_z^{III} = \sum_{m=1}^{\infty} A_m^{III} Z_0\{\gamma_m^{III} r\}\ \exp(j\omega t)\ \sin(\beta_m z) \tag{17}$$

$$E_\theta^{III} = j\omega\mu_0 \sum_{m=1}^{\infty} \frac{1}{\gamma_m^{III}} A_m^{III} Z_0'\{\gamma_m^{III} r\}\ \exp(j\omega t)\ \sin(\beta_m z) \tag{18}$$

where $Z_0\{\gamma_m^{III} r\} = J_0\{\gamma_m^{III} r\} Y_0'\{\gamma_m^{III} r_W\} - J_0'\{\gamma_m^{III} r_W\} Y_0\{\gamma_m^{III} r\}$; A_m^{II}, B_m^{II} and A_m^{III} are the field constants in different analytical regions, identified by given superscript, respectively. $\gamma_n^I \left[= \left(k^2 - \beta_n^2\right)^{1/2}\right]$, $\gamma_m^{II} \left[= \left(k^2 - \beta_m^2\right)^{1/2}\right]$, and $\gamma_m^{III} \left[= \left(\varepsilon_r k^2 - \beta_m^2\right)^{1/2}\right]$ are the radial propagation constants in regions I, II, and III, respectively (**Figure 8**). The axial phase propagation constants β_n in region I and β_m in regions II and III are defined in the same manner as for model-3 [20].

2.3.3 Circular waveguide loaded with alternate dielectric and metal vanes

This model (model-8) is similar to model-5 except the region II filled with dielectric of relative permittivity ε_r between the two consecutive metal vanes [21]. For the sake of analysis, the structure may be divided into two regions; (i) the central cylindrical vane-free free-space region I: $0 \le r < r_V$, $0 \le \theta < 2\pi$, and (ii) the dielectric filled region II between two consecutive metal vanes: $r_V \le r \le r_W$, $< \theta < 2\pi/N$ (**Figure 9**). The relevant (axial magnetic and azimuthal electric) field intensity components in the regions I and II may be given by (11)–(14), in which the radial propagation constant is given as: $\gamma^{II} = \left(\varepsilon_r k^2 - \beta^2\right)^{1/2}$ [21].

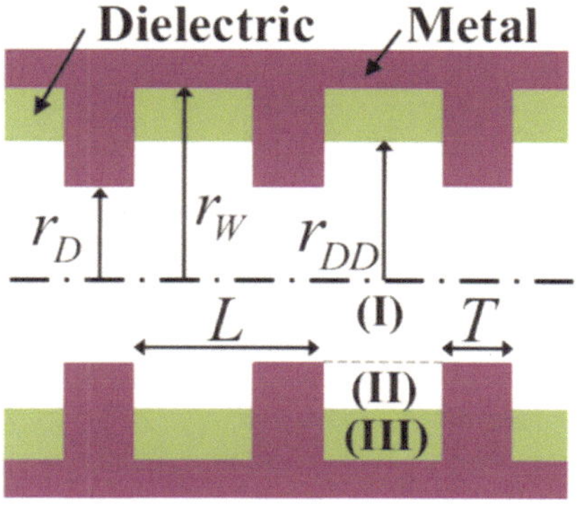

Figure 8.
Circular waveguide loaded with dielectric and metal discs with the hole radius of metal discs lesser than that of dielectric discs [20].

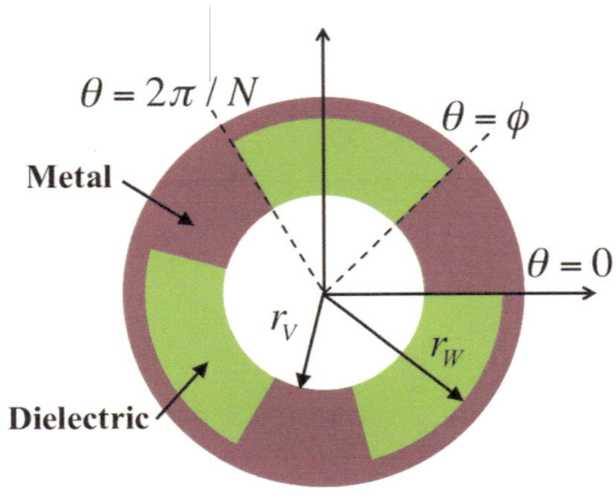

Figure 9.
Circular waveguide loaded with alternate dielectric and metal vanes [21].

3. Boundary conditions

One of the general boundary conditions comprising all the considered models is due to the null tangential electric field intensity at the inner surface of the metallic waveguide that is represented as [2, 5, 9–21]:

$$E_\theta = 0|_{r=r_W} \quad 0<z<\infty \tag{19}$$

Model-1: The relevant electromagnetic boundary conditions for model-1 may be written in the mathematical form as [7]:

$$H_z^I = H_z^{II}|_{r=r_L} \quad 0<z<\infty \tag{20}$$

$$E_\theta^I = E_\theta^{II}|_{r=r_L} \quad 0<z<\infty \tag{21}$$

The boundary conditions (20) and (21) state the continuity of the axial component of magnetic and the azimuthal component of electric field intensities at the interface, $r = r_L$, between the regions I and II (**Figure 2**) [7].

Model-2: The relevant electromagnetic boundary conditions for model-2 (**Figure 3**) may be written in the mathematical form as [7]:

$$H_z^I = H_z^{II}, \quad 0<z<\infty|_{r=r_C} \tag{22}$$

$$E_\theta^I = E_\theta^{II}, \quad 0<z<\infty|_{r=r_C} \tag{23}$$

The boundary conditions (22) and (23) state the continuity of the axial component of magnetic and the azimuthal component of electric field intensities at the interface, $r = r_C$, between the regions I and II (**Figure 3**) [7].

Model-3: The relevant electromagnetic boundary conditions for model-3 (**Figure 4**) may be written in the mathematical form as [2, 5, 9–13]:

$$H_z^I = H_z^{II} \quad 0<z<L-T|_{r=r_D} \tag{24}$$

$$E_\theta^I = \left\{ \begin{array}{ll} E_\theta^{II} & 0<z<L-T \\ 0 & L-T\le z\le L \end{array} \right|_{r=r_D} \tag{25}$$

The boundary conditions (24) and (25) state the continuity of the axial component of magnetic and the azimuthal component of electric field intensities at the interface, $r = r_D$, between the regions I and II, and the null azimuthal component of electric field intensity at the disc hole metallic surface (**Figure 4**) [2, 5, 9–13].

Model-4: The boundary conditions (24) and (25) are also true for the model-4 (**Figure 5**) at the interface, $r = r_D$, between the regions I and II. The additional boundary conditions at the interface, $r = r_{BH}$, between the regions II and III may be written as [2, 9, 14, 15]:

$$H_z^{II} = H_z^{III} \quad 0 \le z \le (L - T - T_{BH})/2\Big|_{r=r_{BH}} \tag{26}$$

$$E_\theta^{II} = \begin{cases} E_\theta^{III} & 0 \le z \le (L - T - T_{BH})/2 \\ 0 & (L - T - T_{BH})/2 \le z \le (L - T + T_{BH})/2 \end{cases}\Bigg|_{r=r_{BH}} \tag{27}$$

The boundary conditions (26) and (27) state the continuity of the axial component of magnetic and the azimuthal component of electric field intensities at the interface, $r = r_{BH}$, between the regions II and III, and the null azimuthal component of electric field intensity at the disc hole metallic surface (**Figure 5**) [2, 9, 14, 15].

Model-5: The relevant electromagnetic boundary conditions for model-5 (**Figure 6**) may be written in the mathematical form as [16–18]:

$$H_z^{I} = H_z^{II} \qquad \phi \le \theta \le 2\pi/N\Big|_{r=r_V} \tag{28}$$

$$E_\theta^{I} = \begin{cases} 0 & 0 \le \theta < \phi \\ E_\theta^{II} & \phi \le \theta \le 2\pi/N \end{cases}\Bigg|_{r=r_V} \tag{29}$$

The boundary conditions (28) and (29) state the continuity of the axial component of magnetic and the azimuthal component of electric field intensities at the interface, $r = r_V$, between the regions I and II, and the null azimuthal component of electric field intensity at the vane tip metallic surface (**Figure 6**) [16–18].

Model-6: The relevant electromagnetic boundary conditions for model-6 (**Figure 7**) are given by (24) and (25), same as for model-3 (**Figure 4**) [19].

Model-7: The relevant electromagnetic boundary conditions (24) and (25) are also true for the model-7 (**Figure 8**) at the interface, $r = r_D$, between the regions I and II. The additional boundary conditions at the interface, $r = r_{DD}$, between the regions II and III may be written as [20]:

$$H_z^{II} = H_z^{III} \qquad 0 < z < L - T\Big|_{r=r_{DD}} \tag{30}$$

$$E_\theta^{II} = E_\theta^{III} \qquad 0 < z < L - T\Big|_{r=r_{DD}} \tag{31}$$

The boundary conditions (30) and (31), respectively, state the continuity of the axial magnetic and the azimuthal electric field intensities at the interface, $r = r_{DD}$, between the disc occupied free space region II and disc-occupied dielectric region III (**Figure 8**) [20].

Model-8: The relevant electromagnetic boundary conditions for model-8 (**Figure 9**) are given by (28) and (29), same as for model-5 (**Figure 6**) [21].

4. Dispersion relations

In general, the field intensity components (**1**)–(18) contain unknown field constants. In order to establish relations between these unknown field constants, the relevant field intensity components are substituted into the respective boundary conditions. Further, the algebraic manipulations of the obtained relations between these field constants eliminate all the field constants, and it results in a characteristic relation of the model known as the dispersion relation. The dispersion relations of various considered models are:

Model-1 [7]:

$$\begin{aligned}&\gamma^{II}J_0'\{\gamma^I r_L\}\left(J_0\{\gamma^{II}r_L\}Y_0'\{\gamma^{II}r_W\} - Y_0\{\gamma^{II}r_L\}J_0'\{\gamma^{II}r_W\}\right)\\ &-\gamma^I J_0\{\gamma^I r_L\}\left(J_0'\{\gamma^{II}r_L\}Y_0'\{\gamma^{II}r_W\} - Y_0'\{\gamma^{II}r_L\}J_0'\{\gamma^{II}r_W\}\right) = 0\end{aligned} \tag{32}$$

Model-2 [7]:

$$\begin{aligned}&\gamma^{II}J_0'\{\gamma^I r_C\}\left(J_0\{\gamma^{II}r_C\}Y_0'\{\gamma^{II}r_W\} - Y_0\{\gamma^{II}r_C\}J_0'\{\gamma^{II}r_W\}\right)\\ &-\gamma^I J_0\{\gamma^I r_C\}\left(J_0'\{\gamma^{II}r_C\}Y_0'\{\gamma^{II}r_W\} - Y_0'\{\gamma^{II}r_C\}J_0'\{\gamma^{II}r_W\}\right) = 0\end{aligned} \tag{33}$$

Model-3 [2, 5, 9–13]:

$$\det\left|M_{nm}J_0\{\gamma_n^I r_D\}Z_0'\{\gamma_m^{II}r_D\} - Z_0\{\gamma_m^{II}r_D\}J_0'\{\gamma_n^I r_D\}\right| = 0 \tag{34}$$

where

$$M_{nm} = \frac{\gamma_n^I\beta_m^{II}\left(1-(-1)^m\exp\left[-j\beta_n^I(L-T)\right]\right)}{\gamma_m^{II}\left[\beta_m^{II} - \exp\left(-j\beta_0^I L\right)\left(\beta_m^{II}\cos\left(\beta_m^{II}L\right) + j\beta_n^I\sin\left(\beta_m^{II}L\right)\right)\right]}. \tag{35}$$

Model-4 [2, 9, 14, 15]:

$$\det\left|M_{nm}J_0\{\gamma_n^I r_D\}\left[J_0'\{\gamma_m^{II}r_D\} + \xi Y_0'\{\gamma_m^{II}r_D\}\right] - J_0'\{\gamma_n^I r_D\}\left[J_0\{\gamma_m^{II}r_D\} + \xi Y_0\{\gamma_m^{II}r_D\}\right]\right| = 0 \tag{36}$$

where

$$\xi = \frac{\gamma_p^{III}J_0'\{\gamma_m^{II}r_{BH}\}Z_0\{\gamma_m^{III}r_{BH}\} - \gamma_m^{II}J_0\{\gamma_m^{II}r_{BH}\}Z_0'\{\gamma_m^{III}r_{BH}\}}{\gamma_m^{II}Y_0\{\gamma_m^{II}r_{BH}\}Z_0'\{\gamma_m^{III}r_{BH}\} - \gamma_m^{III}Y_0'\{\gamma_m^{II}r_{BH}\}Z_0\{\gamma_m^{III}r_{BH}\}} \tag{37}$$

Model-5 [16–18]:

$$\begin{vmatrix} P_{g-1} & Q_{g-1,g} & Q_{g-1,g+1}\\ Q_{g,g-1} & P_g & Q_{g,g+1}\\ Q_{g+1,g-1} & Q_{g+1,g} & P_{g+1}\end{vmatrix} = 0 \tag{38}$$

where

$$\begin{aligned}P_{v'} = &\left(J_{v'}'\{\gamma^{II}r_V\} - \frac{J_{v'}'\{\gamma^{II}r_W\}}{Y_{v'}'\{\gamma^{II}r_W\}}Y_{v'}'\{\gamma^{II}r_V\}\right)\left(\frac{2\pi}{N} - \varphi\right)\\ &-\frac{2\pi}{N}\frac{\gamma^{II}}{\gamma^I}\frac{J_{v'}'\{\gamma^I r_V\}}{J_{v'}\{\gamma^I r_V\}}\left(J_{v'}\{\gamma^{II}r_V\} - \frac{J_{v'}'\{\gamma^{II}r_W\}}{Y_{v'}'\{\gamma^{II}r_W\}}Y_{v'}\{\gamma^{II}r_V\}\right)\end{aligned} \tag{39}$$

$$Q_{v',v} = \left(J_v'\{\gamma^{II}r_V\} - \frac{J_v'\{\gamma^{II}r_W\}}{Y_v'\{\gamma^{II}r_W\}}Y_v'\{\gamma^{II}r_V\}\right)\frac{1-\exp j(v'-v)\varphi}{j(v'-v)} \tag{40}$$

Model-6 [19]:

The dispersion relation for model-6 (**Figure 7**) is same as that for model-3 (**Figure 4**) and is given by (34) through (35) with interpretation of radial propagation constant in dielectric filled region as $\gamma_m^{II} = \left(\varepsilon_r k^2 - \beta_m^2\right)^{1/2}$ [19].

Model-7 [20]:

$$\det\left|M_{nm} J_0\{\gamma_n^I r_{MD}\}\left[J_0'\{\gamma_m^{II} r_D\} + \xi Y_0'\{\gamma_m^{II} r_D\}\right] - J_0'\{\gamma_n^I r_D\}\left[J_0\{\gamma_m^{II} r_D\} + \xi Y_0\{\gamma_m^{II} r_D\}\right]\right| = 0 \tag{41}$$

$$\xi = \frac{\gamma_m^{III} J_0'\{\gamma_m^{II} r_{DD}\} Z_0\{\gamma_m^{III} r_{DD}\} - \gamma_m^{II} J_0\{\gamma_m^{II} r_{DD}\} Z_0'\{\gamma_m^{III} r_{DD}\}}{\gamma_m^{II} Y_0\{\gamma_m^{II} r_{DD}\} Z_0'\{\gamma_m^{III} r_{DD}\} - \gamma_m^{III} Y_0'\{\gamma_m^{II} r_{DD}\} Z_0\{\gamma_m^{III} r_{DD}\}} \tag{42}$$

Model-8 [21]:

The dispersion relation for model-8 (**Figure 9**) is same as that for model-5 (**Figure 6**) and is given by (38) through (39) and (40) with interpretation of radial propagation constant in dielectric filled region as $\gamma^{II} = \left(\varepsilon_r k^2 - \beta^2\right)^{1/2}$ [21].

5. Dispersion characteristics

One may clearly observe that the shape of dispersion characteristics of the model-1 (**Figure 2**) and model-2 (**Figure 3**) change with change in relative permittivity of the dielectric material (**Figures 10** and **11**). The cutoff frequency decreases with increase of the relative permittivity of the dielectric material. The increase of relative permittivity of the dielectric material depresses the dispersion characteristics of the model-1 and model-2 and more at higher value of phase propagation constant. The analytical dispersion characteristics are found within 3% of that obtained using HFSS (**Figures 10** and **11**).

Periodic loading a circular waveguide by the metal annular discs (model-3) brings out alternate pass and stop bands with their respective higher and lower

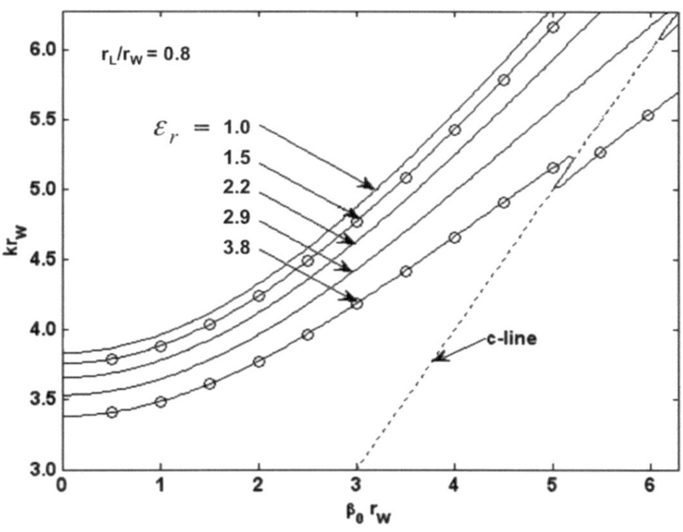

Figure 10.
TE_{01}-mode dispersion characteristics of circular waveguide with dielectric lining on metal wall taking ε_r as the parameter. The characteristics with $\varepsilon_r = 1$ (special case) represents the dispersion characteristics of conventional circular waveguide. Circles represent results obtained using HFSS.

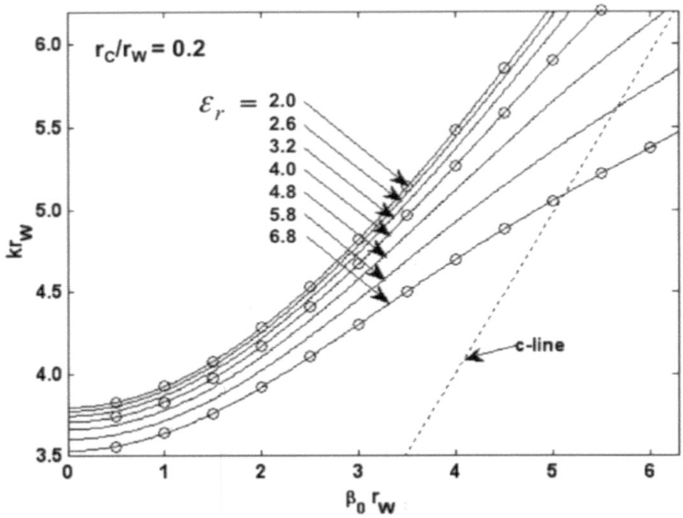

Figure 11.
TE_{01}-mode dispersion characteristics of circular waveguide with coaxial dielectric rod and taking ε_r as the parameter. Circles represent results obtained using HFSS.

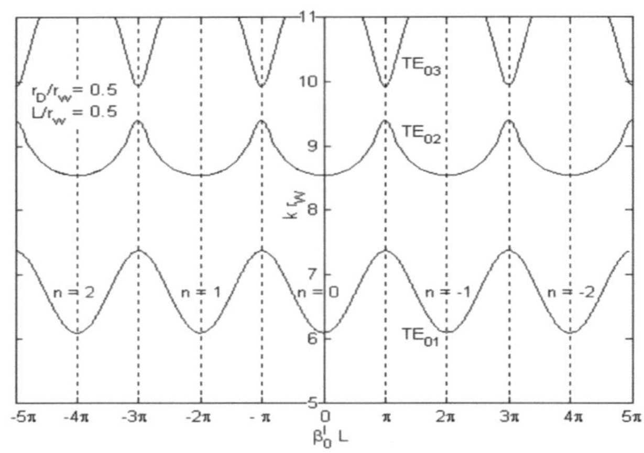

Figure 12.
Pass and stop band characteristics of the infinitesimally thin metal disc-loaded circular waveguide (including higher order harmonics $n = 0, \pm 1, \pm 2, \pm 3, \pm 4, \pm 5$; $m = 1, 2, 3, 4, 5, 6, 7, 8, 9, 10, 11$) [9, 10].

cutoff frequencies [1–3, 5, 9]. The dispersion characteristics taking horizontal axis as normalized phase propagation constant become periodic with the periodicity of $\beta_0^I L = 2\pi$ for a given mode (TE_{01}, TE_{02} and TE_{03}). The normalized passband (kr_W scale) for the TE_{02} mode is narrower than that of the TE_{01} mode. Similarly, the normalized stopband above the TE_{02} mode is narrower than that above the TE_{01} mode (**Figure 12**). The RF group velocity (slopes of dispersion plot) is positive (fundamental forward wave mode, $n = 0$) for the TE01 or TE_{02} mode, it is negative (fundamental backward wave mode) for the TE_{03} mode (**Figure 12**). The dependencies of the structure dispersion characteristics, for typical mode TE_{01}, on the disc-hole radius (**Figure 13**), the structure periodicity (**Figure 14**) and the finite disc thickness (**Figure 15**) are studied. Further, with the increase of either of the parameters, namely, the disc-hole radius and the structure periodicity, the lower and upper edge frequencies of the passband of the dispersion characteristics both decrease, though not equally. This lead to decrease or increase of the passband

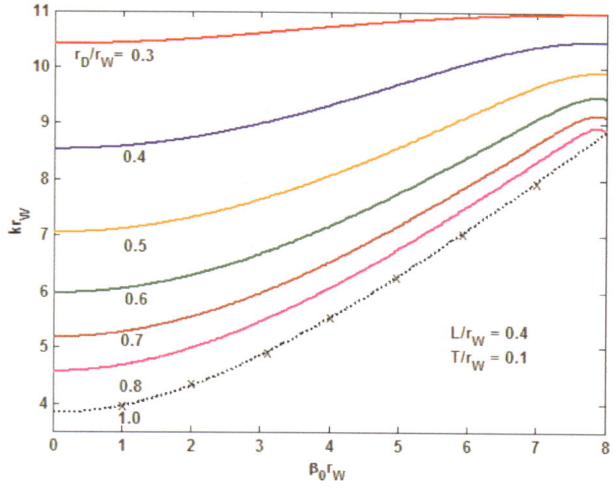

Figure 13.
TE_{01}-mode dispersion characteristics of the conventional disc-loaded circular waveguide (solid curve) taking the disc-hole radius as the parameter. The broken curve with crosses refers to a smooth-wall circular waveguide [2, 5, 11].

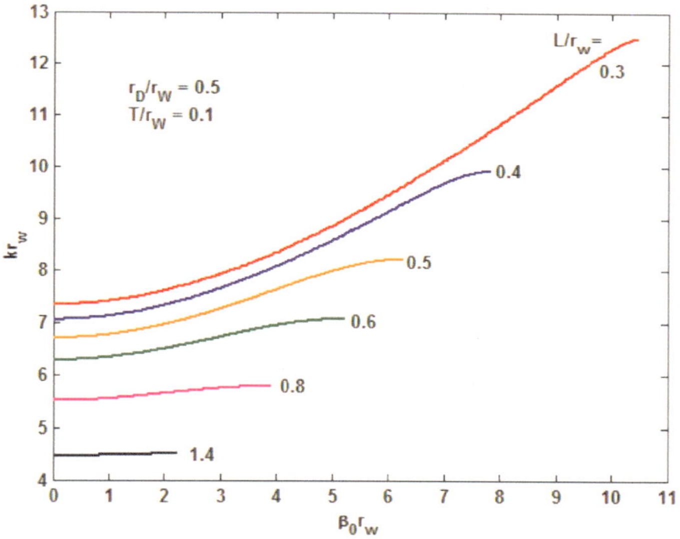

Figure 14.
TE_{01}-mode dispersion characteristics of the conventional disc-loaded circular waveguide (solid curve) taking the structure periodicity as the parameter [2, 5, 11].

according as the disc-hole radius decreases or the structure periodicity decreases, with the shift of the mid-band frequency of the passband to a higher value for the decrease of both the parameters (**Figures 13** and **14**) [2, 5, 11].

Although the disc-hole radius and the structure periodicity tailor the dispersion characteristics, the later one found to be more effective that the former one for widening the frequency range of the straight-line section of the characteristics. Reducing the structure periodicity can increase the frequency range of the straight-line section, however it accompanies with shift in waveguide cutoff (**Figure 14**). This wider straight-line section of the dispersion characteristics may be utilized for wideband coalescence with cyclotron wave (beam mode line) to result a wideband performance of a gyro-TWT. Thus, reducing the structure periodicity (**Figure 14**)

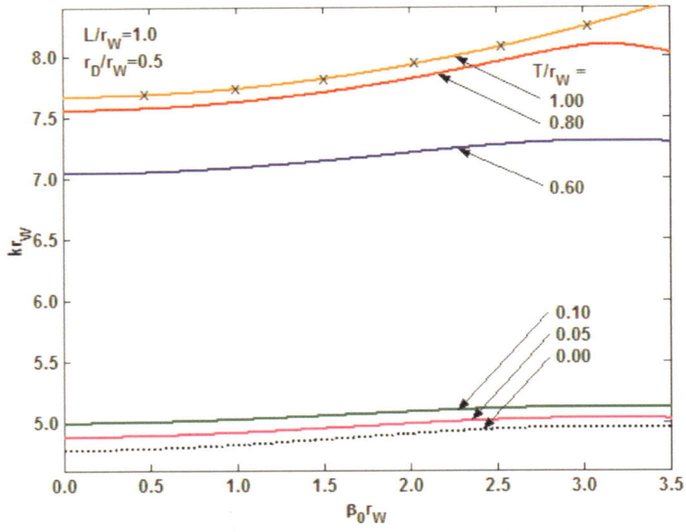

Figure 15.
TE_{01}-mode dispersion characteristics of the conventional disc-loaded circular waveguide taking the disc thickness as the parameter. The broken curve refers to the infinitesimally thin metal disc-loaded circular waveguide [2, 5, 11].

and increasing the disc-hole radius (**Figure 13**) may widen the device bandwidth. However, such broadbanding of coalescence is accompanied by the reduction of the bandwidth of the passband of the structure as well (**Figures 13** and **14**) [2, 5, 11]. The decrease of the disc thickness decreases both the lower and upper edge frequencies of the passband such that the passband first decreases, attains a minima and then increases; and the mid-band frequency of the passband as well as the start frequency of the straight-line section of the dispersion characteristics reduces (**Figure 15**). The shape of the dispersion characteristics depends on the disc thickness, though not as much as it does on the disc-hole radius or the structure periodicity (**Figures 13-15**) [2, 5, 11].

Similar to the conventional disc-loaded circular waveguide (model-3), both the hole-radii (bigger and smaller) of the interwoven-disc-loaded circular (model-4, **Figure 5**) waveguide tailor the dispersion characteristics. The lower- and the upper-cutoff frequencies decrease with increase in hole-radii, such that the passband increases and decreases with decrease of bigger and smaller hole-radii, respectively (**Figures 16** and **17**). Similar to the conventional disc-loaded circular waveguide (model-3), the structure periodicity of the interwoven-disc-loaded circular waveguide is the most effective for the increasing the passband and tailoring the dispersion characteristics (**Figure 18**). Neither, the extent of passband changes nor dispersion tailors with variation of disc-thickness of bigger-hole-disc, however, the mid frequency of the passband shifts to higher frequency with increase of disc-thickness of bigger-hole-disc (**Figure 19**). This nature may be used for shifting the operation band in the passive components or in order to optimizing the beam-wave interaction in designing a gyro-TWT with the interwoven-disc-loaded circular waveguide. In addition to tailoring the dispersion characteristics, required for designing a broadband gyro-TWT, the model-4 shows an interesting characteristic. The passband increases with increase as well as with decrease of disc-thickness of smaller- hole-disc with reference to that of bigger-hole-disc, however, the shift of the passband occur towards higher and lower frequency side, respectively, with increase and decrease of disc thickness of smaller-hole-disc with reference to that of bigger-hole-disc (**Figure 20**). Thus, the structure periodicity (**Figure 18**) and the

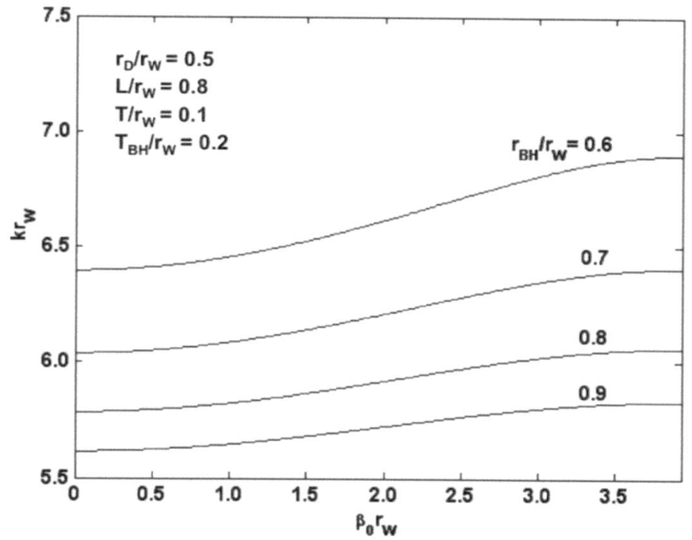

Figure 16.
TE_{01}-mode dispersion characteristics of the interwoven-disc-loaded circular waveguide taking the bigger disc-hole-radius as the parameter [9, 14, 15].

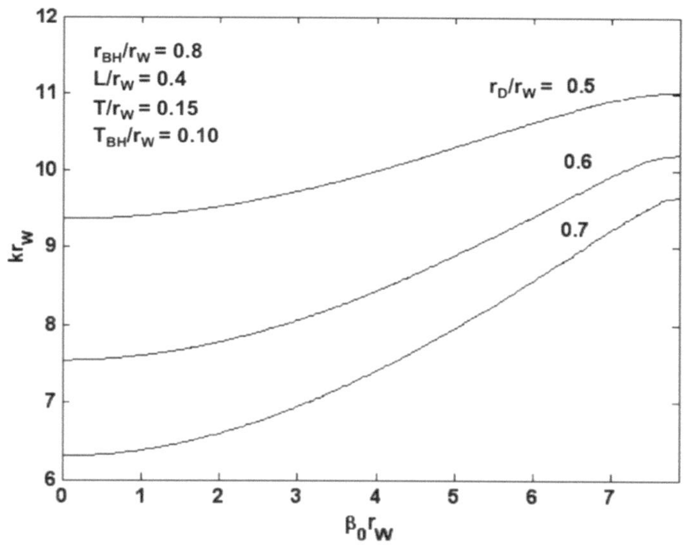

Figure 17.
TE_{01}-mode dispersion characteristics of the interwoven-disc-loaded circular waveguide taking the smaller disc-hole-radius as the parameter [9, 14, 15].

disc-thickness (**Figure 19**) of bigger-hole-disc of the interwoven-disc-loaded circular waveguide are, respectively, the most and the least sensitive parameter for controlling the passband as well as shape of the dispersion characteristics [14, 15].

Axial metal vane loading to a smooth-wall circular waveguide (model-5, **Figure 6**) forms an azimuthally periodic structure, which does not shape its dispersion characteristics, however the insertion of the metal vanes in to the circular waveguide shifts the waveguide cutoff frequency to a higher value [16–18] (**Figures 21–23**). Specifically, the increase of either of the vane angle (**Figure 22**) and the number of metal vanes (**Figure 23**) and the decrease of vane-inner-tip

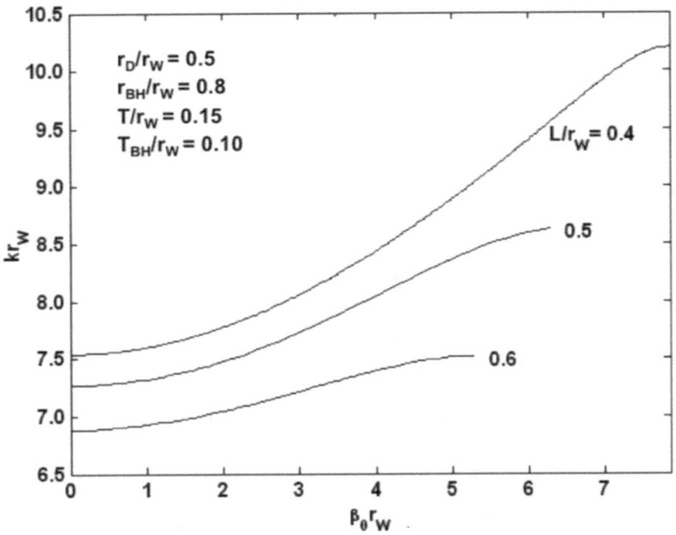

Figure 18.
TE_{01}-mode dispersion characteristics of the interwoven-disc-loaded circular waveguide taking the structure periodicity as the parameter [9, 14, 15].

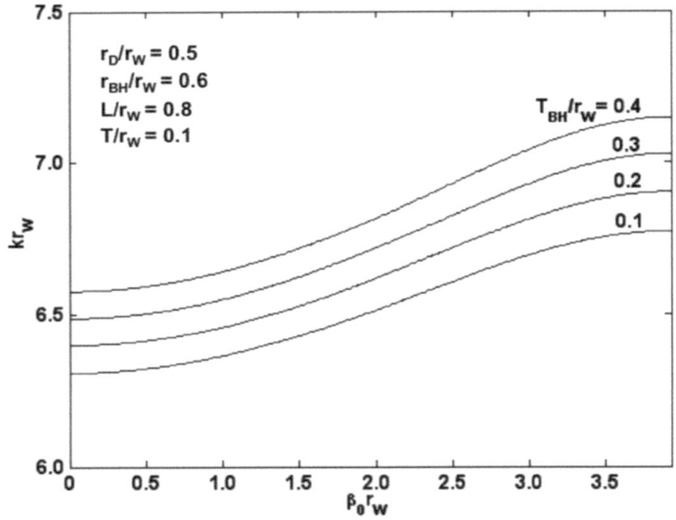

Figure 19.
TE_{01}-mode dispersion characteristics of the interwoven-disc-loaded circular waveguide taking the disc-thickness of bigger-hole-disc as the parameter [9, 14, 15].

radius (**Figure 21**) increases the waveguide cutoff frequency, and none of the parameters tailors the dispersion characteristics.

For the model-6, the variation of relative permittivity of the dielectric discs changes the lower and upper cutoff frequencies of the passband (**Figure 24**). Two lowest order azimuthally symmetric (TE_{01} and TE_{02}) modes are typically considered to study the performance of this model. With the increase of relative permittivity, the passband continuously decreases for the TE_{01} mode, and first decreases and then increases for the TE_{02} mode (**Figure 24**). Also, the variation in relative permittivity shapes of the dispersion characteristics of the structure, for both the

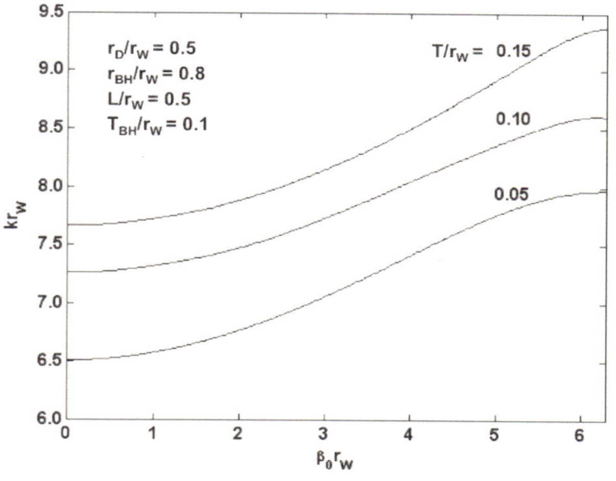

Figure 20.
TE_{01}-mode dispersion characteristics of the interwoven-disc-loaded circular waveguide taking the disc thickness of smaller-hole-disc as the parameter [9, 14, 15].

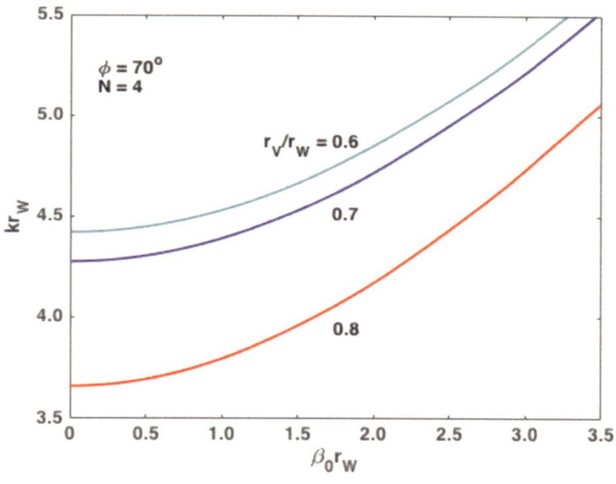

Figure 21.
TE_{01}-mode dispersion characteristics of the metal-vane-loaded circular waveguide taking the vane-inner-tip radius as the parameter [16–18].

modes TE_{01} and TE_{02}, however more significantly for the latter mode (**Figure 24**). The TE_{01} mode of the structure.

Exhibits fundamental forward wave (positive) dispersion characteristics irrespective of the value of relative permittivity (**Figure 24(a)**), however, the TE_{02} mode exhibits fundamental forward (positive) and backward (negative) wave dispersion characteristics, respectively, at higher and lower values of relative permittivity. This suggests that an appropriate selection of the value of relative permittivity in this structure would yield a straightened TE_{02} mode $\omega - \beta$ dispersion characteristics near low value of phase propagation constant for wideband coalescence with the beam-mode dispersion line and consequent wideband gyro-TWT performance (**Figure 24(b)**). Thus the introduction of the dielectric discs between metal discs in the a conventional metal disc-loaded waveguide, with lower values of relative permittivity for the TE_{01} mode and with higher values of relative permittivity for the TE_{02} mode enhances the frequency range of the straight line portion of

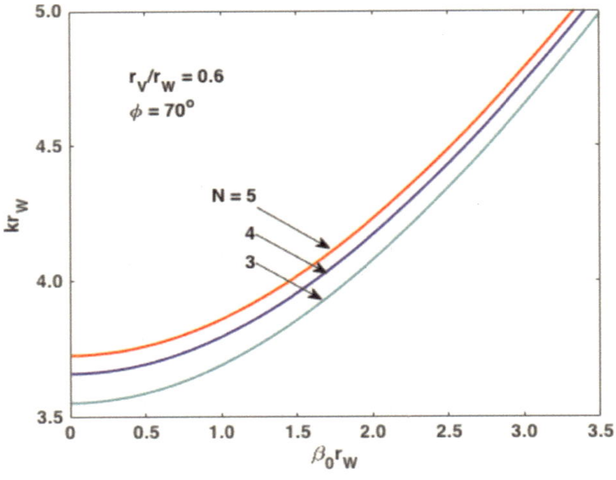

Figure 22.
TE_{01}-mode dispersion characteristics of the metal-vane-loaded circular waveguide taking the vane angle as the parameter [16–18].

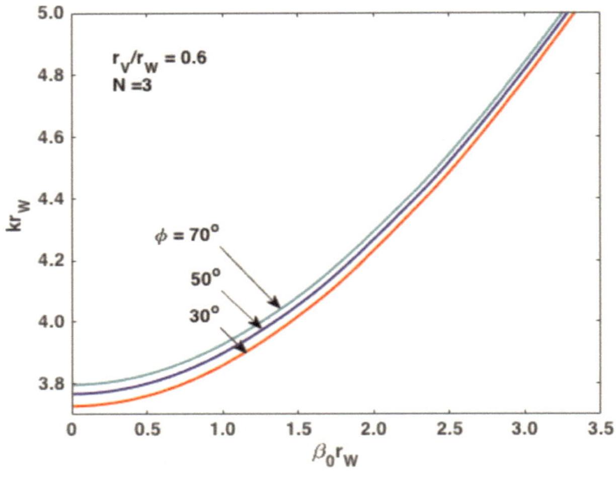

Figure 23.
TE_{01}-mode dispersion characteristics of the metal-vane-loaded circular waveguide taking the number of metal vanes as the parameter [16–18].

the $\omega - \beta$ dispersion characteristics, desired for wideband gyro-TWT performance (**Figure 25**) [19].

The lower and upper cutoff frequencies of the passband vary with thickness of dielectric disc $(L - T)/r_W$ taking structure periodicity constant such that the passband decreases with an increase in thickness of dielectric disc for both the TE_{01} and TE_{02} modes. The thickness of dielectric disc tailors the dispersion characteristics, however, more for the TE_{02} than for the TE_{01} mode, and the control is more prominent for thinner dielectric disc (**Figure 26**) [19]. The less effective parameters, the disc-hole radius, in tailoring the dispersion characteristics of a conventional disc-loaded waveguide [2, 5, 9–13], effectively controls the shape of the characteristics after introducing the dielectric discs between the metal discs, while, the control is more for the TE_{02} (**Figure 27(b)**) than for the TE_{01} (**Figure 27(a)**) mode, however the characteristics is little irregular for higher disc-hole radius

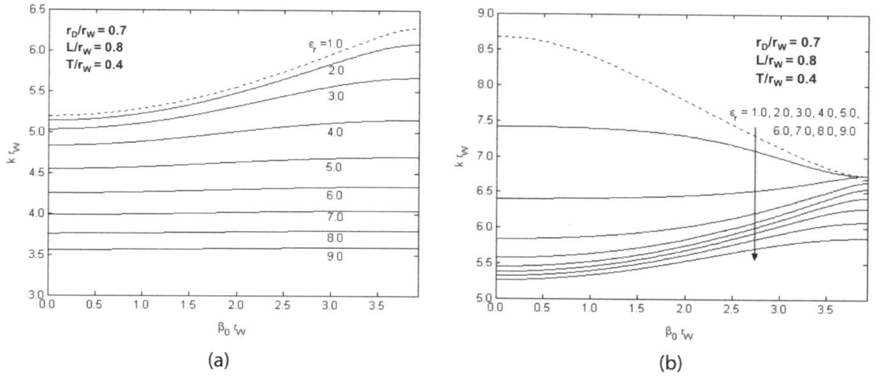

Figure 24.
TE_{01} (a) and TE_{02} (b) mode dispersion characteristics of a circular waveguide loaded with alternate dielectric and metal annular discs (model-6) taking relative permittivity as the parameter. The broken curves refer to a conventional metal disc-loaded circular waveguide [19].

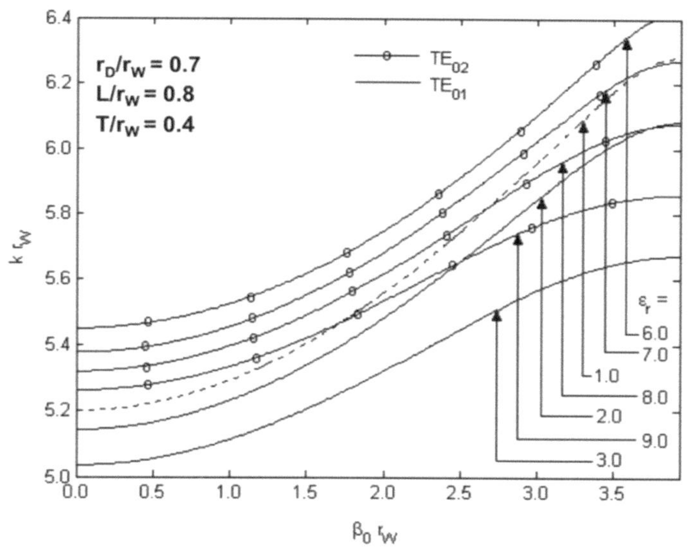

Figure 25.
Dispersion characteristics of a circular waveguide loaded with alternate dielectric and metal annular discs (model-6) for the selected dielectric disc relative permittivity values for the sake of comparison between the modes TE_{01} and TE_{02} with respect to the control of the shape of the dispersion characteristics. The broken curves referring to a conventional metal disc-loaded circular waveguide [19].

(**Figure 27**) [19]. Similar to a conventional all-metal disc-loaded waveguide (model-4), the structure periodicity is the most effective parameter for tailoring the dispersion characteristics of the structure with dielectric discs between the metal discs (model-6), for the TE_{01} and TE_{02} modes, more for the latter. The control of the structure periodicity in straightening the dispersion characteristics, as required for the desired wideband gyro-TWT performance, is enhanced by introducing the dielectric discs in the conventional disc-loaded waveguide, though not enhancing the frequency range of the straight line portion of the dispersion characteristics (**Figure 28**). In this model, a serious care is required while selecting the dielectric material because a heavily dielectric-loaded structure depresses the dispersion characteristics to the slow-wave region (below the velocity of light line in $\omega - \beta$ characteristics).

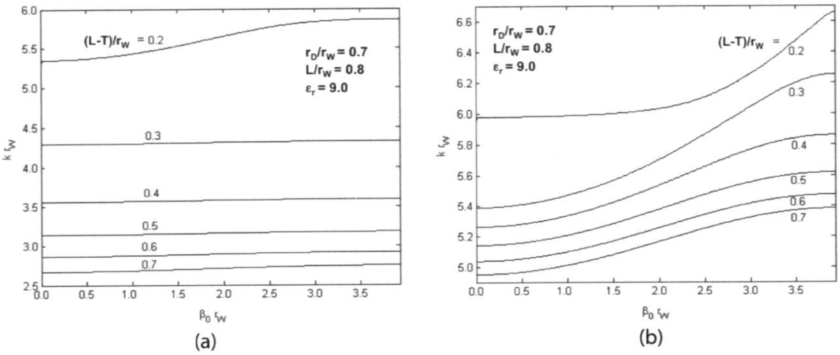

Figure 26.
TE_{01} (a) and TE_{02} (b) mode dispersion characteristics of a circular waveguide loaded with alternate dielectric and metal annular discs (model-6) taking dielectric disc thickness as the parameter for a constant structure periodicity [19].

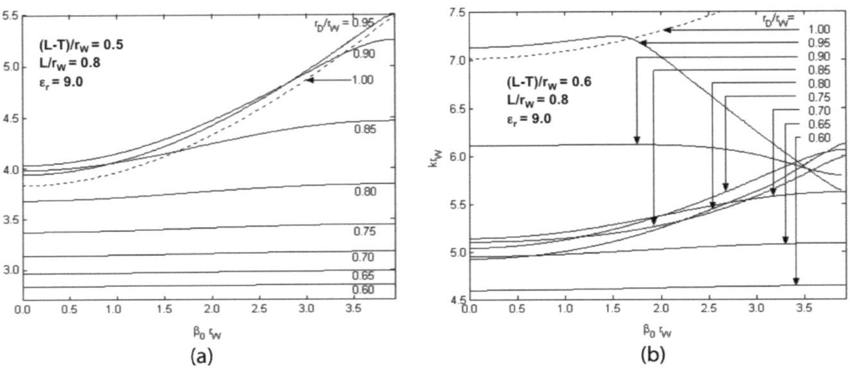

Figure 27.
TE_{01} (a) and TE_{02} (b) mode dispersion characteristics of a circular waveguide loaded with alternate dielectric and metal annular discs (model-6) taking the disc-hole radius as the parameter. The broken curves refer to the special case of a conventional smooth wall circular waveguide [19].

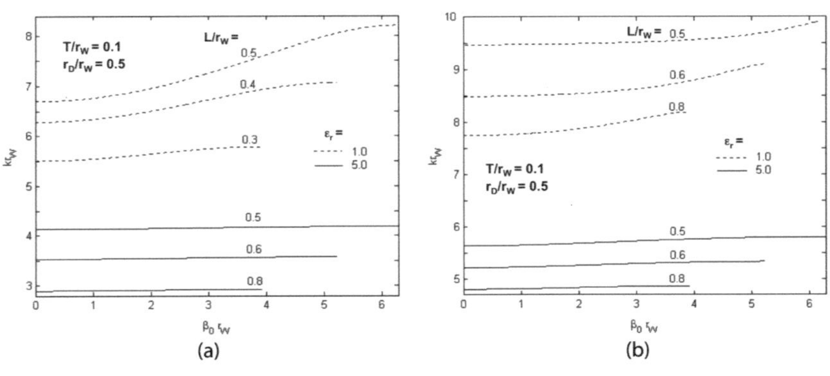

Figure 28.
TE_{01} (a) and TE_{02} (b) mode dispersion characteristics of a circular waveguide loaded with alternate dielectric and metal annular discs (model-6) taking structure periodicity as the parameter. The broken curves refer to a conventional metal disc-loaded circular waveguide [19].

We choose first three lowest order azimuthally symmetric modes in the composite (dielectric and metal) loaded structures for exploring the effect of structure parameter on dispersion characteristics while we chose only the lowest order azimuthally symmetric mode in all-metal variants of the axially periodic structure. The increase of the relative permittivity of the dielectric discs in model-7 reduces the lower and upper cutoff frequencies, however not equally therefore shortens the passband for the TE_{01} mode with shift of the mid-frequency of the passband towards lower value (**Figure 29(a)**). With the increase of the relative permittivity of the dielectric discs, the lower and upper cutoff frequencies shift to lower value for the TE_{01} and TE_{02} modes. The quantitatively the shift in upper cutoff frequency is higher than that of lower cutoff frequency for the TE_{01}, which in turn shortens the passband (**Figure 29(a)**), however, the shift in lower and upper cutoff frequencies are almost equal for the TE_{02} mode, effectively the passband does not change (**Figure 29(b)**). Interestingly for the TE_{02} mode the introduction of dielectric discs converts the fundamental backward mode (the zero group velocity follows to take negative values then again zero and further positive) into fundamental forward mode (the zero group velocity follows to take positive values then again zero and further negative) (**Figure 29(b)**). Thus, introduction of dielectric discs into the conventional disc-loaded waveguide turns the negative dispersion into positive. In absence as well as in presence of the dielectric discs, the TE_{03} mode dispersion characteristics of the disc-loaded waveguide represents the fundamental backward mode, in which the increase of the relative permittivity of the dielectric discs shifts the lower cutoff frequency more than that of upper cutoff frequency, and thus widens the passband for lower relative permittivity. For higher relative permittivity value the lower cutoff frequency remains unchanged and upper cutoff frequency shifts to lower value with the increase of the relative permittivity of the dielectric discs and shortens the passband (**Figure 29(c)**) [20, 21].

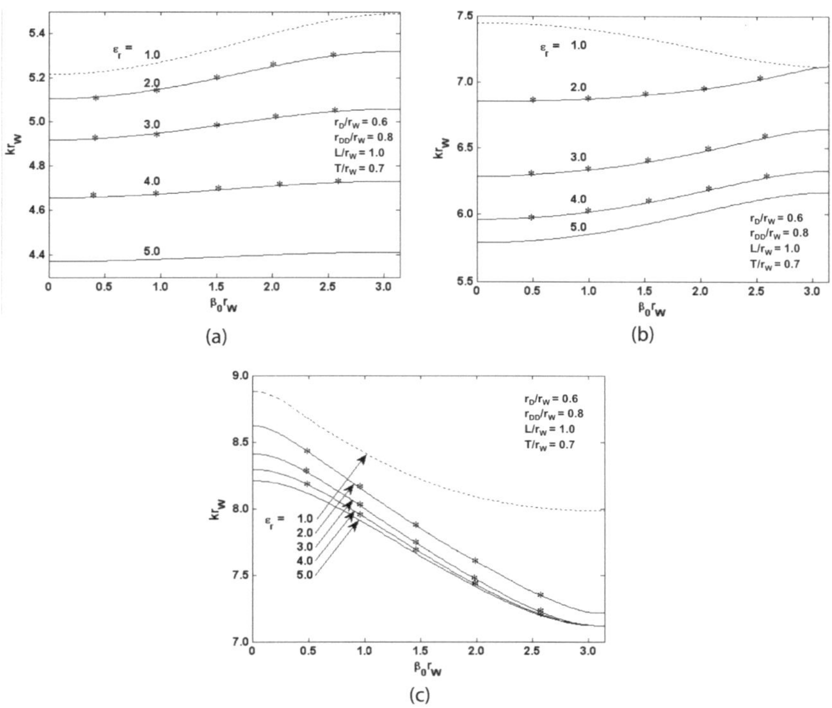

Figure 29.
TE_{01} (a), TE_{02} (b) and TE_{03} (c) mode dispersion characteristics of the alternate dielectric and metal disc-loaded circular waveguide taking relative permittivity of dielectric disc as the parameter [20, 21]. The broken curve refers to the conventional disc-loaded circular waveguide [2, 5, 9–13].

The increase of dielectric disc radius in the model-7 taking constant metal disc radius shifts the lower and upper cutoff frequencies of the TE_{01} mode up, and the passband increases due to lesser the shift in lower cutoff frequency (**Figure 30**). Similarly, the passband of the TE_{02} mode increases with the increase of dielectric disc radius. Here, it is interesting to note that for the taken structure parameters ($r_D/r_W = 0.6$, $L/r_W = 1.0$, $T_{DD}/r_W = 0.3$ and $\varepsilon_r = 5.0$) the frequency shift is maximum for r_{DD}/r_W equal to 0.8–0.9, and minimum for 0.7–0.8 (**Figure 30(b)**). For the lower and higher values of inner dielectric disc radius, the TE_{03} mode dispersion characteristics of the model-7 are fundamental forward (positive) and backward (negative) modes respectively. Thus, there is a possibility of getting straight-line dispersion characteristics parallel to phase propagation constant axis, i.e., zero group velocity line (**Figure 30(c)**) [20, 21]. The increase of periodicity of the alternate dielectric and metal disc-loaded circular waveguide (model-7) reduces the passband and both the lower and upper cutoff frequencies with higher relative reduction in upper cutoff frequency than that of lower (**Figure 31**) for the chosen three azimuthally symmetric modes TE_{01}, TE_{02} and TE_{03}. The TE_{01} and TE_{02} modes are fundamental forward and the TE_{03} is fundamental backward [20, 21]. The change of dielectric disc thickness does not much tailor the dispersion characteristics and the passband (**Figure 32**), however, the decrease of the dielectric disc thickness or the increase of metal disc thickness shifts the passband to lower frequency side for the TE_{01} and TE_{02} modes (**Figure 31(a)** and **(b)**) and least change occur to the TE_{03} mode. In very precise observation, the lower cutoff frequency of the TE_{03} mode is insensitive and the upper cutoff frequency first decreases and then increases with decrease of dielectric disc thickness or with increase of metal disc thickness (**Figure 32**) [20, 21].

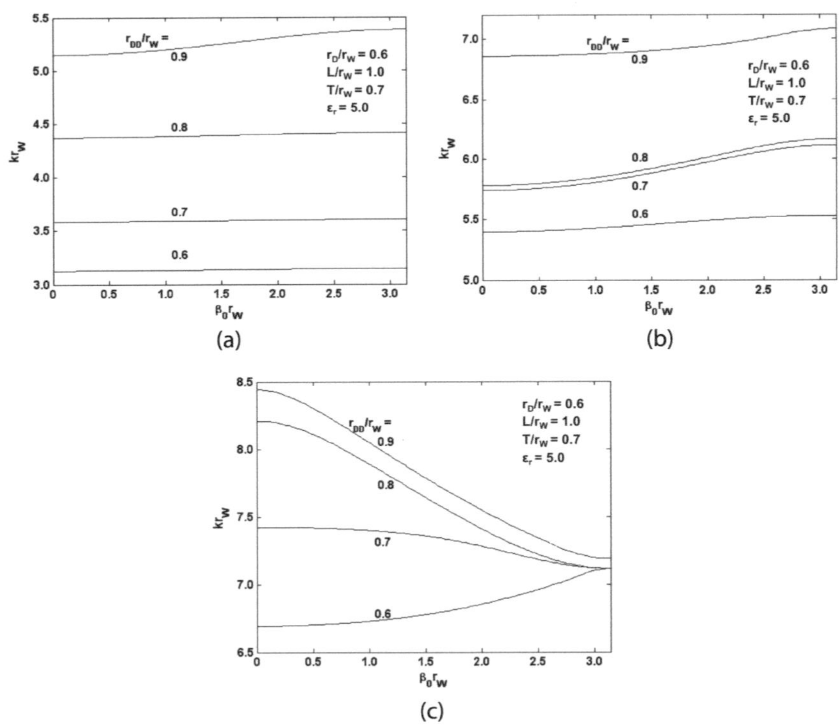

Figure 30.
TE_{01} (a), TE_{02} (b) and TE_{03} (c) mode dispersion characteristics of the alternate dielectric- and metal disc-loaded circular waveguide (model-7) taking inner radius of dielectric disc as the parameter [20, 21].

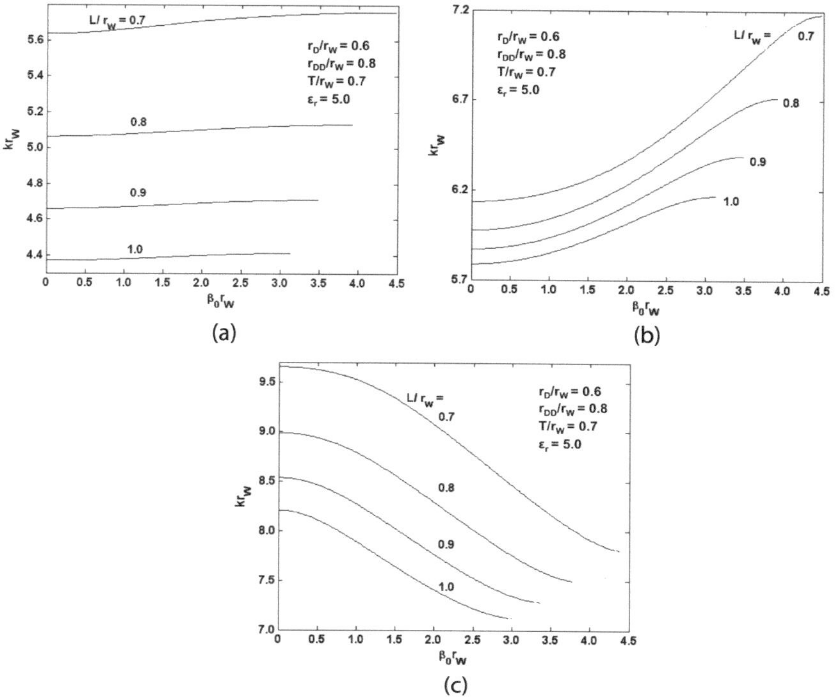

Figure 31.
TE_{01} (a), TE_{02} (b) and TE_{03} (c) mode dispersion characteristics of the alternate dielectric and metal disc-loaded circular waveguide (model-7) taking structure periodicity as the parameter [20, 21].

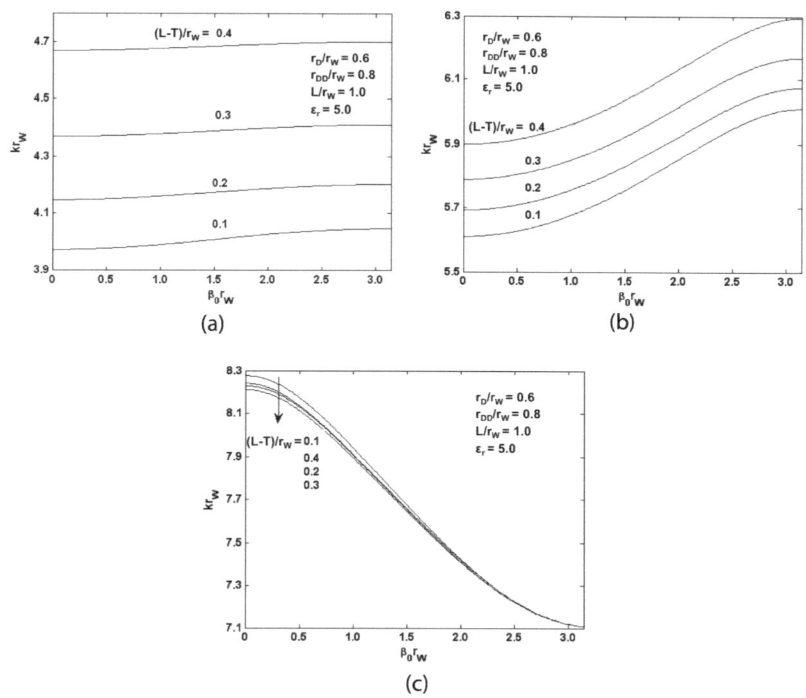

Figure 32.
TE_{01} (a), TE_{02} (b) and TE_{03} (c) mode dispersion characteristics of the alternate dielectric and metal disc-loaded circular waveguide (model-7) taking thickness of dielectric disc as the parameter [20, 21].

Although the geometrical parameters do not tailor the dispersion characteristics of the only metal vane-loaded waveguide [16–18], however, it does for composite-loaded structure (model-8). The radial dimensions (**Figure 33**) are less sensitive in tailoring the dispersion characteristics of the dielectric and metal vanes than their angular dimensions (**Figure 34**), relative permittivity (**Figure 35**) and number of vanes (**Figure 36**). However, the waveguide cutoff (eigenvalue) of the structure depends on all these parameters (**Figures 33-36**). Thus, one may choose angular dimensions, relative permittivity and number of vanes for tailoring the dispersion characteristics and the radial parameter to control the waveguide cutoff frequency [21].

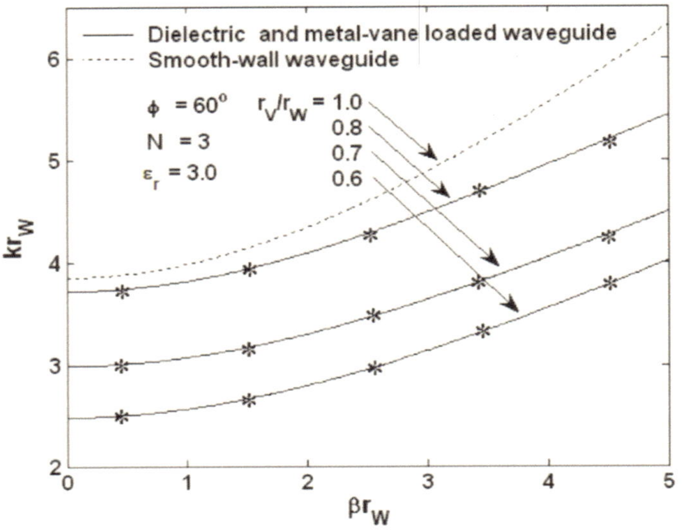

Figure 33.
TE_{01}-mode dispersion characteristics of a circular waveguide loaded with composite alternate dielectric and metal vanes taking the vane inner-tip radius as the parameters. The broken curve refers to a smooth-wall waveguide (free from dielectric and metal vanes) and the star (*) marker refers representative points obtained using HFSS [21].

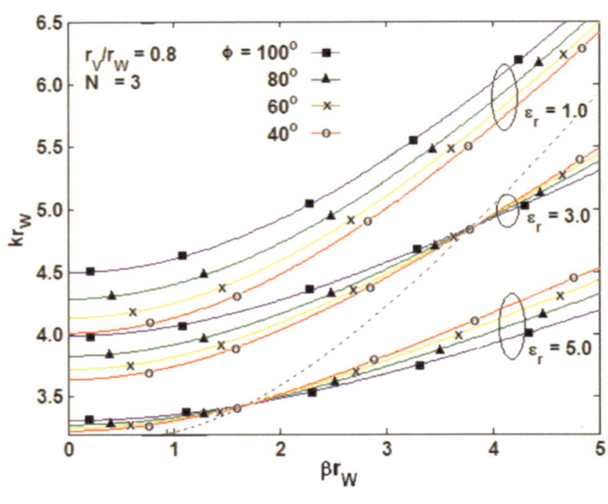

Figure 34.
TE_{01}-mode dispersion characteristics of a circular waveguide loaded with dielectric and metal vanes taking metal vane angle as the parameter. The broken curve represents the locus of the crossover point, which is same as the dispersion characteristics of the smooth-wall circular waveguide of radius r_V [21].

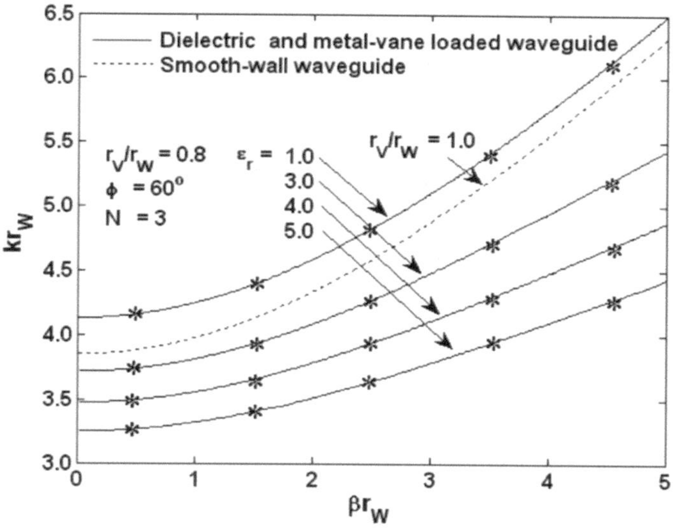

Figure 35.
TE_{01}-mode dispersion characteristics of a circular waveguide loaded with composite alternate dielectric and metal vanes, taking the relative permittivity of dielectric vanes as the parameters, along with the corresponding characteristics of a smooth-wall waveguide (free from dielectric and metal vanes) (broken curve) and the typical representative points of the characteristics obtained by simulation (HFSS) ((star) marker) [21]. $\varepsilon_r = 1$ represents the characteristics of a waveguide loaded with metal vanes alone (**Figure 6**) [16–18].*

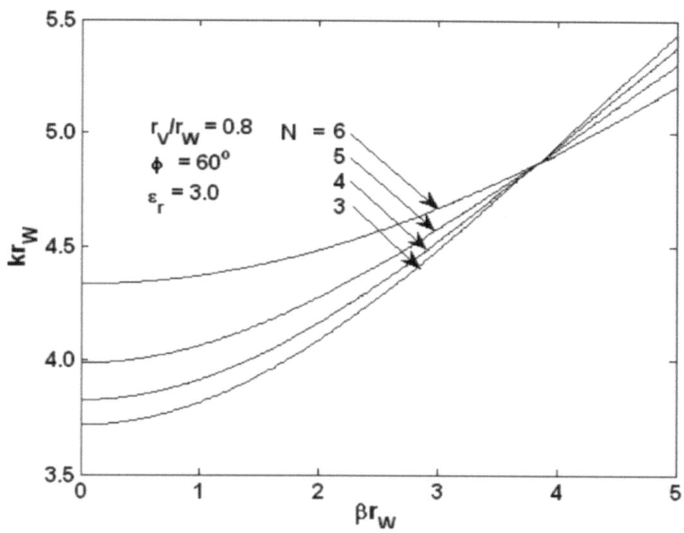

Figure 36.
TE_{01}-mode dispersion characteristics of a circular waveguide loaded with dielectric and metal vanes taking number of vanes as the parameter [21].

A crossover point in the dispersion characteristics appears with varying metal vane angle (**Figure 34**) or varying number of vanes (**Figure 36**). This crossover point sifts to another location for another value of relative permittivity. The locus of such crossover points for varying relative permittivity overlaps with the dispersion plot of a smooth-wall circular waveguide of radius equal to the vane tip radius (**Figure 34**). Below the crossover point, the increase of metal cross-section (either by increasing metal vane angle or by increasing number of vanes) in the cross-section of the waveguide elevates and that of dielectric depresses the dispersion characteristics, and vice versa above the crossover point. This clearly represents the behavior of model-8 as a smooth-wall waveguide of wall radius equal to the vane tip radius at the crossover point and change of metal cross-section (either by changing metal vane angle or by changing number of vanes) in the cross-section of the waveguide affects the dispersion characteristics away from the crossover (**Figures 34** and **36**) [21].

6. Conclusion

The waveguide is inherently a high pass filter with a lower cut-off frequency, a signal of above this frequency will be allowed to propagate or travel through the waveguide. Different kind of loading, such as metal and/or dielectric, has been suggested in the published literature for altering the propagation (dispersion) characteristics. In the present chapter a number of loaded structures are studied with circular cross-sections. All the structure parameters are varied to explore the sensitivity over the dispersion characteristics. In general, the axial periodicity results in periodic dispersion characteristics with a lower and an upper cut-off frequency, which makes the guiding structure a bandpass structure. However, such a characteristic is not reported in literature for the azimuthal periodic structures. The bandpass characteristic arises due to the shaping of the dispersion characteristics. Therefore, the dispersion shaping is only possible with axial periodicity and not with the azimuthal periodicity in all metal structure. However, the azimuthal periodicity in all metal waveguide structure shifts the cut-off frequency over the frequency scale. The dispersion characteristics of various loaded structures have been explored. It has been the interest to study the change in the lower and upper cut-off frequencies and in the passband. The sensitivity of the structure (geometry) parameters on the lower and upper cut-off frequencies and the extent of passband are also included. In case of conventional disc-loaded structure, the periodicity is found to be the most sensitive parameter for dispersion shaping and the disc-hole radius is the most sensitive parameter for shifting the dispersion characteristics over the frequency axis. In interwoven-disc-loaded circular waveguide, the structure periodicity and the disc-thickness of bigger-hole-disc are, respectively, the most and the least sensitive parameter for controlling the passband as well as shape of the dispersion characteristics. Insertion of metal vanes to a circular waveguide does not shape the dispersion characteristics but increases the waveguide cutoff frequency. On the other hand, introduction of dielectric discs into the conventional disc-loaded waveguide turns the negative dispersion into positive. In case of composite vane-loaded structure, the angular dimensions, relative permittivity and number of vanes tailor the dispersion characteristics and the radial parameter controls the waveguide cutoff frequency.

Author details

Vishal Kesari
Microwave Tube Research and Development Centre, Defence Research and Development Organisation, Bangalore, India

*Address all correspondence to: vishal_kesari@rediffmail.com

References

[1] Watkins DA. Topics in Electromagnetic Theory. New York: John Wiley; 1958

[2] Kesari V, Basu BN. High Power Microwave Tubes: Basics and Trends. San Rafael (CA)/Bristol: IOP Concise Physics, Morgan & Claypool Publishers/ IOP Publishing; 2018

[3] Collin RE. Foundations for Microwave Engineering. 2nd ed. USA: Wiley-IEEE Press; 2001

[4] Basu BN. Electromagnetic Theory and Applications in Beam-Wave Electronics. Singapore: World Scientific; 1996

[5] Kesari V. Analysis of Disc-Loaded Circular Waveguides for Wideband Gyro-TWTs. Koln: Lambert Academic Publishing AG & Co.; 2009

[6] Fliflet AW. Linear and nonlinear theory of the Doppler-shifted cyclotron resonance maser based on TE and TM waveguide modes. International Journal of Electronics. 1986;**61**(6):1049-1080

[7] Rao SJ, Jain PK, Basu BN. Broadbanding of gyro-TWT by dispersion shaping through dielectric loading. IEEE Transactions on Electron Devices. 1996;**43**:2290-2299

[8] Choe JY, Uhm HS. Theory of gyrotron amplifiers in disc or helix-loaded waveguides. International Journal of Electronics. 1982;**53**:729-741

[9] Kesari V, Basu BN. Analysis of some periodic structures of microwave tubes: Part II: Analysis of disc-loaded fast-wave structures of gyro-traveling-wave tubes. Journal of Electromagnetic Waves and Applications. 2018;**32**(4): 1-36

[10] Kesari V, Jain PK, Basu BN. Approaches to the analysis of a disc loaded cylindrical waveguide for potential application in wideband gyro-TWTs. IEEE Transactions on Plasma Science. 2004;**32**(5):2144-2151

[11] Kesari V, Jain PK, Basu BN. Analysis of a circular waveguide loaded with thick annular metal discs for wideband gyro-TWTs. IEEE Transactions on Plasma Science. 2005;**33**(4):1358-1365

[12] Kesari V, Jain PK, Basu BN. Analysis of a disc-loaded circular waveguide for interaction impedance of a gyrotron amplifier. International Journal of Infrared and Millimeter Waves. 2005; **26**(8):1093-1110

[13] Kesari V, Jain PK, Basu BN. Modal analysis of a corrugated circular waveguide for wideband potential in gyro-devices. International Journal of Microwave and Optical Technology. 2007;**2**(2):147-152

[14] Kesari V, Keshari JP. Interwoven-disc-loaded circular waveguide for a wideband gyro-traveling-wave tube. IEEE Transaction on Plasma Science. 2013;**41**(3):456-460

[15] Kesari V, Keshari JP. Propagation characteristics of a variant of disc-loaded circular waveguide. Progress in Electromagnetic Research. 2012;**26**: 23-37

[16] Singh G, Ravi Chandra SMS, Bhaskar PV, Jain PK, Basu BN. Analysis of dispersion and interaction impedance characteristics of an azimuthally-periodic vane-loaded cylindrical waveguide for a gyro-TWT. International Journal of Electronics. 1999;**86**:1463-1479

[17] Agrawal M, Singh G, Jain PK, Basu BN. Analysis of a tapered vane loaded broad-band gyro-TWT. IEEE Transactions on Plasma Science. 2001; **29**:439-444

[18] Shrivastava UA. Small-signal theories of harmonic gyrotron and peniotron amplifiers and oscillators [thesis]. Utah: University of Utah; 1985

[19] Kesari V, Jain PK, Basu BN. Modeling of axially periodic circular waveguide with combined dielectric and metal loading. Journal of Physics D: Applied Physics. 2005;**38**:3523-3529

[20] Kesari V, Keshari JP. Analysis of a circular waveguide loaded with dielectric and metal discs. Progress in Electromagnetic Research. 2011;**111**: 253-269

[21] Kesari V. Analysis of alternate dielectric and metal vane loaded circular waveguide for a wideband gyro-TWT. IEEE Transactions on Electron Devices. 2014;**61**(3):915-920

Chapter 14

Terahertz Sources, Detectors, and Transceivers in Silicon Technologies

Li Zhuang, Cao Rui, Tao Xiaohui, Jiang Lihui and Rong Dawei

Abstract

With active devices lingering on the brink of activity and every passive device and interconnection on chip acting as potential radiator, a paradigm shift from "top-down" to "bottom-up" approach in silicon terahertz (THz) circuit design is clearly evident as we witness orders-of-magnitude improvements of silicon THz circuits in terms of output power, phase noise, and sensitivity since their inception around 2010. That is, the once clear boundary between devices, circuits, and function blocks is getting blurrier as we push the devices toward their limits. And when all else fails to meet the system requirements, which is often the case, a logical step forward is to scale these THz circuits to arrays. This makes a lot of sense in the terahertz region considering the relatively efficient on-chip THz antennas and the reduced size of arrays with half-wavelength pitch. This chapter begins with the derivation of conditions for maximizing power gain of active devices. Discussions of circuit topologies for THz sources, detectors, and transceivers with emphasis on their efficacy and scalability ensue, and this chapter concludes with a brief survey of interface options for channeling THz energy out of the chip.

Keywords: terahertz, detector, source, transceiver, silicon

1. Introduction

The significance of terahertz electronics is self-evident for readers of this book. The general consensus among silicon THz circuit designers (!) is that silicon will be the dominant technology for the lower end of the THz spectrum (300 GHz to around 1 THz) in light of recent breakthroughs of silicon circuits in terms of effective isotropic radiated power (EIRP), phase noise, and receiver sensitivity. For many applications, silicon circuits are on par or even superior to III/V compound technologies and optical-based techniques in this frequency range now. This chapter aims to introduce the reader to the fascinating world of silicon THz circuit design through a step-by-step approach: We examine conditions for extracting the most power gain out of a given active device. Popular topologies for silicon sources, detectors, and transceivers are discussed next, and this chapter concludes with a brief survey of THz interface options for efficient energy transfer between circuits and the outside world.

2. How to have THz power and radiate it too

Due to the excessive loss and scarcity of power gain for silicon devices in the THz region, one should strive to extract the most power out of a given device during the whole design phase. This involves making sure that the device is working under the optimum condition (i.e., the device is embedded in the right impedance environment for maximum power gain), the topology of the circuit is optimum for the intended application, and the power is transferred from the circuit through the most efficient interface. This section gives an overview of these areas.

2.1 Power gain maximization for a given active device

The active devices in THz circuits are connected to the rest of the circuits through passive elements, such as capacitors, inductors, and transmission lines. The overall circuit performance is decided both by the active device and these passive elements. Thus, to maximize the circuit performance, a "divide-and-conquer" approach is the logical choice. That is, we first find the "best" active device in a given technology under certain constraints such as power consumption or noise performance. We then decide the best passive network into which the device should be embedded. The problem is there is no such thing as "pure" active device; passive elements are always present in a given active device. Mason [1] has thus defined a figure of merit for active devices:

$$U = \frac{|Y_{21} - Y_{12}|^2}{4(G_{11}G_{22} - G_{12}G_{21})} \tag{1}$$

G_{ij} is the real part of Y_{ij} in Eq. (1). The above FOM is called Mason's invariant U, since it is invariant to passive embedding environments that are linear, lossless, and reciprocal [1, 2].

A device is active if its U is larger than one, which means this device is capable of providing real power. The maximum oscillation frequency (f_{max}) of a device is defined as the frequency where its U equals one, i.e., beyond which frequency it is no longer active. The maximum power gain of this device embedded in the two-port also drops to unity at the maximum oscillation frequency (f_{max}).

U is also the maximum power gain of the device after unilateralization, that is, when Y_{12} is made to zero. Generally speaking, higher U means higher power gain at a given frequency.

For a given two-port shown in **Figure 1(b)**, the power gain is defined as

$$G_p = \frac{P_L}{P_{IN}} = \frac{-0.5 \times \mathrm{Re}(V_2 \times I_2^*)}{0.5 \times \mathrm{Re}(V_1 \times I_1^*)} = \frac{\mathrm{Re}(V_2 \times V_2^* \times Y_L^*)}{\mathrm{Re}(V_1 \times V_1^* \times Y_{IN}^*)} = |A_V|^2 \frac{\mathrm{Re}(Y_L^*)}{\mathrm{Re}(Y_{IN}^*)} \tag{2}$$

P_L and P_{IN} are the real power delivered to the load and to the two-port. V_1, I_1, and V_2, I_2 in **Figure 1** are the voltage and current at port 1 and port 2, respectively. A_V is the voltage gain of the two-port. Y_s and Y_L represent the source and load admittance presented to the device. I_s and Y_s form the Norton equivalent circuit of the signal source.

For an unconditionally stable two-port (it does not oscillate for any passive load and source admittance) at a given frequency, the power gain could be maximized by biconjugate matching at the input and output. Conjugate matching is achieved when the load admittance is equal to the conjugate of the source admittance at a given node; biconjugate matching means that this condition is satisfied both at the input and the output port. For a given two-port, its maximum power gain is

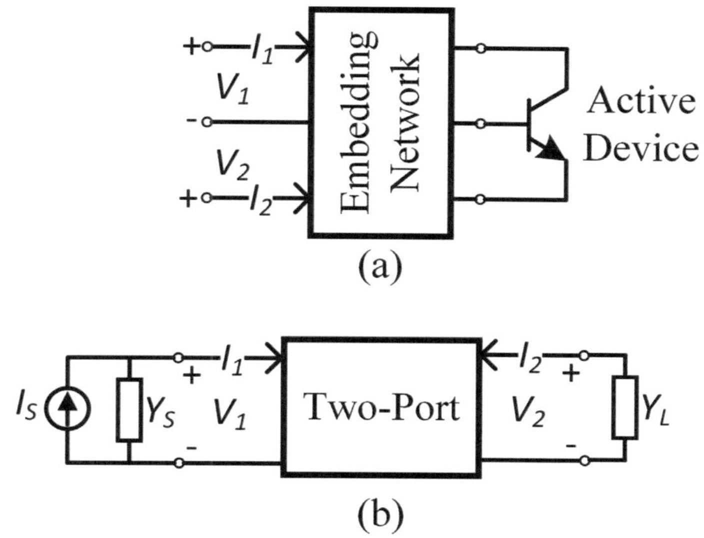

Figure 1.
The two-port representation of an embedded active device. (a) An active device embedded in a linear, lossless, and reciprocal passive network resulting in a two-port and (b) the two-port interfacing to signal source and load.

$$G_{\max} = \left|\frac{Y_{21}}{Y_{12}}\right| \times \left(k - \sqrt{k^2 - 1}\right) \tag{3}$$

where K is the stability factor defined as

$$\frac{2 \times \mathrm{Re}(Y_{11}) \times \mathrm{Re}(Y_{22}) - \mathrm{Re}(Y_{12}Y_{21})}{|Y_{12}Y_{21}|} \tag{4}$$

Biconjugate matching is possible when K is equal or greater than 1. For a given two-port, it is unconditionally stable when the following conditions are simultaneously satisfied:

$$g_{11} \geq 0 \tag{5}$$

$$g_{22} \geq 0 \tag{6}$$

$$K \geq 1 \tag{7}$$

Keep in mind that unlike U, which is invariant to the embedding network, G_{max} is sensitive to its environment. We can modify the embedding environment to make G_{max} larger. It can be shown that the maximum G_{max} for a given device is $\left(\sqrt{U} + \sqrt{U-1}\right)^2$. For a detailed discussion about U and G_{max}, please refer to [3–6] which present ways of designing the embedding network for maximizing G_{max} while maintaining stability under process variations. The basic idea is to utilize feedback to generate negative resistance, such as adding capacitive degeneration to CE (common emitter) amplifiers or adding inductance to the base node of CB (common base) amplifiers.

Another important THz circuit is oscillators. Here, we need to make a distinction between amplifiers and oscillators. The former is one kind of driven circuit, the output of which is controlled by its input. Oscillators belong to the group of autonomous circuit, which generates time-varying signals without time-varying stimulus.

By definition, amplifiers operate below f_{max} to provide power gain. For oscillators, the situation is more complicated: The power gain of active devices within the oscillator should be greater than unity to start the oscillation. As the oscillation amplitude grows, the power gain of the two-port (including device parasitic and loading) gradually compresses to unity when the circuit reaches steady-state oscillation. Thus, it could be argued that devices within oscillators operate at the f_{max} of its embedded two-port in steady-state oscillation. To take the gain compression into account, large-signal parameters should be used for the analysis.

Unlike analog circuit designers who deal exclusively with voltage and current gains, microwave circuit designers are more comfortable with power gain. Momeni [7] has thus shown a refreshing view about the optimum voltage gain and phase shift for a given two-port to oscillate at f_{max}:

$$A_V = \frac{V_2}{V_1} = A_{OPT} = \sqrt{\frac{G_{11}}{G_{22}}} \tag{8}$$

$$\phi = \angle \frac{V_2}{V_1} = \phi_{OPT} = (2k+1)\pi - \angle\left(Y_{12} + Y_{21}^*\right) \tag{9}$$

A_{OPT} and Φ_{OPT} are the optimum voltage gain and phase shift for the two-port at f_{max}. Equations (8) and (9) are derived assuming no clipping to the power rails occurs inside the circuit. If clipping happens, another set of equations apply for A_{OPT}; Φ_{OPT} remains the same [8, 9].

It can be shown that biconjugate matching automatically satisfies Eqs. (8) and (9) at f_{max}.

Under biconjugate matching, G_p reaches unity at f_{max}. Equation (2) can be rewritten as

$$|A_V| = |A_{OPT}| = \sqrt{\frac{\mathrm{Re}\left(Y_{IN}^*\right)}{\mathrm{Re}\left(Y_L^*\right)}} = \sqrt{\frac{\mathrm{Re}\left(Y_{S,OPT}\right)}{\mathrm{Re}\left(Y_{L,OPT}^*\right)}} = \sqrt{\frac{\mathrm{Re}\left(Y_{S,OPT}\right)}{\mathrm{Re}\left(Y_{L,OPT}\right)}} \tag{10}$$

$Y_{S,OPT}$ and $Y_{L,OPT}$ are the optimal source and load admittance for biconjugate matching:

$$Y_{S,OPT} = \frac{|Y_{12}Y_{21}| \times \sqrt{K^2-1} + j \times \left(\mathrm{Im}(Y_{12}Y_{21}) - 2 \times \mathrm{Re}(Y_{22}) \times \mathrm{Im}(Y_{11})\right)}{2\mathrm{Re}(Y_{22})} \tag{11}$$

$$Y_{L,OPT} = \frac{|Y_{12}Y_{21}| \times \sqrt{K^2-1} + j \times \left(\mathrm{Im}(Y_{12}Y_{21}) - 2 \times \mathrm{Re}(Y_{11}) \times \mathrm{Im}(Y_{22})\right)}{2\mathrm{Re}(Y_{11})} \tag{12}$$

By substituting Eqs. (11) and (12) into (10), we have

$$|A_{OPT}| = \sqrt{\frac{G_{11}}{G_{22}}} \tag{13}$$

We are now in the position to derive Φ_{OPT}. First, we have to define the net power flowing into the two-port shown in **Figure 1(b)**:

$$P = V_1 \times I_1^* + V_2 \times I_2^* \tag{14}$$

At f_{max}, the power gain of the two-port drops to unity, which means the real power consumed by the two-port equals the real power generated. Thus, the real part of Eq. (14) equals zero:

$$P_R = |V_1|^2 \times \left[G_{11} + G_{22} \times |A_V|^2 + \mathrm{Re}\left(Y_{12}^* A_V^* + Y_{21}^* A_V\right)\right] = 0 \tag{15}$$

Substituting Eq. (13) into (15), we have

$$2\sqrt{G_{11}G_{22}} + \mathrm{Re}\left[\left(Y_{12} + Y_{21}^*\right)e^{j\phi}\right] = 0 \tag{16}$$

Since the Mason's invariant equals unity at f_{max}, Eq. (1) could be rearranged as

$$\frac{\left|Y_{12} + Y_{21}^*\right|^2}{4G_{11}G_{22}} = 1 \tag{17}$$

Equation (16) thus equals

$$2\sqrt{G_{11}G_{22}} \times \left[1 + \cos\left(\phi + \angle\left(Y_{12} + Y_{21}^*\right)\right)\right] = 0 \tag{18}$$

We have

$$\phi_{OPT} = (2k+1)\pi - \angle\left(Y_{12} + Y_{21}^*\right) \tag{19}$$

Since the above derivation is restricted to f_{max}, it would be interesting to observe the possible deviations of Eqs. (8) and (9) with respect to the two-port's voltage gain and phase shift under biconjugate matching when operating at frequency below f_{max}. A SiGe HBT transistor is used as an example. The emitter width and length of the transistor is 0.12 and 2.5 μm, and the emitter current density is biased for peak f_{max}. The source and load admittance are adjusted for biconjugate matching under each frequency evaluated (**Figures 2** and **3**).

It is clear that Eqs. (8) and (9) are only strictly valid at f_{max}, but the optimum phase shift calculated with Eq. (9) tracks reasonably well with the results obtained with biconjugate matching over a wide frequency range.

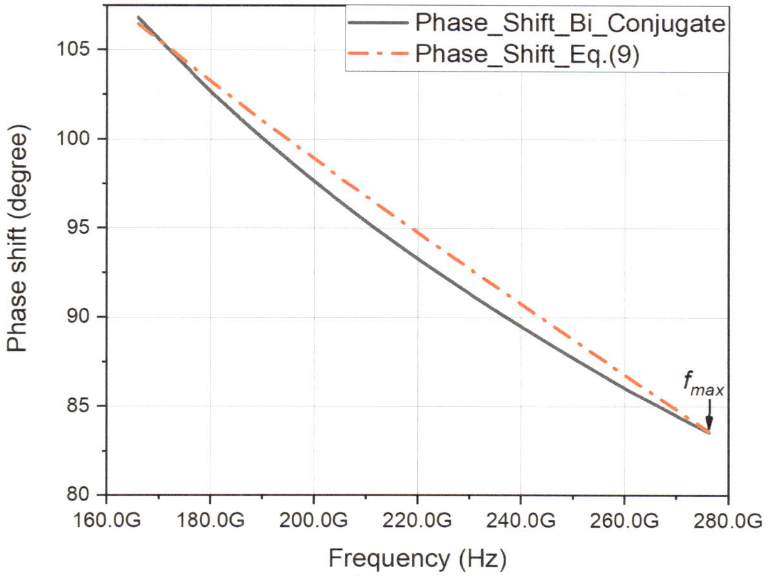

Figure 2.
Comparison of the phase shift of a SiGe HBT transistor under biconjugate matching and the optimum phase shift calculated with Eq. (9).

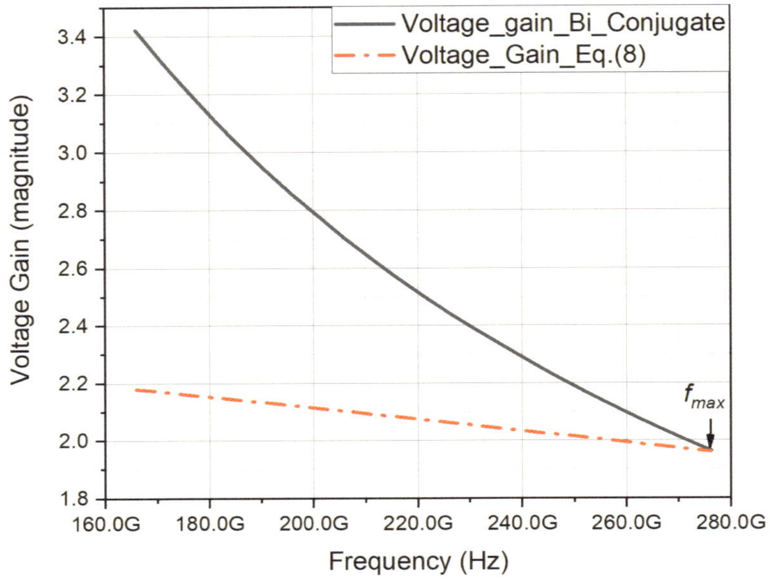

Figure 3.
Comparison of the voltage gain of a SiGe HBT transistor under biconjugate matching and the optimum voltage gain calculated with Eq. (8).

2.2 Circuit topology for THz sources, detectors, and transceivers

Among the many potential benefits offered by THz application, the large bandwidth available is the most obvious one. However, a lot of design issues need to be addressed in order to truly harness this bandwidth potential. We discuss this problem in terms of *SNR* at the receiver:

$$SNR \times B = \frac{P_r}{kTBF} = \left(\frac{\lambda}{4\pi d}\right)^2 \times \frac{P_t G_t G_r}{kTFL} \tag{20}$$

B is the receiver bandwidth. For communications, we would like *B* to scale with frequency. For imaging applications, sometimes *B* is not that important once it reaches certain value as only the range resolution scales with $1/B$. The cross-range resolution scales inversely with wavelength λ. Thus, we lump SNR and B together for trade-offs. P_t and P_r are the power transmitted and received by the transmitter (Tx) and receiver (Rx). G_t and G_r are the gain of the transmitting and receiving antenna. *K* is the Boltzmann constant. *T* is the ambient temperature. *d* is the distance between the transmitter and receiver. *F* is the noise factor of the receiver. *L* represents the loss in the Tx and Rx system, and we assume it scales with $f^{0.5}$to f here.

Assuming constant drive power, P_t for terahertz transmitters approximately follows a $P_t \propto 1/f^2$ relationship since the maximum unilateral gain *U* follows a -20dB/decade slope above $f_{max}/2$. *F* generally scales with f. $SNR \times B$ then scales with $1/f^{5.5}$ to $1/f^6$, which is a really disheartening result. This partly explains why the current silicon THz links are usually demonstrated with link distances ranging from centimeters to meters.

Before leaving this chapter in despair, we can try to manipulate Eq. (20) a little bit further:

$$SNR \times B = \frac{1}{16\pi^2 k} \times \frac{\lambda^2}{Td^2 L} \times \frac{P_t G_t G_r}{F} \tag{21}$$

The first term is by all means beyond our control, and we do not want to change the second term for now. So, what can we do about the last term? It happens that if we were to keep the two-antenna size constants while increasing the frequency, G_t and G_r each come with a nice λ^2 on the denominator. Equation (21) thus equals

$$SNR \times B = \frac{1}{16\pi^2 k} \times \frac{\lambda^2}{Td^2 L} \times \frac{4\pi A_{tp} \varepsilon_t P_t}{\lambda^2} \times \frac{G_r}{F} = \frac{A_{tp} \varepsilon_t P_t}{4\pi k T d^2 L} \times \frac{G_r}{F} \tag{22}$$

A_{tp} and ε_t are the physical area and the aperture efficiency of the transmitting antenna; ε_t is between zero and unity. For active phased arrays, Eq. (22) could be rewritten as

$$SNR \times B = \frac{A_{tp} \varepsilon_t N_t P_{te}}{4\pi k T d^2 L} \times \frac{N_r G_{re}}{F_e} \tag{23}$$

where N_t and N_r are the numbers of transmitters and receivers. P_{te} is the output power for each transmitter. G_{re} is the gain of the receiving antenna for each receiver. F_e is the noise factor of each receiver. For active phased arrays, antenna elements with their corresponding transmitters and receivers are evenly distributed with a pitch of about $\lambda/2$. Thus, N_t and N_r are proportional to $1/\lambda^2$. It is proven that for phased array with lossless combing, the G_r/F term in Eq. (22) scales with N_r [10]. Assuming constant ε_t and G_{re} with respect to f, we see that $SNR \times B$ scales with $f^{-0.5}$ to f^0!

The six orders of magnitude difference of $SNR \times B$ deduced from Eqs. (20) and (23) give us a hint of the size of the design space for silicon THz systems.

2.2.1 THz Sources

When talking about silicon THz sources, a plethora of options is available that varies in functionality, complexity, and performance. For incoherent imaging applications, the most important metrics are output power and efficiency, whereas for spectroscopy, the bandwidth is the most important specification. Perhaps the most demanding application is for THz communications, for which output power, power efficiency, tuning range, phase noise, harmonics, and spurious suppression are all important parameters. This subsection aims to give a brief and incomplete introduction to what has been done in this area.

THz signal can be generated either by frequency multipliers or by on-chip oscillators.

In multipliers, the MOS or bipolar transistor is driven heavily to generate highly nonlinear current. The intended frequency component is then extracted with other components filtered. If efficiency is important, the active device should be conjugate matched for the fundamental and the intended harmonic. The impedance presented to the device at other harmonics is usually short or open circuit to maximize energy transfer between the fundamental and the intended harmonic. But we should not be overzealous about this goal; usually taking care of the first two or three harmonics is enough since the higher harmonics are insignificant. The transistor also has to be biased correctly for maximum harmonic generation. For MOS transistor, the conduction angle is specified. Like power amplifiers, efficient

MOS multiplier works in the class AB, B, or C region depending on the frequency, multiplication factor, and input power. For bipolar transistor, this efficiency is a function of V_{be} (or collector current density) [11].

Relationship of phase noise between the harmonic and the fundamental for multipliers is [11]:

$$S_{out}(\Delta\omega) = N^2 S_{fund}(\Delta\omega) + S_{harm}(\Delta\omega) + S_{amp}(\Delta\omega) \tag{24}$$

where $S_{out}(\Delta\omega)$ and $S_{fund}(\Delta\omega)$ represent the spectral density of phase fluctuations for the harmonic and fundamental signal with $\Delta\omega$ radians offset from the carrier. It is in dBc/Hz form. N is the frequency multiplication ratio. The last two terms in Eq. (24) represent the added noise from the harmonic-generating device and the ensuing amplifiers (if any). The combined value is usually less than 3 dB for reasonably designed circuits.

Multipliers are usually compact and broadband, but they are not as efficient as (well designed) oscillators. A 90–300 GHz transmitter based on distributed quadrupler is designed for spectroscopy and imaging [12]. It resembles the distributed amplifier (DA) in that the input and output capacitance of active device are absorbed in the input and output transmission line. Differential quadrature signal is used to drive two groups of quadrupler diff-pairs, the current of which is then combined to cancel the second harmonic. As another example, quadrupler is used in an 8-element 400 GHz transmitter phased array to replace power amplifier [13]. This also simplifies phase shifter design since the fundamental signal only needs to be shifted within 90 degrees as the phase shift is multiplied by four.

In oscillators, the transistor is made unstable by intentionally introducing positive feedback around it. Steady-state oscillation occurs at the frequency where the open-loop transfer function equals -1 (Barkhausen's criteria). Since the f_{max} of most silicon devices is below 300 GHz, harmonic generation is employed. The fundamental oscillation frequency is usually around 100 GHz for better phase noise, since larger oscillation amplitude and hence better SNR are easier to obtain at lower frequencies.

A high efficiency and scalable 4 × 4320 GHz oscillator array is built in SiGe BiCMOS technology [9]. The oscillator shown in **Figure 4(a)** oscillates at 160 GHz and is optimized for optimum transistor voltage gain and phase shift as discussed in Section 2.1. Y_1 and Y_2 represent the source and load admittance for the transistor. A transmission line with impedance Z_0 and electrical length θ_{TL} is used to introduce

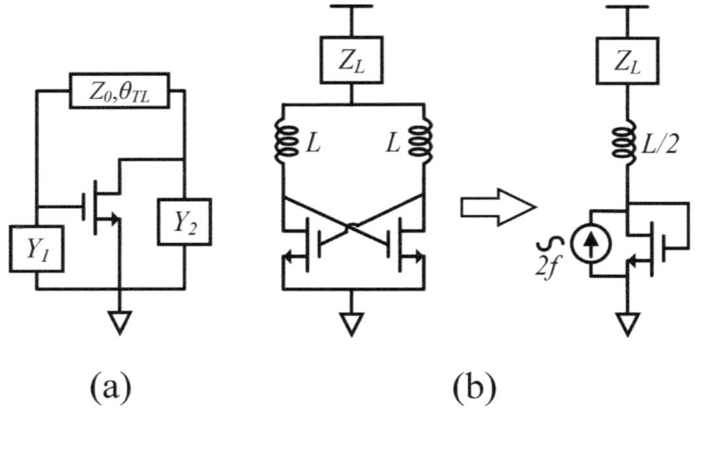

Figure 4.
Topology of (a) self-sustained oscillator and (b) conventional cross-coupled oscillator and its equivalent circuit at second harmonic.

feedback. The transmission line spans approximately a quarter wavelength at the second harmonic to transform the relatively small impedance of the gate node to high values at the drain node. Since harmonic signals are in the current form, boosting the output impedance at the harmonic frequency substantially improves output harmonic power. The DC-to-THz radiation efficiency is 0.54%. Early reports of THz oscillators based on push-push oscillators are less efficient than this partly due to the insufficient output impedance at second harmonic. As is shown in **Figure 4(b)**, the cross-coupled pair in a push-push oscillator is effectively a diode-connected transistor at second harmonic.

Frequency tuning of oscillators is usually done by varying the capacitance of varactor in the oscillation tank. Higher oscillation frequency translates to smaller capacitance, which is problematic for small varactors as its parasitic capacitance would swamp the variable capacitance. This would severely constrain the oscillator's tuning range.

Shown in **Figure 5(a)** and **(b)** are the cross section of a NMOS varactor and its small signal model with its source AC grounded. To increase its Q factor, the minimum channel length is usually employed, which further increases the overlap capacitance (C_{ov}) between the two terminals.

A straightforward way to increase the tuning ratio of the varactor is to place an inductor to partially absorb the parasitic capacitance. A 300 GHz differential Clapp push-push VCO with 8.5% tuning range and phase noise of −85 dBc at 1 MHz offset is reported in [14]. Its simplified schematic and equivalent small signal circuit for calculation of the input impedance seen at the base is shown in **Figure 6**. Note that the base resistance, the depletion capacitance between base and collector, and the output resistance of the transistor are ignored.

The input impedance seen from the base is

$$Z_B = -\frac{1}{j\omega C_{pi}}\left(1 + \frac{g_m + j\omega C_{pi}}{1/R_{var} + j\omega C_{eff}}\right) \tag{25}$$

where C_{eff} is

$$C_{eff} = C_{\text{var}} \times \left(1 - \frac{1}{\omega^2 L_{\text{deg}} C_{\text{var}}}\right) = C_{\text{var}} \times \left(1 - \frac{\omega_0^2}{\omega_{OSC}^2}\right) \tag{26}$$

R_{var} represents the loss in varactor; its conductance is $1/Q$ of the varactor at the evaluation frequency of the quality factor. The resonant frequency ω_0 for C_{var} and L_{DEG} is set below the main oscillation frequency.

The equivalent series resistance and capacitance for Z_B is

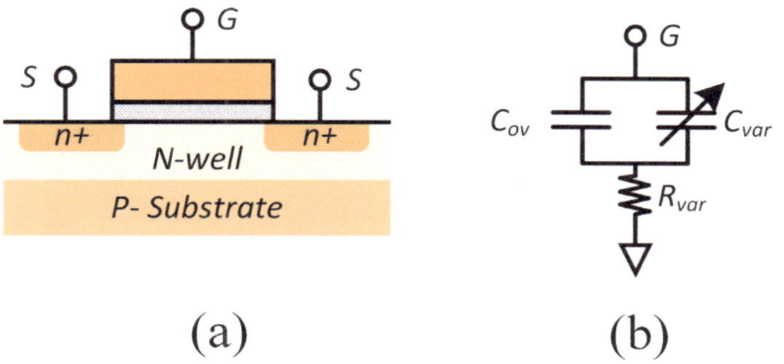

Figure 5.
Cross section of NMOS varactor (a) and equivalent circuit with grounded source (b).

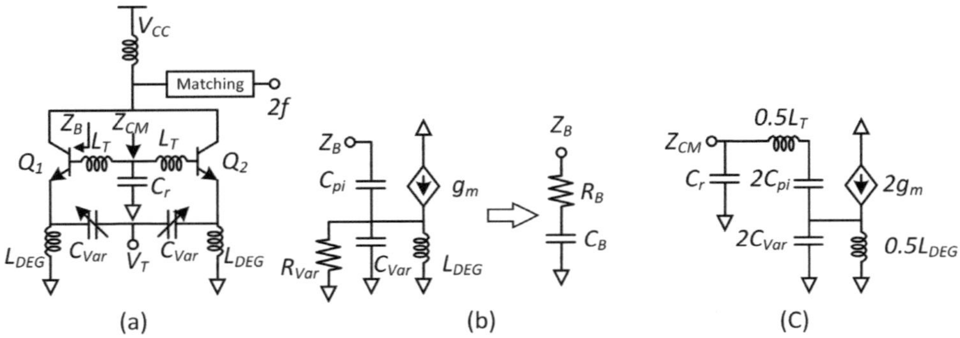

Figure 6.
Push-push VCO with common-mode resonance: (a) schematic, (b) base input impedance for differential mode, and (c) input impedance for common mode.

$$R_B = \frac{1/R_{\text{var}} - g_m C_{eff}/C_{pi}}{1/R_{\text{var}}^2 + \omega^2 C_{eff}^2} \approx -\frac{g_m}{\omega^2 C_{eff} C_{pi}} \tag{27}$$

$$C_B = \left(1 + \frac{g_m/R_{\text{var}} + \omega^2 C_{pi} C_{eff}}{1/R_{\text{var}}^2 + \omega^2 C_{eff}^2}\right) \times \frac{1}{C_{pi}} \approx \frac{C_{eff} \times C_{pi}}{C_{eff} + C_{pi}} \tag{28}$$

Another interesting property of the circuit is that the oscillation frequency for common mode is intentionally set to the second harmonic. C_r in **Figure 6(a)** and **(c)** forms a series resonator with the tank, raising the second harmonic voltage seen at the base significantly. This boosts the second harmonic generation. The output power for the two versions of the VCO is 0.6 dBm and 0.2 dBm, respectively.

As is evident from Eq. (27), the negative resistance seen at the base of the capacitively degenerated transistor could be used to mitigate the loss of the varactor [15]. This property is used in [16] to build a 300 GHz triple-push VCO. The tuning range is 8%, and the phase noise is −101.9 dBc/Hz at 1 MHz offset for the 100 GHz main loop. That translates to a phase noise of −80.28 dBc/Hz at 1 MHz offset for 300 GHz assuming noiseless multiplication.

An interesting observation is that inductors actually have better quality factor than varactors at higher frequency. A carefully designed inductor has a Q of 15–20 at 100 GHz, whereas the quality factor for varactor is around 2–5 at that frequency. It would be nice if we can replace the varactor with a high-quality metal-insulator-metal (MIM) or metal-oxide-metal (MOM) capacitor; we then need to figure out how to tune the inductance of these nice inductors. **Figure 7** shows one such circuit [17] which is also based on Clapp oscillators.

We know that inductance at the base generates a negative resistance and a positive inductance seen from the emitter, as LNA designers can attest to [18]. Careful derivation of Z_E leads to the following results [14]:

$$Z_E \approx \frac{1}{g_m} - \frac{\omega^2 L_{BB}}{\omega_T} + \frac{\omega^2 R_{BB}}{\omega_T^2} + j\omega\left(\frac{R_{BB}}{\omega_T} - \frac{1}{g_m \omega_T} + \frac{\omega^2 L_{BB}}{\omega_T^2}\right) \tag{29}$$

Thus, we can tune the inductance by varying the g_m of the transistor. For bipolar transistors, g_m equals the emitter current divided by the thermal voltage V_T (26 mV in room temperature).

The impedance at the emitter is mapped to the resonant tank through the transformer formed by L_E and L_T in **Figure 7(b)**. The currents of the two active inductors are controlled by tail current source Q_5. The second harmonic current is extracted through L_C, and the degeneration resistor R_{EE} is used to improve

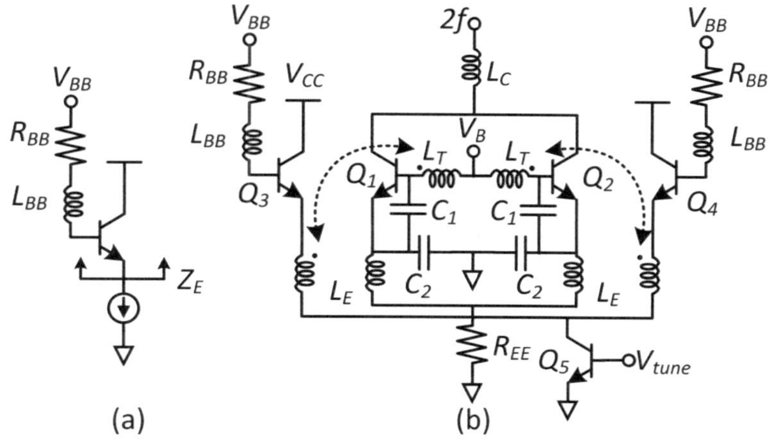

Figure 7.
Schematics for (a) tunable active inductor and (b) complete VCO.

common-mode rejection. Two versions of this VCO with different oscillation frequency are built; the tuning range are 3.5 and 2.8%, respectively. The harmonic power suffers due to the inclusion of R_{EE}, the output power at 201.5 and 212 GHz are −7.2 and−7.1 dBm, respectively. But the phase noise performance is very good, with −87 and −92 dBc at 1 MHz offset for the two VCOs. This translates to −83.5 and −89 dBc at 1 MHz offset if they oscillate at 300 GHz, which is comparable to the circuit shown in **Figure 6** that generates 0 dBm at 300 GHz.

This raises interesting question as smaller output power usually means inferior phase noise performance. One possible explanation is that the noise current at the second harmonic in Q1 and Q2 of **Figure 6** generates large noise voltage at the emitter since these nodes are open circuit due to resonance, thus amplifying the noise current at second harmonic. It should be noted that the phase noise could be improved substantially by breaking the noise current path at second harmonic [19].

A salient feature of Clapp oscillator is the inherent isolation of the load from the tank. This helps to preserve the quality factor of the tank, leading to a better phase noise and less load pulling (variation of oscillation amplitude and frequency caused by load variation). The problem of low output impedance at second harmonic is also mitigated in this topology as the base is isolated from the drain.

One starts to wonder if there is a way to further improve the phase noise performance. We know there is a trade-off between noise performance and power consumption, but we are kind of stuck here: We need larger transistor to burn larger power, but the larger capacitance of the device means smaller inductors in the tank, which complicates the design and ultimately degrades the Q. We need a larger design space here.

One way to do that is to build an array of N oscillators and lock them together. Theoretically the phase noise would drop by N as the SNR increases by N, and the output power would also increase by N. Better still, if we distribute them evenly with a certain pitch (normally $\lambda/2$) and radiate the power out collectively, the energy would focus in certain direction, improving the EIRP by N^2. Tousi et al. shows one such design [20]:

Each individual oscillator shown in **Figure 8** is a cross-coupled push-push oscillator. It is designed for optimum fourth harmonic generation by making sure that the gate is isolated from the drain at the fourth harmonic. The second harmonic is rejected by the narrow band on-chip antenna. Each oscillator is coupled to other oscillators as shown in **Figure 8(a)** through active phase shifters. **Figure 8(a)** forms

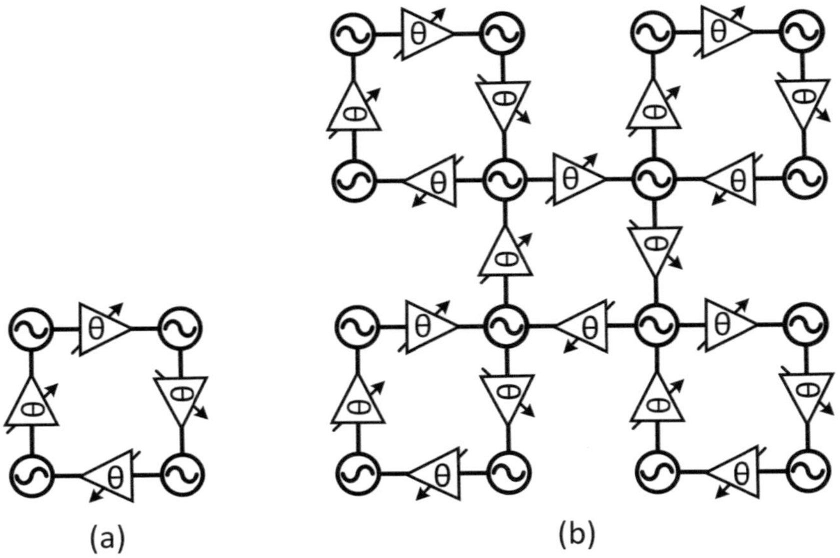

Figure 8.
Block diagram for (a) a unit cell and (b) 2-D oscillator lattice.

a unit cell through which a 2-D oscillator lattice could be formed as shown in **Figure 8(b)**. One nice feature of this design is the use of phase shifter as coupling elements, as this provides us with a new way of tuning the frequency of these injection-locked oscillators [21]:

$$\frac{\Delta f_0}{f_0} = k \sin\left(\Delta\phi(t)\right) \tag{30}$$

Equation (30) is derived from the Adler's equation under locked conditions. Δf_0 is the frequency difference of the injected signal and the free-running frequency f_0 of the slave oscillator. $\Delta\phi(t)$ is the phase difference. K is a factor relating to the quality factor of the oscillation tank, amplitude of the injected signal, and the free-running oscillation amplitude of the tank.

Since the total phase shift through the loop in **Figure 8(a)** is $2k\pi$, the phase shift through the oscillator plus that of the phase shifter is constrained to a set of fixed values. The oscillation frequency of the whole array is changed if we apply a common shift to all the phase shifter in **Figure 8(b)**. Interesting thing happens when we tune the phase shifter connected with an individual oscillator. If we apply a common shift to the four shifters, this disrupts the local loop momentarily as the instantaneous phase of the affected oscillator jumps to a new value to accommodate the change. This ability of changing the phase of radiating elements individually turns this design into a phased array. Measurements show that the beam could be steered $\pm$50 degree in the E plane and $\pm$ 45 degree in the H plane. The frequency tuning range is 2.1%. The peak EIRP is 17.1 dBm, and the phase noise is $-$93 dBc at 1 MHz offset for this 338 GHz array, which show substantial improvements over single oscillators.

A brief comparison between multipliers and harmonic injection-locked VCOs is also in place: The former is generally compact and wideband. They add negligible phase noise if designed properly as dictated by Eq. (22). The biggest issue is the harmonics, which leads to annoying LO spurs that leads to spurious emission and corrupts received signals. The latter comes with higher efficiency and much higher harmonic rejection ratio, but the bandwidth is limited. The close-in phase noise is dominated by the source just like the multiplier, but the far-out phase noise is dominated by the VCO. For mm-wave frequency synthesizers, the two options generally achieve comparable phase noise performances [11]. Since we have to use multipliers to get from mm-wave to THz anyway, this same conclusion holds for THz. For communication applications, it is advisable to use PLL-locked or injection-locked VCOs to generate relatively high LO frequency and use multipliers to boost it to THz frequency (N-push VCOs does this in one place).

2.2.2 THz detectors

THz detectors utilize the nonlinearity of active devices to directly rectify THz signal to DC. A lot of devices could be used, like diode-connected NMOS transistor [22], CE (common emitter) [23] or CB (common base) connected [24, 25] SiGe HBT, CMOS-compatible Schottky diode [26], or P+/n-well diode [27]. THz reception with detectors is incoherent, that is, only the amplitude information is recovered at the receiving side, which limits THz detectors almost exclusively to incoherent THz imaging applications. The strength of THz detectors lies within their simplicity: They do not need LO (local oscillator) signal to do the THz down-conversion. The received THz signal self-mix themselves to DC through even-order nonlinearity of the device. This makes scaling extremely easy, as only low-frequency routing is needed, whereas LO-driven mixer needs cumbersome and power-hungry LO tree which quickly becomes unmanageable when the array gets large. A 1024 pixel NMOS detector array [22] in 65-nm CMOS process showcases the impressive scalability of THz detectors.

However, this flexibility comes with a price: The gain and noise performance of detectors is quite limited, and the specification “responsivity” and “noise equivalent power (NEP)” are used in place of conversion gain and noise figure. The responsivity is defined as the voltage output divided by received power, and NEP is defined as the output noise voltage density divided by responsivity.

The bandwidth of imaging application is usually below 1 MHz; thus, technologies with lower 1/f noise corner frequencies like SiGe HBT or P+/n-Well diode are preferred. The 1/f noise corner frequencies for SiGe HBT and P+/n-well diode are below 1 kHz and 10 kHz, respectively, whereas for NMOS transistor or Schottky diodes, the numbers are well above 1 MHz. For SiGe HBTs, it is shown that CB-connected topology has higher responsivity than CE-connected topology when operating above f_{max} [24].

The principle of THz imaging with detectors largely follow their optical counterpart: They use THz lenses to do the focusing. The problem is that THz wavelength is 2–3 orders of magnitude larger than visible lights, thus large and bulky THz optics are required for reasonable imaging resolutions. They require a lot of effort to set up the imaging setup with the invisible THz radiations [25]. This is the innate deficiency with incoherent imaging. Coherent imaging with THz transceivers could get rid of those optics.

2.2.3 THz transceivers

For transceivers working at lower frequencies, the transmitter and receiver are usually integrated on one chip, and they share the common RF port through switches or duplexers (bandpass filters tuned for simultaneous transmit and receive on different bands). Up to now, neither option is satisfactory for fully integrated silicon THz circuits.

One solution for integrated THz transceiver is to share the antenna and figure out ways to isolate the Tx (transmitter) and Rx (receiver). Park et al. have shown a fully integrated 260 GHz transceiver based on shared leaky-wave antenna [28]. The leaky-wave antenna resembles a lossy transmission line (TL); thus, the Tx and Rx ports could be placed on either end of the antenna. When the transmitter is working, the receiver is turned off and terminates the TL on its side. The same holds true for the receiving mode. The problem with the leaky-wave antenna is that they are relatively long (1.2 mm or 2.5 λ in this design). Statnikov et al. [29] have shown a fully integrated 240 GHz frequency-modulated continuous wave (FMCW) radar transceiver based on shared dual-polarization antenna. A quadrature hybrid coupler is used as a polarizer for the dual-polarization antenna and duplexer for the Tx and Rx. Isolation of the Tx and Rx depends on the orthogonality of left hand circular polarized (LHCP) and right hand circular polarized (RHCP) waves. The Tx and Rx interface with two orthogonal port of the branch-line coupler and are isolated from each other. In Tx mode, the branch-line coupler excites the LHCP mode of the antenna. When the transmitted wave hits a target and bounces back, it changes to RHCP and is subsequently routed to the receiver through the coupler. This scheme is not directly applicable for point-to-point communications, just like frequency-division duplexing (FDD)-based transceivers could not communicate directly with each other.

Another solution is to use two antennas. For FMCW radars, the leakage from the TX to the Rx results in strong interferences around DC [30]. This raises the noise floor in the range spectrum. With area permitting, the Tx and Rx antenna should be separated further apart for better isolation. The measured crosstalk between the two antennas with a separation of about 1.8 mm in a 160 GHz FMCW radar transceiver is below 31 dB [31]. This isolation might be adequate for FMCW radar applications, but it is still wanting for communications.

Transceiver-based THz imaging makes coherent imaging possible, as both the magnitude and phase information of the signal from targets are retained. With both information available, it is possible to get rid of bulky THz optics by sampling the THz field directly and do the focusing digitally. The THz field is usually sampled on a 2-D plane with different THz frequencies; this is fulfilled by raster scanning a FMCW transceiver (or the sample). For a given point in space, the round-trip phase delay from the transceiver to that point is a function of its position and sampling frequency. By raster scanning the transceiver or the sample under different frequencies, its phase delay variation is orthogonal to every other point in the sampling space. This forms the basis for the 3-D imaging through the back-projection algorithm. 3-D imaging based on SiGe FMCW transceivers is reported by several groups [32, 33], showcasing the great potential for low-cost THz imaging applications.

For communication applications, the modulation scheme plays a major role in deciding the transceiver architecture. Low-complexity modulation schemes like on-off keying (OOK) and binary phase shift keying (BPSK) lead to robust and power-efficient design, but the spectrum efficiencies are relatively low. Modulation schemes like 32 QAM and 128 QAM lead to much higher spectrum efficiency, but they are quite demanding on linearity and phase noise performance, and they require image-rejection architectures as the spectra of QAM are asymmetric around the carrier. The upper sideband (USB) and lower sideband (LSB) of the spectra become each other's own image when converted to baseband, and image-rejection is needed to avoid signal corruption. Image-rejection modulation/demodulation is difficult in THz range as I/Q mixers are required. It is very difficult to guarantee phase and amplitude matching for the I/Q LO signal for adequate image-rejection at THz frequency.

A 210 GHz fundamental transceiver chipset with OOK modulation is demonstrated in a 32-nm SOI CMOS process [34]. Ideally speaking, power amplifier (PA)-based fundamental operation is more power-efficient than frequency multipliers.

This helps to boost efficiency of the whole system as PAs are usually the most power-hungry circuits in transceivers. Perhaps the most difficult part of this design is controlling the oscillator pulling effect. Since the PA works at the same frequency as the on-chip VCO, significant coupling could occur between PA and VCO. The injection-locking effect would impact the phase noise performance heavily. The on-chip antenna used in this design only makes things more difficult. To improve the VCO performance, a stacked cross-coupled VCO topology is used to boost oscillation amplitude, improving its robustness in response to interferences.

A 240-GHz direction-conversion transceiver in SiGe BiCMOS technology is demonstrated with BPSK capability. BPSK is a constant-envelope modulation, which means the PA could be driven to saturation for better power efficiency. The spectra of BPSK modulation are symmetric around its carrier (symmetrically modulated), making direct conversion easier to implement as no image-rejection is needed. A 30 GHz LO signal is supplied to this transceiver, and on-chip $\times 8$ multipliers are used for the 240 GHz LO generation. This helps to alleviate the detrimental effect of LO spurs caused by multipliers since they are separated by twice the baseband bandwidth (15 GHz). An on-chip antenna with 1-dB bandwidth of 33 GHz is achieved partly due to the local back etching (LBE) technology used. The silicon substrate below the antenna is removed, resulting in a low-loss air cavity below the antenna. The transceiver link is tested with 15 cm separation, and an impressive 6-dB bandwidth of 35 GHz is obtained. A 25 Gbps wireless link is demonstrated by this transceiver with no equalization. One problem with direct conversion using no I/Q demodulation is that the demodulated signal's SNR is dependent on the phase difference between the Tx LO and Rx LO. A phase shifter is used in this test in case manual tuning is required to boost the SNR.

A 300 GHz QPSK transmitter for dielectric waveguide communication is demonstrated in a 65-nm CMOS process [35]. Again, off-chip LO signal is used to drive on-chip frequency multipliers. The targeted data rate is 30 Gbps, which translates to around 20 GHz baseband bandwidth for QPSK assuming a roll-off factor of 0.3. Thus, the off-chip LO signal frequency is set to 45 GHz. An on-chip quadrature modulator is used to modulate the baseband data to an IF frequency of 135 GHz. It is further shifted by a double-balanced mixer to 315 GHz. Such a high IF alleviates the need for image-rejection mixers since the image frequency falls completely out of band. A 30 Gbps QPSK is demonstrated with on-chip probing.

A 230 GHz direct-conversion 16-QAM 100-Gbps wireless link is demonstrated with a communication distance of 1 meter [36]. The I/Q mixer directly interfaces with on-chip antenna to avoid bandwidth limitation introduced by LNA. On-chip LO multiplier chain is used to convert the external 13.75–16 GHz LO to 220–256 GHz. The baseband bandwidth is around 14 GHz; this poses challenge as the spacing of LO spurs is comparable to this bandwidth. This leads to spurious modulation that overlaps with desired signal. Nevertheless, 100 Gbps with an EVM of 17% is demonstrated.

A 300 GHz 32-QAM and 128-QAM transmitter with 105-Gbps data rate is demonstrated in a 40 nm CMOS process [37]. As there is no PA available, an array of eight square mixers (i.e., mixing through the second-order nonlinearity) is power combined at the output stage. A heterodyne topology is used, and the LO frequencies for the two up-conversion stage are both set at 135 GHz. The IF frequency for the first stage is around 10 GHz, and high-pass filtering is used to suppress the LSB by approximately 10 dB. Single-balanced mixer is used in the first stage to intentionally leak LO signal to the second stage. The second-order nonlinearity of NMOS transistor is used to mix the (IF+LO) signal with LO leakage to obtain the desired intermodulation signal (IF+2LO). Unwanted second harmonics of LO and IF signal is canceled at the output rat-race balun. On-chip probing validates the operation;

the 32-QAM modulation with an EVM of 8.9% is achieved with 105 Gbps. No on-chip antenna is used as this chip is intended to drive high-power THz devices like traveling-wave tubes.

2.3 THz interface for efficient power transfer

The THz interface serves as a gateway between the circuit and the outside world. The efficiency of this interface greatly impacts the performance of the overall system. A simple derivation of the transceiver's link budget would highlight the importance of this interface (**Figure 9**):

$$P_R = P_T - IL_2 - IL_0 - IL_1 \tag{31}$$

where P_R is power received at the receiver, P_T is the output power of the transmitter, and IL_2 and IL_1 are the loss for THz interface at the transmitter and the receiver. IL_0 is the loss associated with the propagation medium. If the medium is free space, which is often the case, IL_0 equals

$$IL_0 = 20 \times \log_{10}\left(\frac{\lambda_0}{4\pi d}\right) + G_{TX} + G_{RX} \tag{32}$$

λ_0 is the free-space THz wavelength, d is the propagation distance, and G_{TX} and G_{RX} are the gain of antenna employed at the transmitter and receiver. Note that Eqs. (31) and (32) are in dB form. Also keep in mind that Eq. (32) is only valid under far-field conditions, that is, the radiation field seen by the receiver is not reactive and could be approximated by plane waves. Common criterion for far-field condition is that the receiver is separated from the transmitter by at least $2D^2/\lambda$, where D is the maximum overall dimension of the transmitting antenna.

To maximize P_R, we have to minimize the loss at the interface and increase the gain of the antennas. This section covers both areas.

For THz silicon chips, grounded coplanar waveguide (GCPW) is the most prevalent medium for on-chip THz routing. It is a combination of coplanar waveguide with microstrip line. This configuration have several merits: First, the ground plane of the microstrip line shields the signal from the electrically thick silicon substrate, which reduces loss and prevents the signal from leaking into substrate modes; the coplanar waveguide makes interface with outside world easier, be it through flip-chip bonding or on-chip probing since the ground conductor lies in proximity with the signal trace.

A 90–300 GHz transmitter and a 115–325 GHz receiver are flip-chip bonded to a liquid-crystal polymer (LCP) substrate [12]. This connects the chip to the 100–280 GHz Vivaldi antenna on the LCP substrate. Such wideband antenna is extremely difficult to realize on-chip. As another example, a CMOS 300 GHz transmitter chip is flip-chip bonded to a GCPW-to-WG transition module [38] implemented on a

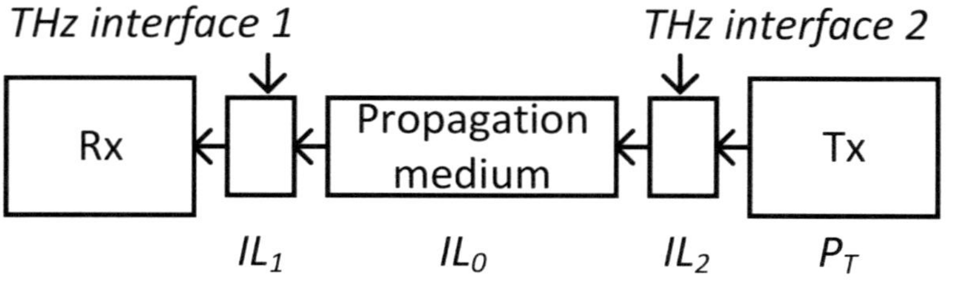

Figure 9.
Link budget analysis of a THz transceiver system.

low-cost glass epoxy PCB. Once transitioned to waveguide interfaces, the chip could be interfaced to a plethora of THz components like horn antennas and high-power amplifier modules. The packaging loss is 8 dB, which includes the transition loss (flip-chip bonding and the GCPW-to-WG), impedance mismatch, and loss in the epoxy material. It should be noted that gold stud bumping is used in both cases, which is compatible with conventional wire bonders and is quite convenient for R&D labs.

An effective way to lower transition loss is to radiate the THz signal directly from chip. Since the high permittivity silicon substrate readily traps the THz radiation, there are basically two lines of thoughts regarding on-chip antenna design. One is to accept this coupling and take this into account while designing antennas; the other is to eliminate the substrate mode altogether.

The first approach tries to make use of the electrically thick antenna to improve the antenna bandwidth. Metal reflector are placed underneath the substrate to reflect energy back, which is often the top layer of the PCB under the chip. To make sure the reflected power adds constructively with the surface radiation, the chip should be $(2k + 1)/4\lambda$ thick [34, 39, 40]. The phase delay of the reflected wave increases linearly with frequency, which limits the bandwidth. Artificial magnetic conductor (AMC) reflector in place of the solid metal reflector on the PCB is used to compensate this phase shift [41], which extends the on-chip antenna bandwidth substantially. The antenna efficiency is within 0.5dB of the peak efficiency between 200 and 300 GHz. Note that the chip should be $1/2\lambda$ thick in this case as AMC introduces zero phase shift at the center frequency.

Although promising, allowing reflections within the substrate increases couplings between antenna elements and on-chip passive components. This makes circuit performance sensitive to antenna location and substrate dimensions (both lateral dimension and chip thickness), which is undesirable for designing arrays.

There are two widely adopted approach to eliminate the substrate mode. The first approach involves attaching a high-k dielectric lens on the back side of the chip, and the antenna radiates through the lens [22, 42–44]. This approach offers high directivity and improves efficiency. The most obvious drawback is the need for nonstandard packaging, which could be quite costly. An intuitive way of understanding why this structure can eliminate the substrate mode can be found in [8].

The second approach is more straightforward: The substrate is shielded from the radiating element using one or a few low-level metal layers. This eliminates the possibility of coupling with substrate modes, but it limits the bandwidth and radiation efficiency severely since the radiating element is only a few microns away from the ground due to the restriction of silicon backend process. The radiating element thus forms a high Q tank with the ground, and the bandwidth of the antenna is on the order of a few percent [13, 20, 25]. The obvious way to increase bandwidth and radiation efficiency is to lower the Q, which could be accomplished by increasing the volume of the resonant tank. One solution is to add a dielectric superstrate above the radiating element [45, 46], which diverts some of the electric fields from the dielectric layer below the radiating element, increasing the resonance volume considerably. The radiation efficiency increases with the superstrate till the onset of the $TE1$ mode, which limits the thickness of the superstrate to $\lambda_0/4\sqrt{\varepsilon_r - 1}$ where ε_r is the dielectric constant of the superstate. Dielectric resonator antenna (DRA) is another option [47, 48]. This antenna composes of a dielectic resonator on top of the chip and a feed element in the top metal layer of the chip. The feed element is used to excite resonance inside the resonator, through which the THz signal radiates. The size of the resonator and relative position of feed could be adjusted for intended oscillation frequency and resonance mode. For a 270 GHz microstrip antenna, the gain is enhanced by 4 dB, and the 3-dB gain bandwidth is extended by 100% when an dielectric resonator is used [49].

For phased array applications, the patch antenna with ground shield is the most straightforward approach.

3. Summary

Silicon THz circuit design is an active research area open to innovations on multiple levels. We need better passive components, better circuits, and the most important thing is we need to come up with better ways of building THz arrays. Scaling is the key for significantly boosting the performance of silicon THz systems as we venture into this last untapped spectrum [50].

Author details

Li Zhuang*, Cao Rui, Tao Xiaohui, Jiang Lihui and Rong Dawei
Key Lab of Aperture Array and Space Applications, He Fei, China

*Address all correspondence to: lizhuang@mail.ustc.edu.cn

References

[1] Mason SJ. Power gain in feedback amplifier. Transactions of the IRE Professional Group on Circuit Theory. 1954;**1**(2):20-25

[2] Gupta MS. Power gain in feedback amplifiers, a classic revisited. IEEE Transactions on Microwave Theory and Techniques. 1992;**40**(5):864-879

[3] Spence R. Linear Active Networks. London, UK: Wiley; 1970

[4] Khatibi H, Khiyabani S, Afshari E. A 173 GHz amplifier with a 18.5 dB power gain in a 130 nm SiGe process: A systematic Design of High-Gain Amplifiers above $f_{max}/2$. IEEE Transactions on Microwave Theory and Techniques. 2018;**66**(1):201-214

[5] Khatibi H, Khiyabani S, Afshari E. A 183 GHz desensitized unbalanced Cascode amplifier with 9.5-dB power gain and 10-GHz band width and −2 dBm saturation power. IEEE Solid-State Circuits Letters. 2018;**1**(3):58-61

[6] Wang Z, Heydari P. A study of operating condition and design methods to achieve the upper limit of power gain in amplifiers at near-fmax frequencies. IEEE Transactions on Circuits and Systems I: Regular Papers. 2017;**64**(2): 261-271

[7] Momeni O, Afshari E. High power terahertz and millimeter-wave oscillator design: A systematic approach. IEEE Journal of Solid-State Circuits. 2011; **46**(3):583-597

[8] Han R, Afshari E. A CMOS high-power broadband 260-GHz radiator array for spectroscopy. IEEE Journal of Solid-State Circuits. 2013;**48**(12): 3090-3104

[9] Han R et al. A SiGe terahertz heterodyne imaging transmitter with 3.3 mW radiated power and fully-integrated phase-locked loop. IEEE Journal of Solid-State Circuits. 2015; **50**(12):2935-2947

[10] Lee JJ. G/T and noise figure of active array antennas. IEEE Transactions on Antennas and Propagation. 1993;**41**(2): 241-244

[11] Wang CC, Chen Z, Heydari P. W-band silicon-based frequency synthesizers using injection-locked and harmonic triplers. IEEE Transactions on Microwave Theory and Techniques. 2012;**60**(5):1307-1320

[12] Chi T, Huang MY, Li S, Wang H. A packaged 90-to-300GHz transmitter and 115-to-325GHz coherent receiver in CMOS for full-band continuous-wave mm-wave hyperspectral imaging. In: International Solid-State Circuits Conference. 2017. pp. 304-305

[13] Yang Y, Gurbuz OD, Rebeiz GM. An eight-element 370–410-GHz phased-array transmitter in 45-nm CMOS SOI with peak EIRP of 8–8.5 dBm. IEEE Transactions on Microwave Theory and Techniques. 2016;**64**(12):4241-4249

[14] Ahmed F, Furqan M, Heinemann B, Stelzer A. 0.3-THz SiGe-based high-efficiency push–push VCOs with >1-mW peak output power employing common-mode impedance enhancement. IEEE Transactions on Microwave Theory and Techniques. 2018;**66**(3):1384-1398

[15] Heydari P. Millimeter-wave frequency generation and synthesis in silicon. In: 2018 IEEE Custom Integrated Circuits Conference (CICC). 2018. pp. 1-49

[16] Chiang PY, Wang Z, Momeni O, Heydari P. A silicon-based 0.3 THz frequency synthesizer with wide locking range. IEEE Journal of Solid-State Circuits. 2014;**49**(12):2951-2963

[17] Chiang P-Y, Momeni O, Heydari P. A 200-GHz inductively tuned VCO With-7-dBm output power in 130-nm SiGe BiCMOS. IEEE Transactions on Microwave Theory and Techniques. 2013;**61**(10):3666-3673

[18] Schmid RL, Coen CT, Shankar S, Cressler JD. Best practices to ensure the stability of sige HBT cascode low noise amplifiers. In: Proceedings of the IEEE Bipolar/BiCMOS Circuits and Technology Meeting. 2012. pp. 2-5

[19] Murphy D, Darabi H, Wu H. Implicit common-mode resonance in LC oscillators. IEEE Journal of Solid-State Circuits. 2017;**52**(3):812-821

[20] Tousi Y, Afshari E. A high-power and scalable 2-D phased array for terahertz CMOS integrated systems. IEEE Journal of Solid-State Circuits. 2015;**50**(2):597-609

[21] Bhansali P, Roychowdhury J. Gen-Adler: The generalized Adler's equation for injection locking analysis in oscillators. In: 2009 Asia and South Pacific Design Automation Conference. 2009. pp. 522-527

[22] Al Hadi R et al. A 1 k-pixel video camera for 0.7-1.1 terahertz imaging applications in 65-nm CMOS. IEEE Journal of Solid-State Circuits. 2012; **47**(12):2999-3012

[23] Uzunkol M, Gurbuz OD, Golcuk F, Rebeiz GM. A 0.32 THz SiGe 4×4 imaging array using high-efficiency on-chip antennas. IEEE Journal of Solid-State Circuits. 2013;**48**(9):2056-2066

[24] Al Hadi R, Grzyb J, Heinemann B, Pfeiffer UR. A terahertz detector array in a SiGe HBT technology. IEEE Journal of Solid-State Circuits. 2013;**48**(9): 2002-2010

[25] Li Z, Qi B, Zhang X, Zeinolabedinzadeh S, Sang L, Cressler JD. A 0.32-THz SiGe imaging array with polarization diversity. IEEE Transactions on Terahertz Science and Technology. 2018;**8**(2):215-223

[26] Han YR, Zhang Y, Kim D, Kim Y, Shichijo H, Kenneth O. Terahertz image sensors using CMOS Schottky barrier diodes International SoC Design Conference, pp. 254–257, 2012

[27] Ahmad Z, Kenneth O. THz detection using P+/n-well diodes fabricated in 45-nm CMOS. IEEE Electron Device Letters. 2016;**37**(7): 823-826

[28] Park J, Kang S, Thyagarajan SV, Alon E, Niknejad AM. A 260 GHz fully integrated CMOS transceiver for wireless chip-to-chip communication. In: 2012 Symposium on VLSI Circuits (VLSIC). 2012. pp. 48-49

[29] Statnikov K, Sarmah N, Grzyb J, Malz S, Heinemann B, Pfeiffer UR. A 240 GHz circular polarized FMCW radar based on a SiGe transceiver with a lens-integrated on-chip antenna. In: Proceedings of European Radar Conference (EuRAD). 2014. pp. 447-450

[30] Park J, Kang S, Niknejad AM. A 0.38 THz fully integrated transceiver utilizing a quadrature push-push harmonic circuitry in SiGe BiCMOS. IEEE Journal of Solid-State Circuits. 2012;**47**(10):2344-2354

[31] Hitzler M et al. Ultracompact 160-GHz FMCW radar MMIC with fully integrated offset synthesizer. IEEE Transactions on Microwave Theory and Techniques. 2017;**65**(5):1682-1691

[32] Jaeschke T, Bredendiek C, Pohl N. 3-D FMCW SAR imaging based on a 240 GHz SiGe transceiver chip with integrated antennas. In: The German Microwave Conference (GeMIC). 2014. pp. 1-4

[33] Hamidipour A, Feger R, Poltschak S, Stelzer A. A 160-GHz system in package for short-range mm-wave applications. International Journal of Microwave and Wireless Technologies. 2014;**6**(3–4): 361-369

[34] Wang Z, Chiang PY, Nazari P, Wang CC, Chen Z, Heydari P. A CMOS 210-GHz fundamental transceiver with OOK modulation. IEEE Journal of Solid-State Circuits. 2014;**49**(3):564-580

[35] Zhong Q, Chen Z, Sharma N, Kshattry S, Choi W, Kenneth KO. 300-GHz CMOS QPSK transmitter for 30-Gbps dielectric waveguide communication. In: 2018 IEEE Custom Integrated Circuits Conference (CICC). 2018. pp. 1-4

[36] Rodríguez-Vázquez P, Grzyb J, Heinemann B, Pfeiffer UR. A 16-QAM 100-Gb/s 1-M wireless link with an EVM of 17% at 230 GHz in an SiGe technology. IEEE Microwave and Wireless Components Letters. 2019; **29**(4):297-299

[37] Takano K et al. 17.9 A 105Gb/s 300GHz CMOS transmitter. In: 2017 IEEE International Solid-State Circuits Conference (ISSCC). 2017. pp. 308-309

[38] Takano K et al. 300-GHz CMOS transmitter module with built-in waveguide transition on a multilayered glass epoxy PCB. In: 2018 IEEE Radio and Wireless Symposium (RWS). 2018. pp. 154-156

[39] Kang S, Thyagarajan SV, Niknejad AM. A 240 GHz fully integrated wideband QPSK transmitter in 65 nm CMOS. IEEE Journal of Solid-State Circuits. 2015;**50**(10):2256-2267

[40] Thyagarajan SV, Kang S, Niknejad AM. A 240 GHz fully integrated wideband QPSK receiver in 65 nm CMOS. IEEE Journal of Solid-State Circuits. 2015;**50**(10):2268-2280

[41] Zhong Q, Choi W, Miller C, Henderson R, Kenneth KO. A 210-to-305GHz CMOS receiver for rotational spectroscopy. Digest of Technical Papers - IEEE International Solid-State Circuits Conference. 2016;**59**:426-427

[42] Filipovic DF, Gearhart SS, Rebeiz GM. Double-slot antennas on extended hemispherical and elliptical silicon dielectric lenses. IEEE Transactions on Microwave Theory and Techniques. 1993;**41**(10):1738-1749

[43] Statnikov K, Ojefors E, Grzyb J, Chevalier P, Pfeiffer UR. A 0.32 THz FMCW radar system based on low-cost lens-integrated SiGe HBT front-ends. In: European Solid-State Circuits Conference. 2013. pp. 81-84

[44] Pfeiffer UR et al. A 0.53 THz reconfigurable source module with up to 1 mW radiated power for diffuse illumination in terahertz imaging applications. IEEE Journal of Solid-State Circuits. 2014;**49**(12):2938-2950

[45] Edwards JM, Rebeiz GM. High-efficiency elliptical slot antennas with quartz superstrates for silicon RFICs. IEEE Transactions on Antennas and Propagation. 2012;**60**(11):5010-5020

[46] Ou YC, Rebeiz GM. Differential microstrip and slot-ring antennas for millimeter-wave silicon systems. IEEE Transactions on Antennas and Propagation. 2012;**60**(6):2611-2619

[47] Mongia RK, Bhartia P. Dielectric resonator antennas—A review and general design relations for resonant frequency and bandwidth. International Journal of Microwave and Millimeter-Wave Computer-Aided Engineering. 1994;**4**(3):230-247

[48] Li C, Chiu T. 340-GHz low-cost and high-gain on-Chip higher order mode dielectric resonator antenna for THz applications. IEEE Transactions on

Terahertz Science and Technology. 2017;**7**(3):284-294

[49] Hou D, Chen J, Yan P, Hong W. A 270 GHz × 9 multiplier chain MMIC with on-chip dielectric-resonator antenna. IEEE Transactions on Terahertz Science and Technology. 2018;**8**(2):224-230

[50] Afshari E. "The Last Untapped Spectrum: Should Industry Care About Terahertz?" 2016. [Online]. Available from: https://sscs.ieee.org/event-recording-and-slides-now-available-the-last-untapped-spectrum-should-industry-care-about-terahertz-presented-by-ehsan-afshari

Chapter 15

Semiconductor Surface State Engineering for THz Nanodevices

Irving Eduardo Cortes-Mestizo, Edgar Briones, Leticia Ithsmel Espinosa-Vega and Victor Hugo Mendez-García

Abstract

This chapter is dedicated to study the semiconductor surface states, which combined with nanolithography techniques could result on remarkable properties of advanced nanodevices suitable for terahertz (THz) signal detection or harvesting. The author presents the use of low-dimensional semiconductor heterostructures for the development of the so-called self-switching diodes (SSDs), studying by simulation tool key parameters in detail such as the shape and size of the two-dimensional electron gas system. The impact of the geometry on the working principle of the nanodevice and the effects on current-voltage behavior will be described in order to acquire design guidelines that may improve the performance of the self-switching diodes when applied to low-power square-law rectifiers as well as elements in rectennas by appropriately setting the size of the components.

Keywords: surface states engineering, self-switching diode, L-shape, V-shape, W-shape, rectenna

1. Introduction

The terahertz frequency range of the electromagnetic spectrum represents a technology gap between the electronic and optical well-stabilized technologies. Thus, the obtention of systems designed to generate, detect, and process THz radiation still is of scientific attention by the full of potential applications. Meanwhile sources available nowadays are typically expensive, fallible, and voluminous. It is possible to extend the electronic and optic concepts, for the THz radiation with millimetric and micrometric wavelength, respectively. However, the use of electronic in the THz range needs small circuits and high-frequency cutoff, while optical requests deal with the dimensions necessaries to work in the THz range [1].

Electronic has been one of the most advanced technologies applied in the modern life, being imperative to extend the electronic concepts such as resistance, diode, and transistor to operate in the THz range with the aim to fill the so-called THz gap by small, efficient, and low-cost technologies. The use of electronic devices in THz implies both high carriers' mobility and short flight time of carriers between electrodes. The nanodevices are one of the most promising alternatives to the development of the THz technology. Recently, the semiconductor nanodevices have been employed for the generation and detection of THz radiation [1–6] showing an enormous potential for their application.

A wide variety of physics phenomena has been used to generate THz radiation by the manipulation of materials systems. THz radiation generation technology can be roughly classified into two categories: photonic and electronic. Phenomena such as the photo-Dember effect [2], Gunn oscillations [3], quantum cascade lasers [4], and difference-frequency mixing in nonlinear crystals [5] have been used to generate THz through the electronic approach. In these techniques, the semiconductor material has a crucial role, and its properties can be further improved by the use of nanostructures. On the other hand, the most applied THz detectors are classified into bolometers and photoconductors. The widely used bolometer is based on temperature variations caused by THz radiation absorption [6].

Important challenges arouse in the development of novel semiconductor nanodevices tuned in THz range to be operated at room temperature (RT). The nanodevice performance at 300 K is reduced principally for the high-impedance nature of this type of technology and the low carrier mobility. In order to improve the RT properties of nanotechnology semiconductor devices, very sophisticate deposition technology has been used where the layer thickness must be controlled as thin as less than one monolayer.

Molecular beam epitaxy (MBE) has been one of the most popular techniques to the fabrication of semiconductor nanostructures used in the development and improvement of nanodevices, achieving faster and efficient electronic and optoelectronic devices like solid-state lasers [7], solar cells [8], and high-electron-mobility transistors (HEMTs) [9]. MBE has allowed the expansion of low-dimensional semiconductor systems, where carriers are restricted to move into two, one, or zero spatial dimensions, becoming an attractive field for researchers due to the possibility to obtain quantum effects that are promisors to fill the so-called THz gap. For example, quantum dots (zero-dimensional restriction) are capable to emit radiation in the THz region as result of their energy states, which depend on its physical dimensions and act as recombination centers for carriers generated in the GaAs layers within the structure when they are exited with a femtosecond optical pump at 800 nm [10]. Additionally, MBE allows for the construction of semiconductor structures with atomically abrupt and flat interfaces that in conjunction with band-structure engineering permit the obtention of high-purity heterostructures embedding a two-dimensional electron gas (2DEG) [11]. In the heterostructures with 2DEG, carriers are confined to a very narrow layer of thickness smaller than De Broglie wavelength which can travel without scattering caused principally by impurities. The large mean free path of electrons in 2DEG is translated into high mobility in the order of 14.4×10^6 cm^2/Vs at low temperature. This high mobility is not possible to be obtained in traditional bulk materials [12].

The use of the modulation-doped semiconductor heterostructures with high-mobility electrons confined to two dimensions in field-effect transistors originated the so-called high-electron-mobility transistor (HEMT). In these devices, the low-temperature mobility was improved over time, leading to innovations in solid-state physics, high frequency, and low-noise electronics [9, 12]. The speed of transistors has been increased over time. HEMTs have demonstrated the highest frequency of operation over 1.5 THz when a combination of techniques such as reducing gate length, material purity, and doping schemes are used in their fabrication [11–13].

High carrier mobility can be proportionated by the appropriate design of modulation-doped heterostructure, where the 2DEG is present. Short flight time of carriers between electrodes is originated by size miniaturization where semiconductor devices enter into the ballistic transport regime, allowing them to be used in high-speed applications. Nevertheless, the current miniaturization technology reaches nanometric dimensions where the extent of surface atoms on semiconductor materials represents an important percent of the whole device, making the performance of the nanodevices be affected by these outmost atoms. Surface physics plays an important role for the functioning and performance of semiconductor nanodevices, being imperative to understand and predict how surface states, surface charge, and near-surface depletion regions can be manipulated in order to engineer nanoscale devices designed to operate at THz frequencies.

1.1 The semiconductor surface

The characteristics and properties that characterize a semiconductor material such as carrier concentration, band structure, or bandgap result from assuming that translational symmetry of infinite bulk crystal is present in all the material volume. However, when the three-dimensional symmetry in the occurrence of a surface is interrupted, it makes the bulk assumptions inadequate. Therefore, to understand the mode that surface properties can be manipulated to produce nanodevices capable to operate at THz range, it is necessary to understand the different behavior adopted by the surface and the effects over the whole semiconductor characteristics.

The sudden termination of the translational symmetry of the crystal lattice at surface originates surface atoms with unpaired electrons. In **Figure 1(a)** the ideal surface atomic arrange is exhibited. In this case, the topmost atomic bonds of surface atoms acquire the same configuration than bulk atoms, in a perfectly periodic two-dimensional surface. The surface bonds assume that lattice cell keeps repeating such if a virtual lattice is deposited above them. Atoms in an ideal surface present similar properties than those atoms from beneath layers in the bulk structure, exhibiting the condition of flat band, where the conduction band (E_C) and valence band (E_V) have the same energy position contrasted with the bulk region. This state is illustrated in **Figure 1(b)**.

The ideal surface is not usually exhibited in III–V semiconductors because free surface bonds involve excessively high-energy states. Surface bonds use both mechanisms to reduce energy: surface reconstructions and reaction with atoms in the atmosphere, usually oxygen. After, the three-dimensional symmetry of surface lattice is changed in the last nanometers. This condition is schematized in **Figure 2(a)**. Surface reconstruction and surface reactions produce electronic states. These states have no equivalent in the band structure of the bulk crystal. Consequently, in real surface the existence of surface states N_S should be considered, positioned inside the semiconductor bandgap, which generally serve as traps for free carriers.

The presence of N_S, sometimes called midgap states, at surface causes important effects on the carrier's distribution and Fermi-level (E_F) position relative to the E_C

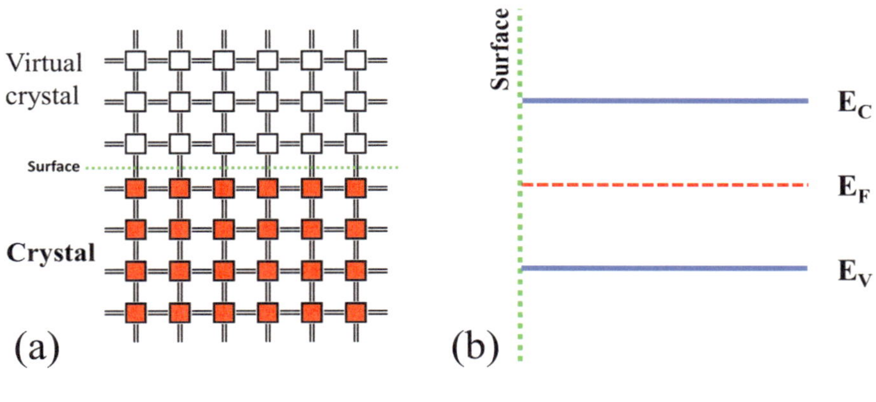

Figure 1.
Ideal surface representation of (a) crystal structure and (b) band diagram.

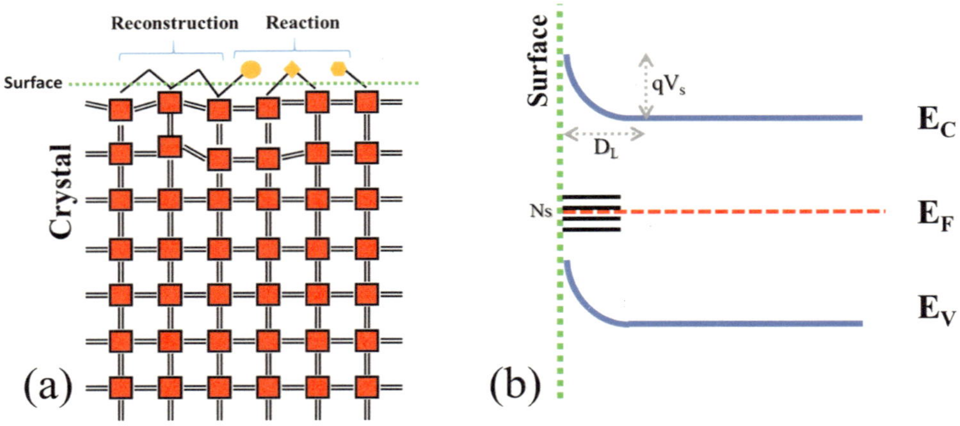

Figure 2.
Real surface representation of (a) crystal structure with surface atom reconstruction and reaction. (b) Band diagram with surface state effect.

and E_V near the surface. The N_S position inside the bandgap offers low energy levels in comparison with E_C and E_V for carriers, and two types of N_S can be recognized according to the charge that is acquired when occupied. Donor-type surface state is obtained when Ns exhibited a positive charge after filled in. On the other hand, acceptor-type surface states are obtained if filled by electrons [14].

In n-type semiconductor, when electrons migrate from their hypothetical bulk positions toward the acceptor-like N_S, two phenomena occur. First, the ionized host atoms lead to the formation of a depletion region, D_L, with a space charge layer. The charge neutrality principle is accomplished when a charge accumulation in a thin region of the surface, n_{SS}, is created by the filling of the N_S. This redistribution of charge is governed by Poisson's equation [15, 16], propitiating the formation of a built-in electric field (E_S) and a voltage (V_S) close to the surface related by

$$V_S = E_S \frac{D_L}{2}. \tag{1}$$

The existence of N_S in the bandgap produces the so-called effects of Fermi-level pinning and the band bending at surface by the formation of a surface barrier with height of $\Phi = qV_S$.

With the purpose to avoid surface effects, a lot of methods have been employed, e.g., altering the chemical composition of the surface modifies the energy of the N_S [16]. It is crucial to recall that band bending is produced by n_{SS} accumulated at semiconductor-environment interface, being possible to modulate the charge density at surface by illumination that would generate electron-hole pairs, which are recombined at surface. Another form to modulate the carrier's population in N_S is by applying an electric field that will exert a selective attraction/repulsion process over the carriers. Accordingly, in surface state engineering, it is imperative to understand and predict how N_S, n_{SS}, and D_L can be manipulated in order to engineer nanoscale devices capable to operate in the THz region.

2. The self-switching diode

The THz systems for detection are roughly divided into two groups according to the detection of amplitude and phase (coherent heterodyne systems) or only amplitude of the signal (noncoherent forward detection systems). The most applied

THz detectors are the bolometers, Schottky barrier diodes, pair-braking detectors, hot electron mixers, and field-effect transistors [17]. Special interest in detector lies in the diode concept, since it is one of the more useful components in electronic engineering, allowing current flow only in one direction.

The diode concept has a very large variety of applications that stimulate the application of semiconductors in electronics. Nevertheless, the cut-off frequency is still under improvements today. For THz detection, ultrafast metal-insulator-metal (MIM) and metal-insulator-insulator-metal (MIIM) tunnel barriers or p-n semiconductor junctions (reverse Esaki tunnel diodes) have been used [18, 19]. Nonetheless, they show low rectification capacity due to their poor diode-like behavior.

One alternative to get a diode device able to operate in the THz regimen is the self-switching diode (SSD) which was first proposed and developed by Song [20]; this device is schematized in **Figure 3(a)**. The properties and performance of the SSD are dependent on both the characteristics of the heterostructure and the grooves etched on it. First, the two-dimensional electron gas (2DEG) created in the heterostructure and illustrated in **Figure 3(b)** is the core of the SSD device, and it is possible to increase the mobility by an adequate layer sequence design. The modulation-doped heterostructures are an excellent example of the importance of the 2DEG properties, since they lead to the development of the high-mobility devices such as the HEMT. The second important element in the SSD structure is the insulating grooves, which can be realized by standard electron beam lithography and wet etching [20].

Due to the groove fabrication process, the symmetry of the crystal structure in the trenches wall is broken, creating N_S. The filling process of N_S produces the formation of a lateral n_{SS} in the trench walls and therefore a D_L inside the nanochannel. The working principle of the SSD is the modulation of the lateral n_{SS} by the bias applied in the ohmic contacts of the device. The current only flows when the anode contact has a positive voltage value with respect to the cathode contact and the grooves force the current to flux only by the nanochannel. In the SSD fabrication process, only one step of lithography is necessary, reducing the cost of production and the use of complex mask alignment steeps [21].

The SSD device has been demonstrated using heterostructures where the 2DEG is presented such as systems based on InAlAs/InGaAs and AlGaAs/GaAs multilayer structures [21, 22]. The working principle of the device is not dependent on the 2DEG properties, allowing to be fabricated in bulk materials like silicon [23] and transparent semiconductors like ZnO and indium tin oxide (ITO) [24, 25], and graphene-based SSD has been also demonstrated [26]. Those devices have exhibited

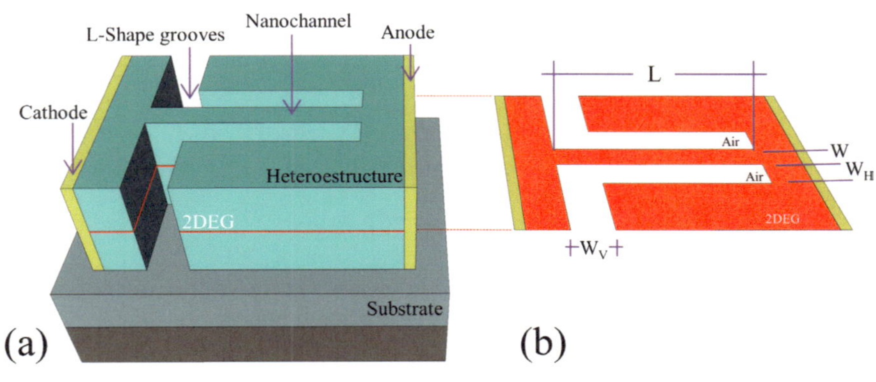

Figure 3.
(a) Self-switching diode layout structure, based on a 2DEG-containing heterostructure. (b) Top view of two-dimensional SSD L-shape topology. The parameters and the nomenclature to define the L-shape SSD geometry are channel length L, channel width W, vertical W_V, and horizontal W_H groove thickness.

the ability to detect extremely weak signals without applied bias [27, 28]; their high sensitivity [29] and their capability to operate in the terahertz regime [27, 30] make the SSD concept a rewarding tool that has opened a broad range of applications that support the THz gap filling.

2.1 Simulation analysis details

The SSD schematized in **Figure 3** represents a real device where it is appreciated that basically the SSD is made with two L-shape grooves etched on a heterostructure with a 2DEG. Hence, it is possible to analyze the SSD performance from two points of view: the heterostructure (one-dimensional) and the groove (two-dimensional) geometry in order to avoid three-dimensional numerical analysis required to simulate the electrical behavior of the SSD device.

The heterostructure component of the SSD can be modeled by Schrödinger-Poisson equation, considering the material and layer sequence in order to get the band profile and calculate the electron distribution in the samples [16]. This type of analysis is useful to determinate the properties of a 2DEG in the heterostructure, which gives the carriers high mobility required to operate in THz. **Figure 4(a)** shows the layer sequence of an InGaAs-/InAlAs-based heterostructure, for which the SSD concept has been experimentally proved [20]. The sequence, material, and thickness of each layer are used as input parameters of the 1D model, else the delta doping (δ-doping) concentration.

Figure 4(b) exhibits the 1D simulation output; the profile of the conduction band of the heterostructure of **Figure 4(a)** is plotted. As observed in the simulation, at the interface InGaAs/InAlAs, the E_C bends toward energies under the Fermi level producing a triangular quantum well with one allowed energy state (E_1). When electrons populate the E_1 sub-level, electrons can move freely along two dimensions. Thus, the electrons are confined within a 2D sheet embedded in a 3D heterostructure. High concentration and high mobility are exhibited by the electrons trapped in the triangular quantum well, making the heterostructure suitable for THz devices.

With the aim of determining the SSD current-voltage behavior (I-V), we employed a commercial physically based device simulator in two-dimensional mode, similar to that reported for 2DEG-based InGaAs SSD and silicon-on-insulator structures [23, 30, 31]. In this model, only the 2DEG layer geometry resulted after the lithography process is analyzed. This type of numerical analysis requires a 2DEG input, where the top view of the SSD and electrical properties of

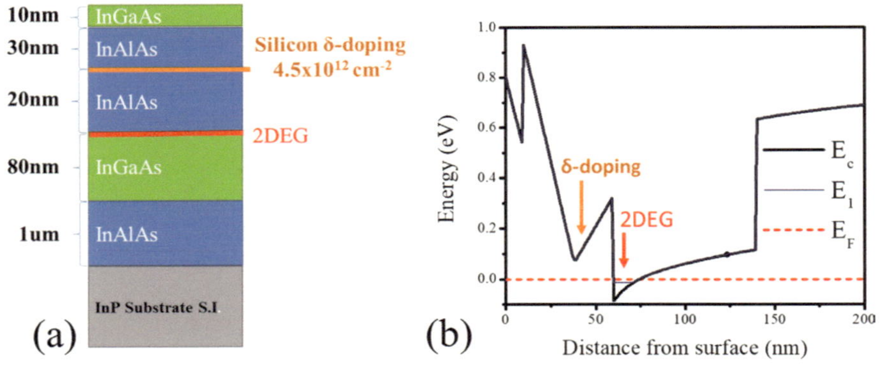

Figure 4.
(a) Layer sequence of the heterostructure analyzed in this work [20]. (b) 1D analysis of conduction band edge energy diagram of the structure at room temperature. The red line indicates Fermi-level position.

the 2DEG are considered as simulation inputs. The SSD layout of **Figure 3(b)** is the schematic top view of an L-shape SSD, showing the geometrical frameworks used in the simulation. The electrodes shown in **Figure 3** have been labeled as anode and cathode according to the convention that in the ideal diode, the carriers flow from anode to cathode.

The background doping and the electron mobility of the 2DEG samples utilized in the simulation were 1×10^{17} cm^{-3} and 12,000 cm^2 V^{-1} s^{-1}, respectively, under no impurity scattering. The dielectric material filling the grooves was air ($\varepsilon_r = 1$), inducing a surface charge density of $n_{SS} = 0.4 \times 10^{12}$ cm^{-2}. The contact resistivity used in the simulation process was 1×10^{-8} Ω cm^2, which is similar to standards that exist for THz devices [32]. Finally, the energy balance model at 300 K was used. The model was employed to perform a systematic geometrical analysis and comparing different types of self-switching structures directed to provide an enabling technology for energy harvesting in the terahertz region. Since the self-switching diodes combine size and electronic transport property effects, the performance of different SSD was analyzed and optimized by changing their shape configuration.

2.2 Self-switching diode working principle

The key to understand the DC performance of the SSD lies on the modulation of the n_{ss} inside the N_S by the built-in electric field, producing a modification in the D_L along the nanochannel. In this way, some authors indicate that the SSD can be understood as a two-dimensional field-effect transistor with gate and drain short circuited where flanges act as a double lateral-gate terminal [28, 33]. The principal mechanism that controls the carrier transport of the SSD is explained in terms of the free carriers that can participate in conduction phenomena and travel through the nanochannel [22]. To expose the SSD performance and their rectifying nature, it is necessary to explore the L-shape device shown in **Figure 5(a)** used as simulation input parameter and termed as the reference geometry. The result of the numerical study is the electron-charge distribution along the 2DEG plane illustrated in **Figure 5(b)** when both electrodes are unbiased.

With no applied bias, simulations show a drastic reduction of the background carrier density inside the nanochannel as a consequence of the electron filling of the N_S created at groove walls. Carrier density decays from the background rate (1×10^{17} cm^{-3}, in orange-red color) to 10^{14} cm^{-3} (cyan color) along the length of the nanochannel and decreases steadily closer to the groove walls, remaining

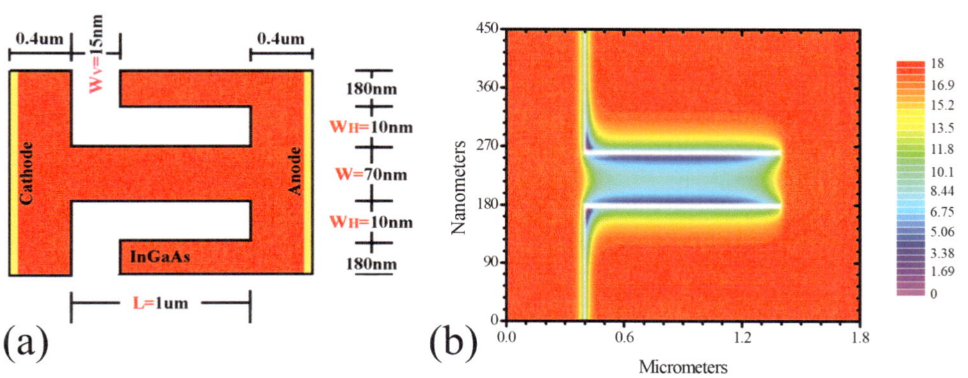

Figure 5.
(a) Top view schematic representation of the proposed L-shape self-switching nanochannels. (b) Electron charge distribution along the 2DEG plane inside the nanostructures under zero bias.

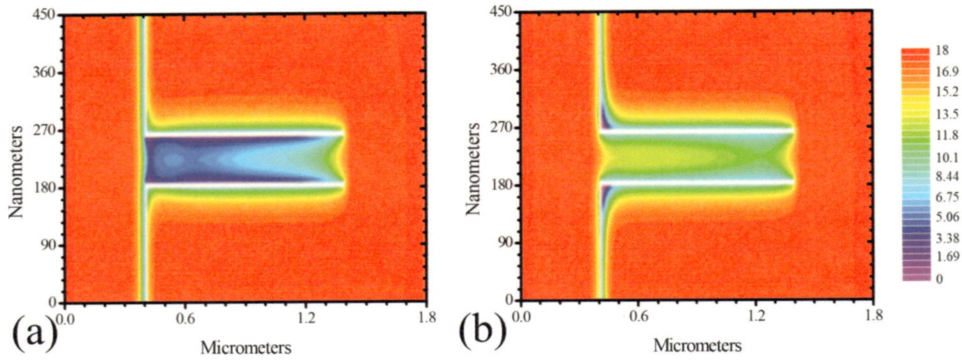

Figure 6.
Electron density distribution (logarithmic scale) for the SSD with the geometry depicted in **Figure 5(a)** *at (a) V = −0.5 (reverse bias) and (b) V = 0.5 (forward condition).*

homogenous all along the L-shape channel, as it is displayed by **Figure 6(b)**. In other words, the D_L created inside the nanochannel limits the free carriers available to participate in conduction phenomena defining high-resistive or not conductive nanochannel.

For reverse bias (V = −0.5 V in the anode, while cathode is set to ground), the charge depletion effects are enhanced along the channel due to the transversal electric field that appears at the groove walls, lowering the carrier density inside channels, according to **Figure 6(a)**. For reverse state, the D_L inside the nanochannel is raised by two orders of magnitude in comparison with the zero-bias case, making a pinch-off condition that prevents any current to flow along the nanochannel due to the lack of carriers. However, reverse current can be established when the bias is high enough to overcome the lateral depletion effects.

Figure 6(b) displays the carrier distribution at V = +0.5 V for the reference L-shape SSD as forward bias which is settled when a positive voltage is applied to the anode. Under forward polarization the n_{SS} in the N_S of the SSD trenches is deflated, reducing the D_L inside the nanochannel. On this condition, the carrier density increases several orders of magnitude ($\sim 10^{15}$ cm^{-3}), close to the background concentration when the D_L is reduced by the transversal electric field. The electron transport along the channel is possible at this condition by the intensifications of the free carrier concentration.

The analysis of the carrier distribution along the 2DEG, where the SSD has been fabricated in addition with the extension/modulation of the D_L inside the nanochannel, indicates that it is possible to control the current flux that travels between electrodes by controlling the free carrier inside the nanochannel. This modulation is originated by the sign of the applied voltage in the anode, making the SSD a rectifier element capable to operate in the THz region by the high-mobility nature of the 2DEG of the heterostructure where the nanometric dimensions of the L-shape grooves are lithographed.

In DC injection mode, the performance of a rectifier element can be simply analyzed by the nonlinearity of their current-voltage curve, the threshold voltage (V_{TH}), and the leakage current present in reverse bias. These properties can be examined by the D_L width inside the nanochannel. The zero-bias simulations indicate that there is a minimum carrier concentration, which propitiates the formation of the D_L necessary to be modulated by the applied bias, being dependent of the groove fabrication process and the nanochannel shape, especially.

Therefore, in the SSD initial design, it is mandatory to achieve a nanochannel with the appropriated carrier concentration at 0 V. As it is going to be analyzed

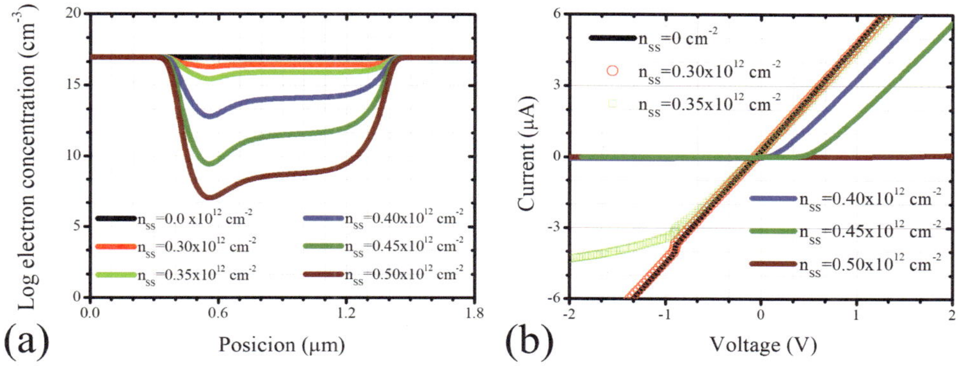

Figure 7.
The effect of the surface charge density (n_{SS}) on (a) the electron concentration along the device and in (b) the current-voltage characteristics.

in the following sections, the nanochannel shape, width, and length ensure that the SSD exhibits diode-like behavior and would provide an estimation of the V_{TH} necessary to turn on the conduction of the SSD. Nevertheless, the N_S are difficult to control in a desire range though they can vary considerably depending on the L-shape groove fabrication techniques such as etching methods, leading to crucial effects on the DC and AC performance of the SSD device.

According to surface physics, the n_{SS} agglomerated in the interface semiconductor environment is a manifestation of N_S. The SSD fabrication determines the available surface states. When the device processing augments the N_S density, more electrons will be used to fill these states, magnifying the D_L inside the nanochannel. Therefore, the rise (reduction) of the n_{SS} has the effect to decrease (increase) the free carriers' density along the channel. **Figure 7(a)** exhibits the carrier distribution along the SSD device from the nanochannel calculated at 0 V when the n_{SS} was varied from 0 to 0.5×10^{12} cm^{-2}. It is clear to see that for the $n_{SS} = 0$, there is not a modification in the free electron concentration along the channel, declining considerably for $n_{SS} > 0.35 \times 10^{12}$ cm^{-2} and enhancing as n_{SS} increases.

Hence, the carrier density inside the nanochannel and their modulation are used to propitiate the rectifier behavior of the SSD device. The current-voltage curve of the SSD-based structure ($n_{SS} = 0.4 \times 10^{12}$ cm^{-2}) is exhibited in **Figure 7(b)** where the diode typical I-V curve is presented. In the case of $n_{SS} = 0$, the I-V is completely lineal; in other words, the absence of N_S produces that the SSD works like a resistor device. When the $n_{SS} = 0.35 \times 10^{12}$ cm^{-2}, the nonlinearity of the I-V curve is exhibited, with a V_{TH} near to zero, a desired value for low-power signal detection. Nevertheless, in this case the reverse bias exhibits an important leakage current. For $n_{SS} > 0.35 \times 10^{12}$ cm^{-2}, the reverse current is removed notoriously, improving the diode-like performance, while an excessive n_{SS} steps the V_{TH} up, improving the nonlinearity and avoiding leakage current.

According to the simulation results, carrier density exhibited by the non-polarized structure indicates that a minimum bias, or voltage threshold, is required in order to propitiate electron conduction. In that case, the lower the carrier concentration is, the higher the V_{TH} results. For short D_L length, low bias is required to turn the SSD on; nevertheless, an important leakage current is present. By modulating the dimensions of the depletion region inside the nanochannel, the L-shape SSDs are found to be suitable for detection of very weak signals, making the SSD concept an important tool in the development of high-frequency integrated circuits since it has been demonstrated that by choosing appropriate geometrical dimensions, the rectification efficiency can be optimized [33, 34].

3. Self-switching diode design guidelines

In order to get a rectifier with cut-off in the THz region by the SSD device, it is important to follow design considerations that may improve the current rectification with low V_{TH} and optimize the SSD up for the best efficiency. These considerations are related to the geometry-depletion region relationship between the SSD dimensions, shape, and the D_L extension. In this section, we evaluate the rectification performance of the L-shape SSD device in DC injection mode by a systematic analysis where the width, W; the length, L; and the vertical and horizontal trench width (W_V and W_H) of the channels are varied in order to get their effect on the I-V behavior of the device. The layout shown in **Figure 5(a)** defines the reference geometries where we have a varied chosen parameter, while the others were fixed.

The simulations were carried out in order to obtain the I-V curves of each geometrical variation with the aim to evaluate the ability of channels to detect the electrical energy of low-voltage THz signals. On this way, the SSDs can be considered as square-law rectifiers that generate electrical power from the incoming THz radiation. In order to understand the size effect on the detection, we use three criteria that define the performance of the SSD. Fist, differential resistance, R_V, of the SSD can be found using the I-V response, key issue for determining the impedance matching between the channels and the signal acquisition system, and it is defined by

$$R_V = \left(\left. \frac{dI}{dV} \right|_{V=0} \right)^{-1} \tag{2}$$

The nonlinearity (N_V), which expresses the diode-like I-V curve behavior, can be determined by evaluating

$$N_V = \left. \frac{d^2 I}{d V^2} \right|_{V=0} . \tag{3}$$

N_V is a main issue for the performance of the rectifier because of higher nonlinearity result in a larger DC output. The impact of R_V and N_V is resumed in the sensitivity, γ_0, of the device, which is calculated by $\gamma_0 = N_V R_V$. R_V, N_V, and γ_0 are usually studied at zero bias, heeding the fact that detection must be performed with no feed bias [35].

3.1 Geometry dimension impact on I-V characteristics

In **Figure 8(a)** the effect of modification in the channel width (W) through I-V curves is presented. It is observed that when W is widened, V_{TH} is reduced as a consequence of the spatial restriction of the electrons that could occupy the surface states inside the grooves in order to maintain charge neutrality. Large D_L indicates high diode resistance and requires high voltage in order to deplete or drain the surface states, producing that R_V goes from 143 MΩ to 233 kΩ when W is broadened from 60 to 80 nm, respectively.

According to Ohm's law, the current is improved as the thickness W is widened by the reduction of R_V. Therefore, by the design of the channel width, it is possible to control the SSD's worth V_{TH} and R_V. Nevertheless, large W generates reverse current at low negative bias, and, as a result, the diode-like character disappears lowering γ_0. Therefore, in order to avoid a reverse current flow and design for a low-threshold voltage V_{TH} device, it is necessary to consider the relationship between the surface charge and W [22, 30].

Figure 8(b) shows I-V characteristics of SSDs with several channel lengths, L, where it is appreciated that V_{TH} shifts to lower voltages when L is reduced due to the

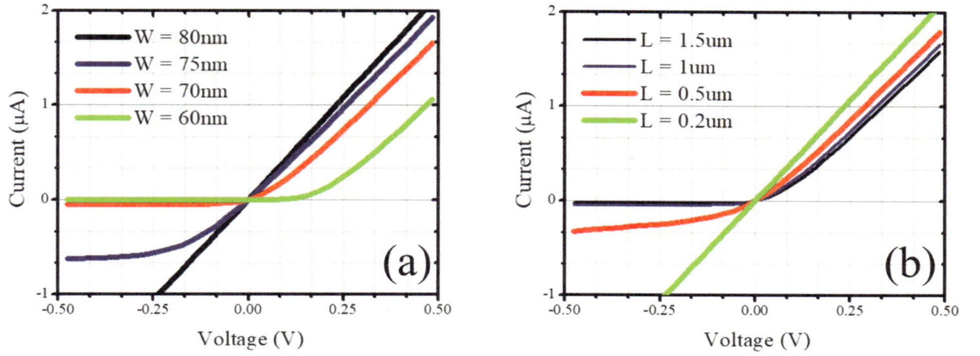

Figure 8.
Current-voltage behavior of the L-shape SSD-based geometry when (a) the channel width (W) parameters are varied from 60 to 80 nm and (b) the channel length (L) is modified from 0.2 to 1.5 μm.

decrease in D_L inside the nanochannel. Small D_L length facilitates ballistic transport; in the meantime that carrier's mean free path is larger than the D_L and reduces R_V. For this case, when L is reduced from 1.5 to 0.5 μm induces a change on R_V from 1.3 MΩ to 423 kΩ. Nevertheless, the reduction of D_L through the reduction of L favors the presence of an important leakage current for reverse bias, affecting the I-V nonlinearity and inducing a reduction in the sensitivity of the devices.

Therefore, for 2DEG-based SSD intended to work at THz frequencies, it is necessary to reduce the carrier's time of flight between electrodes, which can be achieved by improving the electron mobility of the 2DEG, the carrier's mean free path getting better or shortening the L extension, considering that at THz frequencies the electron transport is in the ballistic regime [30]. To design a cost-effective SSD THz device, the election of the appropriated length L for the working frequency in function of the 2DEG mobility is mandatory, keeping γ_0 as high as possible.

The impact of the geometrical dimensions on the I-V characteristic of the SSD was performed by the numeric analysis for all geometric parameters indicated in **Figure 5(a)**, and it is resumed by the nominal sensitivity of the SSD in **Figure 9(a)**. For example, the reduction of the W thickness increases the magnitude R_V while weakens the rectifier performance of the SSD by the apparition of an important leakage current, indicating that the N_V is reduced with this modification. The γ_0 reaches 37 V^{-1} for the narrower W of 60 nm, being reduced for wider W. On the other hand, the largest nanochannel length (L = 1.5 μm) originates that γ_0 = 25 V^{-1}, and conversely the sensitivity is reduced by the reduction of the nanochannel length, rendered in **Figure 9(a)**.

The reduction of D_L width and length is not the only mechanism to modify the DC performance of the SSD. Controlling the magnitude of the electric field that modulates electrons in the N_s would have the effect of improving the I-V behavior. In this task, both the vertical and horizontal width trenches (W_V and W_H, respectively) are the main geometrical elements that can be used to achieve modulation of surface states. It was found that W_V in the range of 5–50 nm did not affect the I-V performance of the SSD and consequentially the γ_0 is constant for these modifications and being reduced with W_V > 50 nm. On the other hand, small changes in W_H affected strongly the DC performance of the SSD; meanwhile variation of W_H from 5 to 30 nm resulted in a variation of V_{TH} from 0 to 0.2 V [33, 34], and γ_0 is optimized for W_H = 30 nm. The effect of W_V and W_H trenches is seen in **Figure 9(a)**.

The SSD ability to rectify signals in the THz region relies in the γ_0 value, which can be improved for nanochannel geometry that exhibits long and wide D_L, amplifying R_V and N_V. Nevertheless, narrow W and large L sizes imply to enlarge the V_{TH}.

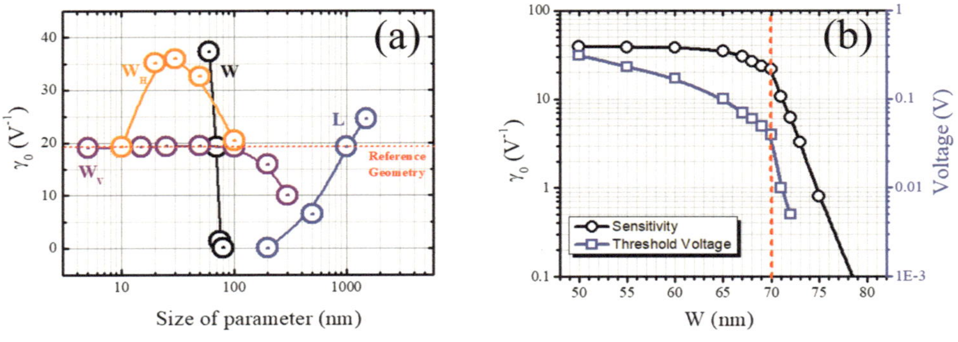

Figure 9.
(a) Sensitivity of the L-shape SSD devices as a function of the size of the geometrical parameters. (b) Effect of modification of W in the V_{TH} *and* γ_0.

The low-power THz radiation detection needs a rectifier with a threshold voltage close to zero, and according to the simulations, the channel width is the element that strongly defined the γ_0 and V_{TH} performance of an L-shape SSD. Numerical simulations were carried out modifying the W to explore the dependence of γ_0 and V_{TH} with the aim to determine the dimensions of W that propitiate $\gamma_0 > 20\ V^{-1}$ y $V_{TH} < 50$ mV (see **Figure 9(b)**). These desirable conditions are stablished in order to overcome the current performance of the best MIM-tunnel barriers, which exhibit sensitivities between 2.5 and 4 V^{-1} [36, 37].

According to the graph in **Figure 9(b)**, it is possible to obtain $\gamma_0 > 20\ V^{-1}$ for $W \leq 70$ nm, improving the rectifier performance of the SSD. On the other hand, when W < 70 nm, the V_{TH} is higher. The best configuration is labeled by the dotted red line at W = 70 nm. For W < 70 nm, the sensitivity of the SSD is improved, expanding the voltage necessary to turn on the device. Conversely, for W values at the right hand of the dotted red line, the threshold voltage necessary to produce a conductive nanochannel is cut down in the same way with that of γ_0. In other words, the geometrical design of the SSD must consider the application where the SSD is going to be used. The design should consider the power of the signal to be detected in order to choose the tolerance margin between γ_0 and V_{TH} that improves the performance of the SSD.

Rendering to numerical results, the SSD devices combine geometrical effects with the 2DEG properties to act like a rectifier, the nanochannel and trench width in conjunction with the appropriate choice of channel length being key parameters. The optimization of W, W_H, and L induces that the nominal sensitivity of the devices raised to measurement around 40 V^{-1}. Self-switching diodes are good candidates for zero-bias detection and harvesting applications as square-law rectifier elements in the rectenna concept. Nevertheless, the high-impedance nature of the SSD will introduce drawbacks that limit the power transfer between the rectifier and the antenna as it will be shown in the next section.

3.2 Rectennas based on self-switching diodes for THz detection

When the L-shape SSD has been optimized to obtain the best rectifier performance, it is possible to use the device in the detection or harvesting of THz electromagnetic signals. In order to evaluate the efficiency of the SSD as rectifier element in the rectenna concept, we analyze the SSDs coupling to gold-made Archimedean spiral antenna lying on the surface of a semi-infinite SiO_2 substrate and designed to operate between 0.25 and 2 THz with a right-hand circular polarization [38]. **Figure 10(a)** illustrates this antenna design. The simulated performance of the designed micrometric

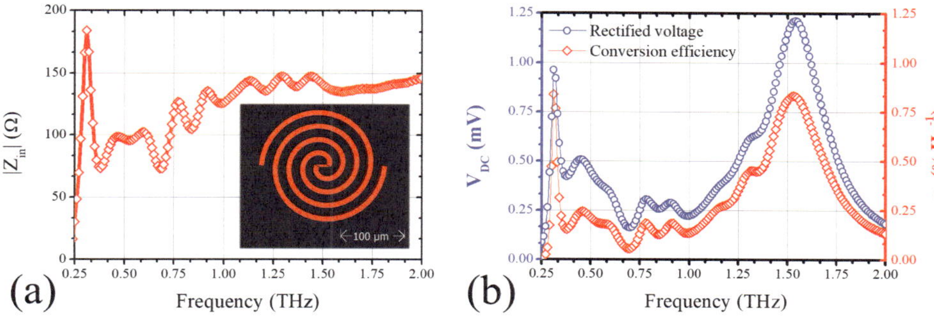

Figure 10.
Performance of L-shape SSD-based rectenna to harvest the optical energy of a terahertz source obtained by the simulation process. (a) Input impedance. Inset shows the Archimedean spiral antenna. (b) Rectified voltage and optical-to-electrical conversion efficiency.

antenna was obtained when it is excited with monochromatic plane waves with an irradiance of 1 W/cm^2 [39] incoming from a precisely tuned terahertz laser. In **Figure 10(a)** the antenna impedance, Z_{in}, is plotted as a first result. In this numeric analysis, we assume that the cut-off frequencies reported for high mobility structures SSD are around 5 THz [22].

The DC voltage generated from the received terahertz radiation by the SSD rectenna is illustrated in **Figure 10(b)**, where a full width at half-maximum parameter of more than 1.8 THz is exhibited with a maximum voltage obtained of 1.25 mV at 1.5 THz. The conversion efficiency of the rectennas is also shown in **Figure 10(b)** as a function of frequency where the total efficiency is found to be ~0.032%. This numerical analysis indicates that in this configuration, the use of L-shape SSDs as rectifiers instead of MIM nanometer diodes enhances the efficiency of the rectennas by a factor of 10^6 [38]. Rectennas based on SSDs are more efficient than those based on MIM barriers due to their I-V characteristic curves, which are more asymmetrical. The relatively low efficiencies (~10^{-2}%) obtained by this SSD rectenna is accredited to the mismatch impedance between the rectifier and the antenna.

From **Figure 10** it is clear that the major voltage and efficiency of the rectenna are obtained when the Z_{in} is raised to about 125 Ω. Higher antenna impedance means a better electrical coupling between the antennas and the diode and more efficient transfer of energy between antenna and rectifier [38]. The coefficient of reflection between both elements is defined by

$$\Gamma = \frac{R_V - Z_{in}}{R_V + Z_{in}}, \tag{4}$$

which exhibits a value around 99.96% for the rectenna system analyzed in this work. In other words, the high magnitude of Γ indicates that only 0.04% of the optical energy is transmitted from the spiral antenna to the L-shape SSD and the rest of the optical power is returned to the spirals [38].

The SSD device performance as rectifier element is modified under unmatched conditions, making necessary to study the R_V behavior on Γ. We use a reference source with a Z_{in} = 50 Ω in order to define the effective sensitivity ($\gamma_{50\Omega}$) to considerer the effects of the unmatched source through the relationship [28]:

$$\gamma_{50\Omega} = \gamma_0\left(1 - |\Gamma|^2\right), \tag{5}$$

where Γ is calculated using the Z_{in} = 50 Ω. The $\gamma_{50\Omega}$ criteria describe the performance of the SSD when their electrodes are short circuited to the antenna arms.

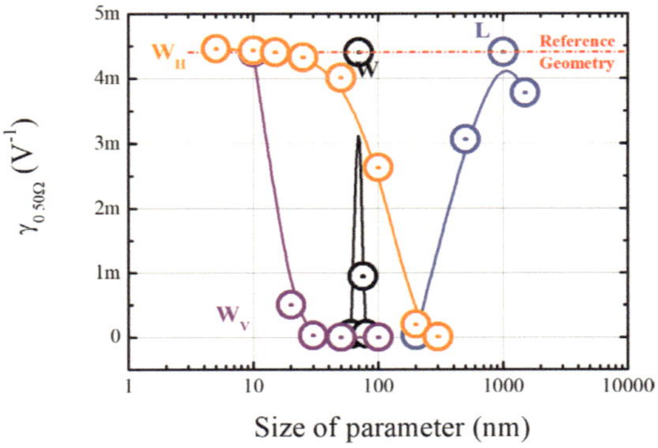

Figure 11.
Effect of the modification of the SSD reference geometry parameters on the effective sensitivity $\gamma_{50\Omega}$.

The trade-off between $\gamma_{50\Omega}$ and R_V is depending on the geometrical parameters; this relationship is illustrated in **Figure 11**.

The numerical results illustrated in **Figure 11** have shown that the reference geometry is the best configuration possible even with the high Γ when they are plugged to a Z_{in} = 50 Ω drive source. This confirms that even when the narrower channels are more sensitive than other structures, they are not the best option to generate and transmit DC electrical power thanks to the high value R_V which heightens the Γ, reducing the power transfer between the antenna and the rectifier. This study indicates that those parameters of the channels that would propitiate to reach higher sensitivities γ_0 are the same with that which produces high $\gamma_{50\Omega}$: the nanochannel width (W), length (L), and the thickness of W_V.

The SSD as square-law detector or rectifier element in energy harvester systems represents an interest alternative, improving the performance of the rectennas based on MIM tunnel barriers, which exhibit very low efficiencies and reduce the technological process required for the rectennas' manufacture, involving only one step of high-resolution lithography. However, to produce terahertz rectennas useful for energy-harvesting applications, an efficient strategy to better match the impedance of the spiral antennas and the self-switching nanodiodes should be incorporated.

4. Self-switching diode shape variations

According to the rectenna performance, a key issue that reduces the use of the self-switching diodes in detection or harvesting applications concerns their high resistance by the nanometric nature of this device. The impedance matching is very difficult, and the losses of the electrical power by reflection will be predominant. The SSD geometry plays the fundamental role defining the DC performance of the device, the design process being a trade-off between geometrical dimensions that exhibits low reverse current in a nonlinear I-V curve with a V_{TH} close to 0 V and, at same time, low channel resistance in conjunction with a high sensitivity quantity. In this section we explore three different SSD shapes to improve their performance as low-power THz rectifier.

The first method to intensify the performance of the SSD is to raise the rectification efficiency by augmenting the numbers of paths through which the carriers travel. In this task, **Figure 12(a)** shows the schematic array of four L-shape SSDs, connected in parallel with geometrical dimensions based on the single SSD of **Figure 5(a)**. The numerical analyses of arrangement for 1, 2, 4, 8, and 16 single

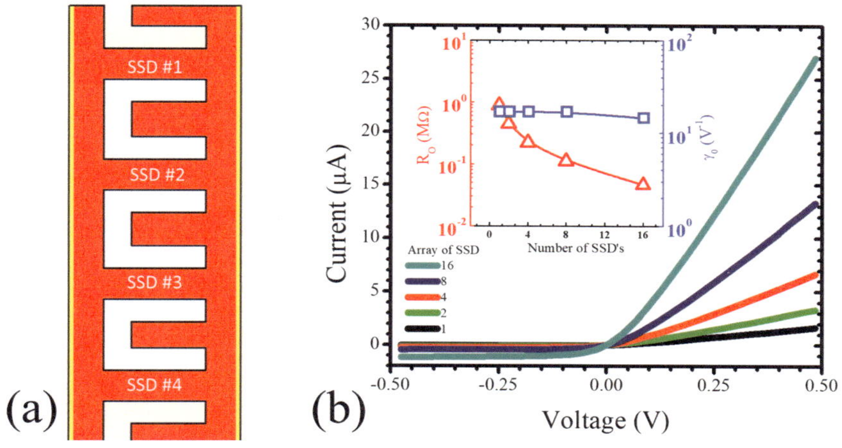

Figure 12.
(a) Top view layout of four L-shape SSDs in parallel array. (b) I-V characteristics of parallel coupled SSD arrangements. The inset shows the R_o and γ_o as functions of the number of SSD elements.

SSDs placed in parallel are exhibited in **Figure 12(b)**. The I-V behavior of the SSD arrays exhibits a small gain of leakage current, while they show threshold voltage of ~0 V. The current is incremented from 1.72 (single SSD) to 28 μA (16 SSD array), representing 16 times larger current in parallel than for a single SSD when the SSD system is biased at 0.5 V.

The augment of SSD elements present in the arrays is translated into a better performance in the current behavior, generating a reduction in the overall resistance, R_O, of the array-based SSD. The changes in device resistance could be considered as $R_O = R_V/N$, where R_V refers to the resistance of the single L-shape element as it is shown in the inset of **Figure 12**. Accordingly, the effect of increasing the number of nanochannels produces R_o changes from 87 to 41 kΩ for a single SSD and 16 SSD arrays, respectively. This reduction in the nanochannel resistance is translated into a reduction in the γ_0 of ~25% as it is plotted in the inset of **Figure 11**. Contrarily, the $\gamma_{50\Omega}$ exhibits a modification from 4×10^{-3} for a single base SSD to 63×10^{-3} V^{-1} for a 16 array-based SSD [33]. It is clear that the major quantity of SSD in parallel results in the reduction of the device resistance, which improves the impedance matching between rectifier elements and antennas by using the appropriate number of SSD [40].

Figure 13(a) exhibits the top view and carrier's distribution for the termed V-shape SSD. In this case, the shapes of the grooves have been modified to reduce the extension of D_L inside the nanochannel by expanding the channel area with the aim to reduce the R_V of the SSD device. According to the SSD working principle, the formation of the D_L along the nanochannel defines the I-V performance of the device. The structure of **Figure 13(a)** and the 2DEG properties described in Section 2.1 were used as input parameters in numerical simulations to analyze the carrier concentration along the V-shape SSD.

When no bias is applied, simulations in **Figure 13(b)** show a drastic reduction of the background carrier density with a width-dependent spatial charge distribution in this shape where electron concentration diminishes close to the channel vortex, reaching populations as low as 6.3×10^{12} cm^{-3} with effective width of 50 nm in a localized region. Reverse polarization is established for V = −0.5 V in **Figure 13(c)**. The pinch-off condition does not occur at −0.5 V occasioned by the small D_L along the channel length, contrasting to the case of the L-shape geometry. Therefore, the carrier distribution can be easily modified to allow the current to flow even at a low reverse bias because of the nanochannel receiving similar carrier concentration for

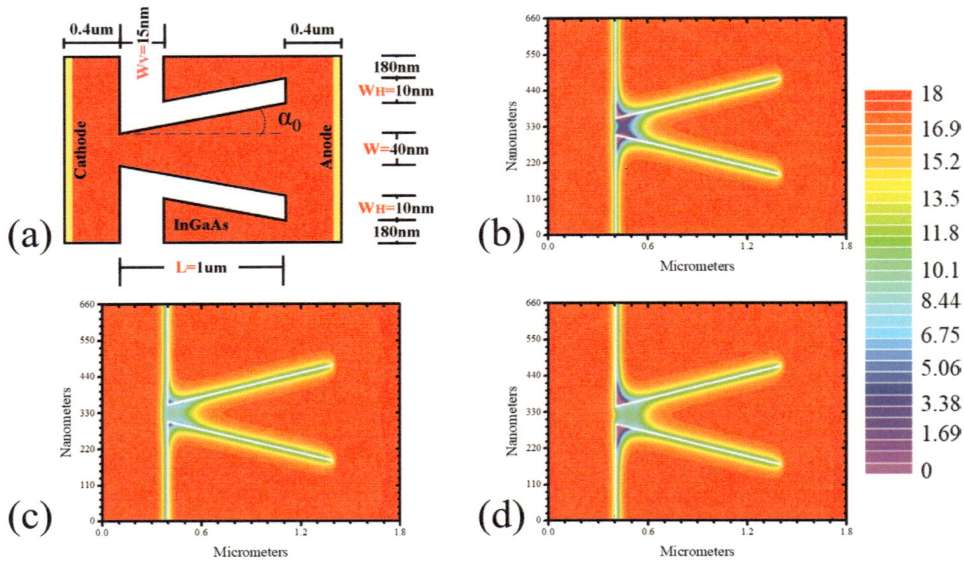

Figure 13.
(a) Schematic representation of the V-shape self-switching diode. Electron charge distribution along the 2DEG plane under (b) zero, (c) reverse (V = −0.5 V), and (d) forward (V = +0.5 V) voltage bias.

reverse and forward condition (at 0.5 V bias), where the redistribution of charge reduces the size of the D_L and enlarges the free carrier concentration. Forward condition is illustrated in **Figure 13(d)**.

Figure 14(a) shows the I-V characteristics for the V-shape SSDs with the reference geometry depicted in **Figure 13(a)** in red solid-line plot where the weak diode-like I-V behavior is appreciated. For this V-shape design, a V_{TH} of ~100 mV and the presence of important current-leakage at reverse breakdown voltage of ~140 mV are presented. Both phenomena are consequence of the short D_L width (along the channel) that is easily filled up when a reverse bias above this voltage is applied. Hence, the D_L extension for the L-shape is larger than the one exhibited by the V-shape SSD, according to **Figures 5(b)** and **13(b)**, respectively. This produces that the V-shape SSD needs a smaller voltage to reach V_{TH} and makes the electrons to flow. Nevertheless, under reverse polarity a leakage current can appear for voltages as low as −0.25 V, reducing the nonlinearity of the device.

The I-V characteristics of V-shape SSDs are sensitive to modifications in their dimensions, as in the case of the L-shape. The performance of the I-V when the channel width, W, is varied is presented in **Figure 14(a)**, while the inset displays the DC behavior when the V leaning angle (α_0) is modified. For this shape, the small D_L between the trenches implies that the diode-like characteristics can be observed only at small W < 45 nm and for thicker size the I-V has a resistor-like behavior. To avoid a reverse current flow and design for a low-V_{TH} V-shape device, to study the relationship between the D_L and α_0 is required. The leaning angle controls the number of free carriers that could occupy the N_S. Therefore, $\alpha_0 > 7°$ is translated in a decay of the D_L width, producing low R_V and V_{TH} in conjunction with poor diode-like behavior. Opposite effect is found to narrow V leaning angles [34] as seen in the inset of **Figure 14(a)**.

Figure 14(b) resumed the impact of the geometry parameters that conform the V-shape SSD by the analysis of the $\gamma_{50\Omega}$ framework where the maximum sensitivity for this analysis is less of $2 \times 10^{-3}\ V^{-1}$ caused by the fact that the low R_V and N_V reduce drastically γ_0. According to **Figure 14(b)** in this shape, all modifications have an important role defining the DC characteristics, producing that the design of this device to be more sensitive to the 2DEG and heterostructure properties. The

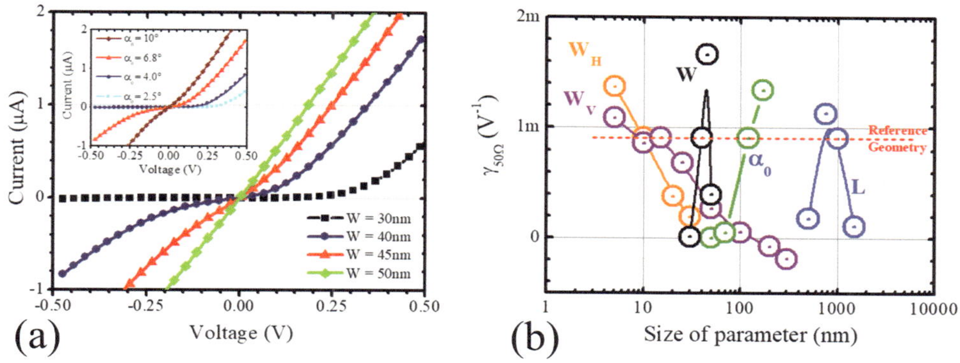

Figure 14.
(a) I-V characteristics of L-shape SSD when the channel width, W, is modified. (b) The inset shows the I-V curves for several leaning angles, α_o.

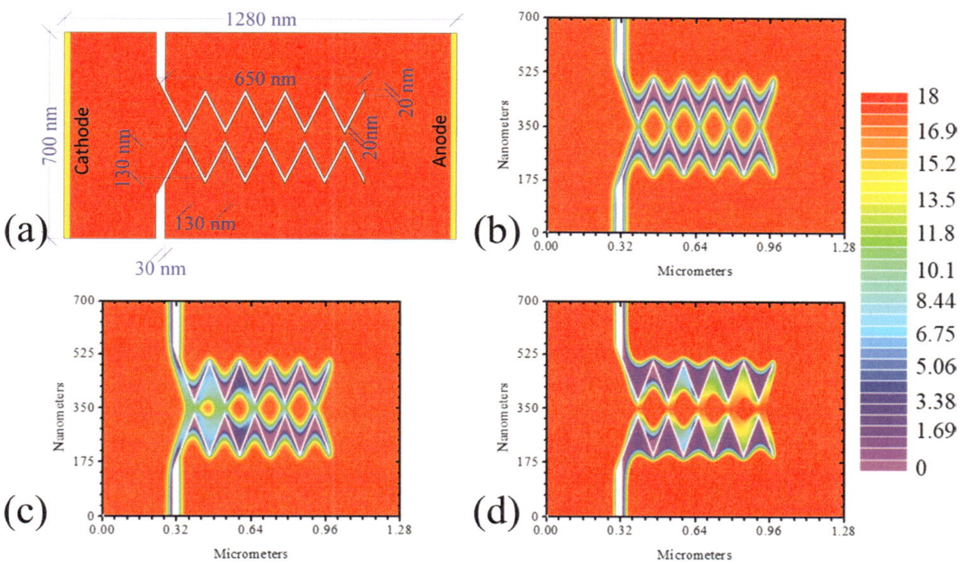

Figure 15.
(a) Two-dimensional scheme of the W-shape SSD where the main geometrical framework is illustrated. Simulation of the W-shape SSD carrier's density for (b) zero bias, (c) reverse bias, and (d) forward bias.

V-shape SSD exhibits poor diode-like characteristics for the 2DEG characteristics used. The fact of the small D_L formed in the channel vortex can be raised if the n_{SS} is strongly elevated or the carrier's concentration is low. Consequently, the L-shape SSD is useless on this condition, being the V-shape the appropriated choice for these situations.

Rendering to previous results, it is necessary to develop a shape of the SSD device that mixed the good rectifier behavior of the L-shape with the low-D_L extension of the V-shape that reduces R_V with the aim to improve the performance of the SSD concept as rectifier. The authors explored the novel SSD shape shown in **Figure 15(a)** labeled as W-shape SSD [40]. In this device, the D_L along the nanochannel has been modified from the completely depleted channel in the L-shape SSD and the less depleted V-shape into a punctual region where the depletion is maximum and easily modified by the electric field applied to electrodes.

For the no-bias situation illustrated in **Figure 15(b)**, the population of electrons inside the nanochannel is enlarged with the vertical distance from the trenches.

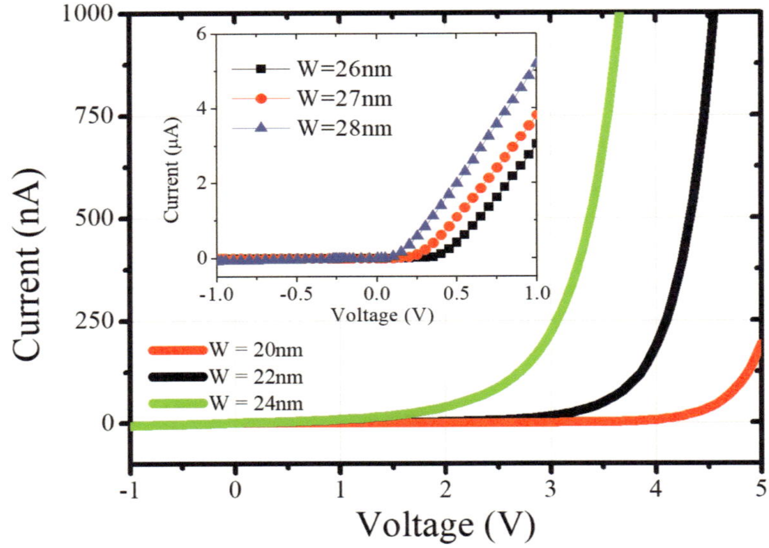

Figure 16.
I-V performance of the W-shape SSD where it is appreciated that the threshold voltage can be modulated by the separation of the grooves.

On the other hand, the electron concentration increases and decreases periodically along the channel, producing an island-like population of electrons with rhombus shape. In this numerical analysis, it is appreciated that the spatial distribution of electrons close to the center of the rhombus is maximum (red color) and it is reduced away to the center of each rhombic electron cluster (cyan color). In this condition the absence of free carriers between the high populated centers of the rhombus can avoid the current flow.

For reverse condition in **Figure 15(c)**, the reduction of the carrier density in the center of the rhombus is obtained, and in the same way, the size of the electron agglomeration is reduced at the left-hand side in contrast with the 0 V case, avoiding free electrons to participate in conduction. When positive voltage is applied to the anode electrode to reduce the D_L, a conducting path close to the chevron corners connects the dots, and a wirelike electron distribution is formed as it is indicated in **Figure 15(d)**, and the current flux can be obtained.

The W-shape SSD current-voltage behavior is dependent on their geometrical dimensions, as in the case of the L- and V-shape devices. Accordingly, the DC performance is optimized by the modification of the shape guidelines, i.e., in the I-V shown in **Figure 16**, the channel width is varied from 20 to 28 nm, being the principal component that determines the DC performance of the SSD as in the case of the L- and V-shape. For the reference geometry with W = 20 nm, the diode-like behavior is present with a negligible negative current, but the V_{TH} is around 4.5 V, undesirable for THz rectification. The slow modification of 1 nm in the channel width modifies strongly the I-V curve in the W-shape SSD.

For instance, the V_{TH} can be tailored by changing the separation of the grooves or the effective channel width from 4.5 to 2.5 V as the channel width is raised from 20 to 24 nm. The inset of **Figure 16** exhibits that when the W-shape SSD is calculated with W > 26 nm, the I-V characteristic behavior turns capable to operate with small power signals owing to the V_{TH} which is near to 0 V. It is important to note the absence of leakage current for reverse bias when the channel width is widened and different behaviors displayed by L- and V-shape designs. The optimization of this device and their performance as rectifier element will be shown in future works.

5. Conclusions

This chapter analyzes the emerging self-switching diode to deal with detection and harvesting of THz radiation from the numerical analysis of their rectifier behavior. The authors have shown the effect of geometry shape and size on the current-voltage performance, finding that the most important values that define the SSD's properties are the channel width and length in conjunction with the trench width for the L-, V-, and W-shape SSDs which are technologically promising by their simplicity in the manufacturing process. We found that the threshold voltage can be tuned to ~0 V, appropriated range for the low power THz signals. The L-shape SSD was analyzed as rectifier element in the rectenna concept where simulations indicate their ability to reach overall efficiencies of ~0.032%, improving the performance of the rectennas based on MIM tunnel barriers for THz applications. The reflection coefficient analysis reveals that one of the problems exhibited by the SSD devices lies in the high-resistance nature of the nanochannel. The reduction of the SSD resistance can be obtained using different shapes which controls the depletion region formed inside the nanochannel. Finally, we conclude that the adequate geometry size and shape of the SSD-based devices must be considered in conjunction with the current development of high-mobility heterostructure and nanolithography process in order to get the ideal rectifier to work in the THz range.

Acknowledgements

The authors acknowledge the financial support from CEMIE-SOL 22, FRC-UASLP, and CONACYT-Mexico through grants INFR-2015-01-255489, CB 2015-257358, PNCPN2014-01-248071, and the Catedras CONACyT (Project No. 44).

Conflict of interest

All authors declare that they have no conflict of interest.

Author details

Irving Eduardo Cortes-Mestizo[1], Edgar Briones[2], Leticia Ithsmel Espinosa-Vega[3] and Victor Hugo Mendez-García[3]*

1 CONACYT-Center for the Innovation and Application of Science and Technology, Autonomous University of San Luis Potosí, San Luis Potosí, Mexico

2 Department of Mathematics and Physics, ITESO Jesuit University of Guadalajara, Tlaquepaque, Mexico

3 Center for the Innovation and Application of Science and Technology, Autonomous University of San Luis Potosí, San Luis Potosí, Mexico

*Address all correspondence to: victor.mendez@uaslp.mx

References

[1] Chamberlain JM. Where optics meets electronics: Recent progress in decreasing the terahertz gap. Philosophical Transactions. Series A, Mathematical, Physical, and Engineering Sciences. 2004;**362**:199-211. DOI: 10.1098/rsta.2003.1312

[2] Maple L, Gow PC, Apostolopoulos V. Simulation of THz generation and propagation from photo-Dember emitters. The Journal of the Optical Society of America B. 2018;**35**(7):1552-1556. DOI: 10.1364/JOSAB.35.001552

[3] Zhao H, Yang L, Zou H, Ma X, Hao Y. Enhancement of negative differential mobility effect in recessed barrier layer AlGaN/GaN HEMT for terahertz applications. IEEE Transactions on Electron Devices. 2019;**66**(3):1236-1242. DOI: 10.1109/TED.2019.2893640

[4] Belkin M, Capasso F. New frontiers in quantum cascade lasers: High performance room temperature terahertz sources. Physica Scripta. 2015;**90**:1-13. DOI: 10.1088/0031-949/90/11/118002

[5] Kojima O, Tarui Y, Shimazu H, Kita T, Majeed A, Ivanov P, et al. Wide frequency tuning of continuous terahertz wave generated by difference frequency mixing under exciton-excitation conditions in a GaAs/AlAs multiple quantum well. Physical Review Applied. 2018;**10**:044035. DOI: 10.1103/PhysRevApplied.10.044035

[6] Ponomarev DS, Lavrukhin DV, Yachmenev AE, Khabibullin RA, Semenikhin IE, Vyurkov VV, et al. Lateral terahertz hot-electron bolometer based on an array of Sn nanothreads in GaAs. Journal of Physics D: Applied Physics. 2018;**51**(13):135101. DOI: 10.1088/1361-6463/aab11d

[7] Norman JC, Jung D, Zhang Z, Wan Y, Liu S, Shang C, et al. A review of high-performance quantum dot lasers on silicon. IEEE Journal of Quantum Electronics. 2019;**55**(2):2000511. DOI: 10.1109/JQE.2019.2901508

[8] Sugaya T. MBE of III-V semiconductors for solar cells. In: Asahi H, Horikoshi Y, editors. Molecular Beam Epitaxy. 1st ed. West Sussex, UK: John Wiley and Sons Ltd; 2019. pp. 265-277

[9] Kuzmik J. Power electronics on InAlN/(In)GaN: Prospect for a record performance. IEEE Electron Device Letters; 2001;**22**:510-512. DOI: 10.1109/55.962646

[10] Daghestani NS, Cataluna MA, Berry G, Ross G, Rose MJ. Terahertz emission from InAs/GaAs quantum dot based photoconductive devices. Applied Physics Letters. 2011;**98**:181107. DOI: 10.1063/1.3586774

[11] Umansky V, Heiblum M. MBE growth of high-mobility 2DEG. In: Henini M, editor. Molecular Beam Epitaxy from Research to Mass Production. 1st ed. Elsevier Inc; 2013. pp. 121-137. DOI: 10.1016/B978-0-12-387839-7.00006-3

[12] Umansky V, de-Picciotto R, Heiblum M. Extremely high-mobility two-dimensional electron gas: Evaluation of scattering mechanisms. Applied Physics Letters. 1997;**71**:683-685. DOI: 10.1063/1.119829

[13] Ajayana J, Nirmal D, Ravichandran T, Mohankumar P, Prajoon P, Arivazhagan L, et al. InP high electron mobility transistors for submillimetre wave and terahertz frequency applications: A review. AEU-International Journal of Electronics and Communications. 2018;**94**:199-214. DOI: 10.1016/j.aeue.2018.07.015

[14] Lüth H. Solid Surfaces, Interfaces and Thin Films. 5th ed. New York,

USA: Springer; 2005. p. 323. DOI: 10.1007/978-3-319-10756-1

[15] Wagner RE, Mandelis A. Intensity dependence of the photoreflectance amplitude in semiconductors. Physical Review B. 1994;**50**:14228-14236. DOI: 10.1103/PhysRevB.50.14228

[16] Cortes-Mestizo IE, Espinosa-Vega LI, Espinoza-Figueroa JA, Cisneros-dela-Rosa A, Eugenio-Lopez E, Mendez-Garcia VH, et al. Determination of the depletion layer width and effects on the formation of double-2DEG in AlGaAs/GaAs heterostructures. Journal of Vacuum Science and Technology B. 2016;**34**:1-6. DOI: 10.1116/1.4942898

[17] Sizov F. THz radiation sensors. Opto-Electronics Review. 2010;**18**:10-36. DOI: 10.2478/s11772-009-0029-4

[18] Bareiss M, Krenz PM, Szakmany GP, Tiwari BN, Kalblein D, Orlov AO, et al. Rectennas revisited. IEEE Transactions on Nanotechnology. 2013;**12**:1144-1150. DOI: 10.1109/TNANO.2013.2281373

[19] Gadalla MN, Abdel-Rahman M, Shamim A. Design, optimization and fabrication of a 28.3 THz nano-rectenna for infrared detection and rectification. Scientific Reports. 2014;4(1):-9. DOI: 10.1038/srep04270

[20] Song AM, Maximov I, Missous M, Seifert W. Diode-like characteristics of nanometer-scale semiconductor channels with a broken symmetry. Physica E: Low-dimensional Systems and Nanostructures. 2004;**21**:1116-1120. DOI: 10.1016/j.physe.2003.11.190

[21] Åberg M, Saijets J, Song A, Prunnila M. Simulation and modeling of self-switching devices. Physica Scripta. 2004;**T114**:123-126. DOI: 10.1088/0031-8949/2004/T114/031

[22] Song AM, Missous M, Omling P, Peaker AR, Samuelson L, Seifert W. Unidirectional electron flow in a nanometer-scale semiconductor channel: A self-switching device. Applied Physics Letters. 2003;**83**: 1881-1883. DOI: 10.1063/1.1606881

[23] Farhi G, Saracco E, Beerens J, Morris D, Charlebois SA, Raskin JP. Electrical characteristics and simulations of self-switching-diodes in SOI technology. Solid-State Electronics. 2007;**51**:1245-1249. DOI: 10.1016/j.sse.2007.07.013

[24] Irshaid YM, Balocco C, Luo Y, Bao P, Brox-Nilsen C, Song AM. Zinc-oxide-based planar nanodiodes operating at 50 MHz. Applied Physics Letters. 2011;**99**:092101. DOI: 10.1063/1.3629995

[25] Kettle J, Perks RM, Hoyle RT. Fabrication of highly transparent self-switching diodes using single layer indium tin oxide. Electronics Letters. 2009;**45**:79-81. DOI: 10.1049/el:20092309

[26] Al-Dirini F, Hossain FM, Nirmalathas A, Skafidas E. All-graphene planar self-switching MISFEDs, metal-insulator-semiconductor field-effect diodes. Scientific Reports. 2014;**4**:1-8. DOI: 10.1038/srep03983

[27] Torres J, Nouvel P, Penot A, Varani L, Sangare P, Grimbert B, et al. Nonlinear nanochannels for room temperature terahertz heterodyne detection. Semiconductor Science and Technology. 2013;**28**:1-6. DOI: 10.1088/0268-1242/28/12/125024

[28] Westlund A, Sangaré P, Ducournau G, Iñiguez-de-la-Torre I, Nilsson PÅ, Gaquière C, et al. Optimization and small-signal modeling of zero-bias InAs self-switching diode detectors. Solid State Electronics. 2015;**104**:79-85. DOI: 10.1016/j.sse.2014.11.014

[29] Balocco C, Kasjoo SR, Lu XF, Zhang LQ, Alimi Y, Winner S, et al. Room-temperature operation of

a unipolar nanodiode at terahertz frequencies. Applied Physics Letters. 2011;**98**:223501. DOI: 10.1063/1.3595414

[30] Mateos J, Vasallo BG, Pardo D, González T. Operation and high-frequency performance of nanoscale unipolar rectifying diodes. Applied Physics Letters. 2005;**86**:212103. DOI: 10.1063/1.1931051

[31] Xu KY, Wang G, Song AM. Electron transport in self-switching nano-diodes. Journal of Computational Electronics. 2007;**6**:59-62. DOI: 10.1007/s10825-006-0048-z

[32] Baraskar AK, Wistey MA, Jain V, Singisetti U, Burek G, Thibeault BJ, et al. Ultralow resistance, nonalloyed ohmic contacts to n-InGaAs. Journal of Vacuum Science and Technology B. 2009;**27**:2036-2039. DOI: 10.1116/1.3182737

[33] Cortes-Mestizo I, Méndez-García VH, Briones J, Pérez-Caro M, Droopad R, McMurtry S, et al. Terahertz harvesting with shape-optimized InAlAs/InGaAs self-switching nanodiodes. AIP Advances. 2015;**5**:117238-117210. DOI: 10.1063/1.4936792

[34] Cortes-Mestizo IE, Briones E, Briones J, Droopad R, Pérez-Caro M, McMurtry S, et al. Study of InAlAs/InGaAs self-switching diodes for energy harvesting applications. Japanese Journal of Applied Physics. 2016;**55**:14304-14307. DOI: 10.7567/JJAP.55.014304

[35] Bean JA, Weeks A, Boreman GD. Performance optimization of antenna-coupled Al/AlO_x/Pt tunnel diode infrared detectors. IEEE Journal of Quantum Electronics. 2011;**47**:126-135. DOI: 10.1109/JQE.2010.2081971

[36] Gadalla MN, Abdel-Rahman M, Shamim A. Design, optimization and fabrication of a 28.3 THz nano-rectenna for infrared detection and rectification. Scientific Reports. 2014;**4**:4270. DOI: 10.1038/srep04270

[37] Dagenais M, Choi K, Yesilkoy F, Chryssis AN, Peckerar MC. Solar spectrum rectification using nano-antennas and tunneling diodes. Proceedings of SPIE. 2010;**7605**:76050E. DOI: 10.1117/12.845931

[38] Briones E, Cortes-Mestizo IE, Briones J, Droopad R, Espinosa-Vega LI, Vilchis H, et al. Efficiency of broadband terahertz rectennas based on self-switching nanodiodes. Journal of Photonics for Energy. 2017;**7**(2):025001. DOI: 10.1117/1.JPE.7.025001

[39] Zhang B, He T, Shen T, Hou Y, Hu Y, Zang M, et al. Conjugated polymer-based broadband terahertz wave modulator. Optics Letters. 2014;**39**(21):6110-6113. DOI: 10.1364/OL.39.006110

[40] Balocco C, Song AM, Åberg M, Forchel A, González T, Mateos J, et al. Nano Letters. 2005;**5**:1423-1427. DOI: 10.1021/nl050779g

[41] Cortes-Mestizo IE, Briones E, Belio-Manzano A, Espinosa-Vega LI, Méndez-García VH. W-shape nanodiode controlled by surface states for THz detection. In: 42nd International Conference on Infrared, Millimeter, and Terahertz Waves (IRMMW-THz); 27 Aug.-1 Sept. 2017. Cancun, Mexico: IEEE; 2017. DOI: 10.1109/IRMMW-THz.2017.8067048